Towards Next Generation Energy Storage Technologies

Towards Next Generation Energy Storage Technologies

From Fundamentals to Commercial Applications

Edited by Minghua Chen

WILEY-VCH

Editor

Prof. Minghua Chen
Harbin University of Science and Technology
School of Electrical and Electronic Engineering
Department of New Energy Materials and Devices
Key Laboratory of Engineering Dielectric and Applications (Ministry of Education)
52 Xuefu Road
Nangang District
Harbin 150080
China

Cover Image: © tommy/Getty Images

Library of Congress Card No.: applied for

British Library Cataloguing-in-Publication Data
A catalogue record for this book is available from the British Library.

Bibliographic information published by the Deutsche Nationalbibliothek The Deutsche Nationalbibliothek lists this publication in the Deutsche Nationalbibliografie; detailed bibliographic data are available on the Internet at <http://dnb.d-nb.de>.

Print ISBN: 978-3-527-35331-6
ePDF ISBN: 978-3-527-84529-3
ePub ISBN: 978-3-527-84530-9
oBook ISBN: 978-3-527-84531-6

Typesetting Straive, Chennai, India

Printed and bound by
CPI Group (UK) Ltd, Croydon, CR0 4YY

C9783527353316_300824

Contents

Preface

The excessive use of nonrenewable fossil energy has triggered energy crises and environmental degradation. To address these issues, replacing nonrenewable fossil energy with renewable energy such as solar, wind, and tidal energy has been proposed and rapidly developed. However, in general, renewable energy cannot provide stable electric output due to its susceptibility to the environment. To harvest renewable energy and achieve output with the required power density, incorporating electrochemical energy storage technologies is believed to be an effective strategy that prevents the wasting of renewable energy. In addition, replacing gasoline-powered cars with electric-powered cars is a promising way to reduce dependence on nonrenewable energy sources, with their electricity supply primarily coming from electrochemical energy storage technologies. Hence, developing electrochemical energy storage technologies is critical to cope with the energy crises and environmental degradation, and hence, they have attracted ever-increasing attention in recent years.

The currently available electrochemical energy storage techniques, dominated by lead–acid batteries, lithium-ion batteries (LIBs), and supercapacitors (SCs), have greatly developed and changed our daily lives. The energy density of state-of-the-art commercial technologies (LIBs) can reach about 180~300 Wh kg^{-1}, which makes them beneficial for applications in electric vehicles and grid-scale energy storage systems. Nevertheless, safety issues (caused by lithium dendrite growth and the use of flammable electrolytes), high cost of raw materials, and the limited energy/power density hinder the widespread application of energy storage technologies in large-scale energy storage systems and long-lasting electric vehicles. To address these challenges, many novel promising energy storage technologies (e.g. solid-state batteries, lithium–sulfur batteries, and lithium–carbon dioxide batteries) based on various electrochemical redox reactions have been proposed recently. These newly emerged technologies are of high energy density, low cost, and high safety and can facilitate a wide range of applications in electric vehicles, grid-scale energy storage systems, and other devices (e.g. power supply of equipment applied in aerospace, deep sea, and polar regions). Substantial work has been devoted to exploring the potential of the aforementioned energy storage technologies, and some of them have stepped into the practical application stage.

To push forward the development of these promising energy storage technologies, this book focuses on the fundamental energy storage mechanisms, critical achievements, and challenges of next-generation energy storage technologies from the perspective of practical applications. This book discusses the recent progress of commercialized energy storage technologies (LIBs and SCs) first and then comprehensively introduces the recent progress of newly emerged energy storage technologies. This book aims to help readers clarify what are the critical issues that hinder the widespread application of these energy storage technologies, translate complex scientific concepts into easily understandable language, and demonstrate the potential of these technologies in practical applications through case studies. In this rapidly evolving field, acquiring up-to-date knowledge is challenging, but we are committed to providing the most up-to-date and comprehensive information to help readers stay informed about the latest developments in energy storage technology.

Lastly, we would like to express our gratitude to all individuals and institutions who have contributed to the writing and publication of this book, as well as to all scientists who have contributed to the development of energy storage technologies. We hope that this book will contribute to the development of energy storage technologies and the goal of achieving clean, sustainable energy.

April 2024

Minghua Chen
Harbin, China

Acknowledgments

The writing of this book was enriched by numerous scientific discussions. I would like to thank my colleagues for their valuable comments. Their knowledge, experience, and skills have greatly enriched this book and have advanced our understanding of next-generation energy storage technologies. We have much gratitude for the many graduate students who have worked on the examples cited and proofread the chapters.

Minghua Chen
Harbin, China

1

Introduction

Minghua Chen[1] *and Yu Li*[2]

[1] *Harbin University of Science and Technology, School of Electrical and Electronic Engineering, Department of New Energy Materials and Devices, Key Laboratory of Engineering Dielectric and Applications (Ministry of Education), 52 Xuefu Road, Nangang District, Harbin, 150080, China*
[2] *Harbin University of Science and Technology, School of Electrical and Electronic Engineering, Department of Electrical Theory and New Technology, 52 Xuefu Road, Nangang District, Harbin, 150080, China*

With ever-increasing concerns about limited fossil fuel reserves (e.g. oil, coal, and natural gas) and the environmental degeneration caused by the emission of harmful gases (e.g. CO_2 and SO_2), there is an urgent need to replace polluted unrenewable energy resources with clean renewable energy resources (e.g. wind and solar energy). Nevertheless, the generation of renewable energy is intermittent, and the electricity output is unstable. Harvesting renewably generated electricity using an electrochemical energy storage system is a practical way to address the intrinsic issues with renewable energy. In addition, removable electrochemical energy storage technologies play a huge role in our daily lives. For example, phones, laptops, and smartwatches we use daily are powered by high-energy-density lithium-ion batteries (LIBs). Some advanced vehicle manufacturing companies (e.g. BYD and Tesla) have launched electric vehicles, in an attempt to decrease harmful gas emissions. Nevertheless, the application of electrochemical energy storage technologies in a wide range of fields is still challenging. The properties of energy storage technologies depend on the applications and the market they are being used. For example, portable energy storage devices and energy storage systems used by people in their daily lives should be energy dense and highly safe. Outdoor energy storage systems in cold regions should possess good low-temperature energy storage performance. Energy storage devices used in regenerative braking should have both high energy density and power density.

To push forward the energy structure upgrade, numerous studies have been dedicated to the investigation of advanced energy storage technologies. Various promising energy storage technologies (e.g. high-safety solid-state batteries, aqueous lithium-ion batteries [ALIBs], aqueous zinc-ion batteries [AZIBs], high-energy-density lithium–sulfur batteries [Li-S batteries], metal–air batteries, and metal–CO_2 batteries) have been developed. These advanced energy storage technologies are designed for different requirements and markets and show

Towards Next Generation Energy Storage Technologies: From Fundamentals to Commercial Applications, First Edition. Edited by Minghua Chen.

promise for next-generation energy storage technologies. In this book, recent advances in energy storage technologies that can be commercialized are thoroughly summarized and discussed in detail, providing the landscape of the state-of-the-art electrochemical energy storage technologies and their widespread practical applications. First, the fundamental knowledge about energy storage technologies will be introduced, and then the practical progress of the existing commercial energy storage technologies and the corresponding energy storage properties, as well as critical challenges, will be presented. Subsequently, energy storage mechanisms, critical issues, design strategies, and practical progress of next-generation energy storage technologies will be discussed. Finally, future trends in these advanced energy storage technologies will be presented. We hope this book can contribute to the development of electrochemical energy storage technologies.

2

Fundamentals of Electrochemical Energy Storage Technologies

Minghua Chen[1] *and Yu Li*[2]

[1] *Harbin University of Science and Technology, School of Electrical and Electronic Engineering, Department of New Energy Materials and Devices, Key Laboratory of Engineering Dielectric and Applications (Ministry of Education), 52 Xuefu Road, Nangang District, Harbin, 150080, China*
[2] *Harbin University of Science and Technology, School of Electrical and Electronic Engineering, Department of Electrical Theory and New Technology, 52 Xuefu Road, Nangang District, Harbin, 150080, China*

2.1 Typical Battery Patterns and Corresponding Functions

Of late, numerous electrochemical energy storage systems have been developed to meet the demands of our daily lives, of which five are commonly used and commercialized: lead–acid batteries, lithium-ion batteries (LIBs), nickel metal hydride (Ni-MH) batteries, nickel–zinc (Ni-Zn) batteries, and supercapacitors. Lead–acid batteries are most commonly used in gasoline cars for engine ignition and running the auxiliary electronics when the engine is off. However, they are harmful to the environment, and replacing them with less expensive, safe, and energy-dense electrochemical energy storage technologies is one of the major challenges faced by researchers. Ni-MH batteries and Ni-Zn batteries are typical alkaline batteries using KOH-based electrolytes. Their energy density is slightly higher than that of lead–acid batteries; however, their output voltage is highly restricted by the limited theoretical thermodynamic stable potential (1.23 V). In addition, alkaline electrolytes will continuously etch the current collector and zinc anode, and thus, cycling lifespans of these batteries can rarely be more than 500 cycles without obvious decay [1]. LIBs have been widely used in portable devices, electric vehicles, and grid-scale energy storage systems due to their high output voltage, high energy density, and long cycling lifespans. The lifespan of advanced LIBs can be more than 10,000 cycles. Although the safety of using LIBs is ensured in various portable devices (e.g. phones and laptops), thermal runaway and explosion can occur in grid-scale energy storage systems and electric vehicles due to the growth of lithium dendrite, which induces internal short circuits and releases substantial heat to ignite flammable electrolytes. Dendrite growth is related to overcharging and over-discharging of batteries. These can be prevented by rational battery management, which can be efficiently controlled in a single

Towards Next Generation Energy Storage Technologies: From Fundamentals to Commercial Applications, First Edition. Edited by Minghua Chen.

cell and hardly controlled in the battery pack [2]. The more the number of LIBs, the higher the thermal runaway risk. Supercapacitor is a power-density-superior electrochemical energy storage device that harvests energy via a rapid physical adsorption/desorption process. However, the energy density of supercapacitors is more than tenfold lower than that of batteries, and thus, they cannot be used as a main power source for the functioning of electric equipment [3].

Nevertheless, all the abovementioned electrochemical energy storage technologies have changed our daily lives and have similar component patterns. In general, all of them have two electrodes. If the electrode materials used are different, they can be classified into cathode and anode, which will be introduced in the following sections. Materials used as electrodes should at least be good electronic conductors. Meanwhile, the cathode and anode need to be separated by electrolytes and separators to avoid direct contact between them. The energy storage mechanism of electrochemical energy storage technologies is mainly based on the electrochemical reactions (reversible redox reaction or intercalation) at cathode and anode. It is noteworthy that the redox reactions happened in electrochemical energy storage devices is quite different with chemical reactions. The primary distinction between an electrochemical reaction and a chemical redox reaction is that, in the former, reduction occurs at one electrode and oxidation occurs at the other, while in the latter, both reduction and oxidation occur at the same electrode. This distinction has several implications. In an electrochemical reaction, oxidation is spatially separated from reduction. Thus, the entire redox reaction is divided into two half-cells. The rate of these reactions can be controlled by externally applying a potential difference between the electrodes, e.g. using an external power supply, a feature that is not present in chemical reactors. Furthermore, electrochemical reactions are always heterogeneous; i.e. they always occur at the interface between the electrolyte and an electrode (and possibly a third phase, such as a gaseous or insulating reactant). Even though half-cell reactions occur at different electrodes, reaction rates are coupled by the principles of conservation of charge and electroneutrality.

2.1.1 Cathode

To date, many electrochemical battery systems have been proposed. The term "cathode" is a relative concept. A cathode in one electrochemical system can be an anode in another system. For example, MnO_2 is a cathode material in Zn-Mn batteries but an anode material in LIBs. Hence, whether a material is a cathode or anode depends on the relative redox potential of the material, which, in an electrochemical device, can be deduced based on standard reduction potential. Standard reduction potential describes the ability of a material to release electrons. The lower the standard reduction potential, the stronger the ability to release electrons. Therefore, when two materials are assembled into an electrochemical system, one with the higher reduction potential is the cathode, and the other one is the anode. Theoretically, any material can be assembled into electrochemical batteries with appropriate electrolytes. The output voltage of an electrochemical

Table 2.1 The standard reduction potential of typical materials.

Half-cell reaction	Potential
$F_2 + 2e^- \leftrightharpoons 2F^-$	+2.87 V
$Pb^{4+} + 2e^- \leftrightharpoons Pb^{2+}$	+1.67 V
$Cl_2 + 2e^- \leftrightharpoons 2Cl^-$	+1.36 V
$Ag^+ + 1e^- \leftrightharpoons Ag$	+0.81 V
$Fe^{3+} + 1e^- \leftrightharpoons Fe^{2+}$	+0.77 V
$Cu^{2+} + 2e^- \leftrightharpoons Cu$	+0.34 V
$2H_2 + 2e^- \leftrightharpoons H_2$	0.00 V
$Fe^{3+} + 3e^- \leftrightharpoons Fe$	−0.04 V
$Pb^{2+} + 2e^- \leftrightharpoons Pb$	−0.13 V
$Fe^{2+} + 2e^- \leftrightharpoons Fe$	−0.44 V
$Nz^{2+} + 2e^- \leftrightharpoons Zn$	−0.76 V
$Al^{3+} + 3e^- \leftrightharpoons Al$	−1.66 V
$Mg^{2+} + 2e^- \leftrightharpoons Mg$	−2.36 V
$Li + 1e^- \leftrightharpoons Li$	−3.05 V

Stronger pulling electron (↑)
Stronger giving electron (↓)

system depends on the potential difference between the cathode and anode. Some electrodes commonly used in diverse electrochemical energy storage technologies are $LiCoO_2$, $LiFePO_4$, $LiMn_2O_4$, $LiNiCoMnO_2$, and $Ni(OH)_2$. All these possess unique crystal structures (e.g. layer structure and sodium super ionic conductor (NASICON) structure) and have sufficient space to store ions and ensure rapid ion transportation (Table 2.1).

2.1.2 Anode

In an electrochemical system, anode is the electrode that reacts at a lower potential among the two electrodes. Metals and graphite are some of the commonly used anode materials in state-of-the-art energy storage technologies. Usually, graphite and other carbon materials are used as host materials to store energy through reversible ion intercalation. The capacity of these materials is usually less than $300\,mAh\,g^{-1}$. In metal anodes, energy storage occurs via reversible ion plating/striping. Theoretically, the capacity of metal anodes is much higher than that of carbon materials. However, in metal anodes, irreversible etching and uneven plating/striping lead to the formation of dendrites and "dead metal," thus presenting them with poor cycling stability and high safety hazard. Designing advanced electrolytes and artificial solid electrolyte interphase (SEI) for ensuring uniform ion plating/striping and protecting irreversible etching is an interesting topic in the field of electrochemical energy storage.

2.1.3 Electrolyte

The electrolyte is an essential component that ensures rapid ionic conductivity inside the battery and prevents direct contact between cathodes and anodes. Electrolytes used in various electrochemical energy storage technologies should generally meet the following basic requirements [4]: (i) high ionic conductivity, at least $1 \times 10^{-3} \sim 2 \times 10^{-2}$ S cm^{-1}; (ii) high thermal and chemical stability, with no separation in a wide voltage range; (iii) a wider electrochemical stability window (ESW) to keep the electrochemical performance stable in a wider voltage range; (iv) good compatibility with other parts of the battery, such as electrode materials, electrode current collectors, and separators; and (v) safe, nontoxic, and nonpolluting.

Based on the energy storage mechanism of electrode materials, various electrolytes have been developed, which can primarily be classified into three categories: aqueous electrolytes, organic electrolytes, and solid-state electrolytes. In aqueous electrolytes, water is used as the solvent. The following are the major advantages of aqueous electrolytes: low cost, high ionic conductivity (compared with organic electrolytes and solid-state electrolytes), and being nonflammable and recyclable. Their primary disadvantage is the limited thermodynamic ESW of water. Theoretically, water decomposes above a voltage of 1.23 V. Considering the overpotential of water splitting, the voltage of aqueous electrolytes is less than 2.0 V. Since energy density is highly dependent on output potential, electrochemical energy storage technologies with aqueous electrolytes usually have much lower energy density than those using organic electrolytes and solid-state electrolytes. Recently, numerous strategies have been developed to make the ESW of aqueous electrolytes wider, e.g. water-in-salt (WIS) electrolytes, deep eutectic solvents, and artificial SEI. In particular, using well-designed WIS electrolytes, the ESW of aqueous electrolytes can be extended to 3 V, as proposed in some previously published studies. However, usually, only electrolytes based on the trifluoromethanesulfonimide (TFSI) anion can barely work above 3 V, and the ESW of aqueous electrolytes using inorganic salts can rarely reach above ~2.5 V in devices. Recent studies suggest that an SEI derived by the decomposition of the TFSI anion is highly porous and dynamically soluble in electrolytes [5]. Since the SEI is not stable, water splitting continuously occurs in LiTFSI-based WIS electrolytes. The continuous generation of gases increases the pressure in the batteries, leading to rapid capacity decay and even explosion. It is proposed that even in high-cost LiTFSI-based electrolytes, electrochemical energy storage devices with aqueous electrolytes can rarely work above 2 V and do not present apparent superiorities over lead–acid batteries [6]. Gel electrolytes are a type of aqueous electrolytes that confine water molecules through strong interactions that can effectively prohibit the decomposition of water at low salt concentrations, which may be a practical way to achieve high output voltage and high energy density in energy storage technologies with aqueous electrolytes. The success of lead–acid batteries can provide a meaningful reference for the development of high-voltage energy storage technologies using aqueous electrolytes.

In organic electrolytes, an organic liquid is used as a solvent to dissolve salts. State-of-the-art LIBs use organic electrolytes, in which $LiPF_6$ is used as the salt and ethylene carbonate (EC)/propylene carbonate (PC)/fluoroethylene carbonate (FEC) is used as the solvent. Each component in the organic electrolytes of LIBs is carefully selected; EC/PC can provide high ionic conductivity with good electrochemical stability and large ESW, while FEC can help form LiF-rich SEI, to ensure stable and rapid ion transportation between the electrode and electrolyte. The critical challenge in using organic electrolytes is safety issues. Dendrites can form even in state-of-the-art technologies. Short circuits induced by dendrites will release a substantial amount of heat to accelerate the decomposition of organic electrolytes and form H_2/O_2 gas, leading to thermal runaway and even explosion.

Since the thermal runaway of LIBs is attributable to the decomposition and poor thermal stability of organic electrolytes, nonflammable and high-mechanical-strength solid-state electrolytes are developed to ensure the safety of LIBs and other batteries. Solid-state electrolytes can be classified into two groups: inorganic electrolytes and polymer electrolytes. Inorganic electrolytes possess sufficient mechanical strength to prevent the formation of metal dendrites and high thermal stability to achieve intrinsic safety. Their ionic conductivity is $1 \times 10^{-2} \sim 10^{-4}\ \mathrm{S\,cm^{-1}}$, which is close to that of liquid electrolytes and meets the criteria for applications in various electrochemical energy storage devices. However, solid–solid electrode–electrolyte interfaces cannot establish an intimate contact, leading to large interface impendence and poor capacity. To ensure good contact, high external pressure is usually required for batteries using solid-state electrolytes, which is hard to achieve in real-world scenarios. Furthermore, the volume expansion of electrodes results in high mechanical strength at electrode–electrolyte interfaces, which leads to the pulverization and falling off of active materials, resulting in poor cycling stability. In contrast, polymer electrolytes are usually soft and can tolerate the volume expansion of active materials and improve the electrode–electrolyte contact, which is conducive to promoting the reversible capacity and cycling lifespans. Nevertheless, polymer electrolytes do not have high enough mechanical strength to prevent short circuits induced by dendrite growth, and their ionic conductivity ($\sim 10^{-6}\ \mathrm{S\,cm^{-1}}$) is insufficient for a high ion transport pathway for electrochemical reactions. Considering all the abovementioned points, combining the advantages of inorganic electrolytes and those of polymer electrolytes can be a promising way to achieve a practicable solid-state electrolyte, and this has been widely investigated in recent years.

2.2 Operating Mechanism of Devices

2.2.1 Potential and Thermodynamics

The working mechanism of an electrochemical battery depends on the intrinsic properties of the materials used, especially thermodynamic properties. As mentioned earlier, a redox reaction of a battery can occur separately as the oxidation reaction takes place at one electrode and the reduction reaction at another, i.e. the whole battery acting as two half-cells. The energy change of the reaction is given by the change in Gibbs free energy for each half-cell reaction:

$$\Delta G = \left(\sum_i s_i \mu_i\right)_{\text{cathode}} - \left(\sum_i s_i \mu_i\right)_{\text{anode}} \tag{2.1}$$

where G refers to the Gibbs free energy, μ_i represents the electrochemical potential of species i, and s_i is the stoichiometric coefficient of species i. If ΔG is negative, electrons will spontaneously flow from the anode to the cathode. This is the driving force for ensuring spontaneous redox reactions in electrochemical batteries. The anode side usually has a lower electrochemical potential, whereas the cathode side has a higher electrochemical potential. The electrochemical potential of different materials may have different reference potentials, e.g. Li/Li^+, Na/Na^+, Ag/AgCl, and Hg/HgO. However, these electrochemical potentials based on different reference potentials can be converted to a standard reference potential.

The maximum amount of work (W_{max}) that can be performed by an electrochemical cell is equal to the product of the cell potential (E°_{cell}) and the total charge transferred during the reaction (nF):

$$W_{\text{max}} = nFE_{\text{cell}} \tag{2.2}$$

W_{max} is expressed as a negative number because work is being done by a system (i.e. an electrochemical cell with a positive potential) on its surroundings.

The change in Gibbs free energy (ΔG) is also a measure of the maximum amount of work that can be performed during a chemical process ($\Delta G = W_{\text{max}}$). Consequently, the relationship between the potential of an electrochemical cell and ΔG can be deduced as follows:

$$\Delta G = -nFE_{\text{cell}} \tag{2.3}$$

A spontaneous redox reaction is therefore characterized by a negative value of ΔG and a positive value of E°_{cell}. When both reactants and products are in their standard states, the relationship between ΔG° and E°_{cell} can be expressed as follows:

$$\Delta G^\circ = -nFE^\circ_{\text{cell}} \tag{2.4}$$

A spontaneous redox reaction is characterized by a negative value of ΔG°, which corresponds to a positive value of E°_{cell}. Thus, equilibrium potential is a function of the intrinsic nature of the species present, as well as their concentrations and, to a lesser extent, temperature. No net current flows at equilibrium, and random thermal collisions between the reactant and product species cause the reaction to occur, sometimes in the forward direction and sometimes in the backward direction. At equilibrium, the rate of the forward reaction is equal to that of the backward reaction. The potential of the electrode at equilibrium is a measure of the electrochemical potential (i.e. energy) of electrons at equilibrium with the reactant and product species. In brief, electrochemical potential can be related to molality m_i and activity coefficient γ_i as follows:

$$\mu_i = \mu_i^\theta + RT \ln m_i\gamma_i \tag{2.5}$$

where μ^θ is independent of concentration, R is the universal gas constant ($8.3143\ \mathrm{J\,mol^{-1}\,K^{-1}}$), and T is the temperature in kelvin.

One can electrically control the electrochemical potential of electrons in an electrode by connecting it to an external power supply, thereby perturbing the equilibrium and driving a reaction. Applying a negative potential to an electrode increases the energy of electrons. This increase in the energy of electrons above the lowest unoccupied molecular orbital of a species in the adjacent electrolyte results in a reduction of that species. This reduction current (flow of electrons into the electrode and from there into the reactant) is also called cathodic current, and the electrode at which this occurs is called the cathode. Conversely, applying a positive potential to an electrode decreases the energy of electrons, causing them to be transferred from the reactants to the electrode. The electrode where such an oxidation reaction occurs is called the anode.

2.2.2 Kinetics of Electrode Reactions

As mentioned earlier, the redox reaction is separated into oxidation and reduction that occur in two electrodes. In galvanic cells, the reduction reaction occurs at the cathode side, and the oxidation reaction occurs at the anode side. The rate of reaction can be expressed using the transition state theory (Figure 2.1). For a reaction to occur, it needs to overcome an energy barrier (Gibbs free energy). Considering this, the rate of reaction can be expressed as follows:

$$k = Ae^{-\frac{\Delta G^*}{RT}}$$

The chemical reaction process on inert electrode surfaces involves the transportation of electrochemically active species (usually ions) from the bulk electrolyte to the region that is very close to the electrode surface, preceding chemical reactions

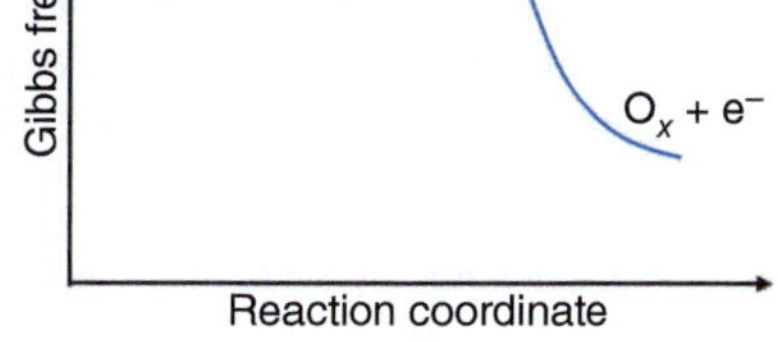

Figure 2.1 Rate of reaction for a chemical process.

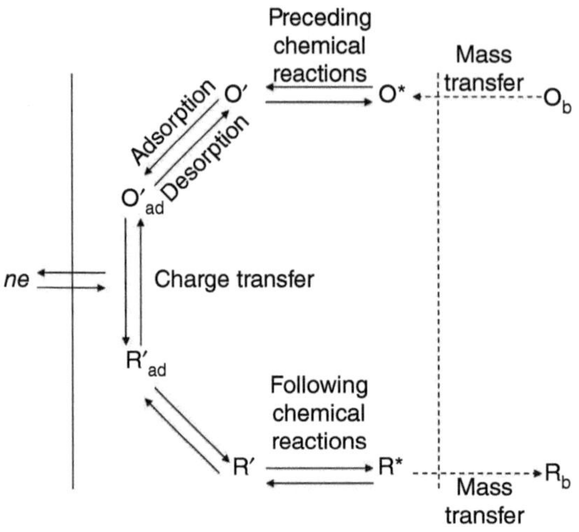

Figure 2.2 Typical charge transfer process at an inert electrode surface.

(not applicable for all electrochemical reactions however), adsorbing on the electrode surface, ensuring charge transfer, desorbing from the electrode surface, following chemical reactions (not applicable for electrochemical reactions however), and migrating to the bulk electrolyte (Figure 2.2). Adsorption, charge transfer, and desorption processes occur rapidly and cannot be separated based on the existing electrochemical technologies or structural characterization technologies. As for an active electrode, the chemical reaction process is different from that of an inert electrode. The charge transfer process can occur at the internal of the electrode. In addition, ions are solvated and undergo a desolvation process before entering the electrode, which kinetically slows down the reaction rate. To make things more complicated, some side reactions occur before the reversible energy storage reaction, which lead to continuous consumption of electrolytes or the formation of SEI.

In addition to the chemical reaction process, the bias voltage acts as a driving force to initiate electrochemical reactions at electrodes and is another essential parameter that affects reaction kinetics. This driving force is termed "surface overpotential" and is denoted as η_s. The rate of a reaction can often be related to surface overpotential using the Butler–Volmer equation, as follows:

$$i = i_0 \left[\exp\left(\frac{\alpha_{ox} nF}{RT} \eta_s \right) - \exp\left(\frac{\alpha_{red} nF}{RT} \eta_s \right) \right] \tag{2.6}$$

The first term on the right-hand side of the equation is the rate of the oxidation reaction, whereas the second term is the rate of the reduction reaction. The difference between these rates is the net rate of reaction. The parameter i_0 is the exchange current density and is analogous to the rate constant used in chemical kinetics. Based on the above equation, the Tafel $I - \eta$ relationship at very large $|\eta|$ can be expressed as follows:

$$\eta_s = \frac{RT}{\alpha_{red} nF} \ln \frac{-i}{i_0} \tag{2.7}$$

$$\eta_s = -\frac{RT}{\alpha_{ox} nF} \ln \frac{i}{i_0} \tag{2.8}$$

This relationship is also called a semi-logarithm relationship. A reaction with a high i_0 value is often called fast or reversible, and $I - \eta$ usually follows a linear relation. In contrast, a low i_0 value approaching disappearance indicates the highly polarized electrode and that the chemical reaction is hardly reversible. The $I - \eta$ relation of a low i_0 value usually is a semi-logarithm relation or similar to the charging behavior of a RC circuit. For a high i_0 value, a high current density can be obtained with a low surface overpotential. In a reaction involving high exchange current density, both the forward and backward reactions occur rapidly, and the net direction of the reaction depends on the sign of the surface overpotential. The exchange current density of the reaction depends on the concentrations of reactants and products, temperature, the nature of the electrode–electrolyte interface, and impurities that may contaminate the surface. Each of these factors can change the value of i_0 by several orders of magnitude; i_0 can range from over 1 mA cm^{-2} to less than 10^{-7} mA cm^{-2}. The parameters α_{ox} and α_{red}, called apparent transfer coefficients, are additional kinetic parameters that inform how an applied potential favors one direction of the reaction over the other. The values of these parameters usually range between 0.2 and 2.

Overpotential is the potential difference above the equilibrium potential that is required to overcome the activation energy of the cell reaction to produce a specified current. Hence, the activation energy can also affect the electrode reaction kinetics. As discussed previously, for a reaction to occur, it needs to overcome an energy barrier (Gibbs free energy). The charge transfer process of electrochemical reactions should overcome an energy barrier as well, and this energy barrier is affected by bias voltage. Considering the half-cell of the reduction reaction as an example, the free energy of reduction species (denoted as R_{ed}) is invariant with voltage, whereas the reduction product (denoted as $O_x + e^-$) is highly affected by voltage. To determine the impact of voltage on electrode reaction kinetics, it is assumed that the effect of voltage on free energy change follows a linear relationship (this is undoubtedly an oversimplification). Using this linear relationship, the activation free energies for reduction and oxidation will be considered to vary as a function of the applied voltage (V) as follows:

$$\Delta G'_{red} = \Delta G'_{red\ no\ voltage} + \alpha FV \tag{2.9}$$

$$\Delta G'_{ox} = \Delta G'_{ox\ no\ voltage} + (1 - \alpha)FV \tag{2.10}$$

The parameter α is called the transfer coefficient, which typically has a value of 0.5. Physically, this provides an insight into the way the transition state is influenced by the voltage. A value of 0.5 means that the transition state behaves midway between the reactants' and products' response to the applied voltage. The free energy on the right-hand side of Eqs. (2.9) and (2.10) can be considered the chemical component of the activation free energy change. It is dependent only on the species involved and not on the applied voltage. Substituting these activation free energy terms into the expressions for the oxidation and reduction rate constants gives the following:

$$k_{red} = Z \exp\left[\left(\frac{-\Delta G'_{\text{red no voltage}}}{RT}\right)\left(\frac{-\alpha FV}{RT}\right)\right] \tag{2.11}$$

$$K_{ox} = Z \exp\left[\left(\frac{-\Delta G'_{\text{red no voltage}}}{RT}\right)\left(\frac{-\alpha FV}{RT}\right)\right] \tag{2.12}$$

These results suggest that the rate constants for a charge transfer process are proportional to the exponential of the applied voltage. So the electrode reaction kinetics can be changed by simply varying the applied voltage. This result provides the fundamental basis for the experimental technique called voltammetry.

Now that the effects of overpotential and activation energy on electrode reaction kinetics are discussed, the following text introduces the effects of mass transfer. Mass transfer is the movement of species in a medium (liquid, gas, or solid). The rate of mass transfer denotes the number of species passing through a unit area within an interval of unit time. Three processes are related to mass transfer, namely diffusion, migration, and convection. Diffusion is the movement of a charged species derived by a gradient of chemical potential (e.g. concentration gradient), which is also called Fick's first law of diffusion. This can be expressed as follows:

$$J_i^{\mathrm{D}} = -D_i \nabla c_i \tag{2.13}$$

Migration refers to the movement of a species driven by a gradient of electric potential, which can be represented as follows:

$$J_i^{\mathrm{M}} = -\frac{Z_i F}{RT} c_i D_i \nabla \phi \tag{2.14}$$

Convection refers to the movement of a species in the fluid flow. Natural convection is induced by density gradient, whereas forced convection is caused by stirring or other hydrodynamic means, as shown below:

$$J_i^{\mathrm{C}} = v c_i \tag{2.15}$$

Hence, mass transfer in an electrochemical cell can be expressed as follows:

$$J_i = J_i^{\mathrm{D}} + J_i^{\mathrm{M}} + J_i^{\mathrm{C}} \tag{2.16}$$

Usually, some measures can be taken so that only one kind of mass transfer contributes predominantly and others are negligible. In a solution, natural convection is the predominant mass transfer mode, and it ensures the uniform concentrations of all species. However, the velocity of the fluid decreases sharply upon approaching a solid surface. Thus, within a very thin layer near the electrode surface, convection can be neglected. Diffusion and migration are predominant in the near surface reaction of electrodes (called the Nernst boundary layer). If the excess of an inert electrolyte is added to a solution and its concentration is much higher than that of the electroactive species, the electric field and migration rate in the solution can be greatly reduced. Thus, diffusion will be the only predominant mass transfer mode within the thin layer near the electrode surface (called diffusion layer).

Upon imposing electrochemical polarization (e.g. potential or current perturbation) to an electrode system, interfacial reactions occur, which render the concentration of species in the vicinity of the electrode different from that of the bulk solution. A concentration gradient then arises, and in the meantime, diffusion takes place, trying to eliminate the concentration gradient. The combined processes of the development of concentration gradient due to interfacial reactions and the diffusion to eliminate the concentration dynamically change the concentration, and the concentration profile gradually extends toward the bulk solution:

$$\frac{\partial c_i}{\partial t} \neq 0 \tag{2.17}$$

Due to natural (or forced) convection, the concentration gradient does not extend unlimitedly toward the bulk, and a steady-state profile of concentration is established finally:

$$\frac{\partial c_i}{\partial t} = 0 \tag{2.18}$$

It is worth noting that the above discussions about the impact of overpotential and activation energy on reaction kinetics do not consider internal reactions of the electrode. However, in electrochemical energy storage technologies, electrode materials participate in the redox reaction through intercalation, conversion, and alloying reactions. Since the bulk phase of electrode materials can participate in the electrochemical reaction, charge transfer and reaction process in the active electrode can significantly affect electrode reaction kinetics. In this regard, the diffusion kinetics of redox species (e.g. Li^+ for LIBs), which can be determined experimentally by diffusion coefficient and theoretically by diffusion barrier, should be taken into account. Hence, to achieve superior energy storage kinetics, both electrodes and electrolytes need to be carefully designed. The typical energy storage processes involved in electrochemical energy storage devices and electrodes are electrochemical double-layer capacitance (EDLC), pseudocapacitance (including underpotential deposition, surface redox pseudocapacitance, intercalation pseudocapacitance), and battery-type faradic reactions (including intercalation, conversion, and alloying), as illustrated in Figure 2.3.

In energy storage devices, EDLC and pseudocapacitor reactions occur at electrode surfaces or near-surface regions, and they occur rapidly compared with battery-type reactions. Battery-type reactions involve bulk electrode reactions, and their reaction kinetics is slower than that of surface-dominated reactions. The electrode process of these reaction mechanisms involves the transport of electrons/ions and possibly contains phase transition. The difference between the capacitive-type electrode process and the battery-type electrode process is that the former primarily occurs at the surface and electrode/electrolyte interface regions, whereas the latter can occur in the bulk of electrode materials. Since electrodes are electrochemically active, various components are defined to describe the equivalent circuit of an electrode and interfaces, including migration of ions within the electrolyte (R_s), adsorption on the electrode surface (R_{ad}, C_{ad}), charge of the electrochemical double layer (C_{dl}), charge of the space charge layer (C_{sc}), ions across the electrode–electrolyte

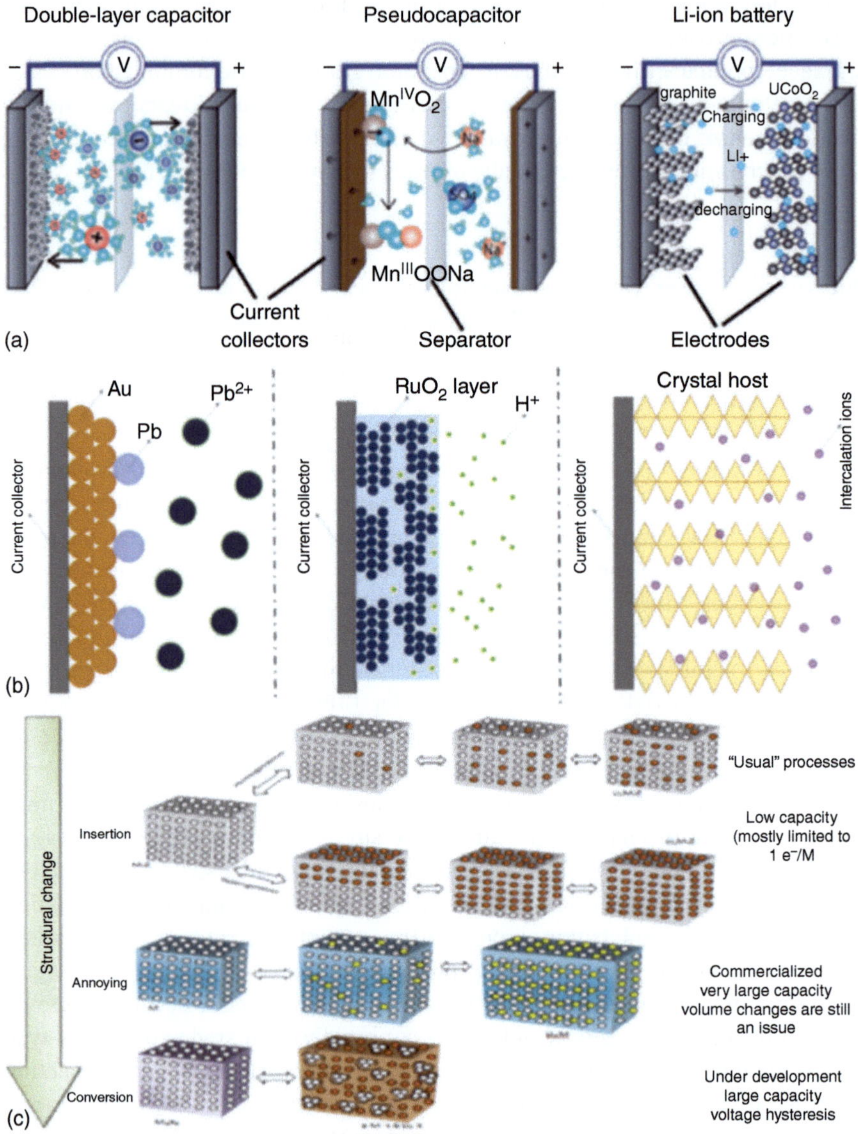

Figure 2.3 Typical process in electrochemical energy storage cells and electrodes. (a) The illustration of electrochemical double layer capacitor, pseudocapacitor, and lithium-ion battery. (b) The diagram of various pseudocapacitors. (c) The energy storage mechanisms of battery-type electrodes. Source: (a) Reproduced with permission from Jost et al. [7]. © 2014/Royal Society of Chemistry. (b) Reproduced with permission from Liu et al. [8]. © 2020/Elsevier. (c) Reproduced with permission from Bensalah and Dowoud [9]. © 2016/Hilaris, CC BY 4.0.

interface ($R_{incorporation}$), charge transfer (R_{ct}), diffusion within the electrolyte inside pores/cavities ($Z_{diffusion}$), bulk diffusion (R_b), diffusion along the grain boundary (R_{gb}), storage of guest atoms in the lattice (C_{chem}), phase transition (C_{chem}), and electron transport (R_e).

In sum, compared with classic electrochemical systems with inert electrodes, the diffusion of ions in the bulk electrode, phase/structural transitions (especially in the long term), and multi-interfaces should be considered in electrochemical energy storage systems with active electrodes.

2.3 Critical Parameters and Design Proposal

To achieve practical applications, electrochemical energy storage technologies should have many properties, such as high energy/power density, intrinsic safety, and long lifespan. The energy density of an electrochemical energy storage technology can be calculated based on the following equation:

$$E = Q \times U \tag{2.19}$$

where Q is the storage charge and U is the output voltage. Hence, high capacity and large voltage window as well as output voltage lead to high energy density. The charge storage ability of a material primarily depends on its theoretical capacity. In general, the theoretical capacity of an electrode material can be calculated using the reversible electron transfer number per mole at the theoretical level and faradic constant. The storage charge of a material can be calculated as follows:

$$Q = n\frac{F}{M \times 3.6} \tag{2.20}$$

where n is the theoretical electron transfer number per mole, F is the faradic constant, and M is the molecular mass of the material. A high electron transfer number and low molecular mass lead to high capacity. The material with the lowest molecular mass that can be used as an active material for electrochemical energy storage is hydrogen, and the theoretical electron transfer number can reach 8 using the redox couple of NH_3/NO_3^-. However, these theoretically high-capacity electrodes have not been practically applied yet because of complicated reaction mechanisms and optimized electrochemical systems. To obtain a high-energy-density technology, the mass between the cathode and anode should be balanced to achieve a good matching. Considering commercial $LiFePO_4$-based LIBs as an example, the capacity of a $LiFePO_4$ cathode and graphite anode is about 140 and 320 mAh g^{-1}, respectively. To store the same amount of charges, the mass of the cathode material should be twofold higher than that of the anode material. Considering the high mass loading required for practical applications, the mass loading of the cathode material can reach 200 mg cm^{-2}, while that of the anode material is about 100 mg cm^{-2}. Such high mass loading of the cathode material leads to a long transport pathway of electrons from the current collector to the cathode surface, and a partially active material in the cathode cannot participate in the redox reaction, leading to poor rate performance and insufficient energy density. In this regard, the closed specific capacity between cathode materials and anode materials is conducive to achieving good matching.

Power density is an important factor in applications that require high power output, such as electric vehicles and energy storage systems for renewable energy. The power density of electrochemical batteries is usually below 500 W kg^{-1}, much less than that of supercapacitors (>10 kW kg^{-1}). Although some power-type LIBs have been commercialized recently, the charging/discharging process can rarely go beyond 3 C, which is insufficient for some applications. Hence, increasing the power density of electrochemical batteries has become a highly researched topic in the field of electrochemical energy storage. High power density means that rapid redox reactions occur in the electrodes, i.e. rate performance. Usually, the ionic conductivity of electrolytes is sufficient to meet the demand for rapid ion diffusion kinetics. The rate performance is primarily determined by ion diffusion kinetics in bulk electrodes. The diffusion of ions from the initial state to the final state occurs through many transition states and overcomes various energy barriers, which indicates that decreasing the diffusion energy barrier can be an effective way to boost ion diffusion kinetics in electrodes. Research shows that heteroatom doping can tailor the electronic structure of a material and decrease its diffusion energy barrier. In addition, rate performance is influenced by operating temperature. The existing commercial LIBs cannot work below 20 °C. At such low temperatures, the ion diffusion barrier in bulk electrodes and the interfacial impendence are significantly increased, leading to poor rate performance.

Cycling lifespan is a critical parameter for applications that serve for a long period, such as energy storage systems and aerospace equipment. The cycling lifespan of an electrochemical battery depends on the structural stability of electrode materials and the chemical/electrochemical stability of electrolytes. In general, for intercalation electrode materials, repeat ion insertion/extraction results in the collapse of the layer structure. This structural degeneration may be caused by several reasons as follows: oxygen evolution may occur at the cathode side, which causes irreversible phase transition; undue ion extraction can induce the Jahn–Teller effect in $LiMn_2O_4$; lithium dendrite growth can occur on the anode side, which can cause "dead Li" formation, leading to irreversible lithium consumption; and so on. In addition, volume expansion may break the SEI on the anode surface, which renders the continuous formation of SEI and consumption of the electrolyte until breakdown. Surface coating and heteroatom doping can effectively prolong the cycling lifespan of electrochemical batteries.

In addition to electrochemical performance under normal conditions, the energy storage performance of electrochemical storage systems in different environments has attracted substantial attention, which are the major issues that should be addressed before being used in wide-ranging applications. State-of-the-art studies primarily focus on the working temperature. Commercial LIBs cannot function at temperatures below −20 °C. Although supercapacitors can work at −40 °C, their energy density is insufficient to be applied as the main power for equipment. Hence, how to achieve high energy/power density below around −40 °C is one of the most important challenges in the field of electrochemical energy storage.

2.4 Common Investigation Technologies

2.4.1 Structural Characterizations

Literature confirms that the nanostructure and crystal structure of a material can significantly affect its electrochemical performance. Briefly, downsizing the nanoparticle size of a material can decrease the distance for ion diffusion between electrodes and electrolytes. Furthermore, decreasing the particle size of a material can increase its specific surface area and contact surface area between electrodes and electrolytes, which can lead to more active materials participating in redox reactions at a relatively high current density. The hierarchical structure of a material further increases its surface area, leading to a high rate performance. Meanwhile, pore structures and hollow structures can provide more space to accommodate the volume expansion of active materials during the charging/discharging process, thus improving structural durability and cycling lifespans of electrodes. With careful design, core–shell composite materials can combine the advantages of different materials, and the shell material can prohibit the structural degeneration of the core material, thus prolonging the cycling lifespans of electrodes. Layered crystal structure materials with large interlayer distances and materials with large frameworks are conducive to rapid ion diffusions, leading to excellent rate performance. It is also worth noting that the structure and components of electrode materials are changed during the continuous charging/discharging process. Hence, understanding the nanostructure and crystal structure of materials is important for the investigation of electrochemical energy storage systems and corresponding materials.

The common structural characterization technologies used in exploring the structure of electrode materials are scanning electron microscopy (SEM), transmission electron microscopy (TEM), atomic force microscopy (AFM), scanning transmission electron microscopy (STEM), Raman spectroscopy, Fourier transform infrared (FTIR) spectroscopy, X-ray photoelectron spectroscopy (XPS), X-ray diffraction (XRD), and Brunauer–Emmett–Teller (BET) test. In general, various electron microscopes are used to analyze the morphology of materials, including nanostructure, size, and phase boundary. These electron microscopes have different resolutions, and the application is also different. For example, SEM is primarily used to determine the nanostructure of materials in the micrometer scale, TEM can reveal the single-particle morphology of materials, and high-resolution TEM can reveal the crystal plane of materials. Spherical-aberration-corrected TEM and STEM can show the atom arrangement of materials, and the results are convincing to identify the doping position of the heteroatom, the existence of the heterointerface of the heterostructure, and the formation of single-atom materials. In addition, electron energy loss spectroscopy (EELS) provides elemental information on the nanometer scale when coupled with TEM and STEM. XRD unravels the crystal structure of materials. The changes in crystal structure induced by heteroatom doping or mechanical strength manifest as the shift of diffraction peaks. Based on Bragg's equation, the shift of diffraction peaks toward a higher degree indicates the decreased corresponding interlayer distance between adjacent crystal planes, and the shift toward a lower degree suggests the increased interlayer distance. XRD has a large database provided by the International Centre for Diffraction Data and is a proven characterization technology. Raman spectroscopy and FTIR can explore the chemical bond environment in materials. Constructing a heterostructure and heteroatom doping can tailor the types of chemical bonds and bond lengths in materials, and thus, Raman spectroscopy and FTIR play a crucial role in ensuring the formation of heterostructures and heteroatom doping. It is also worth noting that Raman spectroscopy is nondestructive and has been used in many fields such as extraterrestrial exploration, testing for pesticide residues, and geological prospecting. XPS estimates the oxidation state and chemical bonds of materials. The existence of vacancies and heteroatoms can affect the binding energy of materials. However, XPS can detect the oxidation states only in the surface region of materials. The chemical bonds and oxidation states of bulk-phase materials can be obtained through X-ray absorption near-edge spectroscopy. It is worth mentioning that extended X-ray absorption fine structure (EXAFS) spectroscopy, a part of X-ray absorption spectroscopy (XAS), presents information about the types and numbers of atoms in coordination with a specific atom and their interatomic distances. Atoms at the same radial distance from a determinate atom form a shell. The number of atoms in the shell is the coordination number. Hence, EXAFS can provide the bulk-phase atomic structure of materials and give more accurate structure information compared with other structural characterization technologies. To ensure the formation of vacancies, electron paramagnetic resonance (EPR, also known as ESR) is a more reliable choice.

Since heterostructure has attracted increasing attention recently, it is necessary to introduce structural characterization methods to identify whether a heterojunction is formed or not. Heterostructure formation can be verified by Raman spectroscopy, XPS, and high-resolution transmission electron microscopy (HRTEM). Once heterostructure formation is confirmed, the band structure of the components of the heterostructure needs to be determined to analyze band bending around the heterointerface. The band structure of ideal semiconductor materials can be described by three parameters, namely Fermi energy, band gap, and electron affinities. Fermi energy can be obtained from UV photoelectron spectroscopy (UPS), and band gap can be estimated using UV-Vis spectroscopy. The energy of the valence band can be identified by valence band XPS. Based on the energy of the valence band and band gap, the energy of the conduction band can be deduced, which is equal to the value of electron affinities, thus establishing the band structure of materials. To estimate the depletion of heterostructure materials, semiconductor type and doping density need to be evaluated. These parameters can be obtained using the electrochemical method (called the Mott–Schottky test) and the Hall effect.

2.4.2 Electrochemical Measurements

Electrochemical measurements are powerful tools for evaluating the energy storage performance of electrode materials and electrolytes, including reversible capacity (or capacitance), rate performance, cycling performance, diffusion coefficient, and charge transfer resistance. Usually, reversible capacity (capacitance), rate performance, and cycling performance are tested using galvanostatic charging and discharging. The capacity of battery-type materials can be calculated using the following equation:

$$Q = \frac{I \times t}{m} \tag{2.21}$$

where Q is the capacity of materials, I refers to the current density used to estimate materials, t is the charging or discharging time, and m is the mass of active materials. The unit of capacity is mAh g^{-1}. Hence, by converting current density to milliamperes, time to seconds, and mass to grams, the capacity of materials can be precisely obtained. Some studies have also used C g^{-1} as the unit of capacity. The relationship between these two units is based on the following specific ratio: $3.6\,\text{C g}^{-1} = 1\,\text{mAh g}^{-1}$. The capacitance of capacitive-type materials can be calculated based on a similar equation:

$$C = \frac{I \times t}{m \times U} \tag{2.22}$$

where U refers to the voltage window of materials. The result is expressed as F g^{-1}. Whether the electrode is a battery type or a capacitor type can be deduced from the cyclic voltammetry (CV) curves of materials. Battery-type materials show sharp redox peaks in their CV curves, while capacitive-type electrodes show approximately rectangle CV curves. The electrode materials discussed in recent studies are hybrid-type materials, which have both battery-type capacity contribution and capacitive-type contribution. To evaluate the contribution of different energy storage behaviors in electrode materials, a power law equation is proposed:

$$I(V) = k_1 v + k_2 v^{1/2} \tag{2.23}$$

where I is the current at a certain potential and v is the scan rate. By plotting $i/v^{1/2}$ versus $v^{1/2}$, the values of k_1 and k_2 can be determined from the slope and the y-axis intercept point of a straight line, respectively. By substituting the values of k_1 and k_2 in Eq. (2.23), one can separately quantify capacitive-type contributions and battery-type contributions. However, it is necessary to remember that this equation is based on an assumption, i.e. no polarization and residual current in the tested materials. Hence, in some studies, the fitted capacitive contribution curves are out of the original CV curves, which is unreasonable. The capacity can be calculated based on CV results. By varying the different conditions (e.g. high/low temperature and external strength), the electrochemical performance of electrode materials and assembled devices can be evaluated in a simulated environment, which can help assess the practicability of the designed materials and devices in real-world applications and the major challenges.

2.4.3 Energy Storage Mechanism Investigations

Once the electrochemical performance and structure of materials are tested, their energy storage mechanism also needs to be investigated to understand why they present such superior or inferior performance. Meanwhile, the design of high-performance electrode/electrolyte materials should be based on an in-depth understanding of their energy storage mechanism, especially during the long-term charging/discharging process. The energy storage mechanism of materials is investigated by the characterization of structural variations of materials during the initial charging/discharging cycles and after long-term cycling. The structural changes in materials can be analyzed by in situ tests (e.g. in situ XRD, in situ Raman spectroscopy, in situ FTIR spectroscopy, and in situ TEM) and ex situ tests. Various X-ray imaging technologies have been developed recently (e.g. transmission X-ray microscopy [TXM], X-ray projection imaging, and scanning transmission X-ray microscopy [STXM]) to observe the three-dimensional changes in materials during the charging/discharging process. Benefiting from these advanced technologies, the understanding of the energy storage mechanism of devices and electrodes/electrolytes has greatly improved. While using these technologies to investigate the designed materials, some factors need to be considered carefully. First, since some electrochemical processes (e.g. adsorption, desorption, and desolvation) occur so quickly, the time resolution of in-situ technologies should match such rapid electrochemical processes. For battery-type electrode materials, a complete charge/discharge process usually takes a few hours or even a few days, which is much higher than that of many in situ characterization technologies. However, for capacitive-type electrodes, the charge/discharge process only takes a few minutes or even seconds. If the signal of the tested materials is not strong enough, the component changes of materials during the reaction cannot be observed. This problem restricts the implementation of in situ Raman spectroscopy in electrochemical energy storage devices because of the weak Raman signal. Fortunately, Li et al. [10, 11] developed various surface-enhanced Raman spectroscopy technologies (e.g. shell-isolated nanoparticle-enhanced Raman spectroscopy [SINERS]) to enhance the Raman signal of materials; using this technology, they observed the configuration and component changes of water on Pt surface. Second, the region size of in situ characterization technologies was investigated. Some technologies can characterize atomic changes during the reaction, while some technologies focus on micrometer-scale changes. So the intention of conducting an in situ test should be clear before testing. Third, the impact of in situ characterization technologies on the structure and components of materials needs to be considered. Taking in situ Raman spectroscopy as an example, if the output power of the laser is too high or the test time is too long, the laser can damage the test materials. Hence, the structure and component changes of materials may not be driven by the electrochemical characteristic. In situ TEM and synchrotron-radiation-based technologies may also result in this issue. In this regard, the test condition should be carefully adjusted. Except for advanced in situ characterization technologies, theoretical computation is a powerful tool for understanding the energy storage mechanism and the enhanced performance of the designed materials. The commonly used theoretical computations are based on the density functional theory through professional software (Vienna Ab initio Simulation Package (VASP) and Materials Studio) and high-performance servers. These computation results can provide an atomic-level understanding of the superiorities of the designed materials and their charge transfer process during the energy storage process [12]. Advances in the development of metal anode materials have triggered the application of a finite element analysis computation method. This computation primarily uses a famous software named COMSOL, which can apply multiphysics on relatively large-scale materials and thus simulate the dendrite growth mechanism and the corresponding impact on electron/electric field distribution, leading to an in-depth understanding of the design principle of metal anodes [13]. Although the energy storage mechanism of various materials has been extensively investigated, understanding their atomic, microsecond, and entire lifespan levels is still a challenge, which should be addressed in future works.

2.5 Common Design Strategies for High-Performance Electrode Materials

One may encounter different types of problems when using these active materials, which can prevent them from giving their best performance. Therefore, various modification strategies have been proposed, which are mainly classified into three types: nanoscale design, conductive composite design, and heteroatom doping. Nanoscale fabrication via bottom-to-up or up-to-bottom methods can effectively reduce the size of electrode materials and achieve a large surface area, which not only increases the contact area between the electrolyte and electrode but also exposes/triggers more active sites to store energy [14]. In addition, the decreased size of electrode materials can result in short transfer paths for ions/electrons, thereby accelerating electrochemical reaction kinetics [15]. The principle behind conductive composite design is as follows: because most of the electrode materials are p-type semiconductors with relatively low electrical conductivity, it is highly necessary to coat a thin conductive layer on them or composite them with a conductive matrix/support [16]. The conductive coating or matrix (e.g. carbon, metal, and conducting polymer) can offer a fast electron transport path, thereby decreasing polarization and promoting performance [17]. However, both nanoscale design and conductive composite design cannot effectively enhance the intrinsic electronic/ionic conductivity in electrode materials. Hence, heteroatom doping can be used to inherently modify or change the electronic/ionic states of active materials [18].

Heteroatom doping usually involves either nonmetal element doping or metal ion doping. Both these types of doping can be used in the case of dual doping. Compared with undoped counterparts, doped electrode materials usually show modified intrinsic electronic/ionic properties [19]. Some interesting phenomena can be observed after heteroatom doping. In one scenario, the doped heteroatom caused an expansion of the crystal cell and enlarged the interlamellar spacing or octahedral/tetrahedral centers, which broadened the transfer path of ions and increased the storage space for sodium ions [20]. In addition, heteroatom doping may cause charge redistribution of crystal cells, especially charge distortion on

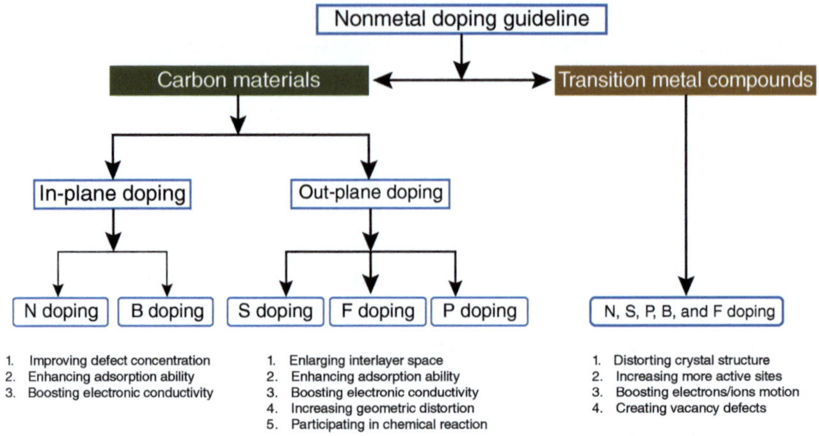

Figure 2.4 Design guideline for nonmetal-doped electrode materials [22]. Source: Reproduced with permission from Li et al. [22]. © 2020/John Wiley & Sons.

surface atoms, and promote electron transport [21]. In other words, heteroatom doping can modulate the intrinsic crystal structure of active materials and affect the charge/ion state, band gap, and crystal structure stability. Of course, energy storage performance is related to doping sites and content.

Generally, the enhanced electrochemical performance of heteroatom-doped materials can be ascribed to the following factors (Figure 2.4): (i) more defects, distortion, and enlarged interlayer space induced by the difference in atomic radii; (ii) enhanced Na ion adsorption and electronic conductivity caused by adjusting electron distribution; (iii) additional active sites originated from heteroatoms; and (iv) synergistic effects between heteroatoms. In nonmetal doping, various plasma treatments to enhance the surface of materials have been developed recently, which can achieve much higher doping content compared with previous doping methods (e.g. chemical vapor deposition and ball milling). In metal ion doping, the dopants are introduced in the host crystal matrix by adding additional metal salts during synthesis reactions (usually hydrothermal reactions), as reported in recent studies. To date, various metal ions such as Ni, Fe, Mg, Ti, Cr, Mo, and Cu have been demonstrated as available dopants for sodium ion battery (SIB) cathodes with enhanced performance. It is worth noting that not all doing is good and sometimes it is harmful. Meanwhile, the effect of doping on electrochemical performance is sometimes unclear, especially doping sites and doping mechanism. Moreover, controllable doping methods are key to achieving optimized performance and need to be developed further. It is expected that using controllable favorable heteroatom doping, intrinsic reaction kinetics can be greatly enhanced, which can lead to enhanced capacity and energy/power density as well as good cycling life. Addressing these issues will promote the practical application of various newly proposed energy storage technologies.

Besides heteroatom doping, a new approach to designing high-performance electrodes has been proposed and developed in recent years: constructing heterostructure. Generally, heterostructure electrodes have higher reversible capacity and better cycling stability compared with their building blocks; some of them even break the limitation of theoretical capacity, demonstrating the superiority of this approach.

In the past condensed matter physics studies, the concept of heterostructure was not originally proposed in the field of energy storage. This concept was first proposed in the field of semiconductor physics by W. Shockley for wide-gap semiconductor emitter applications. According to the definition, a heterostructure consists of different semiconductors with similar crystal structures, similar atomic spacings, and thermal expansion coefficient. The chemical composition and charge distribution change across the interface within the heterostructure, due to different band structures, majority carrier concentration difference, and Fermi-level difference [23]. The performance of the semiconductor heterostructure is much better than individual semiconductor materials due to the synergistic effect. As a generalized definition, a heterostructure consists of two or more types of materials that are physically (mainly van der Waals force) or chemically bonded together, exhibiting complex geometry and indistinct junction interface. It is worth noting that a heterostructure is similar to a composite material with respect to its components; however, focusing issues and design principles are quite different. Generally, a composite material is designed by combining the advantages of two materials, such as coating carbon on a metal compound surface to enhance its electronic conductivity and structural integrity, ignoring the effect of the fundamental nature of building blocks on hybrid materials. In contrast, a heterostructure specifically focuses on the influence of the intrinsic features carried by building blocks (e.g. band structures, alignment styles, semiconductor types, carrier concentration, and Fermi-level difference) on the electronic structure and electric field distribution of whole materials. This is a more reasonable way to reflect on the electronic/ionic environment in hybrid materials. Besides, since charge will redistribute around heterointerfaces, more heterointerfaces can be constructed using heterostructure materials, and the conducive changes undergoing in hybrid materials can be maximized. The composition and structure of heterostructures also play a crucial role in energy storage. From the perspective of the electrochemical energy storage activity of the electrode, the components in a heterostructure could be either metal compounds, carbonaceous materials, or conductive polymers. van der Waals force is common in 2D heterostructures made by stacking different conventional 2D materials, such as graphene, transition metal oxides, chalcogenides, and carbides [24], while other metal compounds and polymers are highly connected by chemical bonds. In this context, heterostructures can be classified into chemically bonded heterostructures and van der Waals heterostructures. According to the location of heterointerfaces and structural features of building blocks, heterostructures can also be classified into core–shell heterostructures, hierarchical heterostructures, island-like heterostructures, and blending heterostructures. Furthermore, heterostructures can also be classified according to their components (silicon-/phosphorus-based [Si-based], phosphorus-based [P-based], metal compounds/metal compounds [MCs/MCs], metal compounds/carbon [MCs/C], MXene [formulated as $M_{n+1}X_nT_x$, where M is

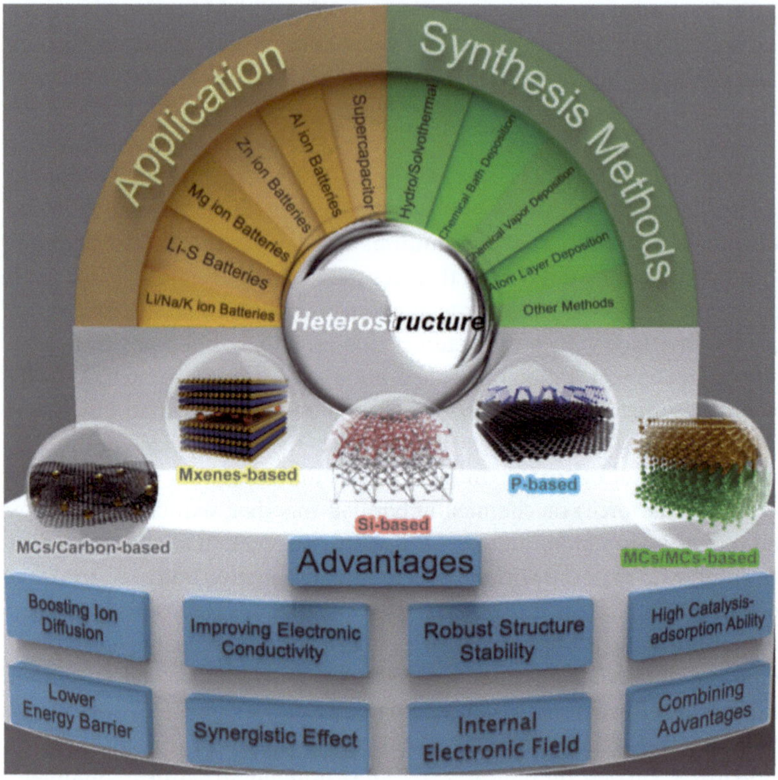

Figure 2.5 A brief introduction of heterostructure materials and corresponding advantages. Source: Reproduced with permission from Li et al. [25]. © 2021/John Wiley & Sons.

a transition metal such as Ti and V; X is C and/or N; and T is a surface termination group OH^-, F^-] heterostructures, and conductive polymer-based types). Generally, the majority of heterostructure electrodes exhibit higher electrochemical storage activity than that of their counterparts due to various synergic effects that heterostructure electrodes might possess (Figure 2.5). First, heterostructure electrodes can incorporate the advantages of different components to neutralize the shortcomings of each component. Second, constructing a heterostructure is an excellent method for enhancing electric conductivity. Electrodes of most of the heterostructure electrochemical storage devices possess refined energy band structures with smaller band gaps, which provide superior conductivity. Third, ion diffusion kinetics can be accelerated due to the introduction of an internal electric field between heterostructure interfaces, which lowers the ion diffusion barrier. Fourth, strong interactions (chemical bands, van der Waals force, and electrostatic force) between multiple phases can enhance the structural stability and prolong the cycling lifespans [26]. Finally, charge redistribution in the building blocks of heterostructures can induce more active sites for energy storage, enhancing the reversible capacity of electrodes.

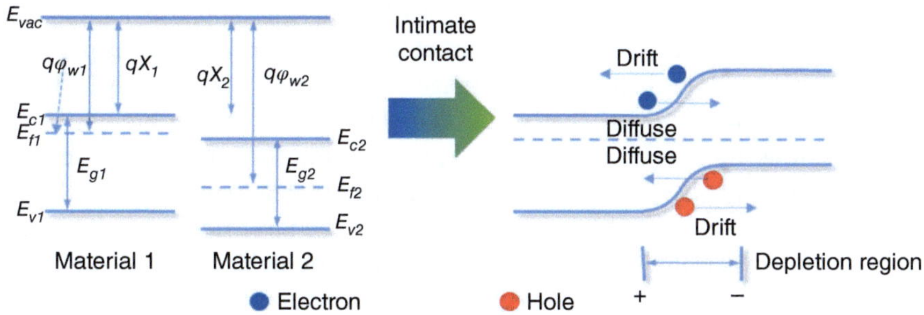

Figure 2.6 Band structure changes after constructing the heterostructure.

The advantages of heterostructure not just lie in simply combining different materials with diverse superiority but are strongly related to the distinct microstructure of heterointerfaces [27]. Once two building blocks are brought into contact, band alignments occur at heterointerfaces spontaneously, which induces charge redistribution near the interfaces (Figure 2.6) until the Fermi level of these materials reaches equilibrium. As a result, electrons and holes accumulate around the heterointerface and are separated by a fully ionized depletion region, forming a certain built-in potential. This built-in potential makes heterostructure materials highly conductive in the forward bias and nearly insulated at the reverse bias. In heterojunctions, conduction bands and valence bands of two building blocks near interfaces bend with a specific band offset during the alignment process, and generally, the bending degree depends on the doping density: the lower the doping density, the larger the band bending. Such bending property in heterostructures may lead to a larger energy barrier for electrons beyond the conduction band offset, which is the electron transport barrier in heterostructures without extreme doping density difference. In addition, the dynamics of the electrons in the semiconductor heterojunctions is also related to band gaps, electron affinities (conduction band minimums), and of course, work functions of combined semiconductors. These unique built-in electric fields and designable electron transport barrier around heterointerfaces bring inspiring electrochemical performance in the field of energy storage, and the related superiorities have also been widely investigated in previous studies.

As mentioned earlier, electrons redistribute around heterointerfaces, so that the partial intrinsic electroneutral building blocks near the boundary are ionized and turn to electric positive or electric negative, which can act as active sites for electrochemical reactions and easily adsorb the oppositely charged carriers. As a result, the positively charged alkali metal ions can be adsorbed at the negatively charged molecules near the heterointerface, storing more electric energy. Investigating the Na^+ storage mechanisms in $CoSe_2$/ZnSe heterostructures, for example, Fang et al. [26] found that the absorbability of $CoSe_2$/ZnSe heterostructures (−2.74 eV) was much higher than that of $CoSe_2$ (−1.23 eV) and ZnSe (−0.28 eV). They claimed that the high absorption of Na ions is due to the redistribution of electrons at heterointerfaces. According to the Bader charge analysis, electrons migrate from the $CoSe_2$ side to the ZnSe side and finally accumulate at the surface of ZnSe.

These negatively charged phases strongly attract Na ions on the phase boundaries close to the ZnSe side, improving the absorbability and energy storage capability of Na ions. Jiang et al. [28] studied the storage behavior of Na ions in Sb_2S_3/SnS_2 heterostructures. They showed that when two semiconductors are brought into contact, electrons transfer from the Sb_2S_3 phase to the SnS_2 phase due to the higher Fermi level of Sb_2S_3, forming an internal electric field pointing from Sb_2S_3 to SnS_2. They proposed that Na ions are attracted to the SnS_2 side and neutralize the accumulated electrons. When the interfacial electric field vanishes, the high concentration gradient at the interface drives the transport of Na ions to both sides of the interface, achieving higher reversible capacity.

In addition to the modified mechanism of the "reservoir model", built-in electric field (BIEF) at heterointerfaces may also be a reason for the improvement in energy storage performance. For energy storage devices such as secondary batteries and supercapacitors, BIEF may accelerate metal ions' diffusion ability by the extra electric field force, and more active materials are involved in energy storage reactions. However, if the electric field accelerates the insertion of ions and remains unchanged upon the charge/discharge process, the extraction of metal ions will be difficult because the direction of diffusion of metal ions is against that of BIEF, leading to the sluggish movability of ions and poor energy storage capability. In contrast, based on experimental results, heterostructure electrodes exhibit better ion diffusion ability than their building blocks, whether in the charging process or discharge process [29]. To address this contradiction, many researchers have been focusing on the working mechanisms of BIEF. Some researchers claim that the direction of the electric field is reversibly changed along with the insertion or extraction of host ions [30]. Xia proposed that the changing of semiconductor types induces electric field inversion [31]. Since the component of active electrode materials is complicated and undergoes variations during the charge/discharge process, the semiconductor types of building blocks are usually difficult to confirm. Nevertheless, this idea strongly inspires researchers to explore the rationality of the direction reversal of the electric field. From a thermodynamics viewpoint, reaction reversibility may be the reason for electric field inversion, especially in three-element heterostructures.

In the case of heterostructures constructed by two building blocks that are composed of the same metal element and different nonmetal elements such as SnS_2/SnO_2, SnS/SnO_2, and $SnSe/SnO_2$, the reversibility of chemical reactions is dominated by nonmetal elements [25]. The lower electronegativity of sulfur and selenium compared with oxygen makes their alkali metal composites easier to reduce, exhibiting higher reversible reactivity. Thus, reversible changed BIEF during the charge/discharge process can be constructed by combining low-Fermi-level metal chalcogenides with high-Fermi-level metal oxides. At the discharge state, electrons at heterointerfaces transfer from metal oxides to metal chalcogenides, hence creating a BIEF to promote the diffusion of Na ions from metal oxides to metal chalcogenides. As for recharge, the extraction process preferentially initiates in the Na_2S or Na_2Se zone due to its higher reactivity than Na_2O. This situation gives rise to Na^+-rich domains in metal sulfides or metal selenides and Na^+-deficient domains in metal oxides. The difference of potential creates a new BIEF in the direction from metal sulfides or metal selenides to metal oxides, thus achieving a reversible changed BIEF. Some researchers have also suggested that using p-type materials and n-type materials with diverse reversibility to construct heterostructures can achieve reversible changed BIEF, such as SnS/SnO_2, Fe_2O_3/FeS_2, Bi_2S_3/Bi_2O_3, and SnS_2/CoS_2. For example in Fe_2O_3/FeS_2 heterostructures, upon discharge, BIEF with an $n \rightarrow p$ direction is formed at interfaces due to the diverse major charge carriers, pushing Na^+ ions from Fe_2O_3 to the neighboring FeS_2. Similarly, due to the higher reversibility of Na_2S, a new BIEF is established upon the charging process, enhancing Na^+ diffusion kinetics in both sodiation and desodiation states.

In the case of heterostructures constructed by one nonmetal element and two different metal elements, the reversibility of chemical reactions can be theoretically calculated. Using theoretical computation, reversibility can be determined by changes in Gibbs free energy in each chemical reaction. For example, Zhang et al. [32] used MoS_2-based heterostructure materials including a conductive Fe_9S_{10} core and carbon coating (Fe_9S_{10}@MoS_2@C) to investigate the working mechanism of BIEF from a thermodynamics standpoint. When placed in contact, electrons move from MoS_2 with a higher Fermi level to the Fe_9S_{10} conductor with a lower Fermi level and develop an electric field in the direction from the MoS_2 layer toward the Fe_9S_{10} core. Using the density functional theory calculation, they compared the reaction reversibility of MoS_2 and Fe_9S_{10} with Na^+ by calculating Gibbs free energy changes, which suggested that the reversibility of Fe_9S_{10} ($\Delta G = -1.49$ eV) was better than that of MoS_2 ($\Delta G = -2.37$ eV). Consequently, when applied as SIB's anode, the Fe_9S_{10} side releases more Na^+ than the MoS_2 side and the different content of Na^+ induces an inversed electric field direction from Fe_9S_{10} to MoS_2, boosting the diffusion ability of Na ions during the recharge process. More recently, a novel idea to explain the working mechanism of BIEF was proposed by Wang et al. that the direction of BIEF remains unchanged during charge/discharge processes. Since the external power supplied was forward-biased to PN-junction upon charging, the space of BIEF would be narrowed [33], which means that the BIEF inhibited the diffusion of ions during the recharge process but the intensity of inhibition was decreased and thus the diffusion of ions could be relatively easier.

Nevertheless, understanding the nature of heterostructures is at an early stage. First, the energy storage mechanism or enhanced mechanism of heterostructures remains controversial. Future studies should figure out the dominant mechanism for the enhanced electrochemical mechanism, how charge redistribution and BIEF affect the entire energy storage process, and what are the most appropriate conditions to achieve the optimized energy storage performance of heterostructures. More solid mechanism studies and experimental studies are needed for a more precise understanding. Recently, machine learning has shown exciting progress in material screening and performance prediction, and COMSOL Multiphysics simulation can help understand the electromagnetic field distribution in electrodes. Meanwhile, numerous material characterization techniques have been explored and applied in energy storage mechanism studies (e.g. in situ characterization techniques, nuclear magnetic resonance, and cryogenic electron microscopy). These advanced theoretical calculation methods and characterization techniques can help researchers enhance the mechanism of heterostructure electrodes. Secondly, the charge redistribution behavior is also related to majority carrier concentration difference, band alignment styles, and the consequent built-in electric field. However, the effect of these factors on the energy storage process is less discussed in the existing literature. Future studies should consider these factors and present optimized situations for different applications. Thirdly, in addition to understanding the fundamental principle behind heterostructures, the influence of structural factors (such as heterointerface locations, interface sizes, coupling forces (chemical bonds or van der Waals force), and transition layer thickness between building blocks) on electron/ion transfer kinetics and energy storage performance is underdeveloped. A more in-depth understanding of these structural factors and fundamental principles can help researchers construct high-performance heterostructures. Fourthly, compared with carbon modification and nanostructure design, heteroatom doping and defect design are rarely discussed in improving the electrochemical performance of heterostructures. However, both heteroatom doping and defect design can adjust carrier concentration and distort local electron distribution [34]. Recently, Song et al. [35] found that defects and dopants show scattering functions to disturb the electron migration route and alleviate the aggregation of electrons at material surfaces. Exploring the potential of heteroatom doping and defect design can help improve the electrochemical performance of heterostructures further. Finally, given practical applications, the existing strategies to producing heterostructure materials usually require a long-time high-temperature annealing process, and thus, the production cost is way higher than that of industry requirements. Meanwhile, the multistep synthesis routes are highly tedious, and the production output is also unsatisfactory. There is no doubt that high yields and low costs are indispensable for commercialization. To this end, exploring kilogram-level outputs and facile one-pot methods is urgently needed to push forward the application of these materials. Microwave-assisted methods can produce pure materials even in seconds. Such powerful techniques can reach ultrahigh temperatures immediately, and diverse metal compounds, as well as their carbon composites, can be successively prepared by manipulating the gas atmosphere in the oven [36]. Therefore, developing microwave-assisted methods may decrease production costs and promote the practical applications of heterostructures.

References

1 Liu, B., Liu, X., Fan, X. et al. (2020). 120 Years of nickel-based cathodes for alkaline batteries. *Journal of Alloys and Compounds* 834: 155185.

2 Kim, T., Song, W., Son, D.-Y. et al. (2019). Lithium-ion batteries: outlook on present, future, and hybridized technologies. *Journal of Materials Chemistry A* 7 (7): 2942–2964.

3 Pacchioni, G. (2019). Superelectrodes for supercapacitors. *Nature Reviews Materials* 4 (10): 625–625.

4 Xia, L., Yu, L., Hu, D. et al. (2017). Electrolytes for electrochemical energy storage. *Materials Chemistry Frontiers* 1 (4): 584–618.

5 Xiao, D., Zhang, L., Li, Z. et al. (2022). Design strategies and research progress for Water-in-Salt electrolytes. *Energy Storage Materials* 44: 10–28.

6 Droguet, L., Grimaud, A., Fontaine, O. et al. (2020). Water-in-Salt electrolyte (WiSE) for aqueous batteries: a long way to practicality. *Advanced Energy Materials* 10 (43): 2002440.

7 Jost, K., Diona, G., and Gogotsi, Y. (2014). Textile energy storage in perspective. *Journal of Materials Chemistry A* 2 (28): 10776–10787.

8 Liu, Y., Jiang, S., and Shao, Z. (2020). Intercalation pseudocapacitance in electrochemical energy storage: recent advances in fundamental understanding and materials development. *Materials Today Advances* 7: 100072.

9 Bensalah, N. and Dawoud, H. (2016). Review on synthesis, characterizations, and electrochemical properties of cathode materials for lithium ion batteries. *Journal of Material Science & Engineering* 5 (4): 1000258.

10 Li, J.F., Huang, Y.F., Ding, Y. et al. (2010). Shell-isolated nanoparticle-enhanced Raman spectroscopy. *Nature* 464 (7287): 392–395.

11 Dong, J.-C., Zhang, X.-G., Briega-Martos, V. et al. (2019). In situ Raman spectroscopic evidence for oxygen reduction reaction intermediates at platinum single-crystal surfaces. *Nature Energy* 4 (1): 60–67.

12 Wu, X., Kang, F., Duan, W. et al. (2019). Density functional theory calculations: a powerful tool to simulate and design high-performance energy storage and conversion materials. *Progress in Natural Science: Materials International* 29 (3): 247–255.

13 Wu, J., Rao, Z., Liu, X. et al. (2021). Composite lithium metal anodes with lithiophilic and low-tortuosity scaffold enabling ultrahigh currents and capacities in carbonate electrolytes. *Advanced Functional Materials* 31 (14): 2009961.

14 Zheng, Q., Liu, W., Li, X. et al. (2016). Facile construction of nanoscale laminated $Na_3V_2(PO_4)_3$ for a high-performance sodium ion battery cathode. *Journal of Materials Chemistry A* 4 (48): 19170–19178.

15 Boebinger, M.G., Xu, M., Ma, X. et al. (2017). Distinct nanoscale reaction pathways in a sulfide material for sodium and lithium batteries. *Journal of Materials Chemistry A* 5 (23): 11701–11709.

16 Liang, X., Chen, M., Pan, G. et al. (2019). New carbon for electrochemical energy storage and conversion. *Functional Materials Letters* 12 (04): 1950049.

17 Stokes, K., Flynn, G., Geaney, H. et al. (2018). Axial Si–Ge heterostructure nanowires as lithium-ion battery anodes. *Nano Letters* 18 (9): 5569–5575.

18 Chowdhury, C., Karmakar, S., and Datta, A. (2016). Capping black phosphorene by h-BN enhances performances in anodes for Li and Na ion batteries. *ACS Energy Letters* 1 (1): 253–259.

19 Li, D., Zhang, L., Chen, H. et al. (2015). Nitrogen-doped bamboo-like carbon nanotubes: promising anode materials for sodium-ion batteries. *Chemical Communications* 51 (89): 16045–16048.

20 Soo Yun, Y. and Jin, H.-J. (2017). Sulfur-doped, reduced graphene oxide nanoribbons for sodium-ion batteries. *Materials Letters* 198: 106–109.

21 Li, X., Wang, H., Robinson, J.T. et al. (2009). Simultaneous nitrogen doping and reduction of graphene oxide. *Journal of the American Chemical Society* 131 (43): 15939–15944.

22 Li, Y., Chen, M., Liu, B. et al. (2020). Heteroatom doping: an effective way to boost sodium ion storage. *Advanced Energy Materials* 10 (27): 2000927.

23 Guo, Q., Chen, N., and Qu, L. (2020). Two-dimensional materials of group-IVA boosting the development of energy storage and conversion. *Carbon Energy* 2 (1): 54–71.

24 Zhang, Y., Xia, X., Liu, B. et al. (2019). Multiscale graphene-based Materials for applications in sodium ion batteries. *Advanced Energy Materials* 9 (8): 1803342.

25 Li, Y., Zhang, J., Chen, Q. et al. (2021). Emerging of heterostructure materials in energy storage: a review. *Advanced Materials* 33 (27): 2100855.

26 Fang, G., Wang, Q., Zhou, J. et al. (2019). Metal organic framework-templated synthesis of bimetallic selenides with rich phase boundaries for sodium-ion storage and oxygen evolution reaction. *ACS Nano* 13 (5): 5635–5645.

27 Song, W., Li, M., Wang, C. et al. (2021). Electronic modulation and interface engineering of electrospun nanomaterials-based electrocatalysts toward water splitting. *Carbon Energy* 3 (1): 101–128.

28 Fang, L., Lan, Z., Guan, W. et al. (2019). Hetero-interface constructs ion reservoir to enhance conversion reaction kinetics for sodium/lithium storage. *Energy Storage Materials* 18: 107–113.

29 Chen, H., Deng, L., Luo, S. et al. (2018). Flexible three-dimensional heterostructured ZnO-Co_3O_4 on carbon cloth as free-standing anode with outstanding Li/Na storage performance. *Journal of The Electrochemical Society* 165 (16): A3932.

30 Zhao, X., Sui, J., Li, F. et al. (2016). Lamellar $MoSe_2$ nanosheets embedded with MoO_2 nanoparticles: novel hybrid nanostructures promoted excellent performances for lithium ion batteries. *Nanoscale* 8 (41): 17902–17910.

31 Xia, T., Zhang, W., Murowchick, J. et al. (2013). Built-in electric field-assisted surface-amorphized nanocrystals for high-rate lithium-ion battery. *Nano Letters* 13 (11): 5289–5296.

32 Jariwala, D., Marks, T.J., and Hersam, M.C. (2017). Mixed-dimensional van der Waals heterostructures. *Nature Materials* 16 (2): 170–181.

33 Zhang, Y., Nie, P., Xu, C. et al. (2018). High energy aqueous sodium-ion capacitor enabled by polyimide electrode and high-concentrated electrolyte. *Electrochimica Acta* 268: 512–519.

34 Liu, X.Y., Zhang, Y.Q., Xia, X.H. et al. (2013). Self-assembled porous $NiCo_2O_4$ hetero-structure array for electrochemical capacitor. *Journal of Power Sources* 239: 157–163.

35 Zhao, Y., He, X., Chen, R. et al. (2018). Hierarchical $NiCo_2S_4$@$CoMoO_4$ core-shell heterostructures nanowire arrays as advanced electrodes for flexible all-solid-state asymmetric supercapacitors. *Applied Surface Science* 453: 73–82.

36 Jia, Y., Ma, Y., Zhu, L. et al. (2019). Hierarchical $NiCo_2O_4$/Bi_2MoO_6 heterostructured nanorod arrays for high-performance supercapacitors. *Materials Letters* 244: 130–133.

3

Lithium-Ion Batteries

Yu Li[1] *and Minghua Chen*[2]

[1]*Harbin University of Science and Technology, School of Electrical and Electronic Engineering, Department of Electrical Theory and New Technology, 52 Xuefu Road, Nangang District, Harbin, 150080, China*
[2]*Harbin University of Science and Technology, School of Electrical and Electronic Engineering, Department of New Energy Materials and Devices, Key Laboratory of Engineering Dielectric and Applications (Ministry of Education), 52 Xuefu Road, Nangang District, Harbin, 150080, China*

3.1 Brief Introduction

The emergence of lithium-ion batteries (LIBs) has greatly accelerated the conversion from nonrenewable fossil energy to renewable clean energy. The concept of LIBs was proposed in 1962 as a primary battery that cannot be recharged, using Li as the anode and MnO_2 as the cathode. The first commercial rechargeable LIB was developed by Moli Energy in 1985, in which Li was the anode and MoS_2 was the cathode. The Li anode stored energy through reversible Li plating/striping, and the MoS_2 cathode stored energy via reversible intercalation and conversion reactions. However, this LIB showed serious safety issues (e.g. thermal runaway and explosion), which could be attributed to the growth of lithium dendrites on the anode surface that induced internal short circuits and released a substantial amount of heat. The next groundbreaking achievement was the development of the graphite anode, which stored energy through reversible lithium intercalation, and hence, lithium dendrites were found to hardly grow on the anode electrode surface. Using lithium cobalt oxide ($Li_{1-x}CoO_2$, LCO) as the cathode material, the first LIB with acceptable safety was commercialized by Sony in 1991. Nevertheless, the reversible capacity of the $LiCoO_2$ cathode was about twofold lower than that of the graphite anode, which meant that to store the same charges, mass loading of the cathode should be at least two times higher than that of the graphite anode theoretically. In addition, the thick electrode increased the diffusion distance from the electrode surface to the current collector, leading to poor rate performance. Therefore, various cathode materials have been developed in the following years. Based on different cathode materials and anode materials, the energy density of LIBs can reach 150–300 Wh kg^{-1}, which is much higher than that of previously reported electrochemical energy storage technologies, such as lead–acid batteries, Ni-MH batteries, and Ni-Zn batteries. Furthermore, the output voltage of LIBs is more than 3 V, which

Towards Next Generation Energy Storage Technologies: From Fundamentals to Commercial Applications, First Edition. Edited by Minghua Chen.

is also higher than that of other electrochemical energy storage technologies. Hence, the emergence of LIBs has brought a renewable and mobilizable energy supply for various applications.

Due to their high energy density and high output voltage, LIBs have been used as the power supply for portable devices, e.g. mobile phones, laptops, and watches. The battery packs of these devices usually consist of 3–12 cells, in parallel or serial connection. The successful application of LIBs in these portable devices is attributable to the following factor: the amount of lithium ions is low, and each battery in a pack can be well controlled by simple battery management systems. However, the application of LIBs in large equipment, such as electric vehicles (EVs) and grid-scale energy storage systems, is difficult. If each LIB is not well controlled, lithium dendrite growth can happen at anode surfaces, leading to thermal runaway and even explosion. With advances in battery management technologies, the safety of LIBs has been greatly improved. Nowadays, EVs have been commercialized, and thermal runaway accidents are rarely reported in recent years. LIBs have also been applied in many other electric mobilities, including pedelecs (electrically assisted bicycles), electric bicycles, electric scooters, hybrid vehicles, electric buses, and trucks. In stationary applications, LIBs are available as mini storage devices with around 2 kWh up to 40 MWh in larger plants.

The superior energy storage performance of LIBs is attributable to their unique energy storage mechanism. As discussed in Chapter 2, LIBs are typical electrochemical energy storage devices consisting of a cathode, anode, separator, and electrolytes. A wide range of commercialized LIBs are based on reversible Li^+ intercalation between cathodes and anodes. Considering the $LiCoO_2$ cathode and the graphite anode as an example, during the discharging process, Li ions are extracted from the interlayer of graphite and migrate across the separator to enter the interlayer of the $LiCoO_2$ cathode. At the same time, electrons flow along the external circuit from the anode to the cathode. During the charging process, Li ions are extracted from the graphite side and intercalated into the cathode. Electrons flow from the $LiCoO_2$ cathode to the graphite anode through the external circuit. So, LIBs are given the name "rocking-chair" batteries because of this energy storage mechanism, as illustrated in Figure 3.1.

Different cell materials are used to produce cylindrical, prismatic, and pouch cells for different applications. Depending on the application, a single battery cell is used or several cells are connected in series in a module. A parallel connection is also possible, depending on the required capacity. Several connected modules form a battery system for automotive applications. To sum up, the major advantages of LIBs are high energy density, long cycling lifespans, and low self-discharge rates. With amazing progress achieved in material science, various materials have been explored for the potential to be used as electrode materials for LIBs, and the energy density and lifespans of LIBs have been greatly improved. Various strategies have been proposed to deal with the tough challenges of uncontrollable lithium dendrite growth on the surface of the lithium metal anode (LMA) and the safety issues of LIBs, further improving the theoretical limits and practical potential of LIBs. The following section will thoroughly summarize the amazing progress made in material

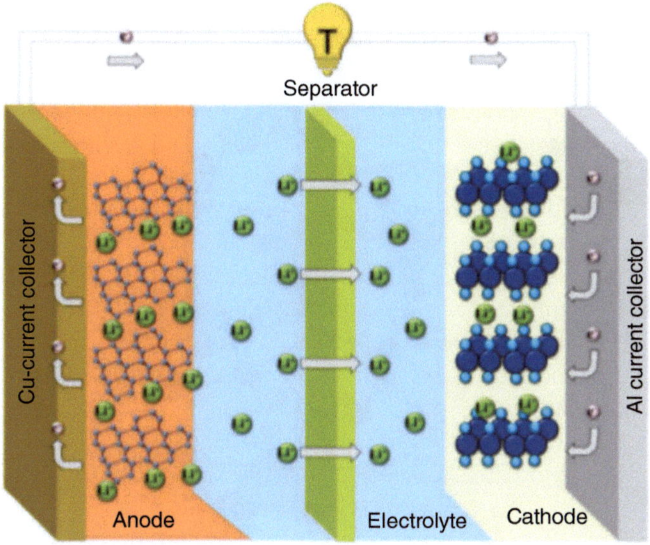

Figure 3.1 Structure and general reaction of LIBs. Source: Reproduced with permission from Qiu et al. [1]. © 2022/Elsevier.

science regarding advanced LIBs and introduce the essential cathode and anode materials that possess attractive practical potential.

3.2 Cathode Materials

3.2.1 $LiFePO_4$

$LiFePO_4$ (LFP) is a natural mineral of the olivine family (triphylite) (Figure 3.2). Arumugam Manthiram and John B. Goodenough identified the polyanion class of cathode materials for LIBs and demonstrated for the first time the possibility of extracting Li from LFP using an insertion method. LFP possesses many advantages such as low raw material costs, nontoxicity, natural abundance of iron, excellent thermal stability, and acceptable specific capacity (170 $mAh\,g^{-1}$), making it suitable for commercialization. The output potential of LFP is about 3.2 V (versus Li/Li^+), which is lower than that of a NiCoMn-based cathode. Although the relatively low output potential of LFP results in low energy density, its electrolyte stability is effectively improved due to the low operating voltage and thus the stability of LIBs. However, LFP exhibits low intrinsic conductivity and hence reduced operating efficiency. Prof. Michel Armand and his coworkers found that adding carbon materials to LFP cathodes significantly improves their performance, making them achieve a practical reversible capacity comparable to the theoretical value (~170 $mAh\,g^{-1}$). LFP rechargeable batteries have a high specific energy (90–170 $Wh\,kg^{-1}$), which is sufficient to power EVs and grid-scale energy storage systems. It is also important to note that some LFP-based batteries offer good cyclic

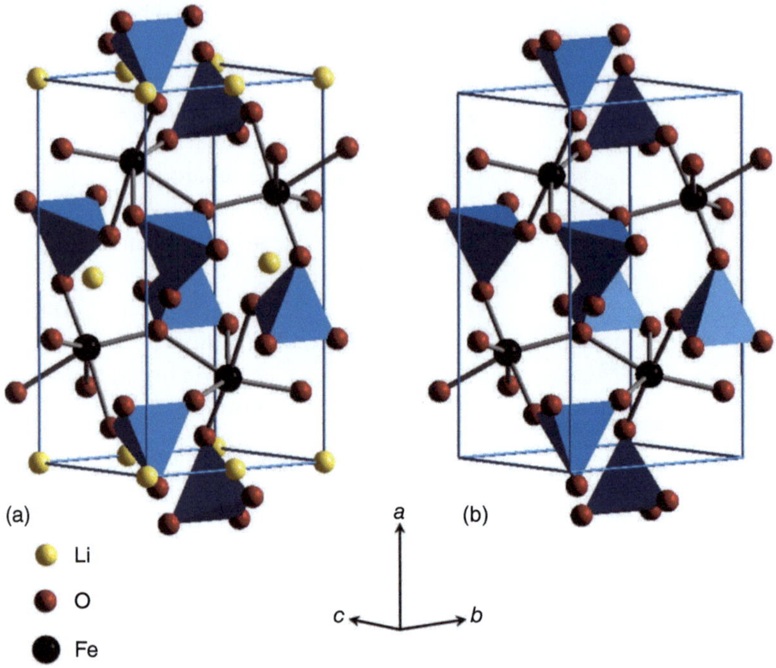

Figure 3.2 Olivine LFP. (a) $LiFePO_4$. (b) $FePO_4$. Source: Reproduced with permission from Castro et al. [2]. © 2022/American Chemical Society.

performance (>3000 cycles) with nominal cell voltage (~3.2 versus Li/Li^+) and even can stably function for more than 10,000 cycles. LFP is a benign substance that does not emit either harmful or toxic gases and is safer for humans and the ecosystem. It is resistant to chemicals (oxygen) and less combustible; therefore, it is less vulnerable to thermal runaway and can withstand temperatures up to 270 °C.

Although LFP has undeniable benefits such as its olivine structure, high theoretical capacity (~170 mAh g^{-1}), safety, low cost, and environmental benignity, there exist a few disadvantages: (i) low intrinsic electronic conductivity (10^{-6} to 10^{-10} S cm^{-1}) and (ii) inadequate Li-ion diffusion kinetics, which limits its performance. The broken octahedral sites of FeO_6 in LFP diminish the speed of Li-ion diffusion across Fe–O–Fe (~10^{-14} cm^2 s^{-1}). To address these issues, various strategies have been explored, such as reducing LFP's particle size, coating a functional surface layer, and heteroatom doping. Reducing the particle size of LFP not only decreases the diffusion distance of Li^+ but alleviates structural variation during the rapid charging/discharging process. However, it creates a large void space between adjacent particles, leading to decreased energy density.

Doping widens the Li^+ migration pathway and enhances the output voltage and working potential of LFP. To develop LIBs with an operating voltage of 3.4 V and above versus Li/Li^+, the redox potential of the cathode materials should be boosted, which can be considerably achieved by ion doping, which is an interesting research topic. Boosting the operating potential, however, is generally associated with a few constraints. To avoid the loss of electrochemical efficiency of the battery, the operating potential must remain within the electrochemical stability window of the electrolyte (i.e. the battery can be recharged and discharged repeatedly without triggering thermal decomposition and redox failure of the electrolyte). Due to variations in free energy during chemical reactions, the electrochemical stability window of the electrolyte becomes shorter than the energy barrier separating the lowest unoccupied molecular orbital and the highest occupied molecular orbital of the electrolyte. Furthermore, high output voltage could induce permanent structural changes in the cathode materials, thereby compromising the electromechanical stability and decreasing the operational lifetime of the LIB. As a result, doping numerous ions requires several factors to be considered in terms of performance and post-doping outcomes. The Li^+ migration route can be shortened by reducing the particle size of LFP; however, this has the drawback of introducing the interface effect, where ions get crowded, resulting in uneven polarization and extensive electron generation; these disadvantages may cause thermal instabilities that necessitate optimization and management.

3.2.2 $LiNi_xCo_yMn_zO_2$

The most frequently examined system of cathode materials consists of layered oxides, with the chemical formula $LiMO_2$ (M = Co and/or Ni and/or Mn and/or Al, Figure 3.3). This system's boundary phases, important binary compounds, and best-known ternary phase $Li_{1-x}(Ni_{0.33}Mn_{0.33}Co_{0.33})O_2$ (NCM) are outlined

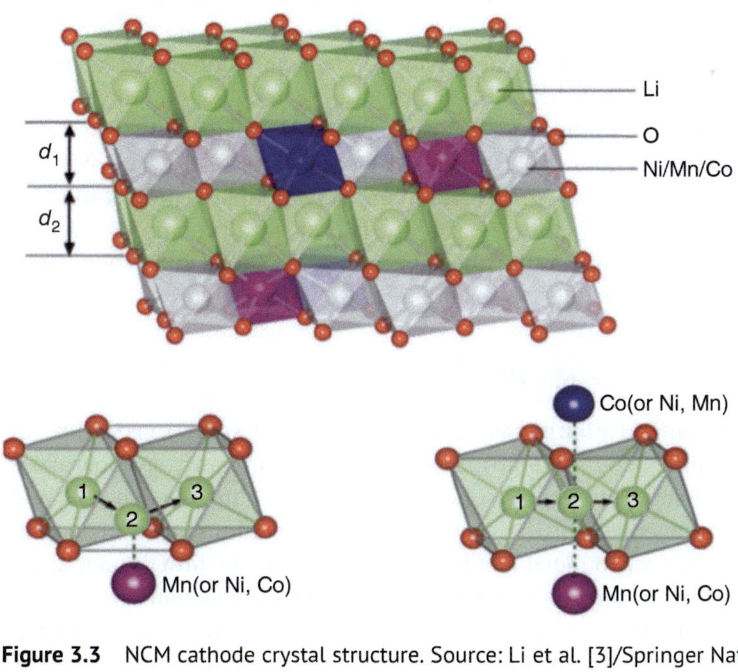

Figure 3.3 NCM cathode crystal structure. Source: Li et al. [3]/Springer Nature/CC BY 4.0.

in the following. LCO has probably been the most widely used cathode material since the market launch of the first rechargeable LIB by Sony in 1991. LCO forms an α-$NaFeO_2$ structure (R-3m). In this structure, cobalt fills the 3a positions and lithium fills the 3b positions. Both are surrounded octahedrally by oxygen on the 6c layers. The two octahedral strata are stacked alternately along the *c*-axis (in the hexagonal setting, along [111] in the rhombohedral cell). Lithium ions in this structure can move within one plane (2D), perpendicular to the stack direction between cobalt octahedral layers. Therefore, they can intercalate and deintercalate the structure. Although the application of LCO is successful, it is not the best cathode material, especially regarding safety and stability. The band structure of LCO and the decrease in the Fermi level caused by charging above 4.6 V versus Li/Li^+ enable the depopulation of 2p- oxygen states, releasing oxygen. Oxygen cannot escape from the cells and, in combination with the organic electrolyte, can react violently by causing flames or even an explosion. Furthermore, the proven solubility of cobalt in commonly used electrolytes might cause the dissolution of cobalt ions from the LCO structure, which might result in capacity loss and ultimately the failure of the cell. Cobalt compounds are relatively expensive because cobalt is very rare in the Earth's crust (30 ppm).

Furthermore, nickel has partially been replaced by cobalt up to a substitution degree of <20% to eliminate the negative effects of the nickel compound without losing its positive characteristics. The resulting compounds in their lithiated form always have developed a structure similar to that of α-$NaFeO_2$. Cobalt substitution results in less mixed occupancy on the lithium layer and consequently in a higher reversible capacity. In addition, the integration of cobalt increases the Li-ion conductivity and the concentration of charge carriers, which increases electrical conductivity.

Further modifications to improve the stability of the compound have been made to lower the 2p- oxygen bands in the band structure, thereby moving them out of the Fermi-level range to minimize oxygen release. One means of achieving this is to incorporate aluminum into the structure. The compound $LiNi_{0.8}Co_{0.15}Al_{0.05}O_2$ (nickel cobalt aluminum, NCA) is available commercially, and in contrast to LNO, it exhibits better stability and cycling capability. Partial substitution of F^- and S^{2-} respectively for the O^{2-} layer could further improve cycling stability because it prevents migration of Ni^{2+} to the Li^+ position.

3.2.3 Spinel (LiM_2O_4; M = Mn, Ni)

$Li_{1-x}Mn_2O_4$ compounds (lithium manganese oxide, LMO spinel) crystallize in the spinel type (space group: Fd-3m), where oxygen atoms form a cubic close packing on the 32e sites (Figure 3.4). Manganese atoms occupy the 16d sites and therefore are coordinated tetrahedrally by six oxygen ions. Lithium ions are located in the 8a positions and surrounded tetrahedrally by oxygen. The $MnO_{6/3}$ octahedra share edges and form a three-dimensional lattice. The 16c positions, which are surrounded octahedrally by oxygen, remain unoccupied and, in combination with the lithium-occupied 8a positions, form a three-dimensional lattice, through

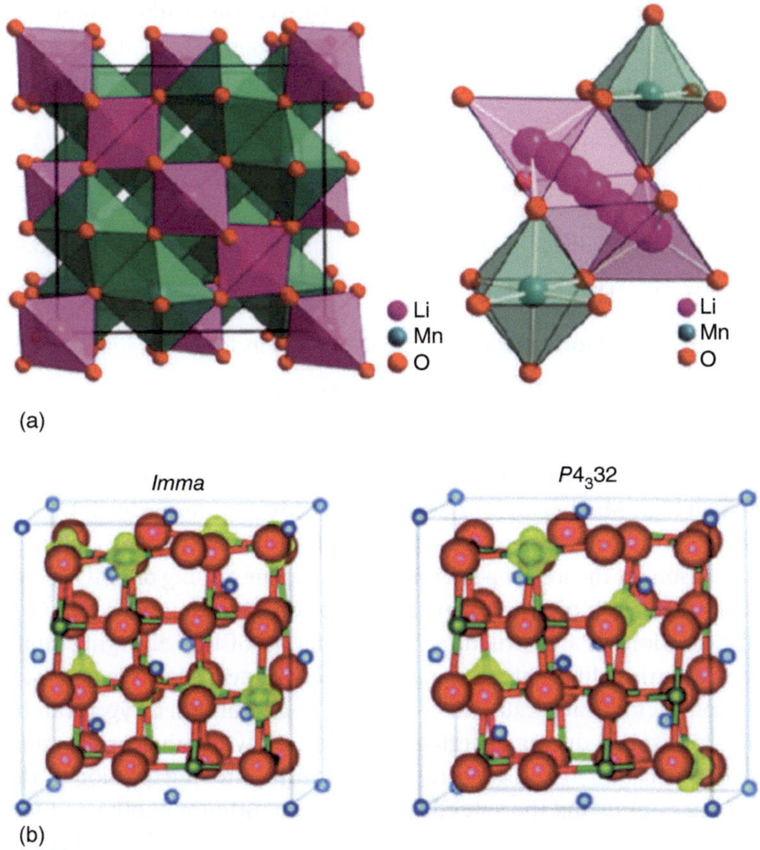

Figure 3.4 Crystal structures of (a) $LiMn_2O_4$ and (b) $LiNi_2O_4$. Source: Reproduced with permission Zhang et al. [4]. © 2013/Elsevier.

which lithium ions can diffuse. Lithium ions can be incorporated into high-lithium compounds such as $Li_{1+x}Mn_{2-y}O_4$ octahedral. Manganese in its lithiated form takes on oxidation states +3 and +4. Trivalent manganese is oxidized during delithiation, and the redox pair Mn^{3+}/Mn^{4+} determines the working voltage of 4.1 V versus Li/Li^+.

LMO and its doping variants are, similar to LCO and NCM, established commercial cathode materials. Their implementation and establishment as first-generation cathode materials are the result of many years of research and development, which aimed to improve the electrochemical characteristics of LMO spinel. Among other things, researchers have focused on the morphology control and surface chemistry of LMO crystallites for decades. LMO exhibits poor cycling behavior as a cathode material, resulting in a continuous decrease in reversible capacity. This characteristic is even more pronounced at temperatures above 50 °C, which can be easily reached in stationary and automotive applications. This behavior can be triggered by multiple factors. Probably, the most important factor is the instability of LMO toward acids. For example, traces of water might react with the conducting salt as shown in Eqs. (3.1)–(3.5), resulting in hydrogen fluoride:

$$LiPF_6 \rightleftharpoons LiF + PF_5 \tag{3.1}$$

$$PF_5 + H_2O \rightarrow POF_3 + 2HF \tag{3.2}$$

$$POF_3 + H_2O \rightarrow HPO_2F_2 + HF \tag{3.3}$$

$$HPO_2F_2 + H_2O \rightarrow H_2PO_3F + HF \tag{3.4}$$

$$H_2PO_3F + H_2O \rightarrow H_3PO_4 + HF \tag{3.5}$$

The resulting hydrogen fluoride reacts with LMO spinel as follows:

$$2LiMn_2O_4 + 4HF \rightarrow 3MnO_2 + Mn_2 + 4F^- + 2Li^+ + 2H_2O \tag{3.6}$$

The Mn^{2+} ions dissolved during the electrolyte reaction can deposit on the cathode as MnO or MnF_2. At the same time, the spinel of the cathode material transforms into deficient high-Mn^{4+} spinel (Li_2MnO_4 and $Li_4Mn_5O_{12}$, respectively). MnO depositions on the anode cause negative effects on the solid electrolyte interface (SEI). A further reason for the loss of performance during the cycling of LMO spinel is the Jahn–Teller instability of Mn^{3+}, which leads to the distortion of the $MnO_{6/3}$ octahedra and distortions of the structure. The formation of two diverging cubic phases is observed during different charging and discharging cycles. Their crystal structures vary considerably and cause microtension in crystallites. Oxygen can be released in the LMO spinel during charging as is the case with other oxidic cathode materials. This causes a loss in capacity and possibly a fire or explosion of the cell. Approaches to minimizing the solubility of Mn^{3+} are lithium doping and formation of phases with a composition of $Li_{1+x}Mn_{2-y}O_4$. Mn^{3+} is partially oxidized to the oxidation state +4, which leads to a decrease in capacity.

3.2.4 Organic Cathodes

Compared with inorganic cathode materials, organic cathode materials have the advantages of high theoretical specific capacities, environmental friendliness, flexible structural designability, high safety, and natural abundance and are a promising class of energy storage materials with broad application prospects. The first commercial rechargeable LIB used $(CF_x)_n$–Li as the cathode material in the early 1970s, which is the prototype concept of organic cathode materials. Afterward, attempts to use fluorinated carbon as the cathode materials in LIBs have never been stopped. However, the subsequent success of intercalation electrode materials has overshadowed the glory of slowly developing organic electrode materials, thereby diverting research attention. In the past 10 years, research on intercalation cathode materials has encountered a bottleneck and gradually slowed down, whereas a series of organic compounds with special structures have been synthesized and their electrochemical performance and reaction mechanism have been investigated. More importantly, the breakthrough has been achieved, providing new ideas for the design of novel organic cathode materials for LIBs and making them research hotspots again. Based on the structural difference, well-studied organic cathode materials can be classified into several types, such as conductive polymers, organosulfur compounds, oxynitride radical compounds, carbonyl-conjugated compounds, and imine compounds. Among them, carbonyl-conjugated compounds have attracted widespread research attention as a novel class of electrochemical redox energy storage materials. Their typical characteristic is a large conjugated system with multiple carbonyl functional groups, which provides them with advantages such as structural diversity, high specific capacity, and fast reaction kinetics. In addition, carbonyl-conjugated organic substances as derivatives of natural biomass extracts have the benefits of low-cost synthesis and large-scale production, which are additional advantages for them as cathode materials. Therefore, compared with other types of organic cathode materials, carbonyl-conjugated compounds are expected to evolve as the next-generation cathode materials for green and sustainable LIBs.

3.3 Anode Materials

3.3.1 Graphite

The electrochemical intercalation of nonsolvated lithium in graphite occurs within a potential range of 0 to 0.25 V versus Li/Li^+. This follows different, well-defined two-phase plateaus with well-defined chemical compounds at the beginning and end of the plateaus. Both phases simultaneously exist throughout the plateaus. Experiments have shown that intercalation steps are measurable and easily discernible because of their color. The hexagonal (ABABAB) and rhombohedral (ABCABC) graphite structures transform into an AAAAAA stack sequence with intercalated lithium during intercalation. Lithium's position is at the center of the C6 rings between two graphene layers [5]. Therefore, the capacity of graphite depends on the number of graphene layers available. The theoretical reversible specific capacity of 372 $mAh\,g^{-1}$ can be achieved if very well-structured graphite (e.g. natural graphite) and slow charging rates (low currents) are used. Figure 3.5 shows that there is a difference between the charging and discharging capacity of the first cycle. This capacity loss results from the electrochemical reactions between lithium ions from the cathode, electrolyte components (organic carbonates, additives, etc.), and the anode surface.

This electrochemical reaction results in a passivation layer between the electrolyte and graphite particles. This layer is called the SEI. The quality of the SEI significantly influences the cycling stability, service life, power, and safety of Li-ion cells. For example, the SEI must exhibit the best possible Li-ion conductivity to achieve the lowest possible overvoltage. At the same time, the SEI should work as a filter for the solvated Li-ion solvent. This prevents the co-intercalation of solvents, which would otherwise destroy the graphite lattice. Furthermore, the SEI should adhere well to the anode particle and be expandable up to a certain degree (ensuring a volume increase without anode particles cracking during intercalation) to prevent further losses during cycling.

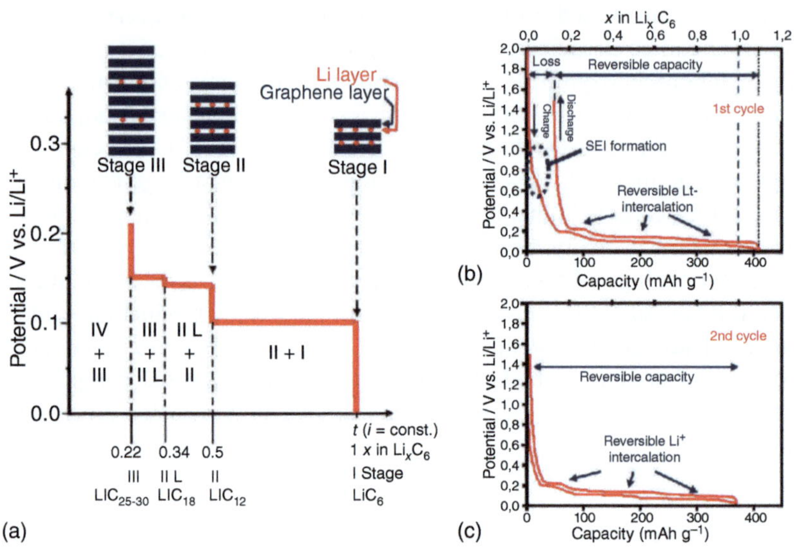

Figure 3.5 Intercalation mechanism of graphite anode. Source: Reproduced with permission from Winter et al. [6]. © 1998/John Wiley & Sons.

The chemistry of the electrolyte and the physical and chemical properties of the anode surface significantly influence the thickness and chemical composition of the SEI. The most important characteristics of the anode surface are specific surface (depending on particle shape, particle size, and porosity) and surface chemistry (e.g. groups such as –COOH, –CO, –OH). Particle shape, particle size, and surface chemistry of the anode also affect the stability and quality of the slurry that is used as a coating for the current collector (copper foil) during electrode production. Round particles flow better during mixing and coating processes and also have a lower specific surface. This means that less binder is necessary for electrode formation, and fewer lithium ions and a lower amount of electrolyte are required for SEI formation during the first charging cycle. Modification of the chemical surface during oxidation/reduction or using process steps such as particle coating of the graphite anode material affects the SEI quality as well as adhesion to the copper foil and the SEI itself.

Regarding anodes, graphite has had a dominating market share of 98% since the invention of LIBs, leaving only a 2% market share for $Li_4Ti_5O_{12}$ (LTO). This is attributable to the fact that graphite has an incomparable balance of relatively low cost, abundance, high energy density (high capacity with low de-/lithiation potential), power density, and very long cycle life. However, the electrochemical performance of graphite needs to be further enhanced to fulfill the increasing demand for advanced LIBs for EVs and grid-scale energy storage stations. In addition, the cycle stability and coulombic efficiency (CE) of graphite need to be further enhanced to fulfill the demand for robust and low-cost large-scale electrochemical energy storage stations. Last but not least, studies on preventing Li dendrite growth and following safety issues by surface engineering of graphite and electrolyte additives are urgently needed to relieve the anxiety regarding the application of graphite in EVs and grid-scale energy storage.

3.3.2 Silicon-Based Materials

As each silicon atom can bind to around four lithium ions, silicon electrodes in LIBs possess a high theoretical gravimetric capacity (4200 mAh g^{-1} for $Li_{22}Si_5$ at 415 °C and 3579 mAh g^{-1} for $Li_{15}Si_4$ at room temperature), making them one of the most promising alternative candidates. Besides, silicon is the second-most abundant element in the Earth's crust, with relatively low redox voltage and remarkable environmental benignity. However, completely replacing traditional graphite electrodes with Si electrodes is still unrealistic because of silicon's low conductivity, insufficient electrode loading, and limited energy density. Furthermore, silicon electrodes also suffer from large volume change and sluggish ion diffusion, resulting in structural pulverization, electrical isolation, and stability decline during repeated charge/discharge processes.

To alleviate these challenges, various Si-based anodes with different modifications have been synthesized. Among these, incorporation of silicon active materials in carbonaceous products (e.g. carbon nanotubes and graphene) is widely used in anode enhancement, such as depositing a silicon layer on the carbon surface, constructing porous silicon/carbon anodes, wedging carbon nanotubes into micron Si, and embedding nanosilicon into a carbon matrix, since carbon materials can not only buffer electrode disintegration but also significantly increase the conductivity of silicon. Nevertheless, the excessive formation of the SEI layer due to the enlarged surface area and unstable structure can partially offset the merits of engineered silicon. Graphite, as a low-cost commercial carbon material with high tap density, low surface area, and stable physical and chemical properties, can significantly reduce unnecessary electrolyte decomposition on the surface of electrodes while maintaining satisfactory anode integrity. Thus, designing an anode that incorporates Si in a mature graphite backbone is expected to be an effective method to alleviate the intrinsic disadvantages of Si-based anodes and attain satisfactory specific capacity and cycling stability.

3.3.3 Lithium Metal

Li metal is considered one of the most promising alternative anode materials due to its low electrochemical potential (−3.04 V versus the standard hydrogen electrode) and an ultrahigh theoretical capacity of 3860 mAh g^{-1}, which is almost 10 times that of a commercial graphite anode (372 mAh g^{-1}). The production of first-generation Li metal batteries (LMBs) consisting of a LMA and a TiS_2 cathode with high energy density and a fast charging rate dates back to the 1970s. However, the application of LMAs in batteries has been impeded by the following problems: (i) Uneven plating/stripping of Li metal results in the growth of Li dendrites. These Li dendrites may pierce the separator and cause internal short circuits and even catastrophic combustion and explosion. (ii) Li metal has a higher chemical and electrochemical activity and thus easily reacts with the liquid electrolyte to form an unstable SEI on the Li metal surface. This persistent reaction causes the rapid consumption of Li metal and electrolyte and results in a low CE and rapid capacity fade. (iii) As a "hostless" material, the relative change in the volume of LMAs is virtually infinite, which may damage the fragile SEI and induce immense internal stress fluctuation of the battery, seriously deteriorating the performance. (iv) The inherent volume variation of repeated Li plating/stripping cycles compromises the mechanical integrity and corresponding passivation functionality of the SEI, resulting in the formation of "dead" Li. These issues have been identified as the major factors for capacity decay and insufficient lifespan.

To date, over 40 years have passed since the advent of exploring viable strategies to address the abovementioned issues of LMAs. Figure 3.6 illustrates the development of strategies for producing dendrite-free LMBs along with their associated materials and structures. In the early stage of the development of LMBs, studies were mainly focused on improving the stability and uniformity of SEIs via the introduction of various electrolyte additives and mechanical barriers. For example, Wen and coworkers modified Li metal by exposing it to nitrogen to form a Li_3N layer. Ex situ-formed protective layers with good adhesion, uniform physical contact with Li surfaces, and acceptable ionic conductivity can suppress the reactions between Li metal and electrolytes. In addition, Zhang and coworkers proposed a self-healing electrostatic mechanism that can fundamentally change the process of Li deposition to avoid Li dendrite formation by adding a low concentration of cations such as cesium and rubidium ions. Some publications have proposed the construction of current collectors for Li metal, and the popularity of this subject has been on the rise since 2016. A strong incentive for advancing LMAs stems from the proposed "lithiophilic" property of some materials. Combined with the understanding of the Gibbs energy formulas for heterogeneous nucleation, a reasonable design philosophy for LMAs has been established, which not only refines the insufficient deficiencies of Li host construction reported in the past, achieving the uniform deposition of Li metal with low overpotential, but also provides much inspiration for other strategies. Thereafter, considerable attention and efforts have been dedicated to the research realm of LMAs. Potential solutions based on Li hosts, electrolyte additives, and artificial protective layers have been validated in LMAs, along with an exponential increase in the number of scientific publications. In addition, assisted by the advanced technique of cryogenic transmission electron microscopy (cryo-TEM), the breakthrough in the direct observation of sensitive battery materials including SEI and Li metal at the atomic resolution level can render a more comprehensive understanding of failure mechanisms in LMAs, which accelerates the research process on LMAs and provides the possibility for constructing ideal LMAs in the future.

3.3.4 Other Anode Materials

Besides the abovementioned anode materials that possess great potential for the practical application of LIBs, other anode materials have been developed recently for different demands of energy storage applications, such as high capacity, long

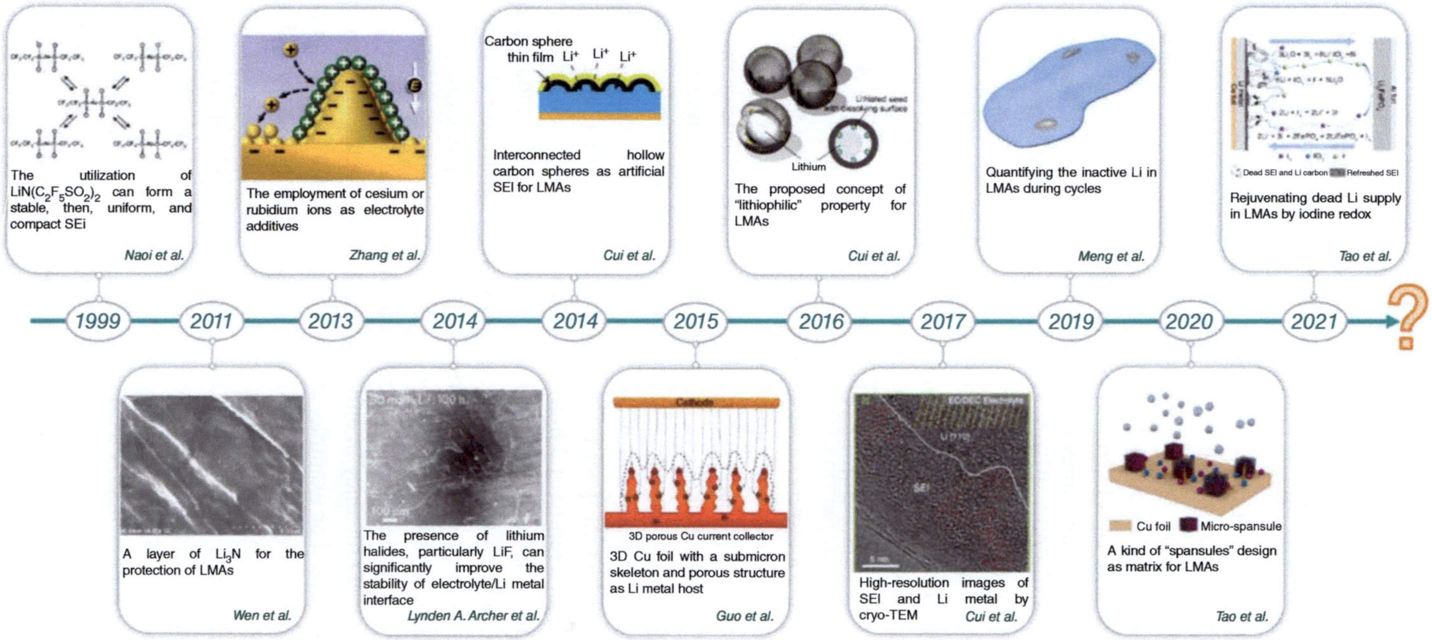

Figure 3.6 Critical development of Li metal anode. Source: Reproduced with permission from Yuan et al. [7]. © 2022/Elsevier.

lifespans, excellent low-temperature performance, and high charging/discharging rate. Metal oxides, sulfides, selenides, phosphides, carbides, nitrides, MXenes, metal-organic frameworks, covalent organic frameworks, sulfur, phosphorus, selenium, tin, bismuth, and so on have been explored as anode materials. These materials can be grouped into three categories based on their redox reaction mechanism: intercalation, conversion, and alloying materials, which is discussed below.

For alloying-type anodes, tin (Sn), antimony (Sb), sulfur (S), phosphorus (P), selenium (Se), silicon (Si), and bismuth (Bi) have been developed. Ultrahigh specific capacity is the major advantage of alloy anodes. The abovementioned metals can contribute more than one electron per mole for reversible energy storage, while six carbon atoms contribute one electron to store energy. Hence, the theoretical capacity of alloy anodes is two to ten times higher than that of carbon anodes. In addition, alloy anodes such as Tin (Sn) alloys have higher onset voltage above Li/Li^+, which can help prevent lithium dendrite growth [8]. These advantages make alloy anodes an interesting research topic with the potential for LIB advancement. Nevertheless, alloy anodes have some drawbacks as well. According to the literature, massive volumetric expansions caused by alloying reactions lead to serious structural degeneration, resulting in rapid capacity decay [9]. Performance degeneration of alloy anodes can be attributable to four reasons: (i) surface oxidation of alloys, which leads to Li loss, and consequently an irreversible loss of capacity; (ii) low Li-ion diffusion rates that render the lithiation/delithiation process irreversible; (iii) continuous formation SEI film on the alloy anode surface due to the huge volume changes of alloy anodes; and (iv) detaching of alloys from the current collector due to the volume expansion of alloy anodes. To address these challenges, various strategies have been proposed, including 3D design, artificial SEI, and hybrid with carbon. Briefly, 3D structures provide sufficient space for the volume expansion of alloy anodes. In the artificial SEI method, a robust and high Li^+ conductive layer is constructed to faceplate alloying reactions. Coating carbon on the surface of alloy anodes provides a mechanical layer to improve the detaching and volume expansion of alloy anodes. From the viewpoint of practical applications, carbon coating is more cost-effective and practicable than others and has been commercialized on a Si anode by Yi Cui and his company (Amprius) [10]. Recently, different hybrid alloying-type materials have demonstrated efficiency in enhancing the electrochemical performance of alloying-type materials. For example, Reddy and Varadaraju [11] designed a LIB anode using a $NbSb_2$ alloy in which antimony reacts with lithium, yielding Li_3Sb. Here, niobium does the work of a buffer substrate. The output and efficiency of the battery are improved as a result of the stress release and reduction in volume variations. Yin et al. [12] developed a Ag-Sn alloy ($Ag_{52}Sn_{48}$ and $Ag_{46}Sn_{54}$), which shows a reversible capacity of $800\,mAh\,g^{-1}$ to be used as an anode. In this setup, Ag serves as an unreactive matrix with buffer action. Mao and Dahn [13] prepared an alloy powder of Sn-Fe-C. In this alloy, $SnFe_3C$ acts as the inert matrix, while Sn_2Fe is the active phase. They showed that when the composite is made with Sn_2Fe and $SnFe_3C$ with compositions of 25% and 75%, respectively, the battery exhibits increased reversible capacities near $1600\,mAh\,cm^{-3}$ and excellent cyclability.

As for conversion-type materials, a substantial number of transition metal compounds (TMCs) have risen to prominence as highly promising anode materials for LIBs due to their attractive compositions alongside a high theoretical specific capacity. Some good examples are magnetite (Fe_3O_4), pyrite (FeS_2), and pyrolusite (MnO_2) [14]. Meanwhile, due to the lower lithium intercalation potential of TMCs than that of graphite anodes, the growth of lithium dendrites usually does not occur in TMC anodes, leading to better operation safety.

Transition metal (TM) oxides (MO_x, where M = TMs, which include iron, cobalt, nickel, copper, and zinc) possess some attractive characteristics that have gained them a lot of interest as top candidates for LIB anode materials. They are nontoxic, possess high power density and theoretical specific capacity, are abundant in nature, and have a low-cost fabrication process. Iron oxide materials (e.g. Fe_3O_4) have lately gained increased recognition as potential anode materials due to their high theoretical capacity (~926 mAh g^{-1}). However, they have some drawbacks. They show a rapid capacity decrease and poor cycling stability due to the pulverization of active materials during the charge/discharge process. In addition, compared with graphite, Fe_3O_4 exhibits poor electron conductivity. This eventually results in a poor rate of performance. Chalcogenides (including SnS, MoS_2, FeS_2, and CoS_2) have stood out as attractive anode materials whose theoretical capacity is quite high due to their Li-ion (and Na) storage capacity by the mechanism of electrochemical conversion [15]. To further improve the cycling stability and rate performance of chalcogenides (especially cobalt sulfide), which are hampered by significant alterations in the bulk dimensions of the material during charging/discharging, highly conductive and stable materials like polyaniline and graphene are applied as coatings on the surface of the sulfides [16]. Metal selenides usually show higher electronic conductivity and interlayer space (if there are 2D materials) than metal oxides and metal sulfides, showing better electrochemical performances.

MXenes as a new anode material for LIBs have gained significant interest due to their high electrical conductivity along with outstanding chemical stability and low Li-ion diffusion impedance [17]. MXenes have a high specific capacity value of 320 mAh g^{-1} at a current of 100 mA g^{-1} after 760 cycles. However, in general, they suffer from restacking during cycling, limiting their Li-ion storage potential. The majority of studies have focused on the Li-ion storage potential of multilayered MXenes, although they suffer from restacking during cycling and perform similarly to graphite. These studies showed that increasing the interlayer spacing, altering the functional groups of MXenes, and synthesizing few-layered MXenes are the most essential strategies to improve the Li-ion storage capabilities of MXenes. MXene-based composites are often manufactured with two goals: limiting MXene restacking and buffering the volumetric expansion and loss of electrical contact of high-capacity anode materials, resulting in a synergistic effect between MXenes and other composite ingredients. The findings of the examined studies show that adding modest concentrations of MXenes to metal oxides, TM dichalcogenides, and silicon can greatly alleviate their difficulties for practical applications in LIBs.

Aluminum niobates have emerged as a better alternative to traditional titanium niobates ($TiNb_2O_7$ and $Ti_2Nb_{10}O_{29}$) to serve as an improved anode material for

LIBs [18]. Despite their attractively high theoretical capacity and impressive electrochemical properties, titanium niobates possess poor charge conductivity due to their reduced rate capability.

Although the electrochemical properties of TMCs are highly attractive, drawbacks such as inherent poor electronic and ionic conductivity, continuous electrolyte decomposition, and relatively large volume expansion (<200%) seriously restrict their application.

3.4 Application and Critical Challenges

3.4.1 Trends in LIBs for Applications

One primary aim for using LIBs is to develop high-capacity materials with higher voltage. Until now, several types of high-voltage cathode materials such as $LiCoO_2$, $LiNi_xCo_yMn_zO_2$, $LiNi_xCo_yAl_zO_2$, and spinel $LiMn_{1.5}Ni_{0.5}O_4$ have gained increasing popularity because of their high specific energy and power densities. To realize high energy density of a cathode LIB, the main strategy is to enhance its capacity and operating voltage according to the following equation: energy density = capacity × voltage. Running cathode materials at a high voltage effectively increases the energy density of the corresponding batteries. For example, researchers have studied $LiCoO_2$ for decades, whose cutoff voltage can be increased from 4.2 to 4.6 V, with improved gravimetric energy density from 547 to 886 Wh kg^{-1}. However, some possible flaws have constrained this effort, including severe structural changes that carry the risk of structural collapse and crack formation. Lithium-rich materials have a higher capacity and voltage. However, after the initial cycle, they pose challenges such as severe oxygen evolution, voltage deterioration, and capacity loss and hence are hard to enforce in the market. Meanwhile, next-generation long-range EVs demand an energy density of 800 Wh kg^{-1}, which is not feasible using current commercial NMC and NCA. Even though $LiMn_{1.5}Ni_{0.5}O_4$ is renowned as a 5 V-class cathode material, it is difficult to prepare because the impurity phase of $Li_xNi_{1-x}O$ deteriorates its electrochemical performance [19]. However, less emphasis is devoted to advancing spinel-structured $LiMn_2O_4$ cathode materials. Considering acceptable specific energy density, high working voltage, safety performance, and cost-effective and robust prospects for sustainable and clean energy, $LiMn_2O_4$ is the most suitable cathode material for LIBs. However, its energy density is too low to support practical applications. Thus, increasing the tap density or upper cutoff voltage of $LiMn_2O_4$ could open new prospects for next-generation LIBs with improved energy and power density.

Massive efforts have been devoted to evaluating the electrochemical behavior of $LiMn_2O_4$ under high voltages (>4.3 V versus Li/Li^+). Xia et al. observed a 10% increase in capacity by increasing the cutoff voltage to 5 V, showing that an excess amount of Li^+ can be further deintercalated and intercalated in the structure. This observation correlates well with that of Tang et al. [20], who showed a reduction in lattice parameters when charged to 5.1 V, confirming the presence of extra Li^+

in the $LiMn_2O_4$ structure. Nevertheless, charging $LiMn_2O_4$ to a higher voltage (>4.3 V versus Li/Li^+) is indeed plagued by large irreversible capacity with rapid capacity fading, which reduces the battery life. The fundamental causes of these issues need to be identified to overcome performance deterioration. Accordingly, some efforts have been made to identify them. In the early 2010s, surface structural changes of $LiMn_2O_4$ were first directly observed under high-voltage cycling. Radical changes in surface chemistry at the atomic level were revealed, coupled with manganese dissolution, resulting in performance deterioration of the battery. Since then, high-voltage stability studies of $LiMn_2O_4$ have become a hot topic of ongoing research. The degradation phenomenon primarily arises at the interface between the cathode and electrolyte, altering the surface structure of the cathode. In addition, interfacial studies have introduced a core challenge, which is deemed necessary to achieve better battery performance.

The challenges of high-voltage $LiMn_2O_4$ include surface distortions, interfacial reaction between electrode and electrolyte, and stress-induced cracking. Complicated structural changes on the surface are highly correlated with Mn dissolution and oxygen evolution, which are complex and not fully understood. Rapid electrolyte oxidation is driven by high charging voltage, allowing the complex and challenging cathode–electrolyte interface reaction, and then compromises the cycle life and safety of $LiMn_2O_4$. Substantial lattice mismatch of $LiMn_2O_4$ at high voltages induces building stress on the surface, which can act as a nucleation site for crack formation, and the spinel structure can collapse.

Nowadays, significant research attention is focused on identifying and developing layered TM oxides as cathode materials. Among numerous layered cathode materials, nickel-rich cathode materials, for example $LiNi_xCo_yMn_zO_2$ (NCM) and $LiNi_xCo_yAl_zO_2$ (NCA), are considered the most promising candidates primarily due to their higher energy density and rate capability. Nickel-rich cathodes with a Ni content above 80% have advantages such as practical capacity (200–250 mAh g^{-1}), along with high gravimetric energy density (~900 Wh kg^{-1}), as they show a considerable improvement over commercial LIB cathodes used worldwide, such as $LiCoO_2$ (140 mAh g^{-1}), $LiNi_{1/3}Co_{1/3}Mn_{1/3}O_2$ (160 mAh g^{-1}), and $LiMn_2O_4$ (120 mAh g^{-1}) [21]. Besides, their lower cost and environmentally friendly production due to the lower Co content endow them with additional fascinating advantages in practice.

Even though considerable progress has been achieved over the past years, these technologically important cathodes have not yet met the criteria for applications in EVs because of their intrinsic physical and chemical shortcomings. Although Ni-rich cathode materials have been applied in EVs, further expansion of market sales is hindered because of their rapid capacity degradation and safety hazard problems. Generally, for Ni-rich cathode materials, there exist several issues and challenges as follows: (i) residual lithium species that are inevitably formed upon the synthesis process and side reactions with the electrolyte; (ii) cation mixing (or Li/Ni mixing) caused by the material preparation and cycling process; (iii) generation of gases (O_2 and CO_2) at the particle surface, which is closely related to surface reconstruction and side reactions with the electrolyte; (iv) surface structure reconstruction upon cycling, including the layered-spinel-rock-salt phase transition along with the formation of lattice oxygen vacancies; (v) microstrain and crack of primary particles formed at high state-of-charge (SOC) and intensive cycling; (vi) fragile thermal stability, which may lead to thermal runaway, leading to serious safety issues; and (vii) TM dissolution due to HF attack. Such detrimental issues seriously impede the further development of Ni-rich-based LIBs as next-generation batteries.

3.4.2 Recycling of LIBs

Nowadays, LIBs have been widely used in electric devices, EVs, and grid-scale energy storage systems. With the increasing application of LIBs, it is reasonable to deduce that a substantial number of LIBs will be produced and retired. How to deal with retired LIBs is one of the biggest foreseeable challenges in the near future. For example, EV sales exceeded one million per year in 2017. Assuming that the average battery pack weight is 250 kg per car, the resultant pack wastes would comprise around 250,000 tons when these vehicles were retired. Considering that the application of LIBs will exponentially grow in the future, the waste materials of LIBs will become a serious challenge. In this regard, recycling of LIBs urgently needs to be investigated.

If tens of millions of EVs are to be produced annually, careful management of the resources consumed by EV battery manufacturing will surely be essential to ensure the sustainability of the future automotive industry, as will a material- and energy-efficient 3R system (reduce, reuse, and recycle). Considering the ever-increasing demand for raw materials for LIBs, recycling high-value materials from retired LIBs can be a promising way to alleviate the anxiety about raw materials and resources. In this regard, waste may also represent a valuable resource. With great efforts, many recycling methods have been developed recently, which are thoroughly discussed below.

The method that was first applied was pyrometallurgical metal reclamation. This method reduces metal compounds to metal alloys (Co, Cu, Fe, and Ni). It can be easily scaled up and is a general approach to recycling metals in waste materials. Since this method has been commercialized, it is considered a good method to recycle metals in consumer LIBs. As metal current collectors aid the smelting process [22], this technique has the important advantage that whole cells or modules can be recycled without splitting LIBs into different components. The products of a pyrometallurgical process contain metallic alloy fraction, slag, and gases. The gaseous products are produced by the decomposition of organic components in LIBs, such as organic electrolytes and polymer binders. The metallic alloy contains different metals that can be separated via hydrometallurgical processes. The slag usually contains the typical metal component in LIBs, including Al, Mn, and Li, which can be reclaimed by further hydrometallurgical processing, but can alternatively be used in other industries (e.g. cement industry). Nevertheless, this method cannot be used to recycle electrolytes, plastics (approximately 40–50 wt%), and lithium salts in LIBs or other components such as lithium salts. Furthermore, the reclamation process will generate toxic gases and require further processes to inhibit the diffusion of toxic gases in the air. Meanwhile, a high-temperature reclamation process requires high energy costs. Unfortunately, this method is a frequently used process for the extraction of high-value TMs (e.g. Ni and Co) [23].

Comminution reclamation is a common method to recycle waste materials. In general, it involves many components, including sieves, filters, magnets, shaker tables, and heavy media, and is based on differences in the properties of waste materials, such as ferromagnetism, hydrophobicity, particle size, and density, which can separate a mixture of lithium-rich solution, plastics, and electrode powders. This method produces electrode coatings in fine fractions of material, and plastics, casing materials, and metal foils in coarse fractions [24]. Generally, magnetic separation processes can separate magnetic materials (e.g. steel casings) in coarse fractions, and density separation processes can separate plastic materials from foils. The fine product is electrode materials, including metal oxides and carbon, which can be separated by froth flotation [25]. Before fine product recycling, polymeric binders in electrodes need to be eliminated to liberate electrode materials (metal oxides and graphite) from current collectors. Based on previous works, the routes to eliminating binders include sonication (because binders are usually soluble in an organic solvent, such as *N*-methyl-2-pyrrolidone [NMP] and dimethylformamide [DMF]) [26], thermal heat treatment (due to the lower thermal stability of binders than that of graphite and metal oxides), or dissolution of the aluminum current collector [27, 28]. However, these methods often require high temperatures (60–100 °C) and are relatively slow (three hours) [29]. While ultrasound can induce faster delamination (1.5 hours), this is still too slow for a continuous-flow process, and the required solvent-to-solid mass ratios of 10 : 1 will not be viable on a commercial scale with these solvents [30].

Hydrometallurgical treatments are developed to recycle LIBs. They use aqueous solutions to leach the desired metals from the cathode material. H_2SO_4/H_2O_2 is the most common combination of reagents in previous works. Significant research has been devoted to identifying the most efficient conditions to achieve an optimal leaching rate, including leaching acid concentration, leaching time, leaching temperature, solid-to-liquid ratio, and addition of a reducing agent. Results showed that adding H_2O_2 can effectively improve leaching efficiency. This shows that H_2O_2 acts as a reducing agent to convert insoluble Co^{3+} materials into soluble Co^{2+} through the following reaction:

$$2LiCoO_2(s) + 3H_2SO_4 + H_2O_2 \rightarrow 2CoSO_4(aq) + Li_2SO_4 + 4H_2O + O_2$$

To date, numerous leaching acids and reducing agents have been investigated. The leached solution may also subsequently be treated with an organic solvent to perform solvent extraction. Once leached, metals may be recovered through several precipitation reactions controlled by manipulating the pH of the solution. Cobalt is usually extracted as either sulfate, oxalate, hydroxide, or carbonate, and lithium can be extracted through a precipitation reaction forming Li_2CO_3 or Li_3PO_4 [31]. Another recycling method proposes the mechanochemical treatment of materials, where electrode materials are ground with a chlorine compound or a complexing agent to produce water-soluble salts of cobalt, which can be separated from insoluble fractions by washing with water [32]. Following an initial work focusing on leaching and remanufacturing of $LiCoO_2$ [33], hydrometallurgical treatments have been applied to recycle other metal oxide materials with multimetal elements (e.g. $LiNiCoMnO_2$). After the leaching process, metal ions are dissolved in solution and recycled. The recycled materials can be used to produce metal oxide electrode materials, thus making it possible to achieve the utilization of waste materials in LIBs.

Although waste materials in LIBs can be recycled using the abovementioned methods, their properties usually change after the recycling process. The recycled materials need further fabrication to produce metal oxide electrode materials, which is quite complex. To avoid the additional fabrication process, various direct recycling methods have been proposed. These methods do not destroy metal oxides, thus avoiding additional fabrication processes and improving recycling efficiency. Due to the crystal structure and element ratio of metal oxides in retired LIBs, recycled metal oxide cathodes require further processing to recover the structure of the active material. Crystal structure can be easily recovered through the high-temperature annealing process. However, the lithium content in directly recycled metal oxide cathodes is usually deficient because materials may not be recovered from batteries in the fully discharged state with the cathodes fully lithiated. Studies published to date have primarily focused on laptop and mobile phone batteries as they are available in huge amounts for recycling. Recycling routes have been outlined recently [34]. Cathode strips, obtained after dismantling spent batteries, are soaked in N-methyl-2-pyrrolidone (NMP) before undergoing sonication. Powders are either regenerated through simple solid-state synthesis with the addition of fresh Li_2CO_3 or treated hydrothermally with a solution containing $LiOH/Li_2SO_4$ before annealing. For high-cobalt cathodes, conventional pyrometallurgical methods or hydrometallurgical methods can recover around 70% of cathode active materials. However, these methods are not suitable for other cathodes, considering the lengthy and expensive purification process. Fortunately, unfavorable purification processes can be avoided using direct recycling methods. This is one of the most important advantages of direct recycling. Although the electrochemical performance of directly recycled cathodes is not as good as the new one, direct recycling methods are promising in recycling lower-value cathodes such as $LiMn_2O_4$ and $LiFePO_4$ [35]. Direct recycling also has the advantage that, in principle, all battery components can be recovered and reused after further processing, except separators. Studies on recycling of graphitic anodes are quite limited compared with those on metal oxide cathodes, due to the lower recovery value of the former. Nevertheless, graphite can be separated through mechanical methods with similar properties to those of pristine graphite, which is more easy to recycle [36].

The major challenge in using direct recycling methods is that the efficiency of direct recycling processes highly depends on the healthy state of LIBs. A low charge state of a LIB means that its lithium content is highly deficient and it needs a lithium compensation process to recover the lithium content in metal oxide cathodes. Considering the high cost of lithium compensation, direct recycling of oxide cathodes in the low charge state will be expensive. There are also potential issues with the flexibility of these methods in handling metal oxides of different compositions. To achieve maximum efficiency, direct recycling processes should be tailored to specific cathode formulations, necessitating different processes for different cathode materials. Furthermore, unknown or poorly characterized LIBs and cathode materials will affect the quality of the recycled cathode materials, which are hard to reuse in LIBs [37].

3.4.3 Critical Challenges

The most difficult challenge for the wide application of LIBs in EVs and grid-scale energy storage systems is thermal runaway and explosion of inferior LIBs in the system. Although using low-voltage cathode materials (e.g. $LiFePO_4$) can effectively prevent thermal runaway, energy density of the systems will be decreased below 200 Wh kg^{-1}, much lower than that of LIBs with high-voltage cathodes (e.g. NCM). Achieving high safety without sacrificing energy density is critical for state-of-the-art LIBs. In addition, even when using a low-voltage cathode, the safety concerns of LIBs cannot be completely resolved due to the flammable and unstable nature of high-voltage organic electrolytes.

In addition to safety issues, energy density of state-of-the-art LIBs almost reaches its limit, based on the theoretical capacity of commercial cathode materials ($LiFePO_4$ and NCM) and anode materials (graphite). To further increase the energy density of LIBs, high-capacity electrodes should be developed. Next-generation anode materials can be silicon/carbon composites and lithium metal. Both of them can reach an ultrahigh theoretical capacity above 3000 mAh g^{-1}. However, silicon anodes show serious volume changes during the charging/discharging process. This serious volume expansion will destroy the SEI layer formed on the silicon surface, thus resulting in continuous SEI formation along cycling. LMAs also suffer from the volume expansion issue, but the challenge for LMAs is beyond that. Due to serious polarization and uneven electric field distribution, lithium ions are unevenly deposited on the anode side, and lithium dendrites start to grow. This uneven striping will induce "dead lithium" formation, decreasing energy density. The continuous growth of lithium dendrites will penetrate the separator and create direct contact with the cathode, inducing an internal short circuit, which will generate a vast amount of heat to accelerate the decomposing of electrolytes and cathodes, finally leading to thermal runaway and explosion. It is presumed that replacing flammable organic electrolytes with inflammable solid-state electrolytes is a practical way to ensure the high safety and high energy density of LIBs. However, the practical application of solid electrolytes in LIBs is still challenging. State-of-the-art solid-state electrolytes and lithium solid batteries will be discussed in detail in Chapter 5.

Meanwhile, the power density of LIBs is insufficient. State-of-the-art power-type LIBs usually use $LiFePO_4$ cathodes because of the low operating potential of $LiFePO_4$, which can avoid electrolyte decomposition, leading to long lifespans. To achieve a high rate of charging/discharging, mass loading of $LiFePO_4$ on the current collector is reduced to decrease the Li^+ diffusion length from the electrolyte to the internal of $LiFePO_4$, which will inevitably decrease the energy density of packed LIBs. Based on the data provided by LIB-producing companies, the energy density of $LiFePO_4$ LIBs is about 180 Wh kg^{-1}. To achieve high power density, the energy density will be decreased to ~120 Wh kg^{-1}. Samsung developed power-type LIBs using NCM as the cathode to improve the energy density of LIBs at a high rate. Although the energy density of these LIBs can reach the performance of energy-type $LiFePO_4$ LIBs, their charging/discharging rate (2 C) is lower than that of $LiFePO_4$ LIBs (6 C). Maintaining a high energy density at a high power density is a critical challenge for LIBs.

Last but not least, state-of-the-art commercial LIBs cannot work below −20 °C, limiting the application of LIBs in low-temperature regions around the world. This limitation seems to attract increasing attention in recent years; however, this limitation has not yet been well addressed. The electrode reaction process of LIBs involves diffusion from bulk electrolytes to the electrode surface region, Li-ion desolvation, electrolyte decomposition (in initial cycles), SEI formation, Li-ion intercalation, and charge transfer. The inferior electrochemical performance of LIBs at low temperatures below −20 °C is related to all the abovementioned electrode reactions. Two significant phenomena have been observed previously, namely increased electrode–electrolyte interface impendence and increased viscosity of bulk electrolytes. Both the components and structures of electrode materials and electrolyte materials should be thoroughly investigated and carefully designed to achieve LIBs that can work at low temperatures.

References

1 Qiu, Y. and Jiang, F. (2022). A review on passive and active strategies of enhancing the safety of lithium-ion batteries. *International Journal of Heat and Mass Transfer* 184: 122288.

2 Castro, L., Dedryvère, R., El Khalifi, M. et al. (2010). The spin-polarized electronic structure of $LiFePO_4$ and $FePO_4$ evidenced by in-lab XPS. *The Journal of Physical Chemistry C* 114 (41): 17995–18000.

3 Li, T., Yuan, X.-Z., Zhang, L. et al. (2020). Degradation mechanisms and mitigation strategies of nickel-rich NMC-based lithium-ion batteries. *Electrochemical Energy Reviews* 3 (1): 43–80.

4 Zhang, T., Li, D., Tao, Z. et al. (2013). Understanding electrode materials of rechargeable lithium batteries via DFT calculations. *Progress in Natural Science: Materials International* 23: 256–272.

5 Zhang, H., Yang, Y., Ren, D. et al. (2021). Graphite as anode materials: fundamental mechanism, recent progress and advances. *Energy Storage Materials* 36: 147–170.

6 Winter, M., Besenhard, J.O., Spahr, M.E. et al. (1998). Insertion electrode materials for rechargeable lithium batteries. *Advanced Materials* 10 (10): 725–763.

7 Yuan, H., Ding, X., Liu, T. et al. (2022). A review of concepts and contributions in lithium metal anode development. *Materials Today* 53: 173–196.

8 Speirs, J., Contestabile, M., Houari, Y. et al. (2014). The future of lithium availability for electric vehicle batteries. *Renewable and Sustainable Energy Reviews* 35: 183–193.

9 Su, L., Jing, Y., and Zhou, Z. (2011). Li ion battery materials with core–shell nanostructures. *Nanoscale* 3 (10): 3967–3983.

10 Zhang, L., Wang, C., Dou, Y. et al. (2019). A yolk–shell structured silicon anode with superior conductivity and high tap density for full lithium-ion batteries. *Angewandte Chemie International Edition* 58 (26): 8824–8828.

11 Reddy, M.A. and Varadaraju, U.V. (2006). $NbSb_2$ as an anode material for Li-ion batteries. *Journal of Power Sources* 159 (1): 336–339.

12 Yin, J., Wada, M., Yoshida, S. et al. (2003). New Ag-Sn alloy anode materials for lithium-ion batteries. *Journal of The Electrochemical Society* 150 (8): A1129.

13 Mao, O. and Dahn, J.R. (1999). Mechanically alloyed Sn-Fe(-C) powders as anode materials for Li-ion batteries: III. $Sn_2Fe : SnFe_3C$ active/inactive composites. *Journal of The Electrochemical Society* 146 (2): 423.

14 Nzereogu, P.U., Omah, A.D., Ezema, F.I. et al. (2022). Anode materials for lithium-ion batteries: a review. *Applied Surface Science Advances* 9: 100233.

15 Chen, H., Ran, M.-Y., Wei, W.-B. et al. (2022). A comprehensive review on metal chalcogenides with three-dimensional frameworks for infrared nonlinear optical applications. *Coordination Chemistry Reviews* 470: 214706.

16 Sajjad, M., Cheng, F., and Lu, W. (2021). Research progress in transition metal chalcogenide based anodes for K-ion hybrid capacitor applications: a mini-review. *RSC Advances* 11 (41): 25450–25460.

17 Aghamohammadi, H., Eslami-Farsani, R., and Castillo-Martinez, E. (2022). Recent trends in the development of MXenes and MXene-based composites as anode materials for Li-ion batteries. *Journal of Energy Storage* 47: 103572.

18 Lou, X., Li, R., Zhu, X. et al. (2019). New anode material for lithium-ion batteries: aluminum niobate ($AlNb_{11}O_{29}$). *ACS Applied Materials & Interfaces* 11 (6): 6089–6096.

19 Zhong, Q., Bonakdarpour, A., Zhang, M. et al. (1997). Synthesis and electrochemistry of $LiNi_xMn_{2-x}O_4$. *Journal of The Electrochemical Society* 144 (1): 205.

20 Tang, D., Ben, L., Sun, Y. et al. (2014). Electrochemical behavior and surface structural change of $LiMn_2O_4$ charged to 5.1 V. *Journal of Materials Chemistry A* 2 (35): 14519–14527.

21 Yi, H., Liang, Y., Qian, Y. et al. (2023). Low-cost Mn-based cathode materials for lithium-ion batteries. *Batteries* 9 (5): 246.

22 Georgi-Maschler, T., Friedrich, B., Weyhe, R. et al. (2012). Development of a recycling process for Li-ion batteries. *Journal of Power Sources* 207: 173–182.

23 Lv, W., Wang, Z., Cao, H. et al. (2018). A critical review and analysis on the recycling of spent lithium-ion batteries. *ACS Sustainable Chemistry & Engineering* 6 (2): 1504–1521.

24 Wang, X., Gaustad, G., and Babbitt, C.W. (2016). Targeting high value metals in lithium-ion battery recycling via shredding and size-based separation. *Waste Management* 51: 204–213.

25 Zhan, R., Oldenburg, Z., and Pan, L. (2018). Recovery of active cathode materials from lithium-ion batteries using froth flotation. *Sustainable Materials and Technologies* 17: e00062.

26 Li, X., Zhang, J., Song, D. et al. (2017). Direct regeneration of recycled cathode material mixture from scrapped $LiFePO_4$ batteries. *Journal of Power Sources* 345: 78–84.

27 Song, D., Wang, X., Zhou, E. et al. (2013). Recovery and heat treatment of the $Li(Ni_{1/3}Co_{1/3}Mn_{1/3})O_2$ cathode scrap material for lithium ion battery. *Journal of Power Sources* 232: 348–352.

28 Chen, J., Li, Q., Song, J. et al. (2016). Environmentally friendly recycling and effective repairing of cathode powders from spent $LiFePO_4$ batteries. *Green Chemistry* 18 (8): 2500–2506.

29 Zhang, Z., He, W., Li, G. et al. (2014). Ultrasound-assisted hydrothermal renovation of $LiCoO_2$ from the cathode of spent lithium-ion batteries. *International Journal of Electrochemical Science* 9 (7): 3691–3700.

30 Nirmale, T.C., Kale, B.B., and Varma, A.J. (2017). A review on cellulose and lignin based binders and electrodes: small steps towards a sustainable lithium ion battery. *International Journal of Biological Macromolecules* 103: 1032–1043.

31 Yang, Y., Zheng, X., Cao, H. et al. (2017). A closed-loop process for selective metal recovery from spent lithium iron phosphate batteries through mechanochemical activation. *ACS Sustainable Chemistry & Engineering* 5 (11): 9972–9980.

32 Wang, M.-M., Zhang, C.-C., and Zhang, F.-S. (2017). Recycling of spent lithium-ion battery with polyvinyl chloride by mechanochemical process. *Waste Management* 67: 232–239.

33 Li, J., Zhao, R., He, X. et al. (2009). Preparation of $LiCoO_2$ cathode materials from spent lithium–ion batteries. *Ionics* 15 (1): 111–113.

34 Shi, Y., Chen, G., Liu, F. et al. (2018). Resolving the compositional and structural defects of degraded $LiNi_xCo_yMn_zO_2$ particles to directly regenerate high-performance lithium-ion battery cathodes. *ACS Energy Letters* 3 (7): 1683–1692.

35 Dunn, J.B., Gaines, L., Sullivan, J. et al. (2012). Impact of recycling on cradle-to-gate energy consumption and greenhouse gas emissions of automotive lithium-Ion batteries. *Environmental Science & Technology* 46 (22): 12704–12710.

36 Sabisch, J.E.C., Anapolsky, A., Liu, G. et al. (2018). Evaluation of using pre-lithiated graphite from recycled Li-ion batteries for new LiB anodes. *Resources, Conservation and Recycling* 129: 129–134.

37 Bolli, C., Guéguen, A., Mendez, M.A. et al. (2019). Operando monitoring of F–formation in lithium ion batteries. *Chemistry of Materials* 31 (4): 1258–1267.

4

Sodium-Ion Batteries

Huihua Li and Hongxu Qv

Harbin University of Science and Technology, School of Electrical and Electronic Engineering, 52 Xuefu Road, Harbin, 150080, China

4.1 Introduction

Lithium-ion batteries (LIBs) are currently used as power sources for various devices ranging from hybrid electric vehicles to consumer electronics, thanks to their high power and energy density. However, despite the great success of LIBs, their widespread implementation in large-scale stationary energy storage applications is restricted due to the limited supply of lithium resources and, particularly, cobalt availability. In this regard, alternative battery chemistries that are cost-effective and based on abundant materials are highly desired for large-scale energy storage applications. Sodium (Na), one of the most abundant resources in the alkali metal family, has been considered a sustainable alternative to lithium (Li) for high-performance, low-cost, and large-scale energy storage devices. Na and Li belong to the same family, so they share many similarities in physical and chemical properties. As shown in Table 4.1, Na provides multiple advantages: (i) low redox potential (−2.71 V versus standard hydrogen electrode), which is only 0.3 V higher than that of Li; and (ii) high abundance in the crust of up to 2.64%. In addition, Na is highly abundant in seawater and easy to exploit. Therefore, sodium-ion batteries (SIBs) are one of the most promising options for developing large-scale energy storage technologies. The study of SIBs started around the 1980s. Although the development of SIBs was slower than LIBs in the early stage due to the huge success of the latter, SIBs have recently gained momentum [1–3]. Nevertheless, several challenges including safety, longevity, cost, and a strong need for enhanced energy and power densities restrict the large-scale application of SIBs [2]. Specifically, SIBs possess an intrinsically lower energy density than LIBs due to the larger ionic radius of sodium and its higher redox potential [4]. Nonetheless, these issues can be (partially) addressed by designing proper electrode materials, allowing for energy density values comparable to those of the lithium-ion technology. Continuous improvements in power, energy density, and safety are required to keep up with the changes in technology.

Towards Next Generation Energy Storage Technologies: From Fundamentals to Commercial Applications, First Edition. Edited by Minghua Chen.

Table 4.1 Comparison of properties of sodium and lithium.

Element	Li	Na
Standard atomic weight ($g\,mol^{-1}$)	6.94	22.9897
Atomic radius (pm)	152	186
Standard electrode potential (V versus SHE)	−3.04	−2.71
Specific capacity ($mAh\,g^{-1}$)	3861	1166
Concentration in the Earth (%)	0.006	2.64

4.2 Energy Storage Mechanism

Batteries are electrochemical devices that operate by converting chemical energy into electrical energy. SIBs typically consist of one or more electrochemical cells, each containing four primary components: negative electrode (anode), positive electrode (cathode), conducting electrolyte, and separator. The working principle of SIBs is illustrated in Figure 4.1 [6]. During charging, sodium ions in SIBs are extracted from the cathode, migrated through the electrolyte, and inserted into the anode, accompanied by electron migration through an extra external wire in the opposite direction, guaranteeing electroneutrality. Upon discharging, sodium ions are released from the anode and reverted to the cathode. Therefore, rechargeable SIBs also can be referred to "rocking chair" or "swing" batteries, analogous to LIBs. However, there are some obvious differences between the two systems. The

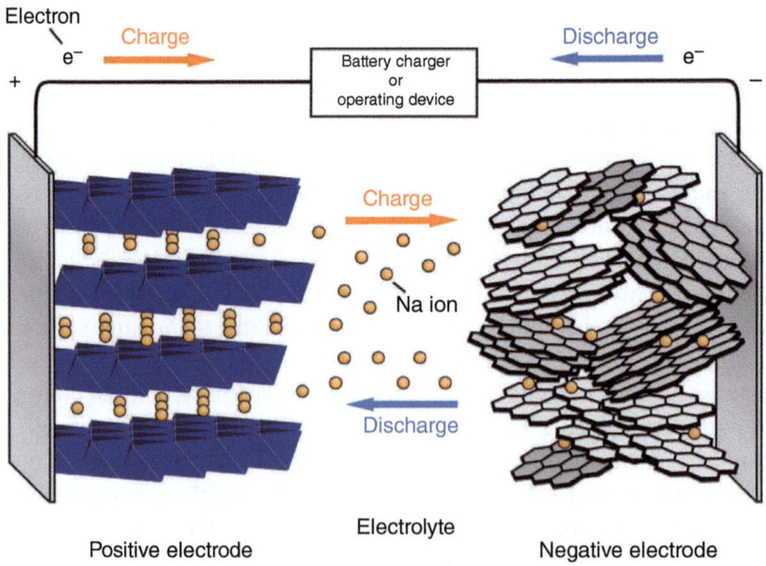

Figure 4.1 Working principle of the sodium-ion cell. Source: Reproduced from Yabuuchi et al. [5]. © 2014/American Chemical Society.

ionic radius of Na^+ ions (1.02 Å) is larger than that of Li^+ ions (0.76 Å), which affects transport properties, interface formation, and phase stability. In addition, sodium has a higher standard electrode potential (−2.71 V versus standard hydrogen electrode compared with −3.02 V for lithium) and is heavier than lithium (23 g mol^{-1} versus 6.9 g mol^{-1} for lithium); thus, compared with LIBs, the specific energy density of SIBs is generally around 20% lower. Nonetheless, due to the growing demand for efficient low-cost energy storage technologies, SIBs have been regarded as an attractive alternative because of the accessibility and availability of sodium. Recently, the breakthrough in material science has guided the design of promising materials for SIB applications, characterized by improved electrochemical performance and cycle life. Compared with LIBs, SIBs are currently at an early stage of development and research; thus further improvements in safety, energy density, and cost are required before successful commercialization and large-scale applications of SIBs.

4.3 Cathode Materials

Cathode materials are the key component in SIBs, which to some extent determine the overall energy density and cost of a battery. An ideal cathode material should fulfill the following requirements [7]: (i) high working potential and high reversible capacity; (ii) excellent structural stability during reversible Na^+ insertion/extraction reactions; (iii) good electronic and Na^+ conductivities; (iv) high electrochemical stability toward the electrolyte; and (v) clean, low-cost, and environmentally friendly synthesis process. Up to now, cathode materials for SIBs can be typically classified into two types, namely layered oxides and polyanionic compounds.

4.3.1 Layered Oxides

4.3.1.1 Classification of Layered Structures

Layered oxides (NaT_MO_2, T_M = Ti, V, Cr, Mn, Fe, Co, Ni, or a mixture of two or three elements) are attractive cathode materials for SIBs due to their simple structure, high theoretical capacity, and ease of synthesis. The most common layered structures can primarily be categorized into two types using the classification proposed by Delmas et al. [8]: P2 type and O3 type, according to the surrounding Na^+ environment and the number of unique oxide layers stacked. The symbols "P" and "O" represent the prismatic and octahedral coordination environment of Na ions, respectively, and "2" and "3" stand for the number of transition metal layers with different types of O stacking in a single cell unit. A schematic illustration of the crystal structures of P2 and O3 phases is shown in Figure 4.2.

P2-type NaT_MO_2 consists of two kinds of T_MO_2 layers and sandwiched Na layers in each unit cell with an ABBA oxygen packing framework [10]. There exist two different types of trigonal prismatic sites for Na^+, which share faces (Na_f) and edges (Na_e) with the T_MO_6 octahedra in the two adjacent slabs, respectively. Na_f (Na_1) contacts the two T_MO_6 octahedra of the adjacent slabs along its face, whereas Na_e (Na_2)

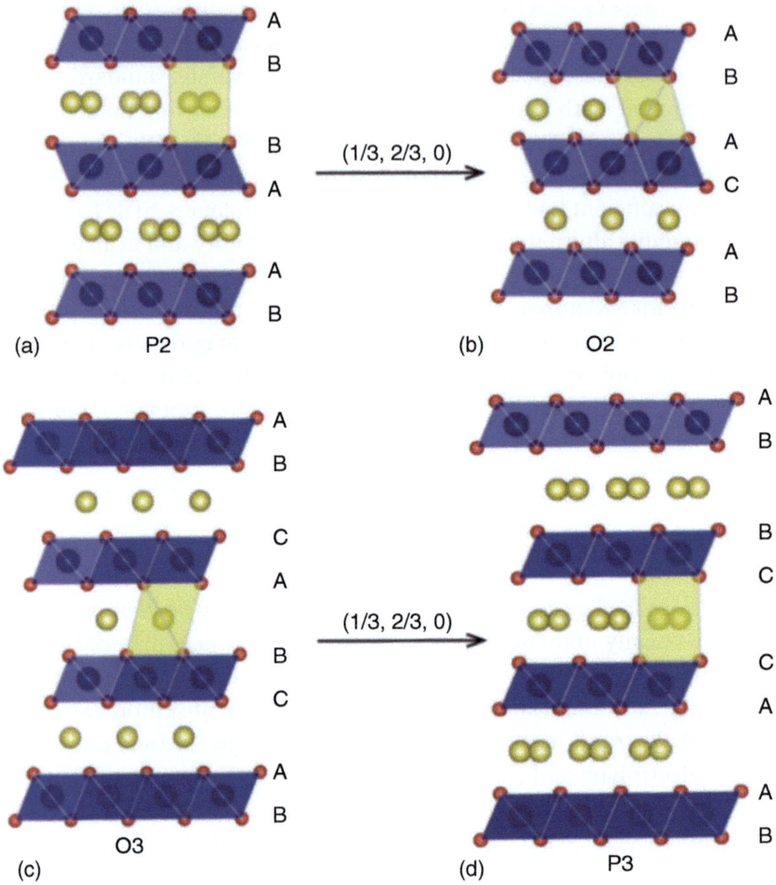

Figure 4.2 Crystal structures of layered Na-containing oxides (Na_xTMO_2) for (a) P2-type, (b) O2-type, (c) O3-type, and (d) P3-type stackings (blue balls: the transition metal; yellow balls: Na ions in the O-type frameworks). Source: Reproduced with permission from Wang et al. [9]. © 2018/John Wiley & Sons.

contacts the six surrounding T_MO_6 octahedra along its edges. These two adjacent Na_f and Na_e sites are not accessible simultaneously because of the large coulombic repulsion between the two adjacent Na ions. P2-type NaT_MO_2 is also classified as a 2H phase with a space group of $P6_3/mmc$ [9].

In O3-type $Na_xT_MO_2$, Na^+ and 3d transition metal ions are situated in separate octahedral sites due to the larger ionic radius of Na ions (1.02 Å) in comparison with 3d transition metal ions in a trivalent state (<0.7 Å). These sites are organized in a cubic close-packed (ccp) oxygen array. O3-type layered phases can be classified as cation-ordered rock-salt superstructure oxides. The edge-shared NaO6 and TMO6 octahedral units are organized in an alternating layered structure that is oriented perpendicular to the [11] crystallographic direction, resulting in the formation of NaO_2 and T_MO_2 slabs, respectively [11]. O3-type compounds have three kinds of AB, CA, and BC-T_MO_2 layers with different O stacking types (Figure 4.2c) to define the unit cell, and Na ions occupy the "octahedral" (O) sites between T_MO_2 layers, creating a typical O3-type layered structure. The structure is also referred to as a 3R phase and has a space group of $R3m$ [5]. Additionally, a prime symbol (′) is appended to the abbreviation when the crystal lattice exhibits in-plane distortion relative to the hexagonal lattice; for example, P′2-type Na_xMnO_2 [12] features an orthorhombic lattice (space group $Cmcm$), and O′3-type $NaNiO_2$ [13] possesses a monoclinic lattice (space group $C2/m$).

However, several major limitations cannot be ignored, including the low voltage, which is normally not higher than 3.5 V, multiple phase transitions, voltage decay upon cycling, and structural instability when exposed to air and, especially, to moisture. These drawbacks strongly limit the energy density and long-term cycling stability of the battery when layered oxide cathodes are used.

4.3.1.2 Challenges

Layered transition metal oxides (LTMOs) have gained great interest as promising candidates for high-capacity cathode materials for SIBs. However, some major drawbacks restrict their commercialization and applications in future electric energy storage systems (ESSs).

Irreversible Phase Transition Generally, during electrochemical cycling, most types of layered Na_xTMO_2 undergo a series of ion rearrangements, structural distortions, and sometimes even complicated phase transitions due to the sliding of TMO_2 layers [14]. P2-type Na_xTMO_2 cathodes can maintain the P2 phase in a certain voltage range until the extraction of extra Na^+, and then, they undergo conversion to the O2 phase due to the gliding ($\pi/3$ rotation) of some TMO_6 octahedra sheets when a specific quantity of Na^+ is extracted, leading to a significant contraction in the crystal structure and decreased interlayer distance (Figure 4.2a,b). The presence of large Na^+ ions in the P2-type phase energetically stabilizes large prismatic sites. As a result, TMO_2 slabs glide (move) to form octahedral sites after Na^+ extraction and by Na^+ exchange, resulting in the formation of a new O2 phase characterized by a unique AB AC AB oxygen stacking (Figure 4.2b). This O2 phase consists of two different TMO_2 slabs with crystallographically distinct oxygen arrangements, namely AB and AC. Vacancies are deliberately introduced between the AB and AC layers. Therefore, the O2-type layered phase is categorized accordingly as having a hexagonal close-packed (hcp) oxygen array, which includes ABA-type and ACA-type oxygen arrangements (Figure 4.2b). In contrast to the previously discussed O3-type phase, which exhibits a ccp oxygen arrangement (ABC-type), only edges are shared for NaO_2 layers and TMO_2 layers on both sides. Based on oxygen packing, the O2-type phase can be characterized as an intergrowth structure combining ccp and hcp arrays. It is worth noting that the nucleation of the O2 phase within a P2 crystallite can occur at multiple locations in the crystal, using the gliding vectors (2/3, 1/3, 0) and (1/3, 2/3, 0). This process results in the formation of extended defects at the boundaries between the different domains [15, 16]. However, the irreversible P2–O2 phase transition of P2-type materials at high voltage causes structural collapse and rapid capacity degradation.

The O3-type Na_xTMO_2 counterparts typically undergo more perplexing phase transitions than P2-type structures upon Na extraction/insertion as follows: O3-P3-P′3-O′3 [17]. After approximately 25% extraction of Na ions, all Na_xTMO_2 compounds undergo an O3–P3 transition [18]. It is noteworthy that these phase transitions, specifically the P2–O2 and O3–P3 transitions, have the potential to alter the mechanism of in-plane Na-ion diffusion. When sodium ions are partly extracted from the O3-type phase, Na^+ at the prismatic site attains energetic stability by forming vacancies, similar to the behavior observed in the P2-type phase. This process is accompanied by the gliding of TMO_2 sheets, allowing the formation of spacious prismatic sites without breaking the TM–O bonds. Consequently, there is a shift in the typical oxygen-stacking pattern from "AB CA BC" to "AB BC CA," leading to the classification of this new phase as a P3-type phase (Figure 4.2d). Notably, the P3-type structure is thermodynamically stable at synthesis temperatures below 800 °C, whereas the P2 phase is the high-temperature phase [19]. Hence, the phase transition from P3/O3 to the P2 cannot be achieved through electrochemical desodiation alone, as these transitions necessitate the breaking or reformation of TM–O bonds via heat treatment.

P2-phase materials always show a good rate capability and cycling stability due to their low diffusion barrier and good structure integrity. O3-type Na_xTMO_2 cathodes normally have high capacity due to their high initial Na content. Various coordination environments and stacking sequences of layered Na_xTMO_2 enhance their diversity to a certain extent. However, the phase transitions and structural evolution that occur during electrochemical cycling can easily induce complex electrochemical behaviors and inevitably cause large expansion and shrinkage of the lattice volume, which significantly deteriorate their cyclic performance and result in rapid capacity degradation, impeding their practical applications. Therefore, mitigating or reducing these irreversible phase transitions is the most effective strategy to improve the electrochemical performance of layered oxides in SIBs.

Sluggish Kinetics Na-based layered oxide cathodes, especially O3-type materials, usually experience sluggish kinetics due to the larger ionic radii of Na^+ than that of Li^+ [20]. As reported by Rojo et al., the sluggish Na^+ kinetics is an intrinsic property of O3-type materials due to the narrower Na interlayer spacing compared with the P2 structure [21]. Activation energy for Na^+ diffusion in the O3 phase is considerably higher (around 800 meV for dumb-bell hops and 200 meV for tetrahedral hops) compared with the P2 phase (approximately 120 meV), even though both phases share the same chemical composition. In the O3 framework, Na^+ should traverse a tetrahedral site to migrate from one octahedral site to another, creating a substantial energy barrier due to the size difference between the large Na ion and the narrow interstitial positions within the tetrahedral sites [22, 23].

Storage Instability Layered cathode materials normally show poor air stability, which may originate from the spontaneous extraction of Na and oxidation of transition metals after exposure to a humid environment, which significantly increases the cost of material storage, transportation, and battery production. Several factors contribute to the underlying reaction mechanism: (i) the introduction of water between TMO_2 slabs introduces additional complexity to this structural family. A prime example of this variability is observed in $Na_xTMO_2{-}y{\cdot}H_2O$ (y: anionic vacancies), which exhibits an exceptionally rich chemical space. These changes have noticeable physical manifestations that are readily visible to the naked eye when the samples are aged under air. For instance, the color evolves from black to brown, and the volume expansion leads to the cracking of pellets [24, 25]; (ii) when exposed to air, layered oxides can also capture CO_2 through a remarkable room temperature process that involves the unprecedented insertion of CO_3^{2-} within the transition metal layers of the Na-based metal oxide, while simultaneously oxidizing the TM ions in the lattice to a higher valence state [26]; (iii) the structural variation observed upon exposure to air has been attributable to the extraction of Na^+ ions from Na layers through either oxidation reactions with water or H^+/Na^+ exchange, resulting in the formation of a protonated phase. It is hypothesized that this process contributes to the increased interlayer spacing between adjacent oxide layers, as the reduced shielding effect by the positively charged Na ions leads to increased interlayer repulsion [27–29]. All the aforementioned mechanisms lead to the formation of electrochemically inactive alkaline species such as Na_2CO_3 and NaOH on the surface of active materials, resulting in battery performance degradation. Moreover, the formed Na_2CO_3 or NaOH acts as an electrical insulator. Considering these unfavorable aspects, it is essential to design air-stable cathode materials with enhanced storage stability. This is crucial for assembling cost-effective SIBs, as the presence of alkali ions in the slurry under ambient conditions can easily damage polyvinylidene fluoride binders and current collectors.

Insufficient Electrochemical Performance Conventional layered cathodes face the challenge of simultaneously achieving high specific energy, excellent rate performance, and long cycle life [7, 30, 31]. Using high-voltage redox couples such as Fe^{3+}/Fe^{4+} and Ni^{2+}/Ni^{4+} to increase the operating voltage of cathode materials, the energy density of layered cathodes can be enhanced while simultaneously maintaining specific capacity. Furthermore, the interface stability of layered oxide cathodes can have an impact on the electrochemical performance of the cell. However, LTMOs continue to encounter numerous interface challenges such as phase transitions within the high voltage range, which lead to surface cracks, and significant side reactions between the cathode material and the electrolyte. These issues contribute to an increase in charge transfer resistance and the dissolution of transition metal ions [32]. In SIBs, the cathode has low sodium-ion conductivity, and electronic conductivity at the cathode material and electrolyte interface is also low, which leads to subpar electrochemical performance. Therefore, layered oxide cathodes often encounter unfavorable, complex irreversible phase transitions, sluggish Na diffusion kinetics, and susceptibility to moisture, leading to rapid capacity decay and limited reversible capacities. Although recent research has significantly improved the cyclability of these electrode materials, there are other issues that need to be addressed to ensure long-term cycling stability, such as the Jahn–Teller effects of

Mn^{3+} and Fe^{3+} migration [33]. Therefore, there is a strong motivation for further advancements in this field.

4.3.1.3 Strategies for Layered Oxide Materials for SIBs

Many strategies have been applied to improve the electrochemical performance of layered oxide materials, such as morphology design, interface engineering, ion doping, and integrated tactics.

Morphological Manipulation Na-based cathodes commonly suffer from sluggish kinetics due to the large ionic radii of Na^+. Implementing appropriate morphology modulation can greatly enhance Na^+ kinetics, thereby improving the rate performance.

1D porous: The construction of 1D porous nanofibers self-assembled with nanoparticles is an effective approach to enhance the diffusion rate of sodium ions, establish oriented electronic pathways, and stabilize the structure, which can improve the rate capability and cycling stability. Liu et al. reported a unique porous P2-$Na_{2/3}Ni_{1/3}Mn_{2/3}O_2$ nanofibers assembled from nanoparticles (20–90 nm) using electrospinning followed by a calcination process [34]. Figure 4.3a,b demonstrates that the presence of porous nanofibers has no influence on the formation of the P2 phase, where sodium ions occupy the prismatic sites. Notably, Figure 4.3c–e, presenting scanning electron microscopy (SEM) and transmission electron microscopy (TEM) images, reveals the well-preserved 1D porous structure even after calcination. High-resolution TEM (HRTEM), selected area electron diffraction (SAED), and energy-dispersive spectroscopy (EDS) mapping results (Figure 4.3f–j) further confirm the successful construction of P2-type $Na_{2/3}Ni_{1/3}Mn_{2/3}O_2$, with all elements uniformly distributed within the nanofibers. Such porous 1D structure assembled by small nanoparticles can effectively accelerate electron/Na^+ transportation, facilitate easy access to the electrolyte, and maintain structural integrity over repetitive sodiation/desodiation processes. Overall, they show significantly improved cycle life (≈81% capacity retention after 500 cycles) as cathode materials for SIBs.

Microsphere structure: A special microsphere structure with a uniform size distribution can achieve better electrodes due to the excellent dispersity with fluidity and generate close-packed arrays, thereby enhancing the volumetric energy density of practical batteries. Lou et al. presented a facile self-templated strategy to synthesize uniform P2-$Na_{0.7}CoO_2$ microsphere cathode materials (Figure 4.4a) [35]. $CoCO_3$ microspheres and Co_3O_4 microspheres show remarkable structural stability upon annealing in air, as demonstrated by magnified field-emission scanning electron microscopy (FESEM) images (Figure 4.4b,c). Additionally, X-ray diffraction (XRD), FESEM, and HRTEM results confirm the attainment of $Na_{0.7}CoO_2$ with uniform microspheres and good crystallinity (Figure 4.4d–g). Notably, this technique has been proved to be successful in obtaining P2-$Na_{0.7}MnO_2$ and O_3-$NaFeO_2$ cathode materials. Benefiting from

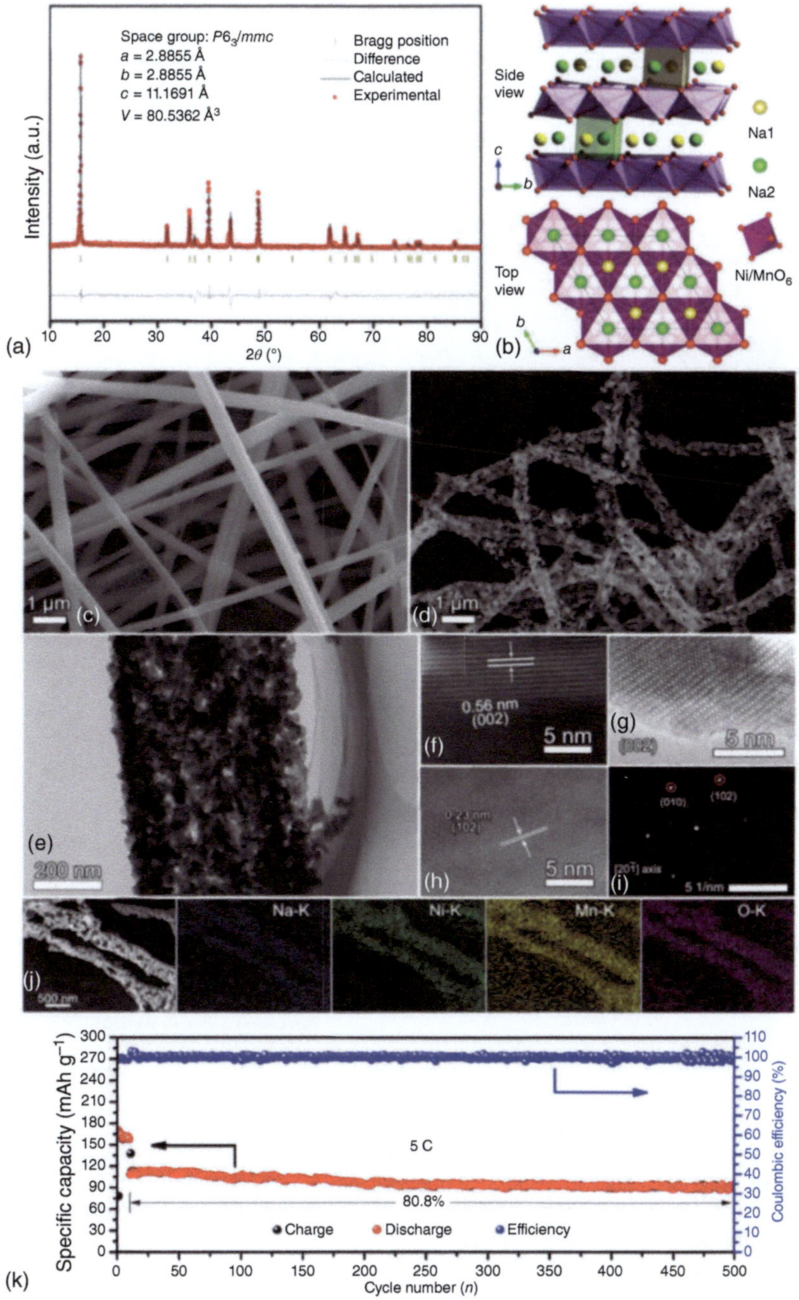

Figure 4.3 (a) Rietveld refined X-ray diffraction pattern with corresponding cell parameters of the P2-$Na_{2/3}Ni_{1/3}Mn_{2/3}O_2$ nanofibers. (b) Schematic of the P2-type crystal structure. SEM images of the (c) electrospun nanofibers and (d) resultant $Na_{2/3}Ni_{1/3}Mn_{2/3}O_2$ nanofibers. (e) TEM image, (f–h) HRTEM images, (i) SAED pattern, and (j) EDS mapping images of $Na_{2/3}Ni_{1/3}Mn_{2/3}O_2$ nanofibers. (k) Long-term cycling performance of $Na_{2/3}Ni_{1/3}Mn_{2/3}O_2$ nanofibers at 5 C. Source: Reproduced with permission from Liu et al. [34]. © 2019/John Wiley & Sons.

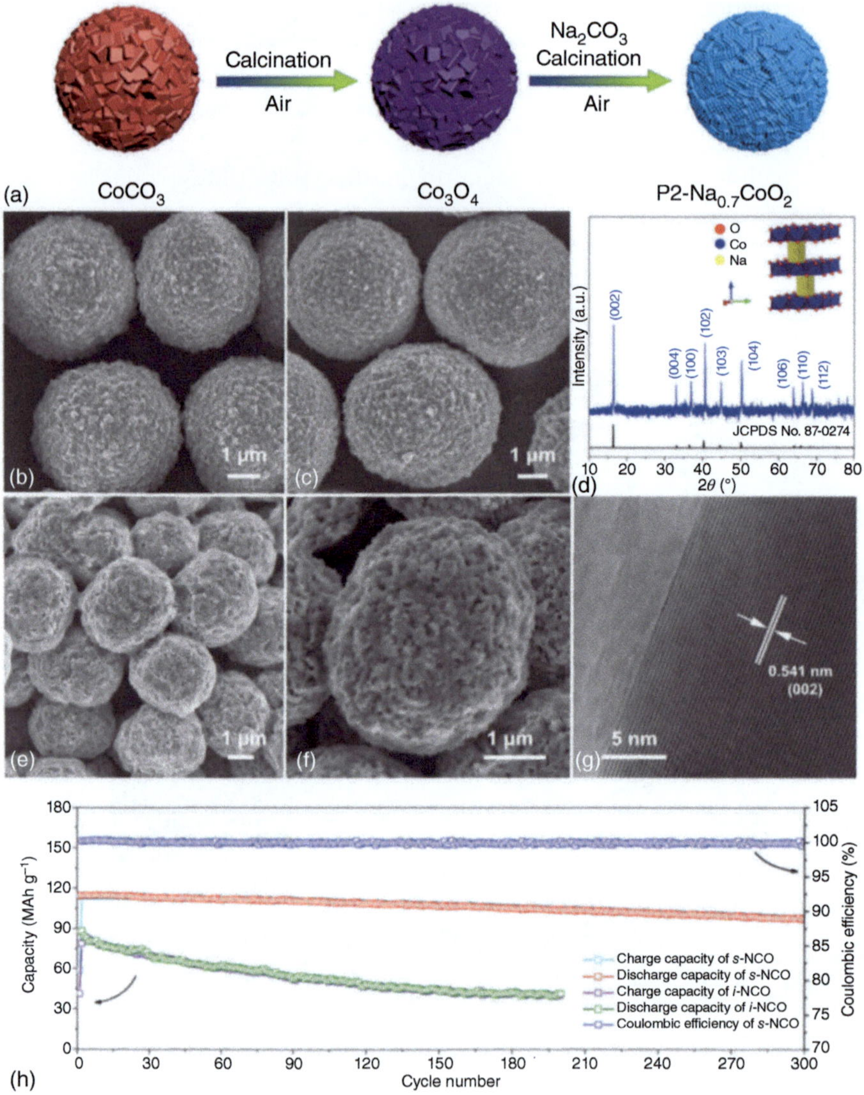

Figure 4.4 (a) Schematic of the two-step synthesis of P2-$Na_{0.7}CoO_2$ microspheres. FESEM images of (b) $CoCO_3$ and (c) Co_3O_4 microspheres. (d) XRD pattern (inset is the corresponding crystal structure of P2-$Na_{0.7}CoO_2$), (e,f) FESEM images, and (g) HRTEM image of P2-$Na_{0.7}CoO_2$ microspheres. (h) Cycling performance of s-NCO and i-NCO at a current rate of 0.4 C and the corresponding coulombic efficiency of s-NCO. (Note: 1 C = 125 mA g^{-1}.) Source: Fang et al. [35]. © 2017/John Wiley & Sons.

the structural advantages, P2-$Na_{0.7}CoO_2$ microspheres show enhanced electrochemical performance for sodium storage in terms of high reversible capacity (125 mAh g^{-1} at 5 mA g^{-1}), superior rate capability, and long cycle life (86% capacity retention over 300 cycles, Figure 4.4h).

The aggregation of secondary microspheres, derived from a multitude of primary crystals with a nanoplate shape, is widely recognized for its contribution to ensuring the retention of microscale particle size distribution in the final product. It is crucial for large-scale manufacturing as it guarantees appropriate physical properties. In addition, the secondary particles, which have a nanoflake structure, efficiently facilitate sodium-ion migration kinetics due to their short diffusion distance and increased electrode–electrolyte contact area. Dou et al. designed and synthesized a novel air-stable O3-type $Na\,[Li_{0.05}Mn_{0.50}Ni_{0.30}Cu_{0.10}Mg_{0.05}]O_2$ cathode material [36]. They synthesized microspherical $Mn_{0.50}Ni_{0.30}Cu_{0.10}Mg_{0.05}CO_3$ precursors via an industrially feasible coprecipitation method by controlling the pH value, feeding rate, and reaction temperature. With the secondary microspheres being aggregated from a large amount of nanoplate primary crystals, the final product $Na\,[Li_{0.05}Mn_{0.50}Ni_{0.30}Cu_{0.10}Mg_{0.05}]O_2$ has a high reversible capacity and remarkable capacity retention of 70.4% after 1000 cycles at 20 C.

Active facet structure: Besides the geometric structure, the rational design of facets is also crucial for enhancing the electrochemical performance. Exposing active facets of {010} in lithium-based LTMOs has been widely reported as an effective strategy to boost their rate performance. Recently, some studies have revealed a similar phenomenon in Na-based LTMO cathodes. The six-sided morphology of the hexagonal crystal, known as the {010} facet, is commonly associated with the electrochemically active plane for Na^+ migration. However, due to the close-packed structure comprising TMO_6 octahedra, the {001} facet can impede Na^+ transportation. Consequently, it becomes essential to increase the ratio and thickness of the exposed {010} facets to optimize Na^+ migration. Mao et al. synthesized O3-$NaNi_{0.5}Mn_{0.5}O_2$ hollow microbars with exposed {010} active facets using the ethanol-mediated coprecipitation method [37]. SEM and TEM images provide additional evidence that the uniform hollow microbar, measuring 2 μm in width and 15 μm in length, is composed of numerous primary nanoparticles arranged in an oriented stacked manner. It is worth noting that the hollow structure plays a significant role in alleviating volume strain, and its formation process is closely associated with the Kirkendall effect. The electrode shows excellent Na^+ diffusion properties while effectively suppressing irreversible phase transitions without introducing additional metal ions. O3-$NaNi_{0.5}Mn_{0.5}O_2$ has a high Na^+ diffusion coefficient, four times higher than that of the contrast sample, and delivers a remarkable rate capability with a specific capacity of 40.0 mAh g^{-1} at 30 C and outstanding cycling stability with a capacity retention rate of 70.0% over 500 cycles at 3.75 C due to the unique hollow structure and well-exposed {010} active facets. Meanwhile, Li et al. successfully designed and synthesized P2-type hierarchical microspheres, where microspheres are assembled by primary nanoplates featuring radially oriented {010} facets [38]. The presence of these well-exposed {010} facets contributes to the exceptional electrochemical properties of the microspheres. Recently, Zhang et al. synthesized highly dispersed hexagonal-prism-like single-crystal P2-type $Na_{0.66}Ni_{0.26}Zn_{0.07}Mn_{0.67}O_2$ with a high proportion of {001} active facets through a combined coprecipitation and molten-salt method [39]. The presence of molten salt at high temperatures provides a quasi-liquid environment that promotes crystal growth and leads to the formation of highly exposed {010} active facets. The synthesized P2-type $Na_{0.66}Ni_{0.26}Zn_{0.07}Mn_{0.67}O_2$ material, characterized by an abundance of exposed active facets, shows improved structural and interfacial stabilities, reduced side reaction, and enhanced Na^+ diffusion kinetics. As a result, it delivers outstanding cyclic stability with a capacity retention of 95.8% after 100 cycles at 100 mA g^{-1} and a good rate capability of 54.8 mAh g^{-1} at 2000 mA g^{-1} with a decent specific capacity. This work presents a novel approach for synthesizing layered oxides with exposed active facets.

Interface Engineering Interface engineering is a well-known and effective modification technique for improving the electrochemical performance of layered oxide cathodes. Furthermore, using fast ion conductor coatings can improve sodium-ion conductivity and promote the stability of the interface between the cathode material and the electrolyte, resulting in improved electrochemical performance. Surface coating is the most commonly used technique for interface engineering, followed by acid treatment. The primary types of coatings include metal oxide, metal fluoride, metal phosphate, conductive polymer, and fast sodium-ion conductor coatings.

Atomic Layer Deposition Coating The metal oxide coating layer, such as MgO, ZrO_2, Al_2O_3, TiO_2, and ZnO, can efficiently minimize undesirable phase transitions and act as a buffer against inherent stress and strain. Kong et al. used MgO to coat Fe- and Mn-based P2-$Na_{0.67}Mn_{0.5}Fe_{0.5}O_2$ [40]. They found that MgO can prevent the side reaction between the electrolyte and the electrode, limit the dissolution of transition metal ions, reduce interface resistance, and enhance e^-/Na^+ migration, improving electrochemical performance. The slab thickness of the TMO_2 layer is further expanded by a small fraction of Mg^{2+} doping, which effectively decreases the migration barrier for Na^+ and prevents the irreversible phase transition in charging–discharge cycles. As a result, the capacity retention ratio of $Na_{0.67}Mn_{0.5}Fe_{0.5}O_2$ is significantly increased from 20% to 72% after 100 cycles at 1 C and from 67% to 78% after 50 cycles. Sun et al. successfully managed to significantly improve the performance of P2-type electrodes using different metal oxide coatings deposited by the atomic layer deposition technique [41]. The rate performance of P2-$Na_{0.66}(Mn_{0.54}Co_{0.13}Ni_{0.13})O_2$ shows significant improvement in the following order: $TiO_2 < Al_2O_3 < ZrO_2$, due to increased fracture toughness and electrical conductivity provided by metal oxide coating layers. Meanwhile, cyclic voltammetry and electrochemical impedance analysis revealed a decrease in charge transfer resistance, indicating that the metal oxide coating layer effectively mitigates undesirable phase transitions and acts as a buffer against inherent stress and strain between the binder, cathode, and current collector, preventing volumetric changes.

Metal phosphates, fluorides, and conductive polymers: In addition to metal oxides, metal phosphates, fluorides, and conducting polymers have also been regarded as promising coating materials for layered oxide cathodes. Zhang et al. fabricated

$AlPO_4$-coated $Na_{0.7}MnO_{2.05}$ materials via precipitation [42]. XRD results showed that a small amount of Al^{3+} can be doped into $Na_{0.7}MnO_{2.05}$ after coating, reducing the Mn^{3+} content and alleviating the Jahn–Teller effect associated with Mn^{3+}. Furthermore, the coating mitigated the dissolution of Mn^{2+} caused by the disproportionation reaction. Moreover, the coated samples show a weaker redox peak pair at 4.25 V/4.1 V than the pristine sample, suggesting that the coating modification effectively inhibits the P2–O3 phase transition in the high-voltage region. Furthermore, researchers have found that $MgPO_4$ can play a role comparable to that of $AlPO_4$ [43]. Jo et al. reported that the β-$NaCaPO_4$ coating layer can effectively improve capacity retention by removing hydrogen fluoride (HF) and H_2O [44]. They used calcium nitrate and phosphoric acid to react with residual sodium-containing species present on the surface, resulting in the formation of an in situ $NaCaPO_4$ coating layer on $Na_{0.67}Ni_{0.33}Mn_{0.67}O_2$ and the dramatic lowering of the residual sodium compounds. This uniform coating layer effectively decreased the presence of non-conductive sodium species on the surface, as confirmed by time-of-flight secondary ion mass spectrometry analysis (Figure 4.5a,b). In addition, $NaCaPO_4$ can react with HF in the electrolyte: $NaCaPO_4 + HF \rightarrow CaHPO_4 + NaF$, serving as a scavenger for $Na_{0.67}Ni_{0.33}Mn_{0.67}O_2$ and protecting electrolyte corrosion. The $NaCaPO_4$ coating has shown sufficient efficacy in preventing crack formation, as confirmed by SEM and TEM observations (Figure 4.5c–f). Metal fluoride has also been proven to be a suitable material. Sun et al. coated gradient $Na[Ni_{0.65}Co_{0.08}Mn_{0.27}]O_2$ with AlF_3 using a dry ball-milling method [45]. AlF_3-coated $Na[Ni_{0.65}Co_{0.08}Mn_{0.27}]O_2$ showed a high specific capacity and stable cycling performance in both half and full cells. The cross-sectional TEM technique was used to analyze the cycled electrode, aiming to identify the reason for the improved electrochemical performance after applying the AlF_3 coating layer. Results showed that the AlF_3 coating is hardly damaged and incurred no significant material loss during cycling, which is beneficial for protecting $Na[Ni_{0.65}Co_{0.08}Mn_{0.27}]O_2$. The AlF_3 coating remained visibly intact even after cycling, as shown in Figure 4.5g. EDS mappings confirm the existence of the AlF_3 coating layer by filling the gap between the two primary particles (Figure 4.5h). Figure 4.5i presents a detailed TEM image of the region highlighted by the green dashed box in Figure 4.5h. It shows that, despite the uncoated appearance, the surfaces are encapsulated by a thin layer (approximately 10 nm) of AlF_3. This observation indicates that the AlF_3 coating extends around the secondary particle, safeguarding the cathode surface. Despite the presence of a skinny layer with a partially collapsed structure at the extreme edge, HRTEM analysis and corresponding fast Fourier transform patterns (Figure 4.5j) of the specified region B in Figure 4.5i showed the well-preserved layered structure in both the particle periphery and the interior. However, as shown in Figure 4.5k–n, the uncoated NRG65 sample showed more surface cracks and a thicker rock-salt-phase layer, suggesting that the AlF_3 coating has enhanced the structural integrity. Liu et al. also reported the positive effect of AlF_3 on improving the electrochemical performance of layered oxide materials [46]. They applied a coating of AlF_3 to $Na_{0.5}Ni_{0.25}Mn_{0.75}O_2$, which

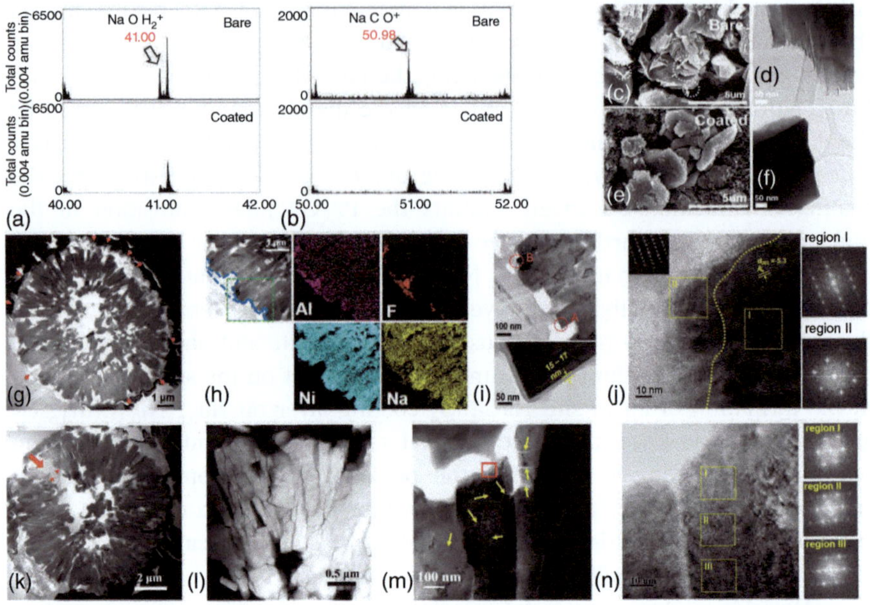

Figure 4.5 TOFSIMS results of pristine bare and $NaCaPO_4$-coated $Na_{2/3}[Ni_{1/3}Mn_{2/3}]O_2$ powders: (a) $NaOH^{2+}$, (b) $NaCO^+$, (c) SEM, and (d) TEM images of the first cycled bare electrode. (e) SEM and (f) TEM of the $NaCaPO_4$-coated $Na_{2/3}[Ni_{1/3}Mn_{2/3}]O_2$ electrode after the first cycle. Source: Reproduced from Jo et al. [44]. © 2018/John Wiley & Sons. (g) Bright-field scanning TEM cross-sectional image of the AlF_3-nanorod gradient $Na[Ni_{0.65}Co_{0.08}Mn_{0.27}]O_2$ (NRG65) cathode particle after 200 cycles (red arrows show the AlF_3 coating layer remaining intact after cycling). (h) Elemental mapping of AlF_3-NRG65 cathode particle after 200 cycles. (i) Bright-field TEM image of the AlF_3-NRG65 cathode after 200 cycles marked by the green dashed box in (h) (the figure below shows the magnified image of the area marked A). (j) HRTEM image of the region marked B in (i) and Fourier transform images of regions I and II. (k–n) TEM image of the corresponding structural analyses on the bare NRG65 cathode particle after 200 cycles. Source: Sun et al. [45]. © 2018/American Chemical Society.

resulted in improved rate performance and cycling stability. Jo et al. reported polypyrrole (PPy)-coated $Na_{0.7}MnO_{2.05}$ hollow microspheres [47]. Studies have shown that the application of a conductive polymer coating improves the electrical conductivity of the electrode material, effectively prevents the dissolution of Mn during cycling, and maintains the stability of the hollow structure. Lin et al. used a copolymer nanolayer of methacrylic acid and acrylonitrile to coat the $Na_{0.67}Li_{0.16}Ni_{0.33}Mn_{0.67}O_{2+\delta}$ cathode [48]. The incorporation of this copolymer nanolayer with robust electron-donating groups allows for the anchoring of transition metal ions through coordinative interactions. This protective mechanism prevents cathode materials from undergoing corrosion by the electrolyte, effectively slowing down the interfacial structural degradation that occurs during cycling.

Ion Doping Cation doping has recently been recognized as a powerful method for controlling complex phase transitions and eliminating the Na^+/vacancy ordered structure, thereby improving electrochemical properties such as cyclic stability and rate capability. In the crystal structure of sodium-based LTMOs, both inert and active metal cations have the potential to occupy the transition metal and Na sites.

Single-metal-ion doping: Due to the similarity in ionic radii between Li ions and transition metal ions, Li ions are often doped into the transition metal site. Nazar et al. reported that the incorporation of 0.2 mol% of Li ions into P2-type $Na_{0.6}MnO_2$ can increase the valence state of Mn to tetravalence and inhibit the P2–P′2 phase transition [49]. In situ XRD measurements reveal suppressed complex phase transitions in pristine $Na_{0.6}MnO_2$ after Li-ion doping. Density functional theory (DFT) calculations suggest that the localized energy levels of oxygen atoms are closer to the Fermi level than those of Mn atoms, indicating that oxide ions may participate in the redox process. Recently, some peer works have also detected the involvement of oxygen redox in the P-type $Na_xLi_yMn_{1-y}O_2$ system using advanced characterization techniques such as neutron pair distribution function analysis, oxygen K-edge soft X-ray absorption spectroscopy (XAS), and high-resolution resonant inelastic X-ray scattering [50]. In contrast, Li-ion doping in O3-type $NaLi_{0.1}Ni_{0.35}Mn_{0.55}O_2$ shows slightly different phenomena compared to the P2 structure [51]. After Li-ion doping, a novel O′3 phase is formed, which effectively mitigates the irreversible O3–P3 phase transformation and minimizes structural degradation, leading to significantly improved electrochemical performance. Furthermore, Oh et al. showed that Li-ion doping in $Na[Li_{0.05}(Ni_{0.25}Fe_{0.25}Mn_{0.5})_{0.95}]O_2$ effectively retards the phase transition from O3 to monoclinic P′3 due to strong bonding between Li and O atoms [52]. The phase transition in the P2 structure is generally presumed to be triggered by the intense electrostatic repulsion between transition metal layers during Na^+ desodiation in the charging process. Recent studies have revealed that Li-ion doping into the P2 structure increases the Na content, successfully preventing the structural transition, as shown in Figure 4.6a. Hu et al. found that the ratio between the interlayer distances of $d_{(O\text{-}Na\text{-}O)}$ and $d_{(O\text{-}TM\text{-}O)}$ serves as a valuable indicator for distinguishing structural competition between P2- and O3-type Na-based LTMOs, leading to the development of high-Na-content P2 materials [53]. Figure 4.6b,c shows that a P2 phase is more likely to form when the ratio of $d_{(O\text{-}Na\text{-}O)}/d_{(O\text{-}TM\text{-}O)}$ exceeds 1.6. Applying this concept, they successfully designed a high-Na-content $Na_{45/54}Li_{4/54}Ni_{16/54}Mn_{34/54}O_2$ material, which showed enhanced structural stability and improved electrochemical performance.

In general, it is presumed that layered oxide cathodes show improved structural stability through Mg-ion doping. Billaud et al. fabricated $Na_{0.67}Mg_xMn_{1-x}O_2$ $(0 \leqslant x \leqslant 2)$ cathodes to investigate the influence of Mg^{2+}-ion substitution for Mn^{3+} ions in $Na_{0.67}MnO_2$ [54]. Replacing two high-spin Mn^{3+} ions with one Mg^{2+} and Mn^{4+} reduces the number of Jahn–Teller active ions. Finally, Mg^{2+} doping

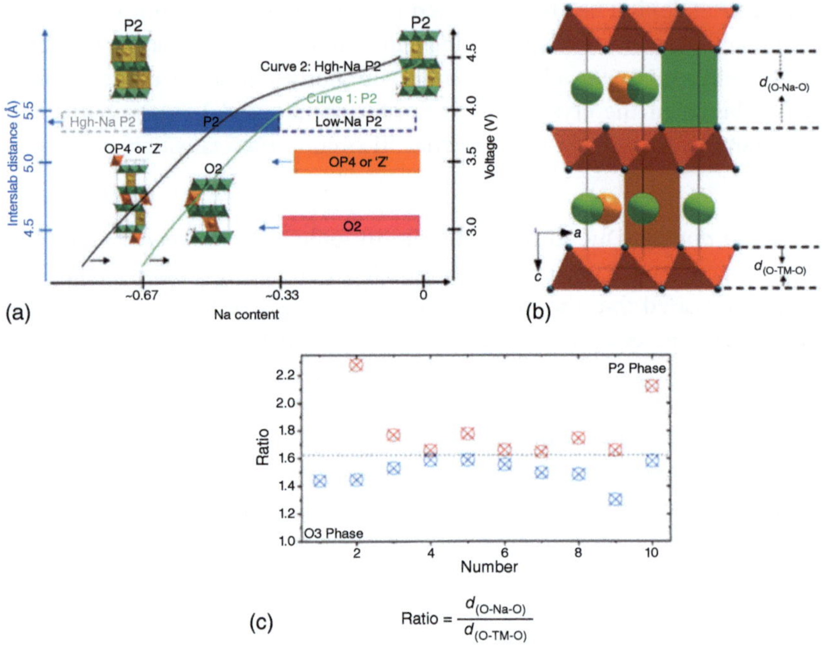

Figure 4.6 (a) The structural evolution mechanism of high-Na P2 oxides during desodiation. (b) Crystal structure of P2-type oxides. (c) The ratio of $d_{(O\text{-}Na\text{-}O)}$ and $d_{(O\text{-}TM\text{-}O)}$ for the typical P2- and O3-type compounds. Source: Reproduced with permission from Zhao et al. [53]. © 2020/American Chemical Society.

enhances cycling stability, resulting in smoother charge–discharge curves and a partial reduction in phase transitions.

Al, being an abundant element, is considered an effective dopant for enhancing the electrochemical performance of layered oxides due to its strong bonding energy with O. Liu et al. fabricated a low-cost cathode, $Na_{0.67}Al_{0.1}Mn_{0.9}O_2$, doped with Al^{3+} ions and systematically investigated the impact of the Al^{3+} dopant [55]. The results showed that Al^{3+} doping reduces the content of Mn^{3+}, mitigates Jahn–Teller distortion, increases Na layer spacing, and suppresses complex phase transitions. As a result, the doped sample showed improved cyclic and rate performance.

Ti^{4+} doping in a $Na_{2/3}Ni_{1/3}Mn_{2/3}O_2$ cathode has been proven to significantly improve the cycling stability and increase the rate of sodium-ion diffusion [56]. Yoshida et al. examined the influence of Ti^{4+} doping in a series of $Na_{2/3}Ni_{1/3}Mn_{2/3-x}Ti_xO_2$ ($0 \leqslant x \leqslant 2/3$) cathodes [57]. Introducing Ti^{4+} effectively curbed the typical Na^+/vacancy ordering and reduced the significant volume fluctuation during the P2–O2 phase transition. As a result, the cycling stability improved, with 88% capacity retention after 20 cycles, and the optimized $Na_{2/3}Ni_{1/3}Mn_{1/2}Ti_{1/6}O_2$ cathode showed a high reversible capacity of 127 mAh g^{-1}.

Similarly, the addition of inert Zn^{2+} to transition metal layers has been recognized as a means to improve the electrochemical performance by stabilizing the structure [58, 59]. Studies by Bai et al. and Wang et al. have provided further evidence that Zn^{2+} doping effectively limits the phase transitions of $Na_{2/3}MnO_2$ during charge–discharge cycles. Intriguingly, Zn^{2+} doping induces the presence of nonbonding O (2p) orbitals, leading to the participation of anionic redox activity.

In recent developments, Nb and Sb ions have emerged as dopants to enhance the electrochemical performance of layered oxides [60]. Active cation doping, such as Fe^{3+}, Co^{3+}, and Cu^{2+}, has also been shown to be effective in improving cycling stability without significantly compromising specific capacity. In contrast, inert cation doping in layered oxides increases structural stability at the expense of specific capacity [61–63]. Additionally, there is growing interest in doping at the alkali metal site, as it offers improved structural stability. Notably, due to the comparable ionic radii of Ca^{2+} (≈1.00 Å) and Na^+ (≈1.02 Å), Ca^{2+} is more likely to occupy the alkali metal site in various layered oxide systems.

Single-nonmetal-ion doping: Nonmetal ion doping, such as F and B ions, has shown promising effects in enhancing the electrochemical performance [64]. Notably, due to the low mass weight of nonmetal elements, nonmetal ion doping can act as a functional primitive with or without minimal reduction in the theoretical capacity. Liu et al. reported a series of F^--doped $Na_{2/3}Ni_{1/3}Mn_{2/3}O_2{-}xF_x$ ($x = 0$, 0.03, 0.05, 0.07) cathodes [65]. Their findings revealed that F^- doping partially reduces the valence of manganese to Mn^{3+}, thereby leading to the formation of active species and improving reversible specific capacity. Moreover, in situ XRD measurements showed that F^- doping can suppress the biphasic reaction at low voltage. As a result, the optimized $Na_{2/3}Ni_{1/3}Mn_{2/3}O_{1.95}F_{0.05}$ cathode showed excellent cyclic stability with a capacity retention of 75.6% over 2000 cycles. The introduction of B ions, which have a high bonding energy with oxygen, theoretically strengthens the structural stability of layered oxides, thereby leading to improved electrochemical properties. Guo et al. incorporated B ions into the transition metal layer of O3-$NaLi_{1/9}Ni_{2/9}Fe_{2/9}Mn_{4/9}O_2$ by forming BO_4 tetrahedra without replacing the transition metal ion [66]. Their results indicated that B-ion doping creates more negative charges around the oxygen sites due to the strong B—O covalent bond, effectively mitigating irreversible oxygen redox. Consequently, the B-ion-doped cathode showed a higher initial reversible specific capacity of 160.5 mAh g^{-1} and good cycling stability with a capacity retention of 82.8% over 200 cycles.

Multiple-ion doping: Multiple-ion doping often shows a synergistic effect of different cations, leading to enhanced stability and improved electrochemical performance compared with single-cation doping. Li et al. devised a co-doped cathode, $Na_{0.67}Mn_{0.8}Ni_{0.1}Mg_{0.1}O_2$, with both Mg and Ni ions [67]. The co-doped cathode demonstrated the more significant d-spacing of the Na layer compared with cathodes with single-Mg- or single-Ni-ion doping, as evidenced by Rietveld XRD refinement results. This effect can be attributable to the cooperative influence of Mg and Ni ions, leading to the contraction of the TMO_2 slab. Remarkably, the co-doped $Na_{0.67}Mn_{0.8}Ni_{0.1}Mg_{0.1}O_2$ cathode showed outstanding rate performance, with high reversible capacities of 110, 66, and 37 mAh g^{-1} at high current densities of 480 (2 C), 1200 (5 C), and 1920 mA g^{-1}

(8 C), respectively. Cheng et al. reported the synergetic doping of Li^+, Mg^{2+}, and Ti^{4+} in P2-$Na_{0.7}Li_{0.03}Mg_{0.03}Ni_{0.27}Mn_{0.6}Ti_{0.07}O_2$, which effectively enhanced the electrochemical performance [68]. Ti^{4+} doping contributes to a high operating voltage, Li^+ doping increases the specific capacity, and Mg^{2+} doping strengthens the crystal structure. The synergistic effect of these dopants results in suppressing the P2–O2 phase transition and converting it into a P2–Z biphasic reaction. The restrained phase transition can be attributable to the increased sliding energy achieved through doping. In addition, the poor air stability of LTMOs, attributable to the spontaneous extraction of Na and oxidation of transition metals in humid environments, significantly increases the costs associated with material storage, shipping, and battery production. In addition to multiple-cation doping at transition metal sites, researchers have proposed dual-site doping, involving both transition metal and Na sites, to improve the electrochemical performance of Na-based LTMO cathodes. A well-designed example is the dual-site-doped O3-type $Na_{0.9}Ca_{0.035}Cr_{0.97}Ti_{0.03}O_2$ cathode proposed by Lee et al., which showed improved electrochemical properties compared with undoped and single-site-doped cathodes [69]. In situ XRD measurements revealed that the dual-site-doped O3-type $Na_{0.9}Ca_{0.035}Cr_{0.97}Ti_{0.03}O_2$ cathode has superior structural reversibility compared with pristine NaCrO. Moreover, the smaller volume changes observed after dual-site doping indicate reduced strain generation, contributing to better cycling stability. The effects of dual-site doping can be summarized as follows: (i) Ti^{4+} doping at the transition metal site increases the d-spacing of Na layers due to the stronger bonding of Ti-O than Cr-O; high-valent Ti^{4+} doping induces Na vacancies, alleviating the migration of Cr ions as well. (ii) Ca^{2+} doping at the alkali metal site also induces Na vacancies, resulting in higher Na^+ diffusion kinetics; moreover, immobile Ca^{2+} ions within the Na layer ensure good structural stability even at high charging potentials, promoting electrochemical reversibility during the charge–discharge process. As a result of dual-site doping, the cathode has a high discharge capacity of 87.8 mAh g^{-1} after 1000 cycles, whereas the undoped sample delivers a specific capacity of only 28.8 mAh g^{-1}.

Various ion doping strategies have been shown to significantly improve the electrochemical performance of layered oxides. It is observed that ion doping strategies typically alter the local chemistry, electronic structure, and phase transition during the insertion/extraction process, leading to improved performance. However, it should be noted that while certain phase transitions are suppressed to a certain extent, lattice changes and intra-grain cracking inevitably occur during long-term cycling, resulting in suboptimal cycle life [59]. Furthermore, different doping elements possess varying atomic weights, interactions, and electrochemical activities, which requires careful and judicious selection and application through comprehensive design. To overcome the challenges of capacity stability, it is possible to develop high-entropy strategies that combine microscale doping of multiple metal elements with interface engineering. This integrated approach aims to improve the overall performance and stability of layered oxide materials.

Integrated Tactics The strategies of morphological manipulation and interfacial engineering that can effectively improve the electrochemical performance of Na-based layered oxides have been discussed in previous sections. However, in some instances, a single modification strategy may not yield the desired level of performance. As a result, researchers have turned to integrated approaches and used multiple manipulation approaches on layered oxides to achieve exceptional results. This integrated approach combines various techniques to enhance the overall performance of materials.

Ion-doping-integrated morphology manipulation: Zhang et al. synthesized a stable P2-$Na_{0.67}Ni_{0.23}$ $Mg_{0.1}Mn_{0.67}O_2$ layered cathode, simultaneously achieving magnesium doping and hierarchical one-dimensional nanostructure composed of a nanoplate subunit assembly (denoted as NNMMO-HNA) using a simple coprecipitation approach that realizes a record voltage of above 3.5 V for sodium-ion full cells [70]. By introducing Mg^{2+} through partial substitution of Ni^{2+}, the harmful P2–O2 phase transition in pure $Na_{0.67}Ni_{0.33}Mn_{0.67}O_2$ resulted in a reduced volume change of about 23%. In situ XRD analysis further revealed that NNMMO-HNA shows a reversible P2–OP4 phase transition, which induces a significantly smaller volume change compared with the P2–O2 phase transition. Moreover, incorporating a unique hierarchical nanostructure can significantly improve the kinetic processes of cathode materials. Research results indicate that NNMMO-HNA has a lower impedance than NNMO-HNA (the sample without the addition of $Mg(CH_3COO)_2 \cdot 4H_2O$), and galvanostatic intermittent titration technique (GITT) results show a higher Na^+ diffusion coefficient for NNMMO-HNA. Consequently, the electrochemical performance of $Na_{0.67}Ni_{0.33}Mn_{0.67}O_2$ is significantly improved by the dual manipulation of Mg^{2+} doping and the integrated hierarchical nanostructure. In addition, the assembled sodium-ion full cells can have excellent performance when combined with a commercial hard carbon (HC) anode. The full cell operates at a high voltage of above 3.5 V and delivers a remarkable energy density of 249.9 Wh kg^{-1} at a power density of 39.6 W kg^{-1}, based on the mass of active cathode and anode materials. Furthermore, the full cell shows decent cycling stability, with capacity retention rates of 72.7% after 50 cycles at 0.1 C and 75.5% after 100 cycles at 1 C.

Coating-integrated doping: A protective coating layer, which acts as a HF scavenger, effectively protects the surface of the cathode particle from exposure to the electrolyte. In addition, doping and substituting cationic elements, including Mg, Ti, Zn, and Li, have been extensively investigated and shown to enhance battery performance by suppressing irreversible, complicated phase transformations. Therefore, an effective strategy involving the combination of coating and doping is used to improve structural stability and provide a novel design approach for high-performance SIB cathode materials. Hwang et al. reported that the electrochemical behavior and structural stability of $Na[Ni_{0.5}Mn_{0.5}]O_2$ within the voltage range of 2.0–4.2 V can be improved by a novel strategy of partial doping of Mg and concurrent coating of MgO, and the schematic of a one-pot synthesis process is shown in Figure 4.7a [71]. Magnesium is a low-toxicity and abundant alkali earth metal with favorable chemical and physical properties, making it an ideal choice for coating and doping purposes. The MgO coating layer on cathode particles effectively protects their surface from side reactions with the electrolyte, especially the erosion of HF. In addition, the homogeneous incorporation of Mg^{2+} into transition metal layers without affecting the original O3-type layer structure significantly improves the structural stability via the reduction in the extent of irreversible multiphase transformation (Figure 4.7b,c). Meanwhile, Zhang et al. adopted a facile method to coat and dope P2-$Na_{0.7}MnO_2$ with a P2-$Na_{0.7}Ni_{0.33}Mn_{0.67}O_2$ layer in a single step to enhance its cycle stability

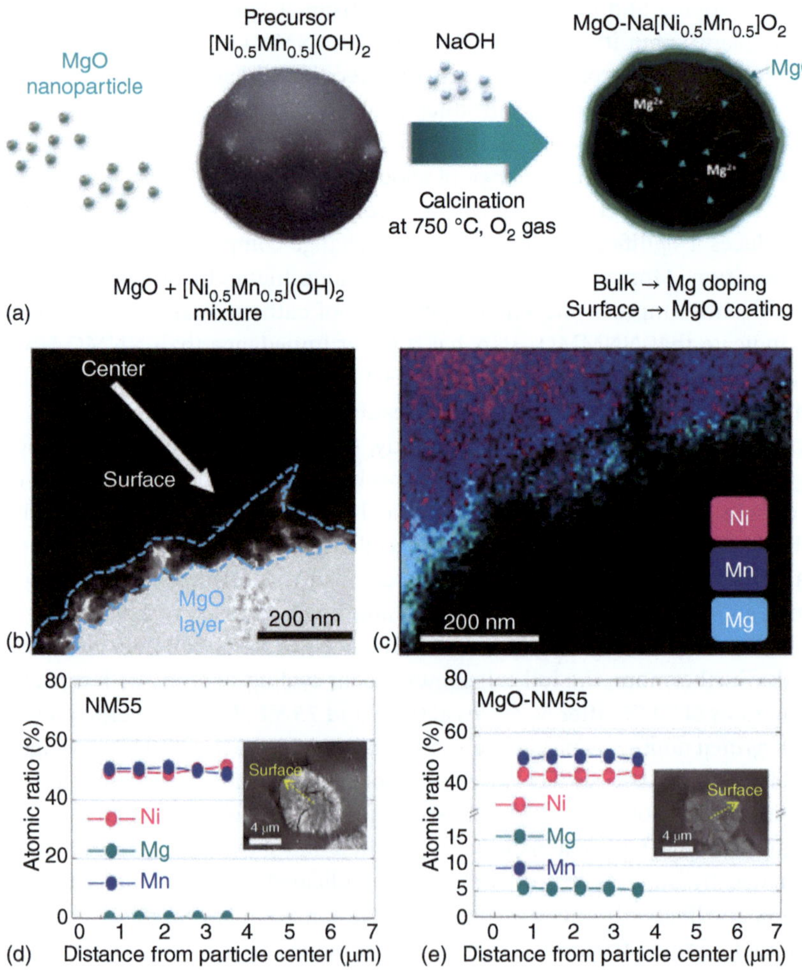

Figure 4.7 (a) Schematic of the synthesis process for preparing the MgO-Na[$Ni_{0.5}Mn_{0.5}$]O_2 (MgO-NM_{55}) cathode. (b) Bright-field TEM image and (c) corresponding elemental mapping of MgO-NM_{55}. EPMA results of (d) NM_{55} and (e) MgO-NM_{55} cathodes. Inset images show the cross-sectional SEM images of NM_{55} and M gO-NM_{55} particles, respectively. Source: Hwang et al. [71]. © 2018/Royal Society of Chemistry.

[72]. Doping Ni into the bulk material yielded $Na_{0.7}MnO_2$ coated with different amounts of $Na_{0.7}Ni_{0.33}Mn_{0.67}O_2$, stabilizing the hexagonal phase. The coating layer provides effective protection against moisture, maintains its stability against air exposure, and suppresses impedance changes during the charge–discharge process. Additionally, the coating layer minimizes lattice strain at the interface, ultimately leading to superior cycling performance. The effective approach of simultaneous coating and doping holds great promise for advancing the development of high-performance cathode materials for SIBs. Furthermore, to achieve precise control over coating and doping levels, extensive optimization is required for parameters such as temperature, duration, atmosphere, calcination techniques, and the composition of mixed materials.

4.3.2 Polyanion-Type Materials

The structures of polyanion-type materials consist of MO_x [M representing a transition metal: Fe, Mn, Co, V, Ti, Cr, etc.] polyhedral and $(XO_4)_m{}^{n-}$ or $(X_mO_{3m+1})^{n-}$[X = S, P, Si, As, Mo, W, etc.] anion groups, which are connected by a corner- and/or edge-sharing manner. Since the oxygen atoms are strongly connected to the transition-metal M and the nonmetal element X by covalent bonds, polyanion-type cathodes can possess high thermal stability, structural stability, and robust crystal frameworks to avoid O_2 release and undertake current attacks [73]. Polyanion cathode materials can generally be categorized into six major types: phosphates, pyrophosphates, fluoride phosphates, sulfates, silicates, and mixed polyanion compounds according to the type of anion groups.

4.3.2.1 Phosphate-Based Polyanion Electrode Materials

$NaMPO_4$ (M = Fe, Mn) Phosphate compounds in SIBs have garnered significant attention due to the commercial success of olivine-structured $LiFePO_4$ (LFP) in LIBs. Therefore, it is reasonable to consider its sodium analogue $NaFePO_4$ (NFP) for SIBs. Furthermore, there is an urgent need to delve deeper into the mechanistic intricacies and optimize the performance of sodium phosphate compounds. NFP was categorized into two distinct phases: olivine (Figure 4.8a) and maricite (Figure 4.8b), which is different from that of LFP [74, 78]. Olivine NFP shows metastability, characterized by a crystal lattice akin to that of LFP, albeit with sodium ions replacing lithium ions. The crystal structure of olivine NFP belongs to the *Pnma* space group [74], which is composed of angularly shared FeO_6 octahedra units connected with PO_4 tetrahedra units by edge sharing, building a one-dimensional sodium-ion migration tunnel similar to a zigzag manner along the *b*-axis. Moreau et al. studied the structural attributes and thermal stability of olivine NFP [75]. In situ XRD analyses (shown in Figure 4.8c) revealed a noteworthy phenomenon wherein the olivine phase undergoes a transition to the maricite phase within an increased temperature range (500–600 °C). Due to the intricacies involved in the direct synthesis of olivine NFP, a rational and widely adopted strategy entails cation exchange, a process facilitated by olivine LFP precursors [75, 79]. Notably, in contrast to the lithium insertion/extraction mechanisms

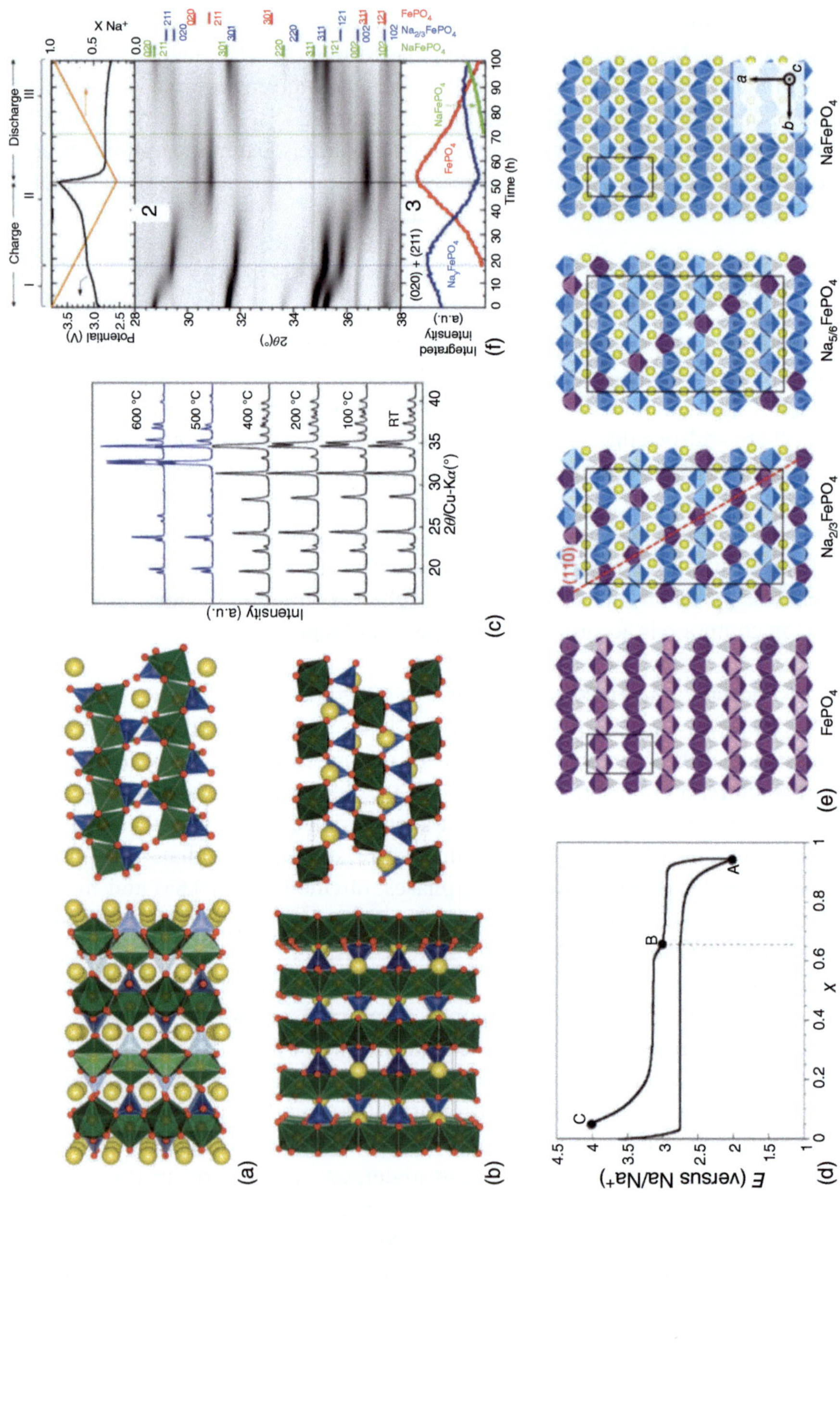
(a)
(b)
(c)
Intensity (a.u.)
600 °C
500 °C
400 °C
200 °C
100 °C
RT
2θ/Cu-Kα(°)
(d)
E (versus Na/Na+)
x
A
B
C
(e)
FePO4
Na2/3FePO4
Na5/6FePO4
NaFePO4
(110)
(f)
Charge
Discharge
Potential (V)
X Na+
2θ(°)
Integrated intensity (a.u.)
Time (h)
(020) + (211)
NayFePO4
FePO4
NaFePO4
Na2/3FePO4

Figure 4.8 Crystal structures of (a) olivine $NaFePO_4$ and (b) maricite $NaFePO_4$, Na atoms (yellow), PO_4 tetrahedra (blue), and FeO_6 octahedra (green). (c) High-temperature in situ XRD pattern showing the phase transition of amphibole to malosite in the $NaFePO_4$ system (400 °C < Tt < 500 °C). Source: Reproduced from Avdeev et al. [74]. © 2013/American Chemical Society. (d) A typical electrochemical curve for the synthesis of $NaFePO_4$ and $Na_{0.7}FePO_4$ in the potentiostatic intermittent titration technique (PITT) mode. Source: Reproduced from Moreau et al. [75]/American Chemical Society. (e) Optimized geometries of the most stable structures for $FePO_4$, $Na_{2/3}FePO_4$, $Na_{5/6}FePO_4$, and $NaFePO_4$. Source: Reproduced from Saracibar et al. [76]/Royal Society of Chemistry. (f) (1) Voltage-versus-time curves of the XRD in situ experiment, and (2) 2-θ-versus-time plot of the XRD patterns comprising a full cycle. (3) Sum of the integrated intensity of the (020) and (211) reflections for each of the phases involved versus time. Source: Galceran et al. [77]. © 2014/Royal Society of Chemistry.

prevalent in LFP, the sodium exchange reaction within the $FePO_4$ lattice (olivine *Pnma*) unfolds in a two-step manner, demarcated by a discernible voltage drop. This phenomenon gives rise to an intermediate phase, denoted as $Na_{0.7}FePO_4$, as shown in Figure 4.8d. This intermediary phase is associated with the ordered arrangement of Na^+ ions within the crystal lattice structure, as expounded upon in prior research [75, 80]. The electrochemical behavior and reaction mechanism of olivine NFP have been extensively analyzed by Cabanas et al. [76, 77]. The emergence of the intermediate-phase $Na_{2/3}FePO_4$ is attributable to the substantial volumetric fluctuation (17.58%) between the two terminal constituents within the system (ranging from NFP to $FePO_4$). A comprehensive phase diagram for the Na_xFePO_4 ($0 \leqslant x \leqslant 1$) system during the progression of the reaction was derived computationally, as shown in Figure 4.8e. Using DFT calculations, it has been ascertained that the stability of Na_xFePO_4 is realized at $x = 2/3$, consequently resulting in a notable increase of 0.16 V increment in the charge profile. Experimental validation through in situ XRD has effectively confirmed that the distinctive asymmetrical voltage profile associated with olivine NFP stems from an alternative reaction pathway, as shown in Figure 4.8f [77, 81]. During the process of desodiation, a uniphasic reaction prevails until the emergence of the $Na_{0.7}FePO_4$ phase at a point of voltage discontinuity. This is succeeded by a biphasic reaction involving Na-rich Na_yFePO_4 and Na-deficient $FePO_4$ phases. Conversely, the sodiation process shows more incredible intricacy, characterized by the concurrent existence of three distinct phases, each characterized by variable miscibility thresholds [77].

Maricite NFP, with thermodynamic stability, has been regarded as electrochemically inactive. The olivine structure undergoes an irreversible transition to the maricite phase upon reaching a temperature of 500 °C [82]. This observation underscores the feasibility of synthesizing the maricite phase through conventional solid-state techniques. The maricite phase in NFP conforms to the *Pnma* space group. It consists of FeO_6 octahedra interconnected through edge sharing, which in turn are linked to PO_4 tetrahedra via corner sharing. From a structural perspective, the conversion from the olivine phase to the maricite phase involves an environmental exchange between Na and Fe atoms. This transformation leads to the elimination of the one-dimensional sodium migration channel. Consequently, the diffusion of Na ions within the maricite phase requires overcoming a substantial energy barrier (with energy barriers of 1.46 and 1.79 eV representing the energies of the most probable pathways), rendering it electrochemically inert, as generally acknowledged [83]. In 2014, for the first time, nanosized maricite NFP was reported to operate as an excellent cathode material for SIBs by Kim et al. [26]. The synthesized nanomaterial showed an impressive capacity of 142 mAh g^{-1} at a C/20 discharge rate and a remarkable capacity retention of 95% after 200 charge–discharge cycles. Notably, the transition from the maricite phase to the amorphous phase of NFP during the initial charging process assumes paramount significance for achieving reversible capacity, as elucidated by the investigations of both Kang et al. and Tao et al. [84, 85]. Kang et al. elucidated that results from quantum mechanical calculations indicate an activation energy of approximately 0.73 eV for amorphous structures along plausible diffusion pathways, a value considerably lower than the energy barriers associated with the NFP maricite structure. Tao et al. [85] revealed the emergence of amorphous structures at the atomic scale using X-ray absorption near-edge and Raman spectroscopy techniques. They proposed a comprehensive structural model showcasing the transition of FeO_6 octahedra into FeO_n polyhedra during the amorphization process. This transformation establishes a pathway for the (de)intercalation of Na ions, leading to significantly improved electrochemical performance. Furthermore, combining experimental findings with theoretical simulations elucidates that reducing crystallite size induces volume contraction due to reduced unit cell parameters. This reduction is attributed to a combination of sodium vacancies and significant changes in bonding characteristics. Fan et al. [86] successfully synthesized downsized NFP nanomaterials (≈1.6 nm) using an electrospinning technique. When incorporated into a complete cell configuration with a pure carbon nanofiber anode, this material showed an impressive energy density of 168.1 Wh kg^{-1} and a significant capacity retention of 87% after 200 charge–discharge cycles.

$NaMnPO_4$ shows both olivine and maricite phases, and extensive research has been conducted on both structures [87]. In the olivine phase, octahedra containing sodium ions share edges, thus forming zigzag chains along the *b*-axis to facilitate the diffusion of Na^+ ions. However, the mobility of Na^+ ions is impeded within the maricite structure, which represents the thermally stable phase. Notably, the olivine phase has unsatisfactory electrochemical performance. Despite using various synthesis methods, including ion exchange, phosphate formate precursor approaches [88], and topotactic molten salt reactions [89], for optimization, investigations into the mechanism of $NaMnPO_4$ remain relatively scarce. A comprehensive understanding of the electrochemical properties of $NaMnPO_4$ would require strategies such as carbon coating.

NASICON-structured materials: Sodium superionic conductor (NASICON) materials constitute a significant class of electrode materials for their exceptional structural stability and superior ionic conductivity. The pioneering NASICON material, $Na_{1+x}Zr_2P_{3-x}Si_xO_{12}$ ($0 \leqslant x \leqslant 3$), was introduced as a solid-state electrolyte (SSE) by Goodenough and colleagues in the 1960s [90]. Due to its capacity to accommodate various alkali metal ions, the $Na_{1+x}Zr_2P_{3-x}Si_xO_{12}$

structure holds great promise as an electrode material. The generic formula for NASICON-structured materials is $Na_xM_2(XO_4)_3$ ($1 \leqslant x \leqslant 4$; M = V, Fe, Ni, Mn, Ti, Cr, Zr, etc.; X = P, S, Si, Se, Mo, etc.). Within this structure, each MO_6 octahedron is connected to three tetrahedral XO_4 units, forming a "lantern" unit, which subsequently assembles into a three-dimensional framework replete with spacious interstices. Both cation M and anion X can be flexibly tuned and replaced within this framework, offering substantial versatility for material engineering [91, 92]. Leveraging its robust structure replete with sodium migration pathways, numerous NASICON materials have been investigated and have shown commendable rate capability and cycling stability.

$Na_3V_2(PO_4)_3$, commonly known as NVP, is an example of NASICON-type materials. Chotard et al. unveiled the existence of four distinct polymorphs of NVP within the temperature range of −30 to 225 °C [93]. Among these polymorphs, the high-temperature phase crystallizes in a rhombic structure belonging to the *R*-3*C* space group. This phase has gained widespread attention due to its straightforward synthesis (one-step method) and commendable performance. Crystallographic investigations have shown the connectivity of $[V_2(PO_4)_3]$ units along the *c*-axis, forming a framework structure by interconnecting PO_4 tetrahedra along the *a*-axis [94]. This structure accommodates two independent Na sites, each occupying different oxygen environments [95]. Na (1) shows sixfold coordination (6b) and is positioned between two $V_2(PO_4)_3$ structural units along the *z*-axis, whereas Na (2) shows eightfold coordination (18e) and resides between two PO_4 tetrahedra. In a fully occupied state, Na sites within $Na_4V_2(PO_4)_3$ divide four Na ions into one group within Na (1) and three within Na (2). Notably, Na (1) is considered non-extractable, and Na^+ ions in Na (1) are deemed more stable. Unfortunately, due to the unstable V^{2+} state and a low potential plateau (1.6 V) stemming from the V^{3+}/V^{2+} redox couple [96], $Na_4V_2(PO_4)_3$ cannot be used as a cathode material. In contrast, $Na_3V_2(PO_4)_3$ has a charge capacity of 118 mAh g^{-1} with a high voltage plateau at 3.4 V, based on the V^{4+}/V^{3+} redox couple. In situ XRD investigations have confirmed a typical two-phase reaction mechanism during cycling [97, 98]. Regarding ion migration pathways within NVP [96], theoretical calculations have identified four potential mechanisms. The associated migration energies are as follows: 0.0904 (*x*-direction), 0.11774 (*y*-direction), and 2.438 eV (path bypassing the octahedra). Furthermore, Chen et al. delved into a concerted ion exchange route known as the Na (1)–Na (2) migration path [99].

In addition to investigating the sodium storage mechanism, extensive efforts have been devoted to mitigate the challenges posed by inferior electronic conductivity and improve the electrochemical performance of NVP [100, 101]. Among the various strategies employed, carbon coating has emerged as a particularly effective approach. Jian et al. used sugar as a carbon source to successfully coat a 6-nm carbon layer onto the surface of NVP [102]. This carbon-coated NVP/C composite has a reversible capacity of 93 mAh g^{-1}, along with improved cycling stability. Recently, the construction of hierarchical structures has proven to be extremely effective in addressing these challenges. The key principles for structural design can be summarized as follows: (i) reducing particle size to shorten the diffusion length of Na^+

ions and (ii) integrating a conductive network to facilitate rapid electron transport while accommodating the volume variations resulting from Na insertion/extraction processes. For instance, Yu et al. employed a facile soft-chemistry method coupled with post-heat treatment to achieve a unique structure comprising carbon-coated nano-NVP embedded within a porous carbon matrix, denoted as (C@NVP)@pC [103]. This composite material showed impressive charge–discharge capabilities within six seconds under an ultrahigh current density of 200C, delivering a specific capacity of 44 mAh g^{-1}. Notably, this capacity even surpasses that of conventional cathode materials such as $LiCoO_2$ and LFP. They also designed similar composites, such as the rGO-framework-hosted NVP nanoparticles (NVP@C@rGO). This material showed excellent rate capability, achieving rates of up to 100C, thanks to enhanced electrode kinetics. Furthermore, the incorporation of metal ions through doping represents another effective avenue for enhancing the electrochemical characteristics of NVP. The systematic investigation of the effects of Fe^{2+}, Mn^{2+}, Al^{3+}, and Cr^{3+} doping in NVP has been carried out by Tirado et al. [104, 105]. Inoishi et al. discovered that $Na_{3.2}V_{1.8}Mg_{0.2}(PO_4)_3$, with Mg^{2+} doping, shows an improved reversible capacity compared with the pure phase due to partial redox involving V^{5+}/V^{4+} [106]. Interestingly, the Mg-rich sample, $Na_{3.5}V_{1.5}Mg_{0.5}(PO_4)_3$, showed a more pronounced plateau corresponding to the V^{5+}/V^{4+} redox reaction.

In addition to vanadium-based NASICON-type materials, research efforts have been extended to explore other single-transition-metal materials, such as $Na_3Fe_2(PO_4)_3$ [107] and $Na_3Cr_2(PO_4)_3$ [108]. $Na_3Fe_2(PO_4)_3$ crystallizes in a monoclinic structure with the *C*2/*c* space group. Rajagopalan et al. investigated the reversible nature of the Fe^{4+}/Fe^{3+} redox couple in $Na_3Fe_2(PO_4)_3$/C, which yielded a discharge capacity of 109 mAh g^{-1} [109]. Liu et al. revealed a two-phase reaction through ex situ XRD analysis [107]. However, it is worth noting that the average potential of $Na_3Fe_2(PO_4)_3$ is relatively low (below 3.0 V), which may pose limitations for its practical application. Yamada et al. studied the electrochemical properties of $Na_3Cr_2(PO_4)_3$ [108]. $Na_3Cr_2(PO_4)_3$ can be characterized by a rhombohedral lattice with the space group *R*-3*c*. Using the Cr^{4+}/Cr^{3+} redox reaction, it shows a flat voltage plateau at 4.5 V, delivering a discharge capacity of approximately 80 mAh g^{-1}. Notably, such an operating voltage is nearly the highest value observed among various cathodes used in SIBs. However, $Na_3Cr_2(PO_4)_3$ grapples with limited capacity and rapid capacity decay, necessitating further optimization for practical applications.

Recently, there has been a growing interest in NASICON-type materials featuring binary transition metals. Given the toxicity and costs associated with vanadium (V), alternative approaches such as V substitution and exploring vanadium-free materials are gaining prominence. The flexible atom arrangement and the synergistic effects of different transition metals make $Na_xM_1M_2(PO_4)_3$ compounds (where M1 and M2 represent distinct transition metals) particularly intriguing for electrochemical applications. Mason et al. substituted half of the vanadium in NVP with titanium (Ti) to synthesize NASICON–$Na_2VTi(PO_4)_3$ [110]. Unlike other vanadium compounds susceptible to degradative oxidation in aqueous systems, NASICON–$Na_2VTi(PO_4)_3$ showed improved cycling stability, retaining 70% of its capacity over 1000 cycles at 10 C [111]. This improvement is attributable to Ti substitution, which effectively prevents structural degradation and vanadium dissolution. This protective effect is probably due to the polarizing influence of Ti atoms and/or the lower redox potential, which reduces the rate of side reactions.

4.3.2.2 Pyrophosphate-Based Polyanion Electrode Materials

Pyrophosphates have garnered significant attention due to their superior electrochemical properties compared with phosphates. One of the notable advantages of pyrophosphates is their remarkable thermal stability, which is mainly attributable to the stability of the $[P_2O_7]$ units [112]. However, it is important to note that the theoretical capacity of pyrophosphates is somewhat diminished due to the increased molecular mass of the respective materials. For instance, while the theoretical capacity of NFP is at 154.2 mAh g^{-1}, that of $Na_2FeP_2O_7$ is reduced to 97.2 $mAhg^{-1}$. Pyrophosphates are typically represented by the formula $Na_2MP_2O_7$, where M can be Fe, Mn, or Co, and these compounds show distinct crystal structures. For instance, cobalt-based pyrophosphates show polymorphic crystal structures, including tetragonal (as depicted in Figure 4.9), orthorhombic, and triclinic phases. The three-dimensional framework formed by $[P_2O_7]$ provides numerous sodium-ion sites and transport channels, offering multiple opportunities for structural evolution during charge–discharge processes.

Iron-Based Pyrophosphate Materials In 2012, Yamada et al. introduced $Na_2FeP_2O_7$ as a promising positive host for sodium insertion, marking the initial discovery of this material [114]. $Na_2FeP_2O_7$, with a triclinic P1 phase, is constructed from $[Fe_2O_{11}]$ dimers formed through the corner sharing of $[FeO_6]$ octahedra and $[P_2O_7]$ groups. This arrangement incorporates both edge sharing and corner sharing, resulting in a three-dimensional framework featuring spacious tunnels along the [011] direction, facilitating the migration of Na^+ ions. Moreover, it possesses large interstitial spaces for accommodating sodium insertion (Figure 4.10a). Notably, $Na_2FeP_2O_7$ can extract

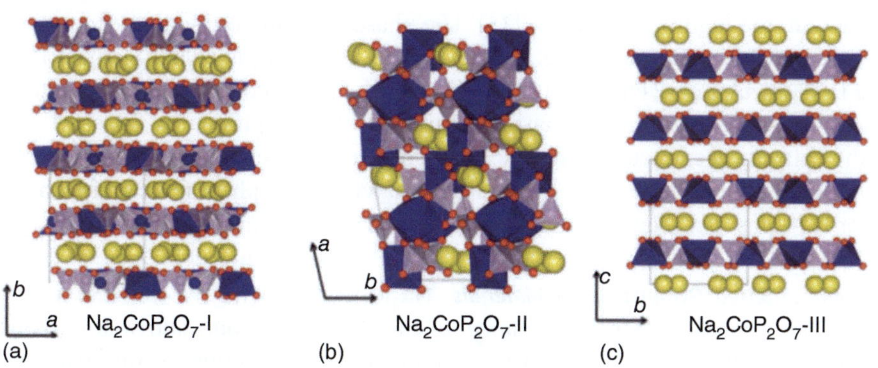

Figure 4.9 Polymorphism in $Na_2CoP_2O_7$: (a) orthorhombic ($P2_1/cn$), (b) triclinic (P-1), and (c) tetragonal ($P4_2/mnm$) forms. Source: Reproduced with permission from Barpanda et al. [113]. © 2012/John Wiley & Sons.

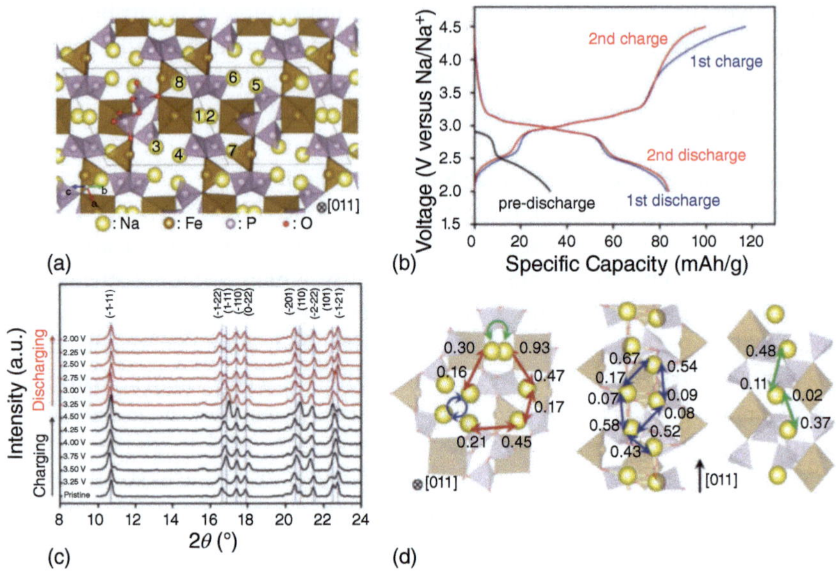

Figure 4.10 (a) Crystal structure of $Na_2FeP_2O_7$. (b) Voltage profiles at a C/20 rate, starting with discharge (pre-discharge). (c) Electrochemical ex situ XRD analysis data of $Na_2FeP_2O_7$ in the first (dis)charging processes. (d) The calculated Na-ion migration barriers for $Na_2FeP_2O_7$. The direction of Na migration and the associated migration barrier (either forward or backward) are denoted by the arrowhead and the accompanying numeric value. A color scheme is employed: red represents 2D migration pathways, while green designates the Na1 and Na2 sites, and blue designates the Na3–Na6 sites that correspond to outward channels. To enhance clarity, the migration barriers along the [011] channel direction (green and blue) are displayed separately in the two rightmost panels. Source: Reproduced with permission from Kim et al. [115]. © 2013/John Wiley & Sons.

only one Na^+ ion, delivering a capacity of approximately 90 mAh g^{-1}. Due to the redox couple Fe^{3+}/Fe^{2+}, $Na_2FeP_2O_7$ shows multistep voltage capacity profiles arising from distinct Na^+ ions (Na1-Na8) in different environments and their sequential extraction (see Figure 4.10b) [114, 115]. The phase behavior and sodium extraction process have been elucidated by Kim et al. [115]. Ex situ XRD analysis (Figure 4.10c) and DFT calculations (Figure 4.10d) reveal that the lower potential plateau around 2.5 V corresponds to a single-phase reaction. During this phase, the first Na^+ ion at the Na1 site is extracted along the [011] channel. Conversely, the higher potential plateau, centered around 3.0 V, is associated with a two-phase reaction in which other Na^+ ions (Na3-Na8) are extracted through 1D and/or 2D pathways.

Manganese-Based Pyrophosphate Materials $Na_2MnP_2O_7$ shows two distinct structures. The layered $Na_2MnP_2O_7$ belongs to the *P*-1 space group and was initially reported by Hwu et al. [116]. Mn_2O_{10} dimers and P_2O_7 groups in this structure share corners and form M-O-P slabs. The Na^+ ions are located between these slabs. $Na_2MnP_2O_7$ shows favorable electrochemical properties within the voltage range of 1.5–4.5 V at a C/20 rate [117]. It delivers a reversible capacity of 90 mAh g^{-1}, with a theoretical capacity of 97 mAh g^{-1}. Notably, despite the typically sluggish kinetic behavior observed in many Mn-based materials, the corner-sharing configuration in the crystal structure of $Na_2MnP_2O_7$ suggests improved accommodation of Jahn–Teller distortions [117]. The charging process involves both single-phase and two-phase reactions. The proposed order of Na^+ extraction within each unit is as follows: half of Na1, all of Na3, and finally half of Na4. Layered $Na_2MnP_2O_7$ can be synthesized at temperatures below 600 °C. However, at 650 °C, it irreversibly transforms into β-$Na_2MnP_2O_7$ [118]. The structure of β-$Na_2MnP_2O_7$ comprises Mn_2O_{10} dimers and P_2O_7 groups in the presence of both edge sharing and angle sharing. This structure features Na^+ tunnels along the [001] direction. β-$Na_2MnP_2O_7$ was initially explored by Yamada et al. as a cathode material for SIBs [119]. It shows a reversible capacity of 80 mAh g^{-1} at a C/20 rate, with an average operating voltage of 3.6 V, attributable to the Mn^{2+}/Mn^{3+} redox couple. A relatively flat voltage plateau characterizes the electrochemical behavior of β-Na_2MnP_2O.

Off-Stoichiometric Iron-Based Pyrophosphate Materials Recent studies led by Nazar et al. have unveiled a series of off-stoichiometric materials within the $Na_{4-\alpha}Fe_{2+\alpha/2}(P_2O_7)_2$ family, where $2/3 \leqslant \alpha \leqslant 7/8$ [120]. Due to the increased availability of reversible sodium ions, $Na_{4-\alpha}Fe_{2+\alpha/2}(P_2O_7)_2$ materials show higher capacities than the stoichiometric counterpart, $Na_2FeP_2O_7$. Specifically, $Na_{3.12}Fe_{2.44}(P_2O_7)_2$ crystallizes into a triclinic structure (space group *P*-1). In contrast to the triclinic structures of $Na_2FeP_2O_7$ and $Na_2MnP_2O_7$, the three-dimensional framework of $Na_{3.12}Fe_{2.44}(P_2O_7)_2$ is constructed from corner-shared centrosymmetric crown units that include $[Fe_2P_4O_{22}]$ and $[Fe_2P_4O_{20}]$ units. Each crown unit consists of two FeO_6 octahedra and two P_2O_7 groups. The theoretical capacity of $Na_{3.12}Fe_{2.44}(P_2O_7)_2$ can reach 117.6 mAh g^{-1}, corresponding to the de-/intercalation of 2.44 Na ions [121], which is notably higher than that of $Na_2FeP_2O_7$ (97.5 mAh g^{-1}). Ex situ XRD patterns collected using a sequential chemical desodiation/sodiation approach confirm that $Na_{3.12}Fe_{2.44}(P_2O_7)_2$ undergoes a single-phase reaction. However, it is essential to note that $Na_{3.12}Fe_{2.44}(P_2O_7)_2$ is sensitive to moisture and CO_2 in the atmosphere. To mitigate this issue, Deng et al. designed a coral-like hierarchical composite of $Na_{3.12}Fe_{2.44}(P_2O_7)_{2/C}$ [121]. This composite shows enhanced surface stability, reduced moisture sensitivity, and a reversible capacity of 107 mAh g^{-1}. Chou et al. proposed a carbon-coated Na-rich $Na_{3.32}Fe_{2.34}(P_2O_7)_2$ material [122]. This material has an exceptionally high initial coulombic efficiency (CE) of nearly 100%. In situ XRD revealed a single-phase reaction with a minimal lattice volume change of approximately 2.19%, which contributes to its structural stability.

Cobalt-Based Pyrophosphate Materials $Na_2CoP_2O_7$ shows three distinct polymorphic phases characterized by variations in the coordination of CoO_4 tetrahedra and CoO_6 octahedra [123]: orthorhombic (*Pna*2$_1$), tetragonal (*P*4$_2$/*mnm*), and triclinic (*P*-1) phases. Yashima et al. conducted an in situ neutron power diffraction study to investigate the temperature-induced phase transition of $Na_2CoP_2O_7$ [124]. At room temperature, the orthorhombic phase transforms into the tetragonal phase, which persists until its melting point is reached at approximately 900 K. Upon melting and subsequent cooling, the triclinic phase is formed. The orthorhombic and triclinic phases of $Na_2CoP_2O_7$ were initially documented by Erragh et al. [125]. The layered orthorhombic $Na_2CoP_2O_7$ structure is constructed from CoO_4 tetrahedra and PO_6 octahedra, with angularly shared oxygen ions forming $[Co(P_2O_7)]^{2-}$ layers oriented parallel to the [001] direction. Each CoO_4 tetrahedron is linked to four adjacent P_2O_7 units via shared oxygen ions. The orthorhombic phase can be easily synthesized using solid-state methods, hydrothermal techniques, and other approaches [126]. The electrochemical properties of the orthorhombic phase of $Na_2CoP_2O_7$ were initially reported by Yamada et al. [126]. In their study, the material exhibited a capacity of 80 mAh g^{-1} at C/20. In contrast to $NaCoPO_4$, which operates at a high voltage of 4.5 V based on the Co^{2+}/Co^{3+} redox couple, $Na_2CoP_2O_7$ shows an average operating voltage of 3 V. This voltage difference is attributed to the weakening of the inductive effect caused by strong covalent bonding in the CoO_4 tetrahedra, as reported by Yamada et al. [126].

4.3.2.3 Fluoride-Phosphate-Based Polyanion Electrode Materials

Iron-Based Fluoride Phosphate Materials The iron-based fluoride phosphate, Na_2FePO_4F, offers improved 2D Na^+ diffusion pathways compared with NFP. Nazar et al. were the first to introduce this framework material of iron-based fluoride phosphate and evaluate its electrochemical properties in LIBs [127]. Orthorhombic Na_2FePO_4F belongs to the *Pbcn* space group [128]. The crystal structure is constructed from bioctahedral units, which consist of face-connected FeO_4F_2 octahedra. This structure encompasses two distinct Na sites: half of the Na atoms (Na1) are positioned in the interlayer region, while the remaining half (Na2) is situated in proximity to the FeO_4F_2 octahedra. The theoretical capacity of Na_2FePO_4F is 124 mAh g^{-1}, corresponding to a single-electron transfer involving the Fe^{2+}/Fe^{3+} redox couple. Studies conducted on NFP materials following chemical oxidation reveal that the Na^+ ions located at the Na1 site remain unchanged, thereby preserving the layered framework. In contrast, the Na^+ ions at the Na2 site are fully extracted during (dis)charging processes [127]. Remarkably, Na_2FePO_4F demonstrates a minimal volume variation of 3.7% throughout these processes.

Cobalt-Based Fluoride Phosphate Materials Cobalt-based fluoride phosphate, Na_2CoPO_4F, shares an isostructural framework with Na_2FePO_4F. Na_2CoPO_4F utilizes the Co^{2+}/Co^{3+} redox couple and has a theoretical capacity of 122.4 mAh g^{-1}. A carbon-modified variant of Na_2CoPO_4F, prepared using the solid-phase method at 600 °C, showed initial charge–discharge capacities of 213 and 71 mAh g^{-1}, respectively, at a rate of 0.05 C, with an irreversible capacity of approximately 142 mAh g^{-1} [129]. This initial irreversible capacity was primarily attributable to electrolyte decomposition resulting from the high-voltage platform of 4.3 V. In another approach, Yang et al. synthesized a Na_2CoPO_4F/C nanocomposite via a spray-drying method [130]. This composite demonstrated a discharge capacity of 107 mAh g^{-1} at 0.5 C. It is worth noting that the incorporation of 2 vol%

fluoroethylene carbonate (FEC), used as an additive electrolyte, further enhances the stability of high-voltage Na_2CoPO_4F battery systems.

Manganese-Based Fluoride Phosphate Materials Manganese-based fluoride phosphate, Na_2MnPO_4F, has a three-dimensional framework structure characterized by the monoclinic $P2_1/n$ space group. Chains of $Mn_2F_2O_8$ are formed by linking the linkage of MnO_4F_2 octahedra via fluorine atoms. The monoclinic structure arises from the connection of the $Mn_2F_2O_8$ chain with PO_4 tetrahedra [131]. Within this structure, Na^+ ions are distributed among two distinct sites adjacent to the fluorine chain, called Na1 and Na4 (or Na2 and Na3). The potential diffusion pathways for Na^+ ions, namely Na1-Na4 and Na2-Na3 chains surrounding the fluorine chain, facilitate the interchain diffusion of Na_2MnPO_4F during the extraction and insertion of Na^+ ions [128]. Initial investigations of Na_2MnPO_4F conducted by Tarascon et al. and Nazar et al. revealed a lack of electrochemical activity, with no reaction observed even in the presence of the oxidant NO_2BF_4 [132]. However, when examining nanoscale Na_2MnPO_4F [133], an initial capacity of 98 mAh g^{-1} was achieved. Furthermore, subsequent studies focused on the preparation of Na_2MnPO_4F/C nanocomposites through the spray-drying method, indicating electrochemical activity [134]. More recently, dual-carbon-confined Na_2MnPO_4F nanoparticles showed a capacity of 122 mAh g^{-1} at 0.05 C and maintained 77% capacity retention after 200 cycles at 0.1 C [135].

Vanadium-Based Fluoride Phosphate Materials Despite the challenges associated with the toxicity and cost of vanadium, the investigation of vanadium-based compounds within polyanion systems continues to attract the interest of researchers. This enduring interest arises from the remarkable properties of these compounds, including their high operating voltage, high energy density, and excellent cycling stability. In fluoride phosphate systems, prominent studies have focused on $NaVPO_4F$, $Na_3V_2(PO_4)_2F_3$, and oxygen-substituted $Na_3(VO_{1-x}PO_4)_2F_{1+2x}$. $NaVPO_4F$, operating with the V^{3+}/V^{4+} redox couple, has a theoretical capacity of 143 mAh g^{-1}. It is worth noting that while some claims have been made regarding the successful synthesis of "monoclinic" $NaVPO_4F$, both XRD patterns and electrochemical performance unequivocally demonstrate that the actual synthesized phase is $Na_3V_2(PO_4)_3$. Thus, the genuinely existing $NaVPO_4F$ adopts a tavorite-like structure with monoclinic symmetry. The tavorite-like $NaVPO_4F$ structure was initially proposed by Masquelier et al. [136]. This structure is similar to tavorite-type compounds such as $LiVPO_4F$. The monoclinic tavorite-like $NaVPO_4F$ structure is characterized by two distinct VO_4F_2 octahedra connected via F atoms, which are connected via an angular linkage to two PO_4 tetrahedra. Traditional solid-state methods encounter difficulties in synthesizing $NaVPO_4F$. However, hydrothermal methods and mechanochemically assisted solid-state methods have been proven effective in its synthesis. Notably, when the raw material ratio of $NaF:VPO_4 = 1:1$, quenching results in a mixture of monoclinic $NaVPO_4F$ and $Na_3V_2(PO_4)_2F_3$, as reported by N.V. Kosova and D. O. Semykina [137]. Furthermore, Li et al. proposed

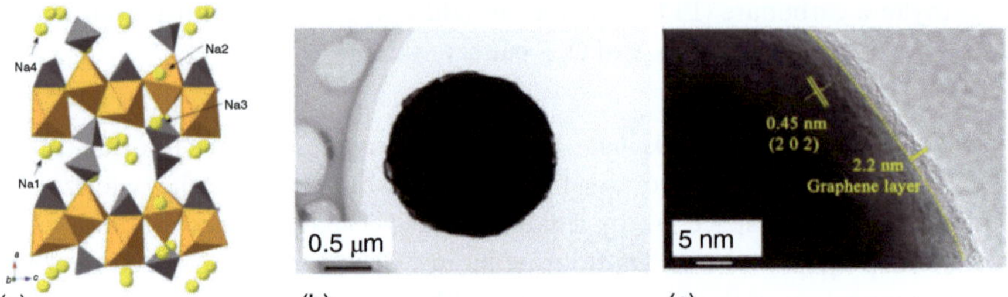

Figure 4.11 (a) Crystal structure of $Na_4M_3(PO_4)_2P_2O_7$; PO_4 and P_2O_7 units are gray, MO_6 units are brown, and Na^+ ions are yellow. Source: Reproduced from Wood et al. [139]. © 2015/American Chemical Society. (b,c) TEM and HRTEM images of $Na_4Fe_3(PO_4)_2P_2O_7$@rGO. Source: Yuan et al. [140]. © 2019/Elsevier.

an electrochemical ion exchange approach for synthesizing tavorite-like $NaVPO_4F$, starting from $LiVPO_4F$ precursors [94].

4.3.2.4 Mixed-Phosphate-Based Polyanion Electrode Materials

Mixed phosphates, such as $Na_4M_3(PO_4)_2P_2O_7$ (where M represents transition metals), have recently gained prominence as promising cathode materials. They offer several advantages, including lower volumetric strain, higher thermal stability, and higher operational potential than other polyanionic materials such as $NaMPO_4$ and $Na_2MP_2O_7$. Nevertheless, like their polyanion-type counterparts, mixed phosphates grapple with challenges related to poor electronic conductivity and low ionic conductivity [138]. The structure of $Na_4M_3(PO_4)_2P_2O_7$ has robust 3D Na^+ transmission channels, formed by MO_6 octahedra, $PO_4{}^{3-}$ groups, and $P_2O_7{}^{4-}$ groups (Figure 4.11a). DFT calculations reveal that these materials undergo a volume change of less than 4% during sodium de-/intercalation, thus endowing them with commendable cycling stability. Considering factors such as cost and environmental impact, iron-based polyanionic compounds are regarded as the most promising cathode materials within the realm of mixed phosphates [141, 142]. For example, $Na_4Fe_3(PO_4)_2P_2O_7$, for instance, shows a theoretical capacity of $129\,mAh\,g^{-1}$ and a reaction potential of 3.2 V versus Na^+/Na, based on the redox reaction involving Fe^{3+}/Fe^{2+}. To tackle the issue of sluggish kinetics, various modification strategies have been used, including nanoparticle granularity control, carbon coating, and ion doping. As shown in Figure 4.11b,c, $Na_4Fe_3(PO_4)_2P_2O_7$ nanoparticles are uniformly dispersed within a 3D rGO layer with a thickness of 2.2 nm. This mitigates volume changes and enhances electrical conductivity, resulting in an impressive rate capability of up to 200 C [140].

4.3.2.5 Sulfate-Based Polyanion Electrode Materials

In the quest for designing high-voltage materials, sulfates hold particular significance due to the stronger electronegativity of $SO_4{}^{2-}$ than other polyanion groups. Alluaudite-type compounds, represented as [$Na_2M(SO_4)_2$·$n$$H_2O$ (where M encompasses transition metals such as Fe, Co, Ni, Cu, Cr, and Mn and n can assume values of 0, 2, 4)], present a veritable treasure trove for SIBs [143]. To date, a multitude of alluaudite-phase $Na_2M(SO_4)_2{\cdot}nH_2O$ materials has been explored, including $Na_2Fe(SO_4)_2$ [144], $Na_2Co(SO_4)_2$ [145], $Na_2Mn(SO_4)_2$ [146], $Na_2Mg(SO_4)_2$ [147], among others. Notably, $Na_2Fe(SO_4)_2$ offers a relatively high operating voltage of 3.25 V (Fe^{2+}/Fe^{3+}) along with a theoretical capacity of 91 mAh g^{-1} [144]. Detailed DFT calculations have revealed that $Na_2M(SO_4)^2$ materials show an extraordinarily small volume change of less than 5% during Na^+ insertion/desertion processes [148]. This property ensures exceptional structural stability over long cycling periods. It is crucial to mention that this structural stability is closely related to the presence of H_2O units within the structure. For instance, the absence of H_2O units can lead to different phase transformation reactions between $Na_2Fe(SO_4)_2$ and $Na_2Fe(SO_4)_2{\cdot}2H_2O$ [149]. An illustrative example is the work by Pan et al., where they reported a cathode material $Na_2Fe(SO_4)_2$ with an operating voltage of 3.6 V (Fe^{2+}/Fe^{3+}) [144]. XRD results showed outstanding thermal stability, with the cathode material remaining stable at temperatures up to 580 °C. Impressively, no hydrate formation was observed even after exposure to the natural environment for two months.

4.3.2.6 Silicate-Based Polyanion Electrode Materials

In comparison with other categories of polyanion cathodes, sodium orthosilicates, specifically Na_2MSiO_4 (where M = Mn, Fe, Co, and Ni), typically offer a higher specific capacity involving a two-electron transfer per formula unit [150]. Similar to lithium orthosilicates, the fundamental building blocks of the Na_2MSiO_4 unit cell consist of NaO_4, MO_4, and SiO_4 tetrahedra. An analysis of bond lengths and bond angles has led researchers to conclude that the stability order of these tetrahedra in Na_2MSiO_4 is as follows: $NaO_4 < MO_4 < SiO_4$ [150]. The robust Si-O bond imparts thermodynamic stability to Na_2MSiO_4 even at temperatures exceeding 1000 °C. Remarkably, the average Si-O bond length experiences minimal change during desodiation, resulting in a volume change of less than 5% even upon full desodiation [151]. One of the most prominent examples among sodium orthosilicate materials is Na_2FeSiO_4, which is primarily due to the abundant resources of the Na-Fe-Si system on the Earth. Na_2FeSiO_4 shows a significant theoretical capacity of 276 mAh g^{-1}, which is attributable to its involvement in two-electron reaction processes.

4.4 Anode Materials

Anode materials are the critical component in SIBs, which to some extent determine energy and power densities, energy efficiency, and cycle lifespan of the battery. An ideal anode material should fulfill the following properties: (i) The de-/sodiation voltage should be as low as possible and ideally should not show large variations in the Na content. (ii) It should be composed of low-atomic-weight elements and be able to accommodate as much Na as possible to reach high reversible gravimetric (mAh g^{-1}) and volumetric (mAh cm^{-3}) capacities. (iii) It should not be soluble in the electrolyte and should not react with electrolyte solvents and sodium salt. (iv) It should possess good electronic and ionic conductivities. (v) Finally, it should

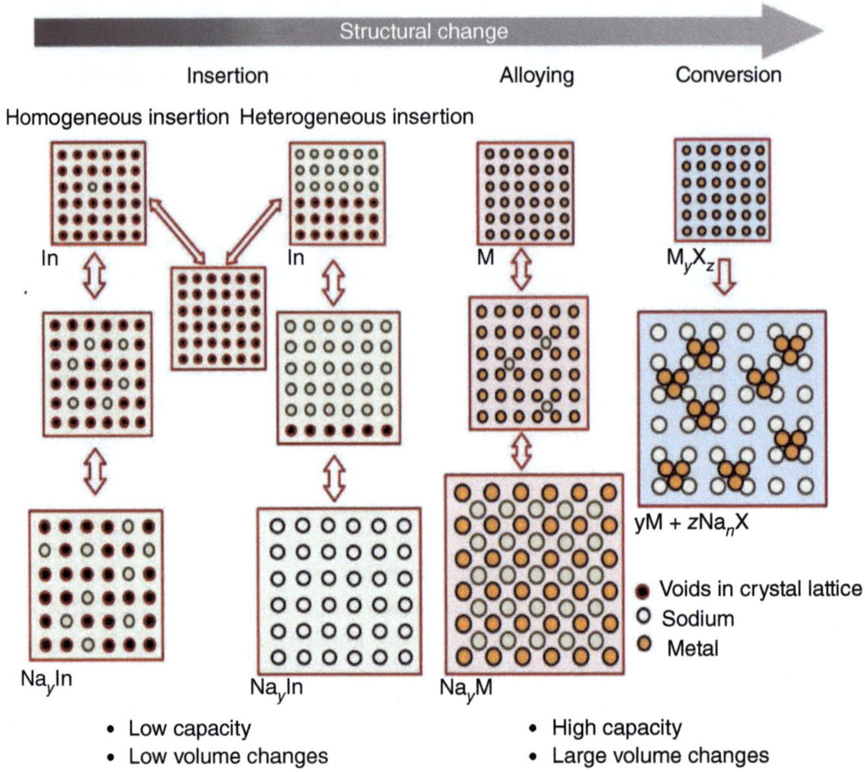

Figure 4.12 Schematic of different sodium storage mechanisms from left to right: insertion, alloying, and conversion. Source: Reproduced from Perveen et al. [152]. © 2020/Elsevier.

be of low cost, safe, and environmentally friendly. Up to now, various anode materials have been explored for SIBs, which generally can be categorized into alloy-type materials (tin [Sn], antimony [Sb], bismuth [Bi], germanium [Ge], etc.), intercalation-based compounds (carbon-based materials, titanium-based materials), conversion-type/conversion-alloying-type anode materials (transition metal nitrides, selenides, sulfides, or phosphides), and organic-type anode materials. The reaction mechanisms occurring during the storage of Na^+ in various classes of anode materials are shown in Figure 4.12.

4.4.1 Alloy-Type Anode Materials

Alloy-type anodes essentially refer to some nonmetallic and metallic elements belonging to groups 14 and 15, such as Si, Ge, Sb, Bi, Sn, and P, which can form alloys with sodium and provide large capacities ranging from 300 to 2000 mAh g^{-1}. The detailed alloying/de-alloying reaction process can be described as follows:

$$M + yNa^+ + ze^- \rightarrow Na_yM \tag{4.1}$$

In contrast to other types of anode materials, the alloying reaction involves multiple steps, which produce various intermediate products before the final phase is obtained. Therefore, energy is required for the solid-state bond breaking of the old phase and the formation of a new phase. Hence, alloy-type anodes suffer from huge volume expansion during ion accommodation, which would lead to the instability of the structure, such as pulverization, resulting in the formation of an unstable solid electrolyte interface (SEI) layer and a poor cycling life.

Several strategies have been developed to overcome the problems related to large volume changes, including nanostructuring, morphological and electrode architectural control, structural engineering, surface engineering, and hybrid composite design. Volume expansion can be minimized by moving from micron-sized to nanosized alloy-type anodes, which can effectively improve the cycling life and reduce the ion diffusion distance for enhanced reaction kinetics.

Silicon (Si): Si has played a pivotal role in the progress of LIBs, but its application in SIBs has been proceeding at a considerably slower pace. Theoretical calculations have indicated that sodium can potentially form a NaSi alloy when paired with Si, offering a remarkable specific capacity of 960 mAh g^{-1} [153]. However, NaSi formation faces significant kinetic barriers, primarily due to the crystalline nature of Si and its inherently poor electrical conductivity, leading to sluggish Na diffusion kinetics and impeding practical research efforts in Si anode development. DFT calculations have revealed that the diffusion of Na^+ ions in crystalline Si involves a high activation energy, which limits the uptake of Na^+ ions in this crystalline structure [154]. Interestingly, during initial cycling, the crystalline structure of Si transforms into an amorphous one, which significantly improves the kinetics of Na-ion insertion. Therefore, the transition from crystalline to amorphous Si structures holds promise. In this regard, Huang et al. introduced a novel approach involving rolled-up amorphous Si nanomembranes, which showed an exceptional rate capability and an ultralong cycle life exceeding 2000 cycles with an impressive 85% capacity retention [155]. This suggests that amorphous Si represents a highly prospective anode material for SIBs. However, it is imperative to address the challenges related to low conductivity and significant volume changes during cycling. The integration of nanoarchitectures and the utilization of conductive matrices are essential strategies to enhance electrochemical performance. A recent study introduced C-Si nanocomposites that show the advantages of having highly conductive ultrathin carbon nanoflakes and nanosized Si particles (with an average size of 9 nm). This innovative electrode design delivered exceptional specific capacities, achieving 378 mAh g^{-1} at a rate of 100 mA g^{-1} and 122 mAh g^{-1} at 200 mA g^{-1}.

Germanium (Ge): Ge has garnered considerable attention as an anode material for LIBs due to its significant lithium storage capacity and fast lithium diffusion kinetics. However, when it comes to SIBs, it is essential to note that the diffusion coefficient of sodium ions within the crystalline structure of Ge is several orders of magnitude slower than that observed for lithium ions. This discrepancy is primarily due to the significantly larger ionic radius of sodium ions than that of lithium ions. The sodiation process of Ge shares similarities with that of Si, particularly in terms of intermediate phases. The insertion of a single sodium atom results in the formation of NaGe as the ultimate crystalline phase in the NaGe system, offering a theoretical capacity of 369 mAh g^{-1} [156]. However, recent simulations have proposed that amorphous Ge has the ability to host more than one sodium atom, indicating a higher theoretical capacity of 576 mAh g^{-1} based on the formation of $Na_{1.56}Ge$ [157]. Similar to Si, crystalline Ge, which performs well in LIBs, does not exhibit favorable electrochemical performance in SIBs. Its characteristic behavior in the crystalline form was investigated by Lu et al. In situ TEM observations revealed the evolution and healing of nanopores on crystalline Ge nanowires during cycling [158]. Furthermore, it has been reported that the alloying reaction is kinetically unfavorable in crystalline Ge due to the higher energy barrier. Meanwhile, the larger sodium ions (1.5 eV versus 0.51 eV for lithium ions) should be overcome when hopping between interstitial sites within the lattice.

Antimony (Sb): Sb has emerged as a promising candidate for SIBs due to its favorable structural characteristics, including an ideal interlayer spacing for Na-ion insertion and a high theoretical specific capacity of 660 mAh g^{-1} [159]. The mechanism of Sb alloy with Na during the charge–discharge process follows Equations 4.2 and 4.3 [160]:

$$Sb + Na^+ + e^- = NaSb \tag{4.2}$$

$$NaSb + 2Na^+ + 2e^- = Na_3Sb \tag{4.3}$$

It is worth noting that most of the mesophases formed during these reactions are amorphous, making precise identification challenging. At the end of the discharge state, Na_3Sb formation competes between hexagonal and cubic-phase polymorphs [161]. Despite the apparent bulk advantages of Sb, its application as an anode material for SIBs remains uncertain. A key challenge is the significant volume expansion (~390%) that occurs during the alloying–de-alloying process, resulting from multiple electrochemical reactions between Sb and Na ions. Another related drawback is the slow kinetics of the alloy reaction [152]. One potential approach to mitigating these challenges is the formation of intermetallic compounds by alloying Sb with inactive elements such as NiSb, Cu_2Sb, $FeSb_2$, and ZnSb, where the inactive metal serves as a mechanical buffer to alleviate volume changes [159, 162]. A binder-free Sb/NiSb alloy was recently reported [163]. The introduction of Ni in this composite improves the conductivity, increases the structural stability, and reduces large volume expansion/contraction. Operando XRD results indicated that Na ions were inserted into the NiSb/Sb lattice, causing atoms to be expelled from the lattice while forming Na_xSb and amorphous Ni. Subsequently, the Na_xSb phase further transformed into hexagonal Na_3Sb. Consequently, the Sb/NiSb anode showed excellent cycling stability and high rate capability, with a reversible specific capacity of 521 mAh g^{-1} after 100 cycles at 200 mA g^{-1} and an initial CE of 86%.

Bismuth (Bi): Compared with other alloy-type anodes, Bi has been considered somewhat less appealing due to its high reaction voltage and relatively low melting point of approximately 270 °C [164]. However, Bi has gained renewed interest as an anode material for SIBs over the past decade, primarily due to its high electrical conductivity and the layered structure that facilitates ion transport within its interlayer spacing. Despite these advantages, the charge storage mechanism of Bi is largely unknown. According to the NaBi equilibrium phase diagram, the alloying of Bi initiates with the formation of NaBi, followed by the generation of Na_3Bi, which theoretically provides a capacity of 385 mAh g^{-1}. Fundamental studies conducted by both Ellis et al. and Sottmann et al. support this proposed alloy reaction pathway [165]. Furthermore, the latter study indicates that the sodiation process of Bi depends on crystallite size. While the possibility of intercalation as a reaction pathway for Bi was considered due to its unique layered structure [166], this theory was challenged by Gao et al. [167]. They argued that the previous misinterpretation of the Bi reaction mechanism stemmed from the metastable nature of Na_3Bi and the easy decomposition of NaBi under ambient conditions. An intriguing report presents a method of transforming conventional bulk Bi materials into a porous structure using a $NaPF_6$-diglyme-carbonate-based solvent through repeated cycling [168]. This modification enhances electrolyte permeability and charge transfer kinetics, resulting in an impressive capacity retention of 94.4% after 2000 cycles. Furthermore, excellent rate capability is achieved, with a minimal capacity reduction observed as the current density increases from 0.05 to 2 A g^{-1}. Moreover, the cycling stability of Bi can be further improved by pairing it with graphite [169]. Extremely small Bi nanoparticles can be effectively stabilized within the graphite layer, leading to a remarkable capacity retention of 90% even at a high rate of 20 C after 10,000 cycles.

Tin (Sn): Sn has garnered significant attention as an anode material for room temperature SIBs due to its remarkable ability to alloy with up to 3.75 Na atoms per Sn atom, resulting in a high theoretical capacity of 847 mAh g^{-1}. The electrochemical storage mechanism of Na within Sn was initially elucidated by Ellis et al., who proposed a sodiation process consisting of four steps, ultimately leading to the formation of $NaSn_3$, NaSn, Na_9Sn_4, and $Na_{15}Sn_4$ phases [170]. Further investigations of the microstructural changes and phase transformations occurring in sodiated Sn nanoparticles, utilizing in situ TEM, suggested a two-step sodiation mechanism. In this mechanism, an amorphous Na-poor phase, denoted as Na_xSn (with $x \approx 0.5$), was formed during the initial step, consuming the pristine Sn material [171]. As sodiation continued, several Na-rich amorphous phases (e.g. Na_3Sn, Na_9Sn) emerged, culminating in the formation of the final crystalline $Na_{15}Sn_4$ phase during the second step. The utilization of Sn nanoparticles allowed for the direct observation of the amorphous phase generated during the first step and showed lower volume expansion than the final crystalline phase. This highlights the advantageous effects of nanostructures combined with material amorphization in accommodating the stress induced by volume changes. Commercial Sn nanoparticles have been extensively tested in aprotic Na cells and showed an impressive capacity of approximately 500 mAh g^{-1} for over 20 cycles, with a first-cycle CE of 86%. These results emphasize that the use of nanoparticles alone may not represent the ultimate solution. However, when combined with appropriate electrode and electrolyte optimizations, nanoparticles, as well as poly(acrylic acid) as a binder and FEC as an electrolyte additive, can lead to superior electrochemical performance [172].

Phosphorus (P): P emerges as a compelling candidate for anodes in SIBs due to its abundance and environmental friendliness. Elemental phosphorus exists primarily in three allotropic forms: white, red, and black. White phosphorus (P4) is highly toxic and flammable under ambient conditions, while black phosphorus is thermodynamically stable at room temperature and pressure. In contrast, red phosphorus (RP) is non-toxic and does not involve in spontaneous combustion [173]. Given the high reactivity of white phosphorus and the demanding synthesis conditions required to obtain black phosphorus, RP, which is easy to prepare and commercially available, has garnered significant attention for use in SIBs. RP offers a theoretical capacity of $2596\,\mathrm{mAh\,g^{-1}}$ (for its highest sodiated phase, Na_3P) and operates at a relatively safe potential ($\approx$0.45 V versus Na/Na^+), with promising increased energy density. RP has a polymeric chain-like structure consisting of covalently bonded P4 tetrahedra. Its low atomic weight, significant sodium uptake capacity, and commercial availability establish RP as one of the most promising anode materials for SIBs. However, amorphous RP suffers from poor electrical conductivity ($<1\times10^{-14}\,\mathrm{S\,cm^{-1}}$) and experiences significant volume expansion (approximately 440%) when fully sodiated to Na_3P. These issues result in sluggish reaction kinetics and rapid capacity degradation. Furthermore, the surface of the resulting Na_3P is highly reactive with the electrolyte, leading to irreversible capacity [174]. Recently, a chemical synthesis method to produce nanostructured RP was reported [175]. The fabricated hollow phosphorus nanospheres show superior electrochemical performance, even with a high P loading ratio (60 wt% of the electrode weight). Mechanical milling has also proven to be an efficient approach for preparing nanostructured carbon composite electrodes [176]. In 2013, Kim et al. prepared an amorphous RP/C composite through ball milling RP and super P carbon in a 7 : 3 weight ratio [177]. This composite demonstrated a reversible capacity of $1890\,\mathrm{mAh\,g^{-1}}$ at a current density of $143\,\mathrm{mA\,g^{-1}}$ ($\approx$0.05 C rate), an operating potential of $\approx$0.4 V versus Na/Na^+, good rate capability ($1540\,\mathrm{mAh\,g^{-1}}$ at a current density of $2.86\,\mathrm{A\,g^{-1}}$, $\approx$1 C rate), and excellent cycling performance with negligible capacity fading over 30 cycles. Successful implementations of nanosized carbon matrices have also been reported for a hand-ground mixture of commercial micro-sized RP with multiwalled carbon nanotubes (CNTs). This composite material showed promising performance, due to the buffering effect of the multiwalled CNTs against the volumetric expansion of P and their improved electron transport properties [178].

4.4.2 Intercalation-Type Anode Materials

By definition, intercalation or insertion mechanism introduces the guest species into the crystal structure without disturbing crystalline parameters such as bond distances, crystal phase, unit cell volume, and distance between crystal phases [152].

The amount of intercalated species is determined by the thermodynamic equilibrium attained at the electrode–electrolyte interface. Upon the insertion process, the charge species are inserted into the crystal lattice, and then, electrons are forced to leave the structure. This means that the structure should include an easily oxidizable and reducible atom. In addition, the structure should also have moderate electronic conductivity. The capacities of insertion-type materials, namely carbon-based materials and intercalation-type metal oxides, are determined by the limited redox sites in the bulk structure and are normally in the range of 200–500 mAh g^{-1} [179]. However, these materials have minor volume changes during Na^+ insertion and extraction, indicating a better tolerance for sodium storage [5]. The detailed insertion process in electrode materials is shown in Figure 4.12 and can be expressed as follows [180]:

$$yNa^+ + ye^- + I_{(\text{intercalation host})} \rightarrow Na_yI \tag{4.4}$$

In recent years, several strategies have been explored to improve Na-ion storage performance. These strategies encompass solvation-ion co-intercalation, heteroatom doping, defect introduction, morphology control, nanostructure design, phase state adjustment, and composite construction.

Among the myriad of potential anode materials, carbon-based materials are consistently the preferred choice due to their exceptional electrical conductivity, low charge–discharge voltage plateau, cost-effectiveness, and environmental friendliness. In addition, carbon-based materials offer a wide range of tunable structures, which not only facilitate the adjustment of their electrochemical performance but also play a pivotal role in designing matrix structures for the fabrication of stable composites. Currently, carbon-based materials are primarily categorized into two groups: graphite and non-graphite materials.

Graphite: Graphite serves as a well-established commercial anode material for LIBs, where lithium ions intercalate into the graphite host, to form binary intercalation compounds denoted as LiC_8. These compounds offer a theoretical specific capacity of 372 mAh g^{-1}. However, it is important to note that graphite does not exhibit reversible Na^+ intercalation to form analogous intercalation compounds. This discrepancy arises from the fundamental nature of the chemical bond between alkali metal atoms and carbon atoms, in which complete charge transfer from alkali metals to carbon atoms occurs, as confirmed by DFT calculations [181]. Nonetheless, the energy released during the transfer of electrons from sodium to carbon atoms is inadequate, resulting in a low plane binding energy for Na ions. This limitation hinders the successful embedding of Na ions into graphite and makes binary Na-intercalated graphite compounds unstable [182]. Expanded graphite, characterized by a larger interlayer distance of 4.3 Å between graphene layers, has been shown to have remarkable electrochemical performance [183]. Jache et al. illuminated the intercalation of solvated alkali ions, a phenomenon often referred to as "co-intercalation," which is achieved through the reduction of graphite, described as follows:

$$C_n + e^- + A^+ + y\,\text{solv} = A^+(\text{solv})_yC_n^- \tag{4.5}$$

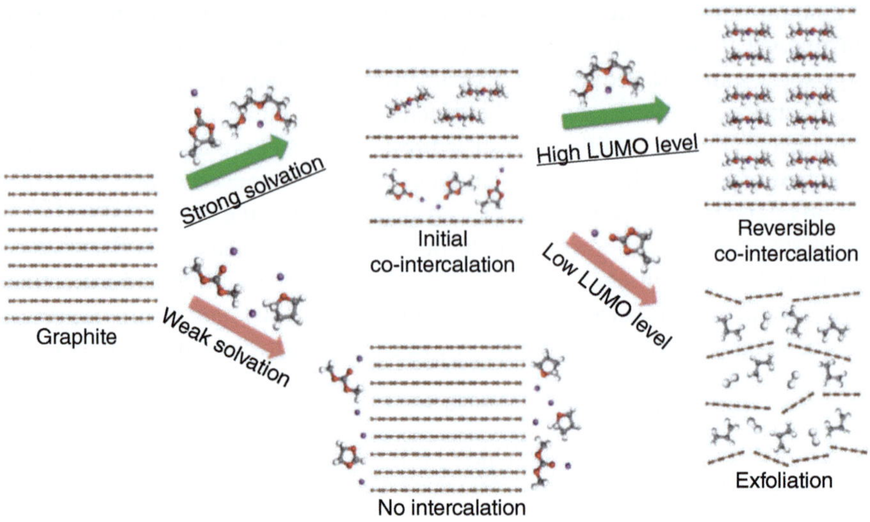

Figure 4.13 Structure of Na solvent co-intercalated in graphite. Source: Reproduced with permission from Yoon et al. [185]. © 2017/John Wiley & Sons.

where *A* and solv represent alkali metal and solvent molecules, respectively, and n and y denote constants [184]. Figure 4.13 provides an overview of the co-intercalation behavior of Na solvent in graphite [185]. The alkali ions can effectively intercalate into graphite through robust solvation to form ternary graphite intercalation compounds, thereby contributing to capacity when these compounds remain stable against oxidation.

In addition, it is widely accepted that the transition from an ester to an ether solvent improves the initial CE, capacity retention, and rate capability of carbon-based anode materials by facilitating the formation of a thin and stable SEI layer. Hu et al. synthesized nitrogen-doped expanded graphite and used it as an anode material in both ether- and ester-based electrolytes. In the ester solvent, they observed an initial discharge capacity of 1702 mAh g^{-1}, while in the ether-based solvent, the delivered capacity was 518 mAh g^{-1}, accompanied by an initial CE of 24.73% and 72.08%, respectively. Remarkably, they showed excellent rate capability, delivering capacities of 125 and 110 mAh g^{-1} at a high rate of 30 A g^{-1}, respectively. These improved results were attributable to the formation of a stable and thin SEI layer, which resulted in low irreversible capacity losses and, consequently, increased CE [186].

Soft carbon: Soft carbon, a type of non-graphitic, graphitizable carbon, is typically produced through liquid- or gas-phase pyrolysis. Its structural layers are nearly parallel to each other, with a slight degree of rotation or twisting. This unique arrangement provides sites for the storage of not only lithium ions but also sodium ions. However, all the soft carbon materials reported so far had limited sodium storage capacities (approximately 200 mAh g^{-1}) and high charging potentials (exceeding 0.5 V), which significantly limited the energy density of SIBs. Pendashteh et al. suggested that the sodium storage performance of soft carbon is closely related to the mechanical treatment of the intermediate or final residue, which influences the structural, textural, and morphological characteristics of the carbon matrix [187]. By optimizing these materials, they achieved a remarkably high reversible specific capacity of over 200 mAh g^{-1} at a C/15 rate (1 C = 372 mA g^{-1}) and excellent high rate capability (approximately 60 mA g^{-1} at 10 C). In addition, defect-rich porous soft carbon nanosheets were developed, where the presence of micropores and defects at the edges synergistically contributed to a significant improvement in specific capacity from 134 to 232 mAh g^{-1}, along with an outstanding rate performance of 103 mAh g^{-1} at 1000 mA g^{-1} for sodium-ion storage [188].

Hard carbon: HC has a distinctive structure characterized by a disordered, randomly stacked arrangement of graphene sheets, resulting in a nanoporous configuration. It is a type of non-graphitic, non-graphitizable carbon material produced through solid-phase pyrolysis from sources such as cellulose, charcoal, coal, and sugar. The process of sodium insertion into disordered carbon materials such as HC and carbon black follows a two-stage mechanism, as proposed by Dahn et al. In the first stage, Na ions intercalate into the disordered, stacked graphene structure, while the second stage, represented by a plateau at lower potentials, corresponds to the filling of Na ions into the nanosized voids within the material's structure (pore filling), as shown in Figure 4.14. This observation aligns with DFT calculations [189]. However, a recent debate arose regarding this mechanism, prompted by the work of Ding et al., who investigated highly porous HC as an anode in SIBs. They observed interlayer enlargement in the plateau region, which they attributed to the intercalation of Na ions [190]. Subsequently, other studies also supported the findings of Ding et al. [191]. Overall, an alternative perspective emerged, suggesting that the slope region might correspond to the adsorption of Na ions at defect sites within the HC structure, followed by the plateau region, which may signify Na-ion intercalation. Toward the end of the discharge process, some Na ions may form clusters or undergo pore filling (Figure 4.14). The highest capacity achieved was 430.5 mAh g^{-1} at a current density of 30 mA g^{-1}, with a CE of 67.7%, reported by Sun et al. for HC derived from biomass, specifically shaddock peel. This material showed exceptional cycling stability, with only a 2.5% capacity degradation after 200 cycles [192]. Wang et al. synthesized HC derived from kelp, featuring interlayer dilation between 3.9 and 4.3 nm. This material showed excellent cycling performance, delivering 205 mAh g^{-1} at 200 mA g^{-1} after 300 cycles and a high rate capability of 96 mAh g^{-1} at 1000 mA g^{-1} [193]. Morphological engineering of carbon materials has also shown promise in increasing the initial capacity due to improved connectivity, larger surface area, and improved structural stability. Zhang et al. prepared micro-/nanostructured HC, resulting in a reversible capacity of 286 mAh g^{-1} at 20 mA g^{-1} after 100 cycles, with an initial CE of 80% [194].

Titanium-based materials: Titanium-based oxides represent a crucial category of intercalation-based anode materials for SIBs. They offer stable redox plateaus, cost-effectiveness, and environmental compatibility. Similar to carbon-based

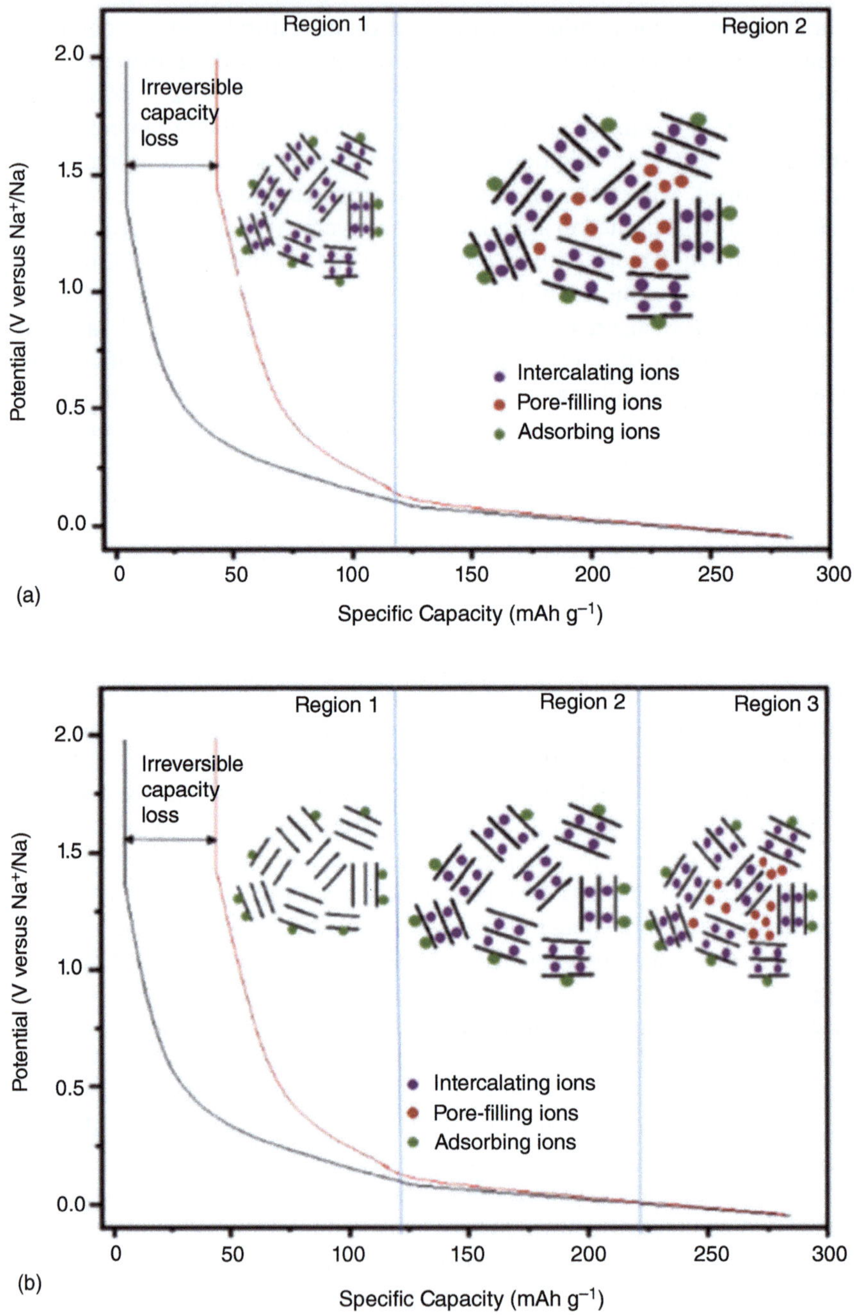

Figure 4.14 Schematic of mechanisms for storage of Na^+ in hard carbon. Source: Reproduced with permission from Perveen et al. [152]. © 2020/Elsevier.

counterparts, titanium-based materials apply an intercalation mechanism to store sodium ions. Prominent examples include TiO_2, $Na_2Ti_3O_7$ with a layered structure, and $Li_4Ti_5O_{12}$ with a spinel structure.

TiO_2 has various polymorphs, including rutile, anatase, bronze (TiO_2-B), and hollandite (TiO_2-H) [195]. In general, bronze TiO_2 is more favorable for Na^+ intercalation than anatase, rutile, and brookite TiO_2 due to its open crystal structure. The crystal structure of anatase consists of TiO_2 octahedra with shared edges [195]. This arrangement creates empty Z-shaped channels and forms a three-dimensional network structure that provides abundant storage sites and diffusion pathways for Na ions. Consequently, anatase TiO_2 exhibits superior electrochemical properties for SIBs. To investigate the influence of crystal structure on TiO_2 anode materials, Liu et al. prepared nanowires of anatase TiO_2 (TiO_2(A)) and bronze TiO_2 (TiO_2(B)) with similar size and surface morphology [196]. They analyzed the Na-ion storage mechanism using XAS and DFT theoretical calculations. While TiO_2(B) features a more open channel than TiO_2(A), both electrodes show similar sodium storage capacities. However, TiO_2(B) shows significantly better stability than TiO_2(A) due to the irreversible formation of Na_xTiO_2 near the surface of TiO_2(A). This finding suggests that the performance of TiO_2 anodes for SIBs can be improved by selecting nanostructures with rough surfaces to mitigate the unnecessary blocking effect of the surface barrier layer.

Among titanium-based materials, $Na_2Ti_3O_7$, with its layered structure, has the lowest working potential. It can reversibly store 2 mol of Na ions per formula unit at an average potential of 0.3 V, resulting in a significant specific capacity of 200 mAh g^{-1}. This characteristic has led to extensive research on SIBs [198]. However, $Na_2Ti_3O_7$ faces challenges such as insufficient ionic conductivity, slow Na-ion migration, and significant structural distortion during Na-ion absorption. These issues ultimately result in poor capacity retention and rate capability. Another noteworthy sodium titanate, $Na_2Ti_6O_{13}$, features a tunnel structure that offers high ionic conductivity, low potential, and robust cycle stability. Unfortunately, its Na-ion capacity is relatively low. To address this limitation, researchers have developed a hybrid structure by combining layered $Na_2Ti_3O_7$ with tunnel-type $Na_2Ti_6O_{13}$ [199]. This hybrid structure combines the excellent Na-ion storage capacity of layered $Na_2Ti_3O_7$ and the high ionic conductivity of tunnel-type $Na_2Ti_6O_{13}$, resulting in commendable cycling stability and outstanding rate performance.

Spinel $Li_4Ti_5O_{12}$ (LTO), recognized as a "zero strain" anode material for SIBs, was introduced by Zhao et al. [200]. LTO has an average voltage of approximately 1.0 V versus Na^+/Na and has a reversible specific capacity of 145 mAh g^{-1}. In contrast to the conventional processes observed in LIBs, the Na-ion storage mechanism in $Li_4Ti_5O_{12}$ follows a unique reaction [201]:

$$2Li_4Ti_5O_{12} + 6Na^+ + 6e^- = LiNa_6Ti_5O_{12} + Li_7Ti_5O_{12} \tag{4.6}$$

Subsequently, DFT calculations confirmed a three-phase reaction mechanism, shedding light on the intricacies of this process.

To further elucidate the phase transformation behavior of the $Li_4Ti_5O_{12}$ anode material during cycling, researchers have used synchrotron in situ XRD

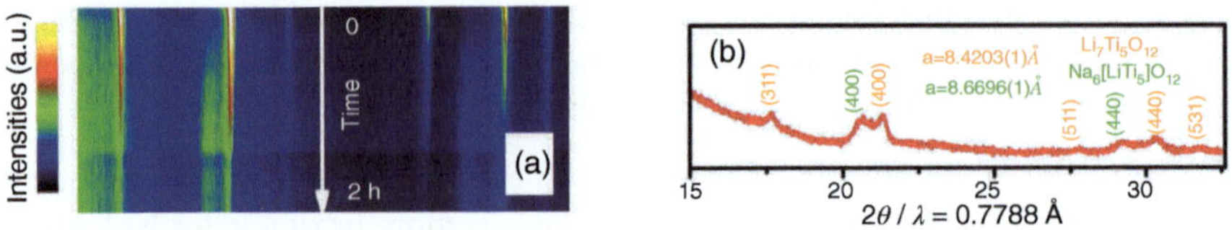

Figure 4.15 (a) Contour plot of peak intensities as a function of reaction time and (b) XRD pattern for final sodiated $Li_4Ti_5O_{12}$. Source: Reproduced from Yu et al. [197]. © 2013/American Chemical Society.

techniques [197]. As shown in Figure 4.15a,b, the peak intensity of $Li_4Ti_5O_{12}$ decreases over reaction time, while a distinct peak set corresponding to Na_6(16c)$LiTi_5$(16d)O_{12}(32e) emerges. This indicates the transformation of $Li_4Ti_5O_{12}$ into a mixture of $Li_7Ti_5O_{12}$ and $Na_6LiTi_5O_{12}$ in a ratio of 46.7 : 53.3. One of the primary challenges with spinel $Li_4Ti_5O_{12}$ anodes is their limited cyclability and rate capability due to poor electron conductivity. This limitation is primarily attributable to the electronic configuration of Ti^{4+}, which is $[Ar]3d^0$. To address this problem, researchers have developed a porous $Li_4Ti_5O_{12}$ nanosheet using cetyltrimethylammonium bromide to control surface morphology. This innovation significantly reduces charge transfer resistance, shortens ion diffusion pathways, and offers more active sites for sodium storage, resulting in reversible specific capacities of 158.9 mAh g^{-1} and 123.2 mAh g^{-1} at 0.1 A g^{-1} and 0.5 A g^{-1}, respectively [202]. Furthermore, composite materials have been explored as an approach to enhancing the performance of $Li_4Ti_5O_{12}$. For example, a sulfur- and nitrogen-doped $Li_4Ti_5O_{12}$/rGO (SN-LTO/rGO) composite was synthesized through a hydrothermal method and subsequent calcination [203]. Compared with the original $Li_4Ti_5O_{12}$ material, SN-LTO/rGO showed lower impedance and faster electrochemical kinetics, underscoring the potential of sulfur and nitrogen doping, as well as composite construction, for electrode modifications in SIBs.

4.4.3 Conversion-/Conversion-Alloying-Type Anode Materials

Conversion materials undergo a process in which metal oxides are chemically transformed into new compounds. As anodes for SIBs, various conversion materials have been extensively studied and can be categorized into oxides, sulfides, selenides, and phosphides. Figure 4.16 presents various conversion anode materials, including their capacity, voltage characteristics, and volume expansion [3]. Depending on the type of metal present in these compounds, such materials can undergo insertion, alloying, and conversion reactions simultaneously. However, their ability to store more than one sodium atom leads to significant volume changes. These challenges can be addressed through the engineering of material morphology, incorporation of carbon materials, or modification of the electrolyte to achieve a stable SEI layer. It is worth noting that the larger ionic radii of sodium ions (Na^+) contribute to the sluggish kinetics observed in these materials.

Oxides are promising anode materials for SIBs, with intermediate voltages relative to Na/Na^+ and volume expansion characteristics that fall between those of sulfides

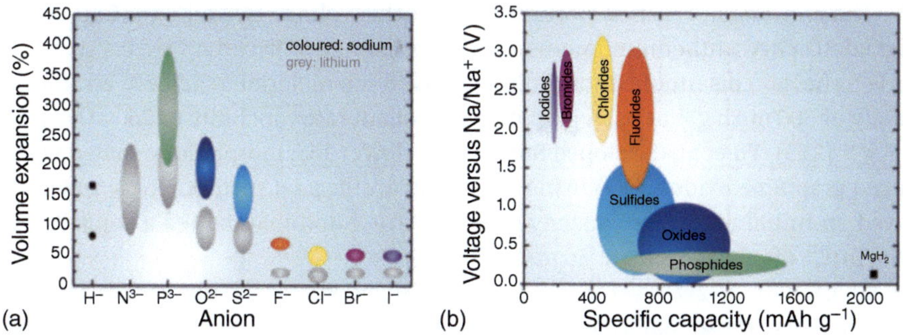

Figure 4.16 Conversion anode materials for SIBs. Source: Reproduced from Nayak et al. [3]. © 2018/John Wiley & Sons.

and phosphides. Research in this area has focused on various types of oxide-based anode materials, including conversion alloying materials such as tin [204, 205] and antimony [206], as well as pure conversion materials such as iron [207], nickel [208], cobalt [209], molybdenum [210], copper [211], and manganese [212]. Nanostructuring of these anode materials is a common approach, because it enhances the electrode–electrolyte interface, promotes the kinetics of Na^+ ions, and mitigates volume changes. For instance, Hariharan et al. synthesized Fe_3O_4 nanoparticles that initially delivered a capacity of 643 mAh g^{-1}. The investigation revealed the presence of crystalline Na_2O embedded within a metallic Fe matrix. Upon desodiation, incomplete oxidation occurred, resulting in a low initial CE of 57% [207].

Metal oxides often suffer from low electronic conductivity and loss of electrical contact due to pulverization during cycling. To address this issue, researchers have explored coating these materials with carbon or incorporating carbon nanomaterials. Lu et al., for example, prepared SnO, SnO_2, and carbon-coated SnO_2 particles. The SnO sample showed the best results and maintained a stable capacity of 530 mAh g^{-1} after 50 cycles, surpassing the theoretical capacity of 398 mAh g^{-1}. This improved performance was attributable to a combination of alloying (55%) and conversion reactions (45%) occurring during Na storage [205].

Metal sulfides have emerged as promising materials for anodes in SIBs, offering advantages over metal oxides. These advantages include high electrical conductivity and weaker M-S bonds than M-O bonds, which enhance the kinetics of the conversion reaction, leading to higher initial CE, improved reversibility, and reduced volume expansion [213, 214]. Consequently, a variety of metal sulfides based on elements such as iron [215], tin [216], cobalt [217], nickel [218], molybdenum [219, 220], zinc [221], titanium [222], and tungsten [223] have been extensively investigated for their potential as anodes in SIBs. The specific mechanism followed by sulfides depends on the type of metal, resulting in either conversion, conversion/insertion, or conversion/alloying reactions. To improve the performance of sulfide-based anodes, nanostructuring, morphology engineering, and carbon coating strategies have been explored. For example, Hu et al. synthesized nanoflowers of MoS_2 with an expanded interlayer spacing, which showed a capacity of 350 mAh g^{-1} at 50 mA g^{-1} and excellent cyclability at an average voltage exceeding

0.4 V. The conversion reaction occurred within this voltage range [220]. Cao et al. designed 3D chrysanthemum microspheres consisting of 2D carbon sheets stacked over FeS sheets. This unique morphology enabled the material to achieve an initial capacity of 500 mAh g^{-1} at 0.2 A g^{-1} and an excellent rate capability of 260 mAh g^{-1} at 1 A g^{-1} [215]. Yu et al. developed SnS_2 with a flower-like morphology coated with reduced graphene oxide (rGO) using a facile hydrothermal method. This material showed an initial discharge capacity of 649 mAh g^{-1} and maintained a capacity of 509 mAh g^{-1} after 100 cycles at 50 mA g^{-1} [224].

Selenides with a 2D layered structure have shown promising properties for anodes in SIBs. These materials offer advantages such as larger interlayer spacing and improved conductivity compared with sulfides, resulting in a better rate performance and CE in SIBs. Various selenides, including SnSe [225] and Sb_2Se_3 (involving conversion alloying mechanisms), as well as $MoSe_2$, $FeSe_2$, ZnSe, and NiSe (undergoing conversion reactions), have been explored as potential anodes in SIBs. To improve the capacities of these materials, various strategies have been used, such as morphology engineering and incorporation of carbon in the form of a conductive network or coating. For example, Ou et al. synthesized a composite of Sb_2Se_3/rGO, which showed a reversible capacity of 682 mAh g^{-1} at 100 mA g^{-1}, coupled with excellent cycling stability. The mechanism involved the intercalation of Na ions into Sb_2Se_3 layers, followed by a conversion reaction between Se and Na^+, resulting in an alloy of Sb with Na ions [226]. Zhao et al. prepared nanoparticles of $FeSe_2$ coated over carbon black, achieving a high capacity of 500 mAh g^{-1} and an excellent rate capability of 250 mAh g^{-1} at 1000 mA g^{-1} [227]. Similarly, Zhang et al. grew nanosheets of $MoSe_2$ on carbon cloth, resulting in a capacity of 452.6 mAh g^{-1} at 0.2 A g^{-1}, with a capacity retention of 85.5% after 100 cycles [228]. Yang et al. developed carbon-supported nickel selenide hollow nanowires, which delivered a reversible capacity of 390 mAh g^{-1} at a 0.2 C rate, along with 97% capacity retention after 100 cycles and excellent rate capability (219 mAh g^{-1} at 5 C) [229].

Metal phosphides are emerging as highly promising materials for SIBs. They feature metals that help mitigate the low electronic conductivity and volume changes associated with phosphorous. Upon full sodiation, the final product consists of Na_3P embedded in a conductive metal matrix that increases conductivity. These metals can be either inactive, such as Co, Mo, Cu, Ni, and Fe, or less active, such as Sn and Se, with respect to Na storage, resulting in minimal volume expansion [230]. Metal active phosphides show a combination of conversion and alloying reactions, leading to the self-healing of cracks that can arise from the latter mechanism during Na storage. The performance of metal phosphides can be further improved through morphology engineering and carbon coating. For example, Liu et al. synthesized yolk–shell Sn_4P_3 nanospheres coated with carbon, with a remarkable capacity of 505 mAh g^{-1} and retaining 360 mAh g^{-1} of the initial capacity at a 1.5 C rate. This enhanced performance can be attributed to void spaces, carbon coating, and alloying/conversion reaction [231]. Furthermore, modifications to the electrolyte can lead to improved performance. Kim et al. prepared Sn_4P_3 with a high capacity of 718 mAh g^{-1} and negligible capacity loss when the FEC additive was used at an average potential of 0.3 V. Mechanistic investigations revealed the formation of Na_3P and Sn through a conversion reaction, which further underwent alloying to form $Na_{15}Sn_4$ during the sodiation process. The starting product is recovered during desodiation via a conversion reaction [232]. In contrast, metal inactive phosphides deliver lower capacity than their metal active counterparts because they exclusively undergo a conversion reaction [233, 234]. However, the capacities of these materials can also be improved by forming composites with carbon-based materials, modifying the electrolyte, or engineering the morphology of the material. For example, Huang et al. synthesized MoP–C, which showed excellent cyclability and a capacity of 398.4 mAh g^{-1} after 800 cycles. Furthermore, it showed exceptional rate performance, with a capacity of 147 mAh g^{-1} at 1600 mA g^{-1} [234].

4.4.4 Organic-Type Anode Materials

Organic materials investigated as potential anodes for SIBs can be broadly categorized into two major classes: small organic molecules (conjugated carboxylate organic compounds) and organic polymers. Small organic molecules show reversible reactions with Na ions at distinct sites, including C-O (comprising quinones, ketones, carboxylates, anhydrides, and imide derivatives), C-N (such as Schiff bases and pteridine derivatives), and N-N (involving azo derivatives), as shown in Figure 4.17 [235]. On the other hand, organic polymers participate in doping reactions, as shown in Figure 4.17, and can be further subcategorized into conjugated conductive polymers, porous organic polymers, organometallic polymers, and organic radical polymers. All these types have demonstrated their

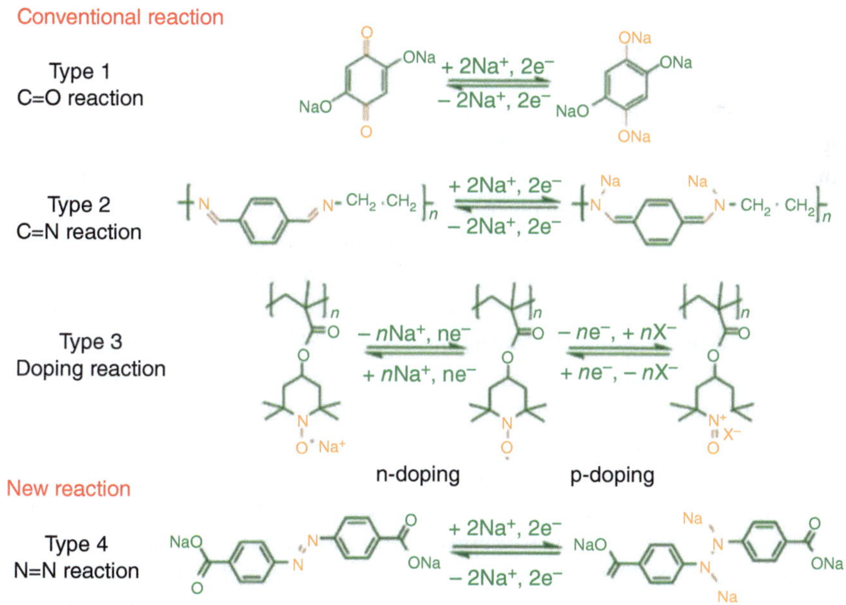

Figure 4.17 Reaction mechanisms for organic electrode materials. Source: Reproduced with permission from Luo et al. [235]. © 2018/John Wiley & Sons.

electroactivity as potential electrode materials for SIBs. The appeal of organic electrode materials lies in their relatively high theoretical specific capacity, abundance in nature, adaptable structural design, and tunable synthesis methods. However, they face challenges such as high solubility in the electrolyte, limited conductivity, and susceptibility to rapid capacity degradation due to the significant volume changes that occur during Na-ion intercalation/deintercalation processes.

Conjugated carboxylate organic compounds are characterized as n-type organics, featuring reversible reactions between C-O and alkali ions, typically showing low potential in the context of Na/Na^+. A recent breakthrough introduced disodium terephthalate ($Na_2C_8H_4O_4$) as a novel organic compound for SIBs, analogous to dilithium terephthalate ($Li_2C_8H_4O_4$) [236]. Disodium terephthalate showed a reversible capacity of 250 mAh g^{-1} at a voltage of 0.45 V, albeit with a modest initial CE of 60%. However, this material suffers from limited rate capability and cycling stability, which is primarily attributable to its propensity to dissolve in organic solvents. This drawback can be addressed effectively by applying a thin layer of Al_2O_3 coating [237]. The thermodynamic and kinetic properties of organic molecules can be tailored through the introduction of various functional groups at the benzenecarboxylate site [238]. Park et al. investigated the electrochemical performance of disodium terephthalate and its derivatives, including amino, bromo, and nitro derivatives. Disodium terephthalate showed a reversible capacity of 290 mAh g^{-1}, while the bromo (electron-withdrawing group), nitro (electron-donating group), and amino (inherently active group) derivatives showed capacities of 300, 302, and 200 mAh g^{-1}, respectively.

Schiff base polymers are organic molecules featuring Schiff-based functional groups ($R_1HC{=}N\text{-}R_2$) that correspond to nitrogenous counterparts of aldehydes and ketones and are achieved by replacing C-O with a C-N-R group. The redox reactions responsible for Na-ion storage occur at the C-N site within Schiff base polymers. These materials have tunable electrical properties, solubility, and processability, making them promising electrode materials for SIBs. The first study in this field was conducted by Castillo-Martínez et al., who reported a reversible capacity of 350 mAh g^{-1} at 26 mA g^{-1}, with a reaction voltage below 1.5 V versus Na/Na^+ [239]. Polyamides are synthesized via the reaction of dianhydrides and diamines with the carbonyl group in polyamides storing Na ions through enolization reactions. Chen et al. synthesized a polyamide polymer with a capacity of 140 mAh g^{-1} at an average potential of 2 V [240]. Polyquinones used in SIBs typically contain sulfur and are prepared through the reaction of sulfur and quinone. They have high solubility in organic solvents, which is partially mitigated through polymerization. Deng et al. used the first quinone poly(anthraquinonyl sulfide) and achieved a reversible capacity of 220 mAh g^{-1}. The rate capability was also excellent, with a capacity of 160 at 6400 mA g^{-1}, and it showed good cyclability with 85% capacity retention at 1600 mA g^{-1} [241]. Conjugated conductive polymers are also used as anode materials for SIBs. These polymers are prepared through p-doping or n-doping, resulting in the polymer carrying either a positive or negative charge, respectively. In n-doped polymers, Na ions are stored. Initially, polyacetylene and polyparaphenylene were used as anodes in SIBs by Shaklette et al. Zhang et al. synthesized a conjugated microporous polymer consisting of 4,7-dicarbazyl [2,1,3]-benzothiadiazole (PDCzBT) through $FeCl_3$ oxidation coupling polymerization. This polymer showed n-type behavior and a reversible capacity of 145 mAh g^{-1} at 20 mA g^{-1} after 100 cycles [242]. Other organic electrodes are derived from natural biomolecules. For example, Wang et al. prepared a juglone/rGO composite with a reversible capacity of 305 mAh g^{-1}, significantly higher than pure juglone (40 mAh g^{-1}) and pure rGO (123 mAh g^{-1}) after 10 cycles [243].

4.5 Electrolyte

The electrolyte is another important component in SIBs, which acts as an ionic conductor for ion transfer between the cathode and anode. Ideally, the electrolyte should offer high ionic conductivity, low viscosity, wide electrochemical stability window (ESW), good thermal stability, low toxicity, and low cost. Besides, it should not react with any other battery components, such as current collectors, separators, and cell packaging materials. With respect to electrochemical considerations, viscosity and ionic conductivity should also be carefully considered. The cell exhibits poor rate capability and cycling performance when using electrolytes with low ionic conductivity and high viscosity. Furthermore, electrolytes should be resistant to massive decomposition into side products, which can lead to the uncontrollable formation of the SEI, resulting in higher polarization, slower transport, loss of charge carriers, and ultimately, cell failure. Due to the similarities between Li and Na, electrolytes in SIBs are formulated with the same strategies already adopted in LIBs [9], which enable the stable formation of SEI to stabilize the anode at low operating potential and avoid the increase in polarization upon cycling [244]. The electrolyte used in SIBs can be classified into two types: liquid electrolytes and solid electrolytes.

4.5.1 Liquid Electrolytes

Liquid electrolytes are commonly used in SIBs due to their significant advantages, including high wettability, low surface tension, low viscosity, and high ionic conductivity within the range of 3×10^{-3} to 2×10^{-2} S cm^{-1} across a wide temperature spectrum. However, they also exhibit notable drawbacks, such as poor physical and chemical stability and potential to facilitate dendrite formation between the electrodes, ultimately leading to short circuits. To mitigate these issues, it is crucial to control the electrode potential and ensure that it remains stable. When the potential of the anode exceeds the highest occupied molecular orbital (HOMO) of the electrolyte, electrolyte oxidation is prevented. In contrast, if it falls below the lowest unoccupied molecular orbital (LUMO) of the electrolyte, no electrolyte reduction occurs. This means that the flow of electrons through the electrolyte is maintained, while only the flow of ions is possible. Thus, the electrode potential of the cell is determined by the HOMO–LUMO gap of the electrolyte. Managing this potential limitation involves kinetic control by maintaining the anode potential at a level too low to be stable.

During the initial charging process, ions from the electrolyte are reduced at the anode and form the SEI layer. This results in a loss of capacity known as irreversible capacity loss. For effective operation, this SEI layer must be stable to prevent further capacity losses due to additional reduction during subsequent cycles. The ideal SEI layer should not dissolve in the solvent but should facilitate the smooth and rapid movement of ions to the electrodes. It should also inhibit the movement of solvent molecules and enhance the interfacial stability of the electrode–electrolyte interface. Any damage to the SEI layer due to side reactions can lead to electrolyte decomposition, increased resistance, heating, reduced energy efficiency, and increased pressure in the SIBs. Therefore, the formation of a stable SEI layer is imperative for smooth operation.

Since the resurgence of interest in SIBs, extensive research efforts have been dedicated to developing suitable electrolytes. Typically, salts dissolved in organic or aqueous solvents are used as electrolytes. Commonly used salts include $NaBF_4$, NaTF, NaTFS, NaFSI, $NaPF_6$, and $NaClO_4$, among others. Organic solvents used include propylene carbonate (PC), ethylene carbonate (EC), dimethylene carbonate (DMC), diethylene carbonate (DEC), or their combinations. However, these electrolytes have not consistently provided promising results, which has led to ongoing research aimed at improving the battery performance and lifespan using additives such as FEC and vinylene carbonate (VC).

One of the earliest studies aimed at identifying a suitable electrolyte for SIBs was conducted by Ponrouch et al. in 2012. They found that a mixture of EC and PC could serve as a superior electrolyte for SIBs, with salts such as $NaClO_4$ and $NaPF_6$ making little or no significant difference [245]. Subsequently, the electrolyte was modified with sodium to $EC_{0.45}$: $PC_{0.45}$: $DMC_{0.1}$, resulting in improved ionic conductivity, lower viscosity, and the formation of a stable SEI layer, ultimately leading to enhanced results [246]. Researchers have also explored other electrolyte options for SIBs, including co-intercalating electrolytes (which participate in redox reactions) and ionic liquids (ILs). Co-intercalating or ether electrolytes, such as diglyme and triglyme, were initially studied with graphite or carbon-based anodes and provided promising results. These solvents facilitated the solvated Na-ion intercalation into the graphitic structure without requiring complete desolvation. Jache et al. were among the first to achieve Na^+ intercalation into graphite, resulting in a reversible capacity of $100\,mAh\,g^{-1}$ after 1000 cycles using sodium triflate in diglyme solvent [184]. Subsequent research from various groups demonstrated that these solvents led to excellent capacity retention and cyclability. The solvents were also applied to non-intercalating materials, such as conversion and alloy materials, with improved outcomes due to the formation of a stable SEI layer. The addition of additives, like FEC, to these electrolytes further improved anode performance, although certain additives previously effective in LIBs, such as ethylene sulfite, VC, and difluoroethylene carbonate, did not perform as effectively in SIBs [247].

ILs represent a well-established field in scientific research, although their application as matrices for Li salts or Na salts remains a relatively less explored avenue. ILs are defined as materials that consist exclusively of ions that remain

Figure 4.18 IL cations Pyr_{13} or BMP (left) and C_2mim or EMIm (right) used in electrolytes. Source: Reproduced from Ponrouch et al. [248]. © 2015/The Royal Society of Chemistry.

in a liquid state below 100 °C. In the context of battery electrolytes, a particularly noteworthy subclass of ILs are aprotic room temperature ILs, often abbreviated as RTILs (hereafter referred to as "ILs"). ILs, in general, have several properties that meet the requirements of electrolyte solvents: they have a wide liquidus range, provide thermal and electrochemical stability, and show negligible to very low vapor pressure, thus rendering them non-flammable – a trait frequently emphasized in ILs. Due to these advantageous characteristics, ILs serve as excellent solvents for ions and inherently possess high ionic conductivities. An IL typically comprises a medium- to large-sized organic cation paired with an anion from the family of weakly coordinating anions (WCAs). Consequently, the SIB electrolyte can be as straightforward as $Na_xCation_{(1-x)}WCA$, with the molar ratio x typically falling within the range of 0.1–0.25. This composition is typically achieved by direct mixing, sometimes with the assistance of heating. Consequently, the system can contain either one or two WCAs. Figure 4.18 presents the two most commonly used IL cations for electrolytes.

The choice of IL cations, as IL anions are typically chosen from WCAs in Na salts (most frequently bis(trifluoromethanesulphonyl)imide (TFSI) or bis(fluorosulfonyl)imide (FSI) anions for RTILs), can impose further limitations on the electrolyte. For example, imidazolium-based $C_n$mim (also known as EMIm, where $n = 2$, BMIm, where $n = 4$, etc.) cations have posed challenges related to the cathodic limit of the ESW, whereas the pyrrolidinium-based Pyr_{1x} family of cations shows better electrochemical stability.

However, there are several primary drawbacks associated with the use of ILs for SIB electrolytes. The major limitation is that nearly all ILs have relatively high viscosities, typically in the order of tens of *cP* at room temperature. Moreover, the viscosity often increases further when doped with the charge carrier, i.e. the Na salt. This phenomenon is attributable to the formation of even stronger ion–ion interactions that complicate the mobility of Na ions, which remains a complex, unresolved issue (some suggest that they move as doubly negatively charged species) [249]. The second primary drawback is the cost of ILs, which depends on the choice of anion. Since WCAs are often the same as those employed in salts, the overall material cost can be high, sometimes even excessive. Opting for FSI-based ILs rather than those based on TFSI can alleviate this concern to some extent, as FSI entails a less cumbersome and therefore less expensive synthesis route. Furthermore, it is worth noting that some studies have used inadequately dried or impure ILs, which cast doubt on certain results and hinder the optimal development of the field.

4.5.2 Solid Electrolytes

Using solid electrolytes in batteries offers several advantages, including enhanced safety, increased energy density, non-flammability, and the potential for employing metallic anodes to augment energy storage capacities. Furthermore, solid electrolytes can serve the dual function of separator and ion transport medium, simplifying cell fabrication and acting as a physical barrier against dendrite formation. However, solid electrolytes also grapple with issues such as poor wettability, low ionic conductivity, and the development of significant interfacial resistance during the charge–discharge process. These challenges should be addressed to facilitate their integration into commercial batteries.

Solid electrolytes can be broadly categorized into gel polymer electrolytes (GPEs) and ceramic electrolytes. GPEs are electrolytes in which a salt solution is immobilized within a polymer matrix, such as polyethylene oxide and polyvinyl alcohol. In this configuration, the salt solution provides ionic conductivity, while the polymer provides mechanical strength. Solvents used in GPEs encompass materials such as EC, PC, and ILs, while salts commonly used include NaTF, NaTFS, NaFSI, and $NaPF_6$, among others [250, 251]. For example, Vignarooban et al. synthesized a GPE based on polyacrylonitrile dissolved in an EC–PC mixture, with $NaClO_4$ as the salt. The optimized composition, consisting of 11% PAN, 12% $NaClO_4$, 40% EC, and 37% PC by weight, showed an ionic conductivity of 4.5 mS cm^{-1}, which is deemed suitable for practical applications. Furthermore, Zheng et al. developed a novel phosphate-based GPE featuring an ionic conductivity of 6.29×10^{-3} S cm^{-1}. This GPE delivered a capacity of 110.7 mAh g^{-1} in a $Na_3V_2(PO_4)_3$/GPE/Na cell, with a capacity retention of 69.2% after 10,000 cycles [252].

Ceramic electrolytes encompass materials such as NASICON and beta-alumina. Historically, these materials have demonstrated satisfactory ionic conductivity at high temperatures but more recently have been modified for integration into room temperature rechargeable SIBs. Some NASICON variants have shown ionic conductivity comparable to that of organic electrolytes. For example, Na_3PS_4, with an ionic conductivity of approximately 2×10^{-4} S cm^{-1}, and $Na_3Zr_2Si_2PO_{12}$, with an ionic conductivity of roughly 5.6×10^{-3} S cm^{-1}, have been reported [253–255]. Notably, Chu et al. developed a chlorine-doped Na_3PS_4 with the formula $Na_{2.9375}PS_{3.9375}Cl_{0.0625}$, which showed an increased ionic conductivity exceeding 0.001 S cm^{-1} at room temperature. This electrolyte gave a specific capacity of 80 mAh g^{-1} in an all-solid-state TiS_2/ ceramic electrolytes/Na cell at a C/10 rate after 10 cycles [256]. Furthermore, Zhang et al. designed a novel composite electrolyte by substituting La^{3+} with Zr^{4+} within the crystal structure of $Na_3Zr_2Si_2PO_{12}$, enhancing the ionic conductivity to 3.4×10^{-3} S cm^{-1} at 25 °C and 1.4×10^{-2} S cm^{-1} at 80 °C. Modified with an IL to improve wetting properties, the $Na_3V_2(PO_4)_3$/IL/ceramic electrolyte/Na system delivered a capacity of 90 mAh g^{-1} at a 10 C rate after 10,000 cycles [257].

A combination of GPE and ceramic electrolyte can be used to prepare hybrid electrolytes that combine the advantages of both materials. For example, Lei et al. designed a hybrid electrolyte comprising crosslinked beta alumina nanowires coated with a poly(vinylidene fluoride-*co*-hexafluoropropylene) GPE. In $Na_3V_2(PO_4)_3$/GPE/Na cells, this hybrid electrolyte showed a capacity retention of 95.3% at 25 °C and 78.8% at 60 °C after 10,000 cycles at an operating rate of 1 C [258].

4.6 Sodium-Ion Batteries at Extreme Temperatures

In recent decades, remarkable progress has been made in comprehending the mechanisms of sodium storage and in the development of high-performance SIBs. These advancements have resulted in SIBs with competitive energy densities, impressive rate capabilities, and exceptional cycling lifespans. These achievements have largely been attributable to the innovative design of new electrode materials and electrolytes, as extensively documented in previous studies [259–261]. However, it is imperative to consider the issue of temperature tolerance when evaluating stationary SIBs. For instance, in the United States, extreme temperatures can vary significantly, reaching as low as −62.2 °C in some regions and as high as 56.7 °C in others. To meet the demands of medium- to large-scale ESSs, SIBs should have a wide temperature tolerance range to ensure optimal functionality under diverse environmental conditions.

Undoubtedly, temperature exerts a profound influence on the kinetics and stability of rechargeable batteries, including SIBs. When these batteries are operated at low temperatures, sluggish electrochemical reactions often occur. This sluggishness leads to significant polarization of the redox reactions, resulting in the rapid approach of cutoff voltages [262]. Consequently, this condition limits the accessible capacity and can even promote the formation of dendrites on the anode. Conversely, high temperatures can enhance reaction kinetics, but they also carry the risk of intensifying the reactions at the electrode–electrolyte interface. In extreme cases, high temperatures can trigger thermal decomposition and even lead to explosions [7]. To circumvent potential SIB failures caused by extreme temperatures, ESSs that rely on rechargeable batteries should incorporate auxiliary temperature control systems. These systems are essential to ensure stable energy storage and release. For example, in some LIBs, nickel foil is used as a self-heating element, which effectively enables rapid warming of the battery system when operated under sub-zero conditions [263, 264]. However, it is important to note that such solutions primarily address the adaptability to cold weather. Furthermore, the introduction of additional accessories into these systems can have adverse effects on the energy density, reliability, and overall cost of stationary ESSs.

Enhancing the inherent temperature tolerance of SIBs requires comprehensive design considerations for all crucial components within SIBs. These components include the electrolyte, cathode, and anode, all of which must be meticulously engineered to operate effectively across a broad temperature range. The design objectives include the facilitation of high intrinsic ion diffusion coefficients to counteract performance losses at low temperatures. Simultaneously, the components should exhibit exceptional thermostability to mitigate the risk of exothermic reactions occurring at high temperatures. For example, Jun Lu and Ya You collaborated to introduce a strategy to prepare low-temperature-adapted sodium-ion electrolytes that enable rechargeable batteries to operate under extremely cold conditions [265, 266]. They demonstrated that the designed electrolyte can achieve stable SEI and high Na^+ mobility at electrode–electrolyte interfaces, through constructing $NaPF_6$ dissolved in strong-solvation and weak-solvation ether solvent mixture. The results affirm that such a temperature-adaptive electrolyte ensures encouraging low-temperature performance in a HC||$Na_{2/3}Ni_{1/4}Cu_{1/12}Mn_{2/3}O_2$ full cell with 90.6% capacity retention over 400 cycles at −40 °C.

There are two principles for all-climate electrodes: (i) preserving sufficient ion diffusivity at low temperatures and (ii) maintaining stable phase structure and electrode–electrolyte interface during charging/discharging at higher temperatures. Due to the low electronic conductivity of NVP, most research focuses on improving the electrochemical performance by adjusting the particle size and coating with conductive substances. For example, Fang et al. used chemical vapor deposition to decorate NVP with a hierarchical carbon framework. This modified NVP showed an impressive reversible capacity of 115 mAh g^{-1} at 0.2 C and maintained a capacity of 38 mAh g^{-1} even at an extremely high discharge rate of 500 C [267]. Capitalizing on the high-rate capabilities of NVP/C composites, Liu et al. investigated the sub-zero performance of their specially designed carbon-coated NVP. This material achieved a discharge capacity of approximately 115.3 and 85.5 mAh g^{-1} at 1 C at temperatures of 55 and −30 °C, respectively [100]. In a different approach, Xu et al. carried out layer-by-layer embedding of NVP into rGO [268]. This composite showed promising capacity retention even at lower temperatures. The NVP@rGO composites delivered discharge capacities of 117.1 and 109 mAh g^{-1} at 1 C at temperatures of 60 and 0 °C, respectively. HC with low graphitization has emerged as a preferred choice for anode material in SIBs because it enables Na intercalation in combination with Na adsorption on isolated graphite layers or filling into nanocavities. Many all-weather full batteries reported so far have used HC as the anode material. In a specific study, Lin et al. conducted a comparative analysis of the room temperature and low-temperature performance of HC in both half and full SIBs [269]. They observed that the HC maintained a stable capacity of approximately 300 and 250 mAh g^{-1} at 25 and −20 °C, respectively. In combination with a $Na_{3.5}V_2(PO_4)_2F_3$ cathode, the full battery had an energy density of 310 Wh kg^{-1}. However, it experienced significant capacity degradation at −20 °C due to the formation of a thicker SEI layer, which increased the resistance and led to a higher Na-ion consumption. To enhance the rate capability, Ponrouch et al. investigated the all-weather performance of a carbon-coated HC in SIBs. This composite material achieved specific capacities of 450 and 265 mAh g^{-1} in a carbonate electrolyte at 75 and −15 °C, respectively [270].

4.7 Other Na-Based Technologies

4.7.1 Sodium Metal Batteries (SMBs)

Because of the limited energy density of SIBs, the development of high-performance electrode materials with high reversible capacity and long-term cycling stability for SIBs is highly needed. On the anode side, metallic sodium has a high theoretical capacity (1166 mAh g^{-1}) and low redox potential (−2.71 V versus standard hydrogen electrode), which could enable high-energy-density sodium batteries [271]. Indeed, sodium metal anodes coupled with high-capacity cathodes, such as sulfur (S), oxygen (O_2), and carbon dioxide (CO_2), have already been explored to realize high-energy sodium-based batteries. However, like Li, the Na anode suffers from several crucial problems such as the inevitable side reaction and the formation of Na dendrites, which lead to cell short circuits and safety concerns [271].

By far, various approaches have been attempted to address these issues. The strategies that can effectively improve electrochemical stability have been summarized in Figure 4.19 [272]. Specifically, these routes include electrolyte manipulation, interfacial engineering between the metal electrode and the electrolyte, electrode architectures that both reduce the current density and serve as effective nucleation substances, and sodium alloys to tune the bulk properties of the metal per se.

Optimizing electrolyte systems: Organic liquid electrolytes play a pivotal role in room temperature SMBs and are critical to achieving highly reversible Na metal anodes [273, 274]. To achieve this, it is essential to control the interfacial side reactions between the electrolyte and the highly reactive Na metal anode. Typically, single linear carbonate-based electrolytes are unsuitable for SMBs due to their tendency to decompose at the Na metal anode [274, 275]. Till now, various solvents,

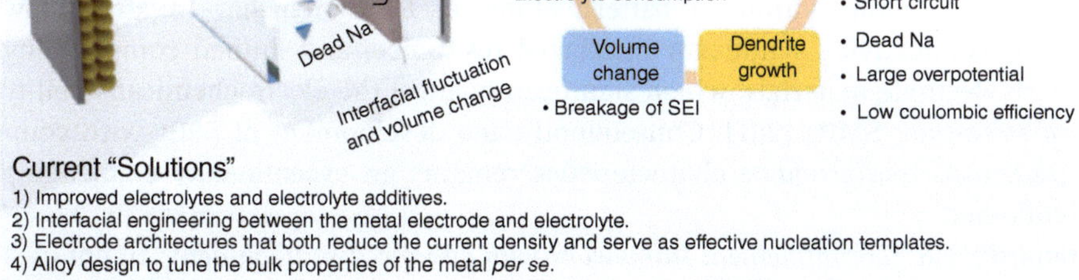

Figure 4.19 Summary of SMBs. Source: Reproduced from Lee et al. [272]. © 2019/American Chemical Society.

salts, and additives have been introduced into the electrolyte to enhance the electrochemical stability of Na anodes [275–277]. Among these additives, FEC stands out as highly effective. It can react with the Na anode and form a NaF-rich SEI layer, which is favorable for stabilizing the Na anode–electrolyte interface [275, 278]. However, the drawback of this approach is the formation of a thick SEI layer, which hinders Na^+ diffusion and increases interfacial impedance, resulting in sluggish Na^+ transport kinetics. In addition, to fully exploit the energy density benefits of high-voltage cathode materials, the development of new electrolytes with high voltage resistance is imperative for optimal performance. This may include exploring highly concentrated electrolytes and IL electrolytes to achieve the desired synergy.

Building artificial SEI layers: The traditional SEI layer is a delicate and vulnerable interface that is formed during the initial cycling of Na anodes. Unfortunately, it tends to fracture due to large volume changes that the Na anode undergoes during plating and stripping cycles [279]. To address this issue and stabilize the highly reactive Na anode, the creation of an artificial SEI layer directly on the Na metal surface becomes imperative [277]. For example, Hu et al. pioneered a low-temperature plasma-enhanced atomic layer deposition technique to construct an ultrathin protective layer of Al_2O_3 (approximately 2.8 nm thick) on the Na metal anode [280]. This artificial SEI layer effectively shields the Na metal from the electrolyte, leading to enhanced electrochemical stability, and can withstand over 450 hours at a current density of $0.25\ mA\ cm^{-2}$ in symmetrical cells using carbonate-based electrolytes. However, it is worth noting that these artificial SEI layers often exhibit inadequate interfacial contact with the Na anode, resulting in increased interfacial resistance and impeded Na^+-ion migration.

Constructing solid-state electrolytes: Given the inherent challenges associated with uncontrolled dendrite formation and persistent side reactions observed in liquid metal batteries, there is an urgent need for their replacement with SSEs in future battery technologies [281]. In the field of solid-state SMBs, SSEs serve two primary functions: (i) acting as a physical barrier to impede the transfer of side products from the cathode to the anode and (ii) providing a mechanical barrier to suppress the growth of sodium (Na) dendrites, which represents a critical safety risk in such systems [274]. Despite their effectiveness in restraining dendrite growth, SSEs are not immune to dendrite-related issues. Na dendrites can find pathways along grain boundaries and voids, eventually leading to internal short circuits within the battery. Moreover, SSEs often have relatively low ionic conductivity, significant interfacial resistance, and limited compatibility with electrode materials, which significantly affect the electrochemical stability of solid-state SMBs [281]. Consequently, the development of SSEs with comprehensive performance characteristics remains an exceptionally challenging endeavor.

Establishing a 3D confinement framework with enhanced sodiophilicity: In contrast to the aforementioned strategies, the use of a 3D confinement framework represents a distinctive approach that fundamentally alters the nucleation and growth dynamics of Na deposition. This novel approach not only provides a 3D porous

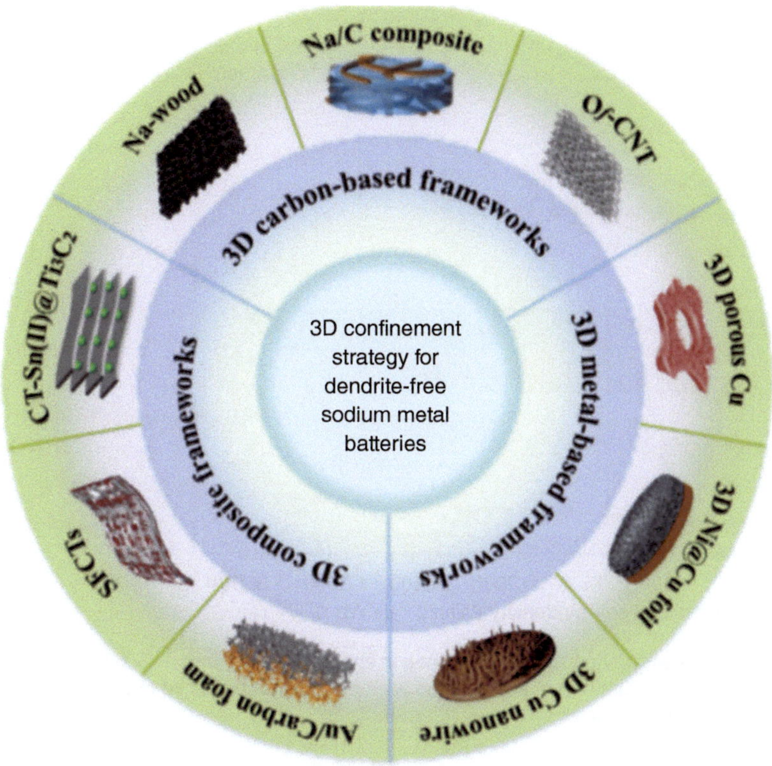

Figure 4.20 Various strategies for dendrite-free SMBs. Source: Reproduced with permission from Li et al. [271]. © 2018/John Wiley & Sons.

host capable of accommodating the significant volume changes associated with the Na anode, but also exerts precise control over the nucleation processes of Na^+ ions, resulting in more stable Na deposition and growth [274]. The 3D porous framework, characterized by a high specific surface area, effectively disperses the local current density and reduces the formation of mossy Na dendrites, as reported in previous studies [273]. Moreover, the introduction of a sodiophilic interfacial layer on the surface of the 3D framework allows for the effective regulation of Na^+-ion flux distribution during the plating process. Various examples of such 3D confinement frameworks are depicted in Figure 4.20, including Na-infused carbonized wood (Na-wood) [282], Na-infused 3D carbon felt (Na/C) [283], oxygen-functionalized carbon nanotube network (O_f-CNT), Na-infiltrated Fe_2O_3-coated carbon textile (SFCT) [284], and Sn^{2+}-pillared Ti_3C_2 MXene (CT-Sn(II)@Ti_3C_2) [285], among others. These frameworks have demonstrated remarkable effectiveness in suppressing the growth of Na dendrites. By constructing an interconnected 3D framework with superior sodiophilicity and exceptional structural stability, the deposition behavior of Na^+ ions can be significantly enhanced, spanning from initial nucleation to final growth stages.

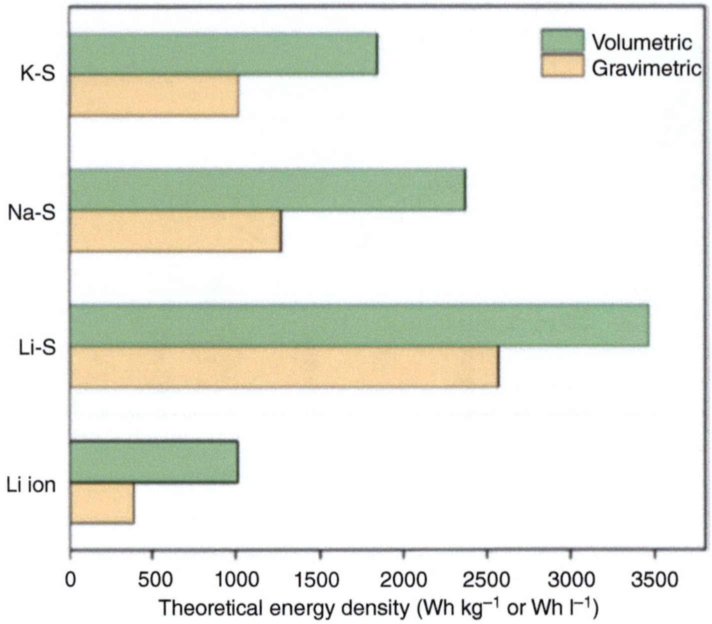

Figure 4.21 Theoretical gravimetric and volumetric energy of alkali metal–S systems based on M_2S (M = Li, Na, K). Source: Reproduced with permission from Lee et al. [272]. © 2019/American Chemical Society.

4.7.2 Sodium Metal Batteries: S, Se, and SSe Systems

In terms of gravimetric energy, the Li-S system currently holds prominence. However, considering the techno-economic advantages of Na versus Li, the Na-S system emerges as a promising avenue for both research and potential commercialization, as illustrated in prior studies [286, 287]. The theoretical energy density of the Na-S system is around 1274 Wh kg^{-1}, significantly higher than that of current LIBs, which is around 387 Wh kg^{-1} (Figure 4.21). In addition, as an alternative or supplement to sulfur (S), selenium (Se) can be employed as a cathode material paired with a Na metal anode. While Se has a lower gravimetric capacity than S (Na_2Se = 678 mAh g^{-1}, Na_2S = 1675 mAh g^{-1}), its volumetric capacity is competitive (Na_2Se = 3250 mAh cm^{-3}, Na_2S =3470 mAh cm^{-3}) due to its high density (4.81 g cm^{-3}) [288]. Typical S-Na and Se-Na cells consist of a Na metal anode; a hybrid electrolyte consisting of carbonates, ethers, or ILs; and S or Se active phases incorporated within a carbon matrix host. During the discharge process, both S and Se are reduced by Na. The discharge reactions at the S and Se cathodes can be expressed as follows [289, 290]:

$$\text{Se cathode : Se} + 2\text{Na}^+ + 2\text{e}^- \leftrightarrow \text{Na}_2\text{Se}$$

$$\text{S cathode : } n\text{S} + 2\text{Na}^+ + 2\text{e}^- \leftrightarrow \text{Na}_2\text{S}_n. \quad (4 \le n \le 8)$$

The ultimately reduced states of S and Se, specifically Na_2S and Na_2Se, share similarities in terms of stoichiometry and equilibrium structure, both adopting a

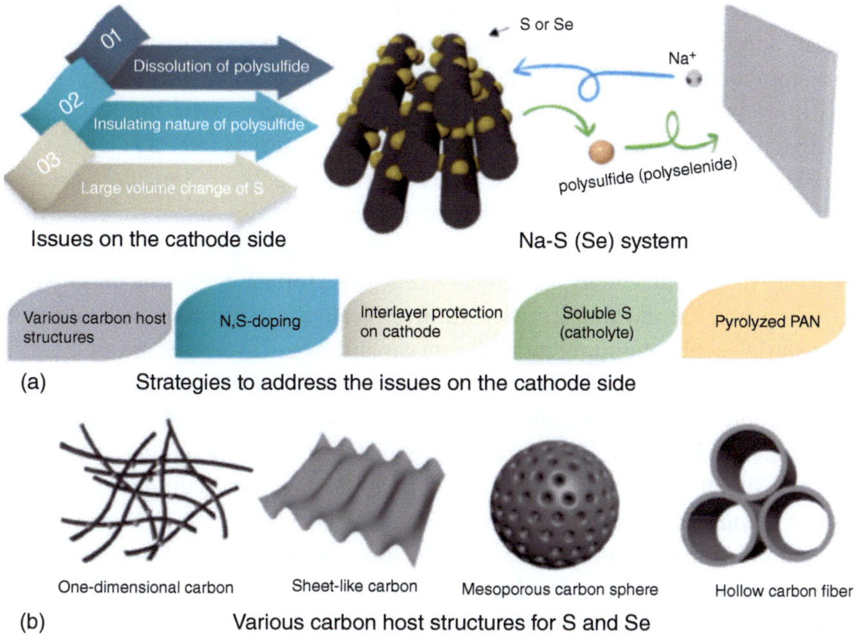

Figure 4.22 (a) Schematic of a Na-S (or Se) system, the electrochemical issues, and solutions. (b) Various carbon host structures incorporate S and/or Se through physical and/or chemical interactions. Source: Reproduced with permission from Lee et al. [272]. © 2019/American Chemical Society.

face-centered cubic configuration. Nonetheless, the intermediate processes of sodiation and desodiation are notably complex, and distinctions emerge between Se and S in this regard. The existence of equilibrium crystalline phases in the final sodiated state is not consistently verified, particularly in nanoconfined architectures.

Several concerns are associated with the S-Na, Se-Na, or their alloy SSe-Na systems: (i) dissolution of polysulfides and polyselenides into the electrolyte during cycling; (ii) electrical insulating properties of polysulfides and polyselenides when they accumulate on the anode due to crossover; and (iii) significant volume changes in S, Se, and SSe during the final stages of sodiation into a solid phase, resulting in localized electrode pulverization (Figure 4.22a) [291]. Intermediate polysulfide and polyselenide anions have high solubility in most organic electrolytes, allowing them to migrate from the cathode and coat the Na metal anode, resulting in surface "poisoning" [292]. This reaction may also impact the formation of the SEI in unexpected ways and require further investigation. Lower-ordered polysulfides can also shuttle back to the cathode and undergo oxidation again, leading to parasitic self-discharge and a decrease in CE. Unoccupied pore space within the host material can mitigate some of the volume changes associated with Na_2Se, Na_2S, etc. However, the required amount of "free" space and the role of pore size distribution (whether mesopores are more effective than micropores or vice versa) remain subjects for future research.

To effectively address these challenges, strategies from the Li-S battery system have been adopted, with varying degrees of success. These strategies encompass a range of approaches, including the use of different carbon hosts with tailored pore size distributions and surface chemistries designed to interact with the Na-Se and Na-S active phases. Notable examples involve carbon materials rich in N and S that promote strong chemical interactions, such as pyrolyzed polyacrylonitrile. Furthermore, polyion blocking interlayers were used to minimize crossover, and electrolytes and additives were adjusted to improve performance (Figure 4.22b) [293, 294]. Considering that the overall energy capacity of a cell is also influenced by the mass of the "inactive" carbon frameworks, a critical future objective is to increase the loading of active materials while maintaining high rate capability and cycling stability.

Incorporating both S and Se into various carbon host materials has emerged as a widely adopted strategy to address the aforementioned challenges, albeit with varying degrees of success. Researchers have explored a variety of one-, two-, and three-dimensional carbon host structures, including carbon nanofibers, carbon nanosheets, and porous carbon spheres. These carbon hosts utilize their hollow spaces and pores to accommodate the significant volume changes associated with S or Se. Furthermore, they effectively mitigate the shuttle effect by entrapping intermediate species within their micropores and small mesopores. However, it is worth noting that the loading capacity of active materials in these pores is inherently limited, resulting in a reduction in overall capacity, particularly in terms of volume. To partially address this limitation, some studies have investigated the impregnation of Se into a monolithic carbon matrix and yielded promising results such as a high areal capacity of 3.39 mAh cm^{-2} and an impressive 98% capacity retention after 150 cycles [295].

Heteroatom doping of the carbon matrix with N represents a well-established strategy aimed at modifying the electron density distribution. This modification serves to increase the binding energy between the carbon matrix and polysulfide ions, as highlighted in previous studies. Specifically, some studies have demonstrated that a high concentration of nitrogen in the carbon matrix effectively traps polysulfides in nanopores via electrostatic interactions. Furthermore, the formation of Na—N bonds has been observed to inhibit undesirable side reactions between polysulfides and the carbonate-based electrolyte [296]. Another effective approach involves the modification of separators to prevent the transition of polysulfides, thereby preventing their contact with the metal anode. For example, Manthiram et al. used various carbon-based interlayers, such as carbon nanofibers, CNTs, and carbon foam, positioned between the sulfur cathode and a glass fiber separator. These interlayers have proven effective in suppressing the initial capacity drop. In addition, cation-selective membranes such as Nafion can fulfill a similar role to carbon-based interlayers in the context of S-Na systems [297]. Leveraging the cation-selective properties of Nafion, Manthiram et al. achieved stable cycling of a polysulfide (Na_2S_6) catholyte, even with the inclusion of solid sulfur powder, for up to 100 cycles while maintaining a high capacity of approximately 600 mAh g^{-1} [298].

Khalil et al. introduced Se_xS_y as a promising cathode material suitable for both lithium metal batteries and SMBs at room temperature [299]. Subsequent studies by the same authors highlighted that $Se_{0.6}S_{0.4}$, when integrated into a carbon nanofiber composite, shows significantly improved cycling stability compared with S incorporated into a carbon nanofiber matrix. This improved stability was attributable to the chemical bonding between S and Se, which effectively inhibits the dissolution of Se_xSe_y species [300]. It is worth noting that the crystalline selenium has significantly higher electrical conductivity than sulfur, with values differing by about 20 orders of magnitude (10^{-5} S cm^{-1} for Se compared to 5×10^{-30} S cm^{-1} for S) [299]. Even in its amorphous state, selenium has a markedly higher electrical conductivity, which is around 10^{-11} S cm^{-1}. Thus, using Se_xS_y as a cathode material can be viewed as a synergistic approach that takes advantage of the gravimetric capacity of sulfur and the superior electrical conductivity of selenium [301]. Lou et al. also reported outstanding electrochemical performance using SeS_2 in combination with pyrolyzed polyacrylonitrile as a cathode material. They reported a remarkable reversible capacity exceeding 800 mAh g^{-1} after 400 charge–discharge cycles [301].

4.8 Summary and Outlook

Over the past decade, the field of SIBs has experienced significant momentum, marked by rapid advances in material development and overall performance. The notable progress suggests that commercialization of SIBs is technically feasible. However, to achieve commercial viability, a market niche should be identified in which SIBs have specific advantages over LIBs or other established battery technologies. Batteries that utilize cost-efficient materials composed of abundant elements represent a potential niche that could gain relevance in the future, particularly as resources and supply chain challenges for LIBs become greater; naturally, this requires a comprehensive and integrated approach.

Regarding cathode materials, polyanionic materials with robust structures and minimal volume changes represent a promising alternative. The family of polyanionic compounds has a high average potential due to the inductive effect of the polyanionic group. Some polyanion materials have been reported to have a high rate capability exceeding 100 C and a long cycle life of thousands of cycles. However, due to the large size of polyanionic groups, these materials often suffer from low electronic conductivity and a moderate capacity below 130 mAh g^{-1}. To obtain electrodes with both high reversible capacity and high rate capability, significant efforts in rational structure design are required, such as carbon decoration [2, 267] and nanosizing [302]. In contrast, layered oxide materials show unique sodium storage properties that differ from those of oxide materials in LIBs. The theoretical capacity of these materials is in the range of 230–250 mAh g^{-1}, depending on the transition metal components. The presence of multiphase transition reactions and effective metal doping/substitution render layered oxides more complex. There is ample room for further exploration to fine-tune the chemical composition to optimize the sodium storage performance, and complicated reaction mechanisms should be studied in more detail. Notably, the sensitivity of most layered oxides to moisture increases the costs of cells, because moisture-free processes are required in both material synthesis and cell preparation. Considering the cost implications of scaling and the intrinsic electrochemical performance of electrodes, O3-$Na_{0.9}Cu_{0.22}Fe_{0.30}Mn_{0.48}O_2$ and P2-$Na_{2/3}Fe_{1/2}Mn_{1/2}O_2$ emerge as promising electrodes among layered oxides. Among various polyanionic compounds, high-capacity $Na_3V_2(PO_4)_3$, $NaVOPO_4$, and low-cost $Na_4Fe_3(PO_4)_2(P_2O_7)$ materials stand out.

Anode materials undergo activation through insertion, conversion, and alloying reactions. Among these, HC emerges as a promising candidate, with a high capacity of about 350 mAh g^{-1}, comparable to the intercalation of Li ions in graphite. This positions HC as a leading candidate for anode material in SIBs, especially considering its successful adoption in the initial LIBs. Therefore, HC could potentially also be the first commercially viable anode material for SIBs. A major problem is the deposition of sodium metal onto the surface of HC at low voltage, which poses safety challenges due to the high reactivity of sodium metal. In addition, the quest for a capacity surpassing 300 mAh g^{-1}, a benchmark below that of graphite anodes in LIBs, is essential to improve the energy density of SIBs. Consequently, alternative anode materials that can deliver high capacity without experiencing sodium metal deposition during discharge are currently the subject of active investigation. Materials relying on conversion and alloying reactions are being specifically explored for this purpose. While conversion reaction materials show high reversible capacities, efforts continue to be made to improve their cycling stability and rate capability. Challenges such as large volume changes and relatively poor electronic conductivity require extensive work, including nanostructuring and carbon decoration, to overcome these limitations. To improve overall performance, significant potential hysteresis should also be addressed. Alloy reaction materials, especially those incorporating Sn, Sb, and P, hold great promise for achieving high-capacity anodes. However, the considerable volume change during sodium uptake leads to rapid capacity degradation. Strategies such as carbon coating, nanosizing, and intermetallic compounds are being explored to improve cyclic stability. Challenges persist in identifying suitable nanostructures and intermetallic compounds to further improve these materials. Ti-based anodes have also attracted attention, but their moderate sodium storage performance requires further improvement. Understanding and mitigating the catalytic reactions between the electrode and the electrolyte is crucial to optimizing their performance. Ongoing research efforts are focused on addressing these challenges to realize the full potential of Ti-based anodes in SIBs.

To date, the specific energy density of SIBs has been estimated primarily based on the weight of the active materials, utilizing calculations with the assumption of 300 mAh g^{-1} for HC as a negative electrode material in full cells. This estimate results in a specific energy density for SIBs with certain cathode materials that reaches or exceeds 300 Wh kg^{-1}. Unfortunately, when taking into account not only the active materials but also all components that make up SIBs, the actual energy density of the first commercial cylindrical 18650 SIBs is considerably lowered to below 100 Wh kg^{-1}. Therefore, it is crucial to contemplate how we can improve the energy density of a SIB system during the development of high-capacity cathode and anode materials, as well as stable electrolytes, including additives that operate in high-voltage regions, with the aim of maximizing overall energy density. Furthermore, there is an urgent need for more systematic studies on surface modification of active materials to minimize side reactions with electrolytes, binders, current collectors, and other components. This intensive progress is essential to propel SIBs toward practical applications in energy storage and highlights the importance of improving the overall energy density of the system.

References

1 Christoph, V., Daniel, B., Marcel, W et al. (2018). A cost and resource analysis of sodium-ion batteries. *Nature Reviews Materials* 3: 18013.

2 Hwang, J.Y., Myung, S.T., and Sun, Y.K. (2017). Sodium-ion batteries: present and future. *Chemical Society Reviews* 46 (12): 3529–3614.

3 Nayak, P.K., Yang, L., Brehm, W. et al. (2018). From lithium-ion to sodium-ion batteries: advantages, challenges, and surprises. *Angewandte Chemie International Edition* 57 (1): 102–120.

4 Shen, L., Wang, Y., Wu, F. et al. (2019). Hierarchical metal sulfide/carbon spheres: a generalized synthesis and high sodium-storage performance. *Angewandte Chemie International Edition* 58 (22): 7238–7243.

5 Yabuuchi, N., Kubota, K., Dahbi, M. et al. (2014). Research development on sodium-ion batteries. *Chemical Reviews* 114: 11636–11682.

6 Peters, J., Peña Cruz, A., and Weil, M. (2019). Exploring the economic potential of sodium-ion batteries. *Batteries* 5 (1): 10.

7 You, Y. and Manthiram, A. (2017). Progress in high-voltage cathode materials for rechargeable sodium-ion batteries. *Advanced Energy Materials* 8 (2): 1701785.

8 Delmas, C., Fouassier, C., and Hagenmuller, P. (1980). Structural classification and properties of the layered oxides. *Physica B+C* 99 (1–4): 81–85.

9 Wang, P.-F., You, Y., Yin, Y.-X. et al. (2018). Layered oxide cathodes for sodium-ion batteries: phase transition, air stability, and performance. *Advanced Energy Materials* 8 (8): 1701912.

10 Wang, C., Liu, L., Zhao, S. et al. (2021). Tuning local chemistry of P_2 layered-oxide cathode for high energy and long cycles of sodium-ion battery. *Nature Communications* 12 (1): 2256.

11 Bhange, D.S., Ali, G., Kim, D.-H. et al. (2017). Honeycomb-layer structured $Na_3Ni_2BiO_6$ as a high voltage and long life cathode material for sodium-ion batteries. *Journal of Materials Chemistry A* 5 (3): 1300–1310.

12 Doeff, M.M., Peng, M.Y., Ma, Y.P. et al. (1994). Orthorhombic Na_xMn0_2 as a cathode material for secondary sodium and lithium polymer batteries. *Journal of the Electrochemical Society* 141: L145.

13 Vassilaras, P., Ma, X., Li, X. et al. (2012). Electrochemical properties of monoclinic $NaNiO_2$. *Journal of the Electrochemical Society* 160 (2): A207–A211.

14 Liu, Q., Hu, Z., Chen, M. et al. (2019). Recent progress of layered transition metal oxide cathodes for sodium-ion batteries. *Small* 15 (32): e1805381.

15 Tournadre, F., Croguennec, L., Willmann, P. et al. (2004). On the mechanism of the P_2–$Na_{0.70}CoO_2$→O2–$LiCoO_2$ exchange reaction—Part II: an in situ X-ray diffraction study. *Journal of Solid State Chemistry* 177 (8): 2803–2809.

16 Paulsen, J.M., Mueller-Neuhaus, J.R., and Dahn, J.R. (2000). Layered $LiCoO_2$ with a different oxygen stacking (O_2 structure) as a cathode material for rechargeable lithium batteries. *Journal of the Electrochemical Society* 147 (2): 508–516.

17 Mu, L., Xu, S., Li, Y. et al. (2015). Prototype sodium-ion batteries using an air-stable and Co/Ni-free O_3-layered metal oxide cathode. *Advanced Materials* 27 (43): 6928–6933.

18 Li, X., Wang, Y., Wu, D. et al. (2016). Jahn–Teller assisted Na diffusion for high performance Na ion batteries. *Chemistry of Materials* 28 (18): 6575–6583.

19 Chen, X., Zhou, X., Hu, M. et al. (2015). Stable layered P_3/P_2 $Na_{0.66}Co_{0.5}Mn_{0.5}O_2$ cathode materials for sodium-ion batteries. *Journal of Materials Chemistry A* 3 (41): 20708–20714.

20 Zhao, C., Lu, Y., Chen, L. et al. (2019). Ni-based cathode materials for Na-ion batteries. *Nano Research* 12 (9): 2018–2030.

21 Katcho, N.A., Carrasco, J., Saurel, D. et al. (2017). Origins of bistability and Na ion mobility difference in P_2- and O_3-$Na_{2/3}Fe_{2/3}Mn_{1/3}O_2$ cathode polymorphs. *Advanced Energy Materials* 7 (1): 1601477.

22 Wang, P.F., Yao, H.R., Liu, X.Y. et al. (2017). Ti-substituted $NaNi_{0.5}Mn_{0.5-x}Ti_xO_2$ cathodes with reversible O_3-P_3 phase transition for high-performance sodium-ion batteries. *Advanced Materials* 29 (19): 1700210.

23 Lv, W.J., Huang, Z., Yin, Y.X. et al. (2019). Strategies to build high-rate cathode materials for Na-ion batteries. *ChemNanoMat* 5 (10): 1253–1262.

24 Cabana, J., Chernova, N.A., Xiao, J. et al. (2013). Study of the transition metal ordering in layered $Na_xNi_{x/2}Mn_{1-x/2}O_2$ ($2/3 \leq x \leq 1$) and consequences of Na/Li exchange. *Inorganic Chemistry* 52 (15): 8540–8550.

25 Stoyanova, R., Carlier, D., Sendova-Vassileva, M. et al. (2010). Stabilization of over-stoichiometric Mn^{4+} in layered $Na_{2/3}MnO_2$. *Journal of Solid State Chemistry* 183 (6): 1372–1379.

26 Duffort, V., Talaie, E., Black, R. et al. (2015). Uptake of CO_2 in layered P_2-$Na_{0.67}Mn_{0.5}Fe_{0.5}O_2$: insertion of carbonate anions. *Chemistry of Materials* 27 (7): 2515–2524.

27 Buchholz, D., Chagas, L.G., Vaalma, C. et al. (2014). Water sensitivity of layered P_2/P_3-$Na_xNi_{0.22}Co_{0.11}Mn_{0.66}O_2$ cathode material. *Journal of Materials Chemistry A* 2 (33): 13415–13421.

28 Monyoncho, E. and Bissessur, R. (2013). Unique properties of α-$NaFeO_2$: de-intercalation of sodium via hydrolysis and the intercalation of guest molecules into the extract solution. *Materials Research Bulletin* 48 (7): 2678–2686.

29 Han, M.H., Gonzalo, E., Sharma, N. et al. (2015). High-performance P_2-phase $Na_{2/3}Mn_{0.8}Fe_{0.1}Ti_{0.1}O_2$ cathode material for ambient-temperature sodium-ion batteries. *Chemistry of Materials* 28 (1): 106–116.

30 Cao, X., Li, X., Qiao, Y. et al. (2019). Restraining oxygen loss and suppressing structural distortion in a newly Ti-substituted layered oxide P_2-$Na_{0.66}Li_{0.22}Ti_{0.15}Mn_{0.63}O_2$. *ACS Energy Letters* 4 (10): 2409–2417.

31 Yang, L., Li, X., Liu, J. et al. (2019). Lithium-doping stabilized high-performance P_2-$Na_{0.66}Li_{0.18}Fe_{0.12}Mn_{0.7}O_2$ cathode for sodium ion batteries. *Journal of the American Chemical Society* 141 (16): 6680–6689.

32 Chen, C., Han, Z., Liang, C. et al. (2022). Electrochemically induced cleavage cracking at twin boundary of sodium layered oxide cathodes. *Acta Materialia* 238: 118212.

33 Lee, E., Brown, D.E., Alp, E.E. et al. (2015). New insights into the performance degradation of Fe-based layered oxides in sodium-ion batteries: instability of Fe^{3+}/Fe^{4+} redox in α-$NaFeO_2$. *Chemistry of Materials* 27 (19): 6755–6764.

34 Liu, Y., Shen, Q., Zhao, X. et al. (2019). Hierarchical engineering of porous P_2-$Na_{2/3}Ni_{1/3}Mn_{2/3}O_2$ nanofibers assembled by nanoparticles enables superior sodium-ion storage cathodes. *Advanced Functional Materials* 30 (6): 1907837.

35 Fang, Y., Yu, X.Y., and Lou, X.W.D. (2017). A practical high-energy cathode for sodium-ion batteries based on uniform P_2-$Na_{0.7}CoO_2$ microspheres. *Angewandte Chemie International Edition* 56 (21): 5801–5805.

36 Deng, J., Luo, W.-B., Lu, X. et al. (2018). High energy density sodium-ion battery with industrially feasible and air-stable O_3-type layered oxide cathode. *Advanced Energy Materials* 8 (5): 1701610.

37 Mao, Q., Gao, R., Li, Q. et al. (2020). O_3-type $NaNi_{0.5}Mn_{0.5}O_2$ hollow microbars with exposed {0 1 0} facets as high performance cathode materials for sodium-ion batteries. *Chemical Engineering Journal* 382: 122978.

38 Li, F., Fan, K., Hou, P. et al. (2021). Boosting the redox kinetics of high-voltage P_2-type cathode by radially oriented {010} exposed nanoplates for high-power sodium-ion batteries. *Small Structures* 3 (1): 2100123.

39 Zhang, F., Lu, Y., Guo, Y. et al. (2023). Highly stabilized single-crystal P_2-type layered oxides obtained via rational crystal orientation modulation for sodium-ion batteries. *Chemical Engineering Journal* 458: 141515.

40 Kong, W., Wang, H., Sun, L. et al. (2019). Understanding the synergic roles of MgO coating on the cycling and rate performance of $Na_{0.67}Mn_{0.5}Fe_{0.5}O_2$ cathode. *Applied Surface Science* 497: 143814.

41 Kaliyappan, K., Liu, J., Xiao, B. et al. (2017). Enhanced performance of P_2-$Na_{0.66}(Mn_{0.54}Co_{0.13}Ni_{0.13})O_2$ cathode for sodium ion batteries by ultrathin metal oxide coatings via atomic layer deposition. *Advanced Functional Materials* 27 (37): 1701870.

42 Zhang, Y., Pei, Y., Liu, W. et al. (2020). $AlPO_4$-coated P_2-type hexagonal $Na_{0.7}MnO_{2.05}$ as high stability cathode for sodium ion battery. *Chemical Engineering Journal* 382: 122697.

43 Wang, Y., Tang, K., Li, X. et al. (2019). Improved cycle and air stability of P_3-$Na_{0.65}Mn_{0.75}Ni_{0.25}O_2$ electrode for sodium-ion batteries coated with metal phosphates. *Chemical Engineering Journal* 372: 1066–1076.

44 Jo, C.-H., Jo, J.-H., Yashiro, H. et al. (2018). Bioinspired surface layer for the cathode material of high-energy-density sodium-ion batteries. *Advanced Energy Materials* 8 (13): 1702942.

45 Sun, H.H., Hwang, J.Y., Yoon, C.S. et al. (2018). Capacity degradation mechanism and cycling stability enhancement of AlF_3-coated nanorod gradient $Na[Ni_{0.65}Co_{0.08}Mn_{0.27}]O_2$ cathode for sodium-ion batteries. *ACS Nano* 12 (12): 12912–12922.

46 Liu, Y., Yang, J., Guo, B. et al. (2018). Enhanced electrochemical performance of $Na_{0.5}Ni_{0.25}Mn_{0.75}O_2$ micro-sheets at 3.8 V for Na-ion batteries with nanosized-thin AlF_3 coating. *Nanoscale* 10 (26): 12625–12630.

47 Lu, D., Yao, Z., Zhong, Y. et al. (2019). Polypyrrole-coated sodium manganate hollow microspheres as a superior cathode for sodium ion batteries. *ACS Applied Materials & Interfaces* 11 (17): 15630–15637.

48 Lin, J., Huang, Q., Dai, K. et al. (2022). Mitigating interfacial instability of high-voltage sodium layered oxide cathodes with coordinative polymeric structure. *Journal of Power Sources* 552: 232235.

49 de la Llave, E., Talaie, E., Levi, E. et al. (2016). Improving energy density and structural stability of manganese oxide cathodes for Na-ion batteries by structural lithium substitution. *Chemistry of Materials* 28 (24): 9064–9076.

50 House, R.A., Maitra, U., Pérez-Osorio, M.A. et al. (2019). Superstructure control of first-cycle voltage hysteresis in oxygen-redox cathodes. *Nature* 577 (7791): 502–508.

51 Zheng, S., Zhong, G., McDonald, M.J. et al. (2016). Exploring the working mechanism of Li^+ in O_3-type $NaLi_{0.1}Ni_{0.35}Mn_{0.55}O_2$ cathode materials for rechargeable Na-ion batteries. *Journal of Materials Chemistry A* 4 (23): 9054–9062.

52 Oh, S.-M., Myung, S.-T., Hwang, J.-Y. et al. (2014). High capacity O_3-type $Na[Li_{0.05}(Ni_{0.25}Fe_{0.25}Mn_{0.5})_{0.95}]O_2$ cathode for sodium ion batteries. *Chemistry of Materials* 26 (21): 6165–6171.

53 Zhao, C., Yao, Z., Wang, Q. et al. (2020). Revealing high Na-content P_2-type layered oxides as advanced sodium-ion cathodes. *Journal of the American Chemical Society* 142 (12): 5742–5750.

54 Billaud, J., Singh, G., Armstrong, A.R. et al. (2014). $Na_{0.67}Mn_{1-x}Mg_xO_2$ $(0 \leq x \leq 0.2)$: a high capacity cathode for sodium-ion batteries. *Environmental Science & Technology* 7 (4): 1387–1391.

55 Liu, X., Zuo, W., Zheng, B. et al. (2019). P_2-$Na_{0.67}Al_xMn_{1-x}O_2$: cost-effective, stable and high-rate sodium electrodes by suppressing phase transitions and enhancing sodium cation mobility. *Angewandte Chemie International Edition* 58 (50): 18086–18095.

56 Wang, P.-F., Yao, H.-R., Liu, X.-Y. et al. (2018). Na^+/vacancy disordering promises high-rate Na-ion batteries. *Science Advances* 4 (3): eaar6018.

57 Yoshida, H., Yabuuchi, N., Kubota, K. et al. (2014). P_2-type $Na_{2/3}Ni_{1/3}Mn_{2/3-x}Ti_xO_2$ as a new positive electrode for higher energy Na-ion batteries. *Chemical Communications* 50 (28): 3677–3680.

58 Bai, X., Sathiya, M., Mendoza-Sánchez, B. et al. (2018). Anionic redox activity in a newly Zn-doped sodium layered oxide P_2-$Na_{2/3}Mn_{1-y}Zn_yO_2 (0 < y < 0.23)$. *Advanced Energy Materials* 8 (32): 1802379.

59 Wang, K., Zhang, Z., Cheng, S. et al. (2022). Precipitate-stabilized surface enabling high-performance $Na_{0.67}Ni_{0.33-x}Mn_{0.67}Zn_xO_2$ for sodium-ion battery. *eScience* 2 (5): 529–536.

60 Shi, Q., Qi, R., Feng, X. et al. (2022). Niobium-doped layered cathode material for high-power and low-temperature sodium-ion batteries. *Nature Communications* 13 (1).

61 Yang, Q., Wang, P.-F., Guo, J.-Z. et al. (2018). Advanced P_2-$Na_{2/3}Ni_{1/3}Mn_{7/12}Fe_{1/12}O_2$ cathode material with suppressed P_2–O_2 phase transition toward high-performance sodium-ion battery. *ACS Applied Materials and Interfaces* 10 (40): 34272–34282.

62 Yao, H.-R., Lv, W.-J., Yin, Y.-X. et al. (2019). Suppression of monoclinic phase transitions of O_3-type cathodes based on electronic delocalization for Na-ion batteries. *ACS Applied Materials and Interfaces* 11 (25): 22067–22073.

63 Wang, L., Sun, Y.-G., Hu, L.-L. et al. (2017). Copper-substituted $Na_{0.67}Ni_{0.3-x}Cu_xMn_{0.7}O_2$ cathode materials for sodium-ion batteries with suppressed P_2–O_2 phase transition. *Journal of Materials Chemistry A* 5 (18): 8752–8761.

64 Feng, Y., Huang, Q., Ding, Z. et al. (2022). Constructing interstitial pillar to manipulating interlamination interaction force: towards high sodium-content P_2/O_3 intergrowth cathodes. *Electrochimica Acta* 433: 141253.

65 Liu, K., Tan, S., Moon, J. et al. (2020). Insights into the enhanced cycle and rate performances of the F-substituted P_2-type oxide cathodes for sodium-ion batteries. *Advanced Energy Materials* 10 (19): 2000135.

66 Xue, T., Geng, G., Han, Y. et al. (2021). Open fire exposure increases the risk of pregnancy loss in South Asia. *Nature Communications* 12 (1): 3205.

67 Li, Z.-Y., Gao, R., Zhang, J. et al. (2016). New insights into designing high-rate performance cathode materials for sodium ion batteries by enlarging the slab-spacing of the Na-ion diffusion layer. *Journal of Materials Chemistry A* 4 (9): 3453–3461.

68 Cheng, Z., Zhao, B., Guo, Y.J. et al. (2022). Mitigating the large-volume phase transition of P_2-type cathodes by synergetic effect of multiple ions for improved sodium-ion batteries. *Advanced Energy Materials* 12 (14): 2103461.

69 Lee, I., Oh, G., Lee, S. et al. (2021). Cationic and transition metal co-substitution strategy of O_3-type $NaCrO_2$ cathode for high-energy sodium-ion batteries. *Energy Storage Materials* 41: 183–195.

70 Peng, B., Sun, Z.H., Zhao, L.P. et al. (2021). Dual-manipulation on P_2-$Na_{0.67}Ni_{0.33}Mn_{0.67}O_2$ layered cathode toward sodium-ion full cell with record operating voltage beyond 3.5 V. *Energy Storage Materials* 35: 620–629.

71 Hwang, J.-Y., Yu, T.-Y., and Sun, Y.-K. (2018). Simultaneous MgO coating and Mg doping of $Na[Ni_{0.5}Mn_{0.5}]O_2$ cathode: facile and customizable approach to high-voltage sodium-ion batteries. *Journal of Materials Chemistry A* 6 (35): 16854–16862.

72 Zhang, J. and Yu, D.Y.W. (2018). Stabilizing $Na_{0.7}MnO_2$ cathode for Na-ion battery via a single-step surface coating and doping process. *Journal of Power Sources* 391: 106–112.

73 Jin, T., Li, H., Zhu, K. et al. (2020). Polyanion-type cathode materials for sodium-ion batteries. *Chemical Society Reviews* 49 (8): 2342–2377.

74 Avdeev, M., Mohamed, Z., Ling, C.D. et al. (2013). Magnetic structures of $NaFePO_4$ maricite and triphylite polymorphs for sodium-ion batteries. *Inorganic Chemistry* 52 (15): 8685–8693.

75 Moreau, P., Guyomard, D., Gaubicher, J. et al. (2010). Structure and stability of sodium intercalated phases in olivine $FePO_4$. *Chemistry of Materials* 22 (14): 4126–4128.

76 Saracibar, A., Carrasco, J., Saurel, D. et al. (2016). Investigation of sodium insertion–extraction in olivine Na_xFePO_4 ($0 \leq x \leq 1$) using first-principles calculations. *Physical Chemistry Chemical Physics* 18 (18): 13045–13051.

77 Galceran, M., Saurel, D., Acebedo, B. et al. (2014). The mechanism of $NaFePO_4$ (de)sodiation determined by in situ X-ray diffraction. *Physical Chemistry Chemical Physics* 16 (19): 8837–8842.

78 Wang, J. and Sun, X. (2015). Olivine $LiFePO_4$: the remaining challenges for future energy storage. *Energy & Environmental Science* 8 (4): 1110–1138.

79 Oh, S.-M., Myung, S.-T., Hassoun, J. et al. (2012). Reversible $NaFePO_4$ electrode for sodium secondary batteries. *Electrochemistry Communications* 22: 149–152.

80 Galceran, M., Roddatis, V., Zúñiga, F.J. et al. (2014). Na–vacancy and charge ordering in $Na_{\approx 2/3}FePO_4$. *Chemistry of Materials* 26 (10): 3289–3294.

81 Gaubicher, J., Boucher, F., Moreau, P. et al. (2014). Abnormal operando structural behavior of sodium battery material: influence of dynamic on phase diagram of Na_xFePO_4. *Electrochemistry Communications* 38: 104–106.

82 Lu, J., Chung, S.C., Nishimura, S. -i. et al. (2013). Phase diagram of olivine Na_xFePO_4 ($0 < x < 1$). *Chemistry of Materials* 25 (22): 4557–4565.

83 Wang, D., Wu, Y., Lv, J. et al. (2019). Carbon encapsulated maricite $NaFePO_4$ nanoparticles as cathode material for sodium-ion batteries. *Colloids and Surfaces A: Physicochemical and Engineering Aspects* 583: 123957.

84 Kim, J., Seo, D.-H., Kim, H. et al. (2015). Unexpected discovery of low-cost maricite $NaFePO_4$ as a high-performance electrode for Na-ion batteries. *Energy & Environmental Science* 8 (2): 540–545.

85 Xiong, F., An, Q., Xia, L. et al. (2019). Revealing the atomistic origin of the disorder-enhanced Na-storage performance in $NaFePO_4$ battery cathode. *Nano Energy* 57: 608–615.

86 Liu, Y., Zhang, N., Wang, F. et al. (2018). Approaching the downsizing limit of maricite $NaFePO_4$ toward high-performance cathode for sodium-ion batteries. *Advanced Functional Materials* 28 (30): 1801917.

87 Boyadzhieva, T., Koleva, V., and Stoyanova, R. (2017). Crystal chemistry of Mg substitution in $NaMnPO_4$ olivine: concentration limit and cation distribution. *Physical Chemistry Chemical Physics* 19 (20): 12730–12739.

88 Koleva, V., Boyadzhieva, T., Zhecheva, E. et al. (2013). Precursor-based methods for low-temperature synthesis of defectless $NaMnPO_4$ with an olivine- and maricite-type structure. *CrystEngComm* 15 (44): 9080–9089.

89 Lee, K.T., Ramesh, T.N., Nan, F. et al. (2011). Topochemical synthesis of sodium metal phosphate olivines for sodium-ion batteries. *Chemistry of Materials* 23 (16): 3593–3600.

90 Goodenough, J.B., Hong, H.Y.-P., and Kafalas, J.A. (1976). Fast Na^+-ion transport in skeleton structures. *Materials Research Bulletin* 11 (2): 203–220.

91 Jian, Z., Hu, Y.S., Ji, X. et al. (2017). NASICON-structured materials for energy storage. *Advanced Materials* 29 (20): 1601925.

92 Delmas, C. (2018). Sodium and sodium-ion batteries: 50 years of research. *Advanced Energy Materials* 8 (17): 1703137.

93 Chotard, J.-N., Rousse, G., David, R. et al. (2015). Discovery of a sodium-ordered form of $Na_3V_2(PO_4)_3$ below ambient temperature. *Chemistry of Materials* 27 (17): 5982–5987.

94 Li, L., Xu, Y., Sun, X. et al. (2018). Fluorophosphates from solid-state synthesis and electrochemical ion exchange: $NaVPO_4F$ or $Na_3V_2(PO_4)_2F_3$? *Advanced Energy Materials* 8 (24): 1801064.

95 Jian, Z., Yuan, C., Han, W. et al. (2014). Atomic structure and kinetics of NASICON $Na_xV_2(PO_4)_3$ cathode for sodium-ion batteries. *Advanced Functional Materials* 24 (27): 4265–4272.

96 Song, W., Ji, X., Wu, Z. et al. (2014). First exploration of Na-ion migration pathways in the NASICON structure $Na_3V_2(PO_4)_3$. *Journal of Materials Chemistry A* 2 (15): 5358–5362.

97 Saravanan, K., Mason, C.W., Rudola, A. et al. (2013). The first report on excellent cycling stability and superior rate capability of $Na_3V_2(PO_4)_3$ for sodium ion batteries. *Advanced Energy Materials* 3 (4): 444–450.

98 Wei, C., Luo, F., Zhang, C. et al. (2019). Voltage window-dependent electrochemical performance and reaction mechanisms of $Na_3V_2(PO_4)_3$ cathode for high-capacity sodium ion batteries. *Ionics* 26 (5): 2343–2351.

99 Wang, Q., Zhang, M., Zhou, C. et al. (2018). Concerted ion-exchange mechanism for sodium diffusion and its promotion in $Na_3V_2(PO_4)_3$ framework. *The Journal of Physical Chemistry C* 122 (29): 16649–16654.

100 Liu, T., Wang, B., Gu, X. et al. (2016). All-climate sodium ion batteries based on the NASICON electrode materials. *Nano Energy* 30: 756–761.

101 Ren, W., Zheng, Z., Xu, C. et al. (2016). Self-sacrificed synthesis of three-dimensional $Na_3V_2(PO_4)_3$ nanofiber network for high-rate sodium–ion full batteries. *Nano Energy* 25: 145–153.

102 Jian, Z., Han, W., Lu, X. et al. (2013). Superior electrochemical performance and storage mechanism of $Na_3V_2(PO_4)_3$ cathode for room-temperature sodium-ion batteries. *Advanced Energy Materials* 3 (2): 156–160.

103 Zhu, C., Song, K., van Aken, P.A. et al. (2014). Carbon-coated $Na_3V_2(PO_4)_3$ embedded in porous carbon matrix: an ultrafast Na-storage cathode with the potential of outperforming Li cathodes. *Nano Letters* 14 (4): 2175–2180.

104 Aragón, M.J., Lavela, P., Alcántara, R. et al. (2015). Effect of aluminum doping on carbon loaded $Na_3V_2(PO_4)_3$ as cathode material for sodium-ion batteries. *Electrochimica Acta* 180: 824–830.

105 Klee, R., Lavela, P., Aragón, M.J. et al. (2016). Enhanced high-rate performance of manganese substituted $Na_3V_2(PO_4)_3/C$ as cathode for sodium-ion batteries. *Journal of Power Sources* 313: 73–80.

106 Inoishi, A., Yoshioka, Y., Zhao, L. et al. (2017). Improvement in the energy density of $Na_3V_2(PO_4)_3$ by Mg substitution. *ChemElectroChem* 4 (11): 2755–2759.

107 Qu, R., Zhang, W., Liu, N. et al. (2018). Antioil Ag_3PO_4 nanoparticle/polydopamine/Al_2O_3 sandwich structure for complex wastewater treatment: dynamic catalysis under natural light. *ACS Sustainable Chemistry & Engineering* 6 (6): 8019–8028.

108 Kawai, K., Zhao, W., Nishimura, S.-i. et al. (2018). High-voltage Cr^{4+}/Cr^{3+} redox couple in polyanion compounds. *ACS Applied Energy Materials* 1 (3): 928–931.

109 Rajagopalan, R., Chen, B., Zhang, Z. et al. (2017). Improved reversibility of Fe^{3+}/Fe^{4+} redox couple in sodium super ion conductor type $Na_3Fe_2(PO_4)_3$ for sodium-ion batteries. *Advanced Materials* 29 (12): 1605694.

110 Mason, C.W. and Lange, F. (2015). Aqueous ion battery systems using sodium vanadium phosphate stabilized by titanium substitution. *ECS Electrochemistry Letters* 4 (8): A79–A82.

111 Wang, H., Zhang, T., Chen, C. et al. (2017). High-performance aqueous symmetric sodium-ion battery using NASICON-structured $Na_2VTi(PO_4)_3$. *Nano Research* 11 (1): 490–498.

112 Niu, Y., Zhang, Y., and Xu, M. (2019). A review on pyrophosphate framework cathode materials for sodium-ion batteries. *Journal of Materials Chemistry A* 7 (25): 15006–15025.

113 Barpanda, P., Nishimura, S.-i., and Yamada, A. (2012). High-voltage pyrophosphate cathodes. *Advanced Energy Materials* 2 (7): 841–859.

114 Barpanda, P., Ye, T., Nishimura, S.-I. et al. (2012). Sodium iron pyrophosphate: a novel 3.0 V iron-based cathode for sodium-ion batteries. *Electrochemistry Communications* 24: 116–119.

115 Kim, H., Shakoor, R.A., Park, C. et al. (2013). $Na_2FeP_2O_7$ as a promising iron-based pyrophosphate cathode for sodium rechargeable batteries: a combined experimental and theoretical study. *Advanced Functional Materials* 23 (9): 1147–1155.

116 Huang, Q. and Hwu, S.-J. (1998). Synthesis and characterization of three new layered phosphates, $Na_2MnP_2O_7$, $NaCsMnP_2O_7$, and $NaCsMn_{0.35}Cu_{0.65}P_2O_7$. *Inorg. Chem.* 37: 5869–5874.

117 Park, C.S., Kim, H., Shakoor, R.A. et al. (2013). Anomalous manganese activation of a pyrophosphate cathode in sodium ion batteries: a combined experimental and theoretical study. *Journal of the American Chemical Society* 135 (7): 2787–2792.

118 Tanabe, M., Honma, T., and Komatsu, T. (2018). Unique crystallization behavior of sodium manganese pyrophosphate $Na_2MnP_2O_7$ glass and its electrochemical properties. *Journal of Asian Ceramic Societies* 5 (2): 209–215.

119 Barpanda, P., Ye, T., Avdeev, M. et al. (2013). A new polymorph of $Na_2MnP_2O_7$ as a 3.6 V cathode material for sodium-ion batteries. *Journal of Materials Chemistry A* 1 (13): 4194–4197.

120 Ha, K.-H., Woo, S.H., Mok, D. et al. (2013). $Na_{4-\alpha}M_{2+\alpha/2}(P_2O_7)_2(2/3 \leq \alpha \leq 7/8$, M = Fe, $Fe_{0.5}Mn_{0.5}$, Mn): a promising sodium ion cathode for Na-ion batteries. *Advanced Energy Materials* 3 (6): 770–776.

121 Lin, B., Zhang, S., and Deng, C. (2016). Understanding the effect of depressing surface moisture sensitivity on promoting sodium intercalation in coral-like $Na_{3.12}Fe_{2.44}(P_2O_7)_2/C$ synthesized via a flash-combustion strategy. *Journal of Materials Chemistry A* 4 (7): 2550–2559.

122 Chen, M., Chen, L., Hu, Z. et al. (2017). Carbon-coated $Na_{3.32}Fe_{2.34}(P_2O_7)_2$ cathode material for high-rate and long-life sodium-ion batteries. *Advanced Materials* 29 (21): 1605535.

123 Sanz, F., Parada, C., Rojo, J.M. et al. (1999). Studies on tetragonal $Na_2CoP_2O_7$, a novel ionic conductor. *Journal of Solid State Chemistry* 604–611.

124 Avdeev, M., Wang, C.-W., Barpanda, P. et al. (2019). Polymorphism and temperature-induced phase transitions of $Na_2CoP_2O_7$. *Inorganic Chemistry* 58 (24): 16823–16830.

125 Erragh, F., Boukhari, A., Elouadi, B. et al. (1991). Crystal structures of two allotropic forms of $Na_2CoP_2O_7$. *Journal of Crystallographic and Spectroscopic Research* 21: 321–326.

126 Barpanda, P., Lu, J., Ye, T. et al. (2013). A layer-structured $Na_2CoP_2O_7$ pyrophosphate cathode for sodium-ion batteries. *RSC Advances* 3 (12): 3857–3860.

127 Ellis, B.L., Makahnouk, W.R.M., Makimura, Y. et al. (2007). A multifunctional 3.5 V iron-based phosphate cathode for rechargeable batteries. *Nature Materials* 6 (10): 749–753.

128 Tripathi, R., Wood, S.M., Islam, M.S. et al. (2013). Na-ion mobility in layered Na_2FePO_4F and olivine Na[Fe,Mn]PO_4. *Energy & Environmental Science* 6 (8): 2257–2264.

129 Kubota, K., Yokoh, K., Yabuuchi, N. et al. (2014). Na_2CoPO_4F as a high-voltage electrode material for Na-ion batteries. *Electrochemistry* 82 (10): 909–911.

130 Zou, H., Li, S., Wu, X. et al. (2015). Spray-drying synthesis of pure Na_2CoPO_4F as cathode material for sodium ion batteries. *ECS Electrochemistry Letters* 4 (6): A53–A55.

131 Kim, S.-W., Seo, D.-H., Kim, H. et al. (2012). A comparative study on Na_2MnPO_4F and Li_2MnPO_4F for rechargeable battery cathodes. *Physical Chemistry Chemical Physics* 14 (10): 3299–3303.

132 Ellis, B.L., Makahnouk, W.R.M., Rowan-Weetaluktuk, W.N. et al. (2009). Crystal structure and electrochemical properties of A_2MPO_4F fluorophosphates (A = Na, Li; M = Fe, Mn, Co, Ni). *Chemistry of Materials* 22 (3): 1059–1070.

133 Wu, X., Zheng, J., Gong, Z. et al. (2011). Sol–gel synthesis and electrochemical properties of fluorophosphates $Na_2Fe_{1-x}Mn_xPO_4F/C$ (x = 0, 0.1, 0.3, 0.7, 1) composite as cathode materials for lithium ion battery. *Journal of Materials Chemistry* 21 (46): 18630–18637.

134 Lin, X., Hou, X., Wu, X. et al. (2014). Exploiting Na_2MnPO_4F as a high-capacity and well-reversible cathode material for Na-ion batteries. *RSC Advances* 4 (77): 40985–40993.

135 Ling, R., Cai, S., Shen, K. et al. (2019). Dual carbon-confined Na_2MnPO_4F nanoparticles as a superior cathode for rechargeable sodium-ion battery. *Ceramics International* 45 (16): 19799–19807.

136 Boivin, E., Chotard, J.-N., Bamine, T. et al. (2017). Vanadyl-type defects in Tavorite-like $NaVPO_4F$: from the average long range structure to local environments. *Journal of Materials Chemistry A* 5 (47): 25044–25055.

137 Kosova, N.V. and Semykina, D.O. (2019). Mechanochemically assisted solid-state synthesis of sodium vanadium fluorophosphates. *Solid State Ionics* 343: 115119.

138 Kosova, N.V. and Belotserkovsky, V.A. (2018). Sodium and mixed sodium/lithium iron ortho-pyrophosphates: synthesis, structure and electrochemical properties. *Electrochimica Acta* 278: 182–195.

139 Wood, S.M., Eames, C., Kendrick, E. et al. (2015). Sodium ion diffusion and voltage trends in phosphates $Na_4M_3(PO_4)_2P_2O_7$ (M = Fe, Mn, Co, Ni) for possible high-rate cathodes. *The Journal of Physical Chemistry C* 119 (28): 15935–15941.

140 Yuan, T., Wang, Y., Zhang, J. et al. (2019). 3D graphene decorated $Na_4Fe_3(PO_4)_2(P_2O_7)$ microspheres as low-cost and high-performance cathode materials for sodium-ion batteries. *Nano Energy* 56: 160–168.

141 Pu, X., Wang, H., Yuan, T. et al. (2019). $Na_4Fe_3(PO_4)_2P_2O_7$/C nanospheres as low-cost, high-performance cathode material for sodium-ion batteries. *Energy Storage Materials* 22: 330–336.

142 Kosova, N.V. and Shindrov, A.A. (2019). Effect of mixed Li^+/Na^+-ion electrolyte on electrochemical performance of $Na_4Fe_3(PO_4)_2P_2O_7$ in hybrid batteries. *Batteries* 5 (2): 39.

143 Driscoll, L.L., Kendrick, E., Knight, K.S. et al. (2018). Investigation into the dehydration of selenate doped $Na_2M(SO_4)_2{\cdot}2H_2O$ (M = Mn, Fe, Co and Ni): stabilisation of the high Na content alluaudite phases $Na_3M_{1.5}(SO_4)_{3-1.5x}(SeO_4)_{1.5x}$ (M = Mn, Co and Ni) through selenate incorporation. *Journal of Solid State Chemistry* 258: 64–71.

144 Pan, W., Guan, W., Liu, S. et al. (2019). $Na_2Fe(SO_4)_2$: an anhydrous 3.6 V, low-cost and good-safety cathode for a rechargeable sodium-ion battery. *Journal of Materials Chemistry A* 7 (21): 13197–13204.

145 Dwibedi, D., Gond, R., Dayamani, A. et al. (2017). $Na_{2.32}Co_{1.84}(SO_4)_3$ as a new member of the alluaudite family of high-voltage sodium battery cathodes. *Dalton Transactions* 46 (1): 55–63.

146 Dwibedi, D., Araujo, R.B., Chakraborty, S. et al. (2015). $Na_{2.44}Mn_{1.79}(SO_4)_3$: a new member of the alluaudite family of insertion compounds for sodium ion batteries. *Journal of Materials Chemistry A* 3 (36): 18564–18571.

147 Bejaoui, A., Souamti, A., Kahlaoui, M. et al. (2018). Spectroscopic investigations on vanthoffite ceramics partially doped with cobalt. *Ionics* 24 (9): 2867–2875.

148 Ri, G.-C., Choe, S.-H., and Yu, C.-J. (2018). First-principles study of mixed eldfellite compounds $Na_x(Fe_{1/2}M_{1/2})(SO_4)_2$ (x=0–2, M = Mn, Co, Ni): a new family of high electrode potential cathodes for the sodium-ion battery. *Journal of Power Sources* 378: 375–382.

149 Barpanda, P., Oyama, G., Ling, C.D. et al. (2014). Kröhnkite-type $Na_2Fe(SO_4)_2·2H_2O$ as a novel 3.25 V insertion compound for Na-ion batteries. *Chemistry of Materials* 26 (3): 1297–1299.

150 Bianchini, F., Fjellvåg, H., and Vajeeston, P. (2017). First-principles study of the structural stability and electrochemical properties of Na_2MSiO_4 (M = Mn, Fe, Co and Ni) polymorphs. *Physical Chemistry Chemical Physics* 19 (22): 14462–14470.

151 Zhu, L., Zeng, Y.-R., Wen, J. et al. (2018). Structural and electrochemical properties of Na_2FeSiO_4 polymorphs for sodium-ion batteries. *Electrochimica Acta* 292: 190–198.

152 Perveen, T., Siddiq, M., Shahzad, N. et al. (2020). Prospects in anode materials for sodium ion batteries – a review. *Renewable and Sustainable Energy Reviews* 119: 109549.

153 Kume, T., Iwai, Y., Sugiyama, T. et al. (2013). NaSi and Si clathrate prepared on Si substrate. *Physica Status Solidi C* 10 (12): 1739–1741.

154 Malyi, O.I., Tan, T.L., and Manzhos, S. (2013). A comparative computational study of structures, diffusion, and dopant interactions between Li and Na insertion into Si. *Applied Physics Express* 6 (2): 027301.

155 Huang, S., Liu, L., Zheng, Y. et al. (2018). Efficient sodium storage in rolled-up amorphous Si nanomembranes. *Advanced Materials* 30 (20): e1706637.

156 Xiao, X., Li, X., Zheng, S. et al. (2017). Nanostructured germanium anode materials for advanced rechargeable batteries. *Advanced Materials Interfaces* 4 (6): 1600798.

157 Jung, S.C., Jung, D.S., Choi, J.W. et al. (2014). Atom-level understanding of the sodiation process in silicon anode material. *The Journal of Physical Chemistry Letters* 5 (7): 1283–1288.

158 Lu, X., Adkins, E.R., He, Y. et al. (2016). Germanium as a sodium ion battery material: in situ TEM reveals fast sodiation kinetics with high capacity. *Chemistry of Materials* 28 (4): 1236–1242.

159 Sarkar, S. and Peter, S.C. (2021). An overview on Sb-based intermetallics and alloys for sodium-ion batteries: trends, challenges and future prospects from material synthesis to battery performance. *Journal of Materials Chemistry A* 9 (9): 5164–5196.

160 Qian, J., Chen, Y., Wu, L. et al. (2012). High capacity Na-storage and superior cyclability of nanocomposite Sb/C anode for Na-ion batteries. *Chemical Communications* 48 (56): 7070–7072.

161 Darwiche, A., Marino, C., Sougrati, M.T. et al. (2012). Better cycling performances of bulk Sb in Na-ion batteries compared to Li-ion systems: an unexpected electrochemical mechanism. *Journal of the American Chemical Society* 134 (51): 20805–20811.

162 Shi, L. and Ding, J. (2021). Sodium storage performance of $FeSb_2$@C composite. *International Journal of Energy Research* 46 (2): 2081–2085.

163 Zheng, X.-M., You, J.-H., Fan, J.-J. et al. (2020). Electrodeposited binder-free Sb/NiSb anode of sodium-ion batteries with excellent cycle stability and rate capability and new insights into its reaction mechanism by operando XRD analysis. *Nano Energy* 77: 105123.

164 Dai, R., Wang, Y., Da, P. et al. (2014). Indirect growth of mesoporous Bi@C core-shell nanowires for enhanced lithium-ion storage. *Nanoscale* 6 (21): 13236–13241.

165 Sottmann, J., Herrmann, M., Vajeeston, P. et al. (2016). How crystallite size controls the reaction path in nonaqueous metal ion batteries: the example of sodium bismuth alloying. *Chemistry of Materials* 28 (8): 2750–2756.

166 Su, D., Dou, S., and Wang, G. (2015). Bismuth: a new anode for the Na-ion battery. *Nano Energy* 12: 88–95.

167 Gao, H., Ma, W., Yang, W. et al. (2018). Sodium storage mechanisms of bismuth in sodium ion batteries: an operando X-ray diffraction study. *Journal of Power Sources* 379: 1–9.

168 Wang, C., Wang, L., Li, F. et al. (2017). Bulk bismuth as a high-capacity and ultralong cycle-life anode for sodium-ion batteries by coupling with glyme-based electrolytes. *Advanced Materials* 29 (35): 1702212.

169 Chen, J., Fan, X., Ji, X. et al. (2018). Intercalation of Bi nanoparticles into graphite results in an ultra-fast and ultra-stable anode material for sodium-ion batteries. *Energy & Environmental Science* 11 (5): 1218–1225.

170 Ellis, L.D., Hatchard, T.D., and Obrovac, M.N. (2012). Reversible insertion of sodium in tin. *Journal of the Electrochemical Society* 159 (11): A1801–A1805.

171 Wang, J.W., Liu, X.H., Mao, S.X. et al. (2012). Microstructural evolution of tin nanoparticles during in situ sodium insertion and extraction. *Nano Letters* 12 (11): 5897–5902.

172 Komaba, S., Matsuura, Y., Ishikawa, T. et al. (2012). Redox reaction of Sn-polyacrylate electrodes in aprotic Na cell. *Electrochemistry Communications* 21: 65–68.

173 Sangster, J.M. (2009). Na-P (sodium-phosphorus) system. *Journal of Phase Equilibria and Diffusion* 31 (1): 62–67.

174 Yabuuchi, N., Matsuura, Y., Ishikawa, T. et al. (2014). Phosphorus electrodes in sodium cells: small volume expansion by sodiation and the surface-stabilization mechanism in aprotic solvent. *ChemElectroChem* 1 (3): 580–589.

175 Zhou, J., Liu, X., Cai, W. et al. (2017). Wet-chemical synthesis of hollow red-phosphorus nanospheres with porous shells as anodes for high-performance lithium-ion and sodium-ion batteries. *Advanced Materials* 29 (29): 1700214.

176 Qian, J., Wu, X., Cao, Y. et al. (2013). High capacity and rate capability of amorphous phosphorus for sodium ion batteries. *Angewandte Chemie International Edition* 52 (17): 4633–4636.

177 Kim, Y., Park, Y., Choi, A. et al. (2013). An amorphous red phosphorus/carbon composite as a promising anode material for sodium ion batteries. *Advanced Materials* 25 (22): 3045–3049.

178 Li, W.J., Chou, S.L., Wang, J.Z. et al. (2013). Simply mixed commercial red phosphorus and carbon nanotube composite with exceptionally reversible sodium-ion storage. *Nano Letters* 13 (11): 5480–5484.

179 Hou, H., Qiu, X., Wei, W. et al. (2017). Carbon anode materials for advanced sodium-ion batteries. *Advanced Energy Materials* 7 (24): 1602898.

180 Etacheri, V., Marom, R., Elazari, R. et al. (2011). Challenges in the development of advanced Li-ion batteries: a review. *Energy & Environmental Science* 4 (9): 3243.

181 Huang, J.-X., Csányi, G., Zhao, J.-B. et al. (2019). First-principles study of alkali-metal intercalation in disordered carbon anode materials. *Journal of Materials Chemistry A* 7 (32): 19070–19080.

182 Raccichini, R., Varzi, A., Passerini, S. et al. (2015). The role of graphene for electrochemical energy storage. *Nature Materials* 14 (3): 271–279.

183 Wen, Y., He, K., Zhu, Y. et al. (2014). Expanded graphite as superior anode for sodium-ion batteries. *Nature Communications* 5: 4033.

184 Jache, B. and Adelhelm, P. (2014). Use of graphite as a highly reversible electrode with superior cycle life for sodium-ion batteries by making use of co-intercalation phenomena. *Angewandte Chemie-International Edition* 53 (38): 10169–10173.

185 Yoon, G., Kim, H., Park, I. et al. (2017). Conditions for reversible Na intercalation in graphite: theoretical studies on the interplay among guest ions, solvent, and graphite host. *Advanced Energy Materials* 7 (2): 1601519.

186 Hu, M., Zhou, H., Gan, X. et al. (2018). Ultrahigh rate sodium ion storage with nitrogen-doped expanded graphite oxide in ether-based electrolyte. *Journal of Materials Chemistry A* 6 (4): 1582–1589.

187 Pendashteh, A., Orayech, B., Suhard, H. et al. (2022). Boosting the performance of soft carbon negative electrode for high power Na-ion batteries and Li-ion capacitors through a rational strategy of structural and morphological manipulation. *Energy Storage Materials* 46: 417–430.

188 Yao, X., Ke, Y., Ren, W. et al. (2018). Defect-rich soft carbon porous nanosheets for fast and high-capacity sodium-ion storage. *Advanced Energy Materials* 9 (6): 1803260.

189 Tsai, P.-C., Chung, S.-C., Lin, S.-K. et al. (2015). Ab initio study of sodium intercalation into disordered carbon. *Journal of Materials Chemistry A* 3 (18): 9763–9768.

190 Ding, J., Wang, H., Li, Z. et al. (2013). Carbon nanosheet frameworks derived from peat moss as high performance sodium ion battery anodes. *ACS Nano* 7 (12): 11004–11015.

191 Wahid, M., Puthusseri, D., Gawli, Y. et al. (2018). Hard carbons for sodium-ion battery anodes: synthetic strategies, material properties, and storage mechanisms. *ChemSusChem* 11 (3): 506–526.

192 Sun, N., Liu, H., and Xu, B. (2015). Facile synthesis of high performance hard carbon anode materials for sodium ion batteries. *Journal of Materials Chemistry A* 3 (41): 20560–20566.

193 Wang, P., Zhu, X., Wang, Q. et al. (2017). Kelp-derived hard carbons as advanced anode materials for sodium-ion batteries. *Journal of Materials Chemistry A* 5 (12): 5761–5769.

194 Zheng, P., Liu, T., and Guo, S. (2016). Micro-nano structure hard carbon as a high performance anode material for sodium-ion batteries. *Scientific Reports* 6: 35620.

195 Aravindan, V., Lee, Y.-S., Yazami, R. et al. (2015). TiO_2 polymorphs in 'rocking-chair' Li-ion batteries. *Materials Today* 18 (6): 345–351.

196 Liu, Y., Zhao, F., Li, J. et al. (2017). Influence of crystal phase on TiO_2 nanowire anodes in sodium ion batteries. *Journal of Materials Chemistry A* 5 (37): 20005–20013.

197 Yu, X., Pan, H., Wan, W. et al. (2013). A size-dependent sodium storage mechanism in $Li_4Ti_5O_{12}$ investigated by a novel characterization technique combining in situ X-ray diffraction and chemical sodiation. *Nano Letters* 13 (10): 4721–4727.

198 Senguttuvan, P., Rousse, G., Seznec, V. et al. $Na_2Ti_3O_7$: lowest voltage ever reported oxide insertion electrode for sodium ion batteries. *Chemistry of Materials* 23: 4109–4111.

199 Wu, C., Hua, W., Zhang, Z. et al. (2018). Design and synthesis of layered $Na_2Ti_3O_7$ and tunnel $Na_2Ti_6O_{13}$ hybrid structures with enhanced electrochemical behavior for sodium-ion batteries. *Advanced Science* 5 (9): 1800519.

200 Zhao, L., Pan, H.-L., Hu, Y.-S. et al. (2012). Spinel lithium titanate ($Li_4Ti_5O_{12}$) as novel anode material for room-temperature sodium-ion battery. *Chinese Physics B* 21: 028201.

201 Sun, Y., Zhao, L., Pan, H. et al. (2013). Direct atomic-scale confirmation of three-phase storage mechanism in $Li_4Ti_5O_{12}$ anodes for room-temperature sodium-ion batteries. *Nature Communications* 4 (1): 1870.

202 Liang, K., He, H., Ren, Y. et al. (2019). Porous lithium titanate nanosheets as an advanced anode material for sodium ion batteries. *Journal of Materials Science* 55 (10): 4372–4381.

203 Liang, K., Huang, X., Hong, X. et al. (2021). Sulfur and nitrogen-doped $Li_4Ti_5O_{12}$/rGO as an anode material for advanced sodium-ion batteries. *Journal of Alloys and Compounds* 857: 158190.

204 Chen, S., Zhou, R., Chen, Y. et al. (2017). Carbon-covered Fe_3O_4 hollow cubic hierarchical porous composite as the anode material for lithium-ion batteries. *Journal of Nanoparticle Research* 19: 1–13.

205 Lu, Y.C., Ma, C., Alvarado, J. et al. (2015). Electrochemical properties of tin oxide anodes for sodium-ion batteries. *Journal of Power Sources* 284: 287–295.

206 Li, K., Liu, H., and Wang, G. (2014). Sb_2O_3 nanowires as anode material for sodium-ion battery. *Arabian Journal for Science and Engineering* 39 (9): 6589–6593.

207 Hariharan, S., Saravanan, K., Ramar, V. et al. (2013). A rationally designed dual role anode material for lithium-ion and sodium-ion batteries: case study of eco-friendly Fe_3O_4. *Physical Chemistry Chemical Physics* 15 (8): 2945–2953.

208 Sun, W., Rui, X., Zhu, J. et al. (2015). Ultrathin nickel oxide nanosheets for enhanced sodium and lithium storage. *Journal of Power Sources* 274: 755–761.

209 Jian, Z., Liu, P., Li, F. et al. (2014). Monodispersed hierarchical Co_3O_4 spheres intertwined with carbon nanotubes for use as anode materials in sodium-ion batteries. *Journal of Materials Chemistry A* 2 (34): 13805–13809.

210 Zhang, X., Fu, C., Li, J. et al. (2017). MoO_3/reduced graphene oxide composites as anode material for sodium ion batteries. *Ceramics International* 43 (4): 3769–3773.

211 Wang, X., Liu, Y., Wang, Y. et al. (2016). CuO quantum dots embedded in carbon nanofibers as binder-free anode for sodium ion batteries with enhanced properties. *Small* 12 (35): 4865–4872.

212 Zhang, Z., Zhao, X., and Li, J. (2016). Facile synthesis of nanostructured MnO_2 as anode materials for sodium-ion batteries. *ChemNanoMat* 2 (3): 196–200.

213 Yu, X.Y. and David Lou, X.W. (2018). Mixed metal sulfides for electrochemical energy storage and conversion. *Advanced Energy Materials* 8 (3): 1701592.

214 Zhou, Q., Liu, L., Huang, Z. et al. (2016). Co_3S_4@polyaniline nanotubes as high-performance anode materials for sodium ion batteries. *Journal of Materials Chemistry A* 4 (15): 5505–5516.

215 Cao, Z., Song, H., Cao, B. et al. (2017). Sheet-on-sheet chrysanthemum-like C/FeS microspheres synthesized by one-step solvothermal method for high-performance sodium-ion batteries. *Journal of Power Sources* 364: 208–214.

216 Zhang, S., Zhao, H., Wu, M. et al. (2018). One-pot solvothermal synthesis 2D SnS_2/CNTs hybrid as a superior anode material for sodium-ion batteries. *Journal of Alloys and Compounds* 737: 92–98.

217 Zhu, X., Jiang, X., Liu, X. et al. (2017). Amorphous CoS nanoparticle/reduced graphene oxide composite as high-performance anode material for sodium-ion batteries. *Ceramics International* 43 (13): 9630–9635.

218 Qin, W., Chen, T., Lu, T. et al. (2016). Layered nickel sulfide-reduced graphene oxide composites synthesized via microwave-assisted method as high performance anode materials of sodium-ion batteries. *Journal of Power Sources* 302: 202–209.

219 Wang, Y.X., Seng, K.H., Chou, S.L. et al. (2014). Reversible sodium storage via conversion reaction of a MoS_2-C composite. *Chemical Communications* 50 (73): 10730–10733.

220 Hu, Z., Wang, L., Zhang, K. et al. (2014). MoS_2 nanoflowers with expanded interlayers as high-performance anodes for sodium-ion batteries. *Angewandte Chemie-International Edition* 53 (47): 12794–12798.

221 Qin, W., Li, D., Zhang, X. et al. (2016). ZnS nanoparticles embedded in reduced graphene oxide as high performance anode material of sodium-ion batteries. *Electrochimica Acta* 191: 435–443.

222 Chaturvedi, A., Edison, E., Arun, N. et al. (2018). Two dimensional TiS_2 as a promising insertion anode for Na-ion battery. *ChemistrySelect* 3 (2): 524–528.

223 Wang, X., Huang, J., Li, J. et al. (2016). Improved Na storage performance with the involvement of nitrogen-doped conductive carbon into WS_2 nanosheets. *ACS Applied Materials & Interfaces* 8 (36): 23899–23908.

224 Yu, Z., Li, X., Yan, B. et al. (2017). Rational design of flower-like tin sulfide @ reduced graphene oxide for high performance sodium ion batteries. *Materials Research Bulletin* 96: 516–523.

225 Kim, Y., Kim, Y., Park, Y. et al. (2015). SnSe alloy as a promising anode material for Na-ion batteries. *Chemical Communications* 51 (1): 50–53.

226 Ou, X., Yang, C., Xiong, X. et al. (2017). A new rGO-overcoated Sb_2Se_3 nanorods anode for Na^+ battery: in situ X-ray diffraction study on a live sodiation/desodiation process. *Advanced Functional Materials* 27 (13): 1606242.

227 Zhao, F., Shen, S., Cheng, L. et al. (2017). Improved sodium-ion storage performance of ultrasmall iron selenide nanoparticles. *Nano Letters* 17 (7): 4137–4142.

228 Zhang, Y., Liu, Z., Zhao, H. et al. (2016). $MoSe_2$ nanosheets grown on carbon cloth with superior electrochemical performance as flexible electrode for sodium ion batteries. *RSC Advances* 6 (2): 1440–1444.

229 Yang, X., Zhang, J., Wang, Z. et al. (2018). Carbon-supported nickel selenide hollow nanowires as advanced anode materials for sodium-ion batteries. *Small* 14 (7): 1702669.

230 Yang, F., Gao, H., Chen, J. et al. (2017). Phosphorus-based materials as the anode for sodium-ion batteries. *Small Methods* 1 (11): 1700216.

231 Liu, J., Kopold, P., Wu, C. et al. (2015). Uniform yolk–shell Sn_4P_3@C nanospheres as high-capacity and cycle-stable anode materials for sodium-ion batteries. *Energy & Environmental Science* 8 (12): 3531–3538.

232 Kim, Y., Kim, Y., Choi, A. et al. (2014). Tin phosphide as a promising anode material for Na-ion batteries. *Advanced Materials* 26 (24): 4139–4144.

233 Fullenwarth, J., Darwiche, A., Soares, A. et al. (2014). NiP_3: a promising negative electrode for Li- and Na-ion batteries. *Journal of Materials Chemistry A* 2 (7): 2050–2059.

234 Huang, Z., Hou, H., Wang, C. et al. (2017). Molybdenum phosphide: a conversion-type anode for ultralong-life sodium-ion batteries. *Chemistry of Materials* 29 (17): 7313–7322.

235 Luo, C., Xu, G.L., Ji, X. et al. (2018). Reversible redox chemistry of azo compounds for sodium-ion batteries. *Angewandte Chemie-International Edition* 57 (11): 2879–2883.

236 Armand, M., Grugeon, S., Vezin, H. et al. (2009). Conjugated dicarboxylate anodes for Li-ion batteries. *Nature Materials* 8 (2): 120–125.

237 Zhao, L., Zhao, J., Hu, Y.-S. et al. (2012). Disodium terephthalate ($Na_2C_8H_4O_4$) as high performance anode material for low-cost room-temperature sodium-ion battery. *Advanced Energy Materials* 2 (8): 962–965.

238 Wang, S., Wang, L., Zhu, Z. et al. (2014). All organic sodium-ion batteries with $Na_4C_8H_2O_6$. *Angewandte Chemie-International Edition* 53 (23): 5892–5896.

239 Castillo-Martinez, E., Carretero-Gonzalez, J., and Armand, M. (2014). Polymeric Schiff bases as low-voltage redox centers for sodium-ion batteries. *Angewandte Chemie-International Edition* 53 (21): 5341–5345.

240 Chen, L., Li, W., Wang, Y. et al. (2014). Polyimide as anode electrode material for rechargeable sodium batteries. *RSC Advances* 4 (48): 25369–25373.

241 Deng, W., Liang, X., Wu, X. et al. (2013). A low cost, all-organic Na-ion battery based on polymeric cathode and anode. *Scientific Reports* 3: 2671.

242 Zhang, S., Huang, W., Hu, P. et al. (2015). Conjugated microporous polymers with excellent electrochemical performance for lithium and sodium storage. *Journal of Materials Chemistry A* 3 (5): 1896–1901.

243 Wang, H., Hu, P., Yang, J. et al. (2015). Renewable-juglone-based high-performance sodium-ion batteries. *Advanced Materials* 27 (14): 2348–2354.

244 Li, K., Zhang, J., Lin, D. et al. (2019). Evolution of the electrochemical interface in sodium ion batteries with ether electrolytes. *Nature Communications* 10 (1): 725.

245 Ponrouch, A., Marchante, E., Courty, M. et al. (2012). In search of an optimized electrolyte for Na-ion batteries. *Energy & Environmental Science* 5 (9): 8572–8583.

246 Ponrouch, A., Dedryvère, R., Monti, D. et al. (2013). Towards high energy density sodium ion batteries through electrolyte optimization. *Energy & Environmental Science* 6 (8): 2361–2369.

247 Lin, Z., Xia, Q., Wang, W. et al. (2019). Recent research progresses in ether- and ester-based electrolytes for sodium-ion batteries. *InfoMat* 1 (3): 376–389.

248 Ponrouch, A., Monti, D., Boschin, A. et al. (2015). Non-aqueous electrolytes for sodium-ion batteries. *Journal of Materials Chemistry A* 3 (1): 22–42.

249 Monti, D., Jónsson, E., Palacín, M.R. et al. (2014). Ionic liquid based electrolytes for sodium-ion batteries: Na^+ solvation and ionic conductivity. *Journal of Power Sources* 245: 630–636.

250 Isa, K.B.M., Othman, L., Zainol, N.H. et al. (2013). Studies on sodium ion conducting gel polymer electrolytes. *Key Engineering Materials* 594–595: 786–792.

251 Vignarooban, K., Badami, P., Dissanayake, M.A.K.L. et al. (2017). Poly-acrylonitrile-based gel-polymer electrolytes for sodium-ion batteries. *Ionics* 23 (10): 2817–2822.

252 Zheng, J., Zhao, Y., Feng, X. et al. (2018). Novel safer phosphonate-based gel polymer electrolytes for sodium-ion batteries with excellent cycling performance. *Journal of Materials Chemistry A* 6 (15): 6559–6564.

253 Hueso, K.B., Armand, M., and Rojo, T. (2013). High temperature sodium batteries: status, challenges and future trends. *Energy & Environmental Science* 6 (3): 734–749.

254 Hayashi, A., Noi, K., Sakuda, A. et al. (2012). Superionic glass-ceramic electrolytes for room-temperature rechargeable sodium batteries. *Nature Communications* 3: 856.

255 Porkodi, P., Yegnaraman, V., Kamaraj, P. et al. (2008). Synthesis of NASICONsA molecular precursor-based approach. *Chemistry of Materials* 20: 6410–6419.

256 Chu, I.H., Kompella, C.S., Nguyen, H. et al. (2016). Room-temperature all-solid-state rechargeable sodium-ion batteries with a Cl-doped Na_3PS_4 superionic conductor. *Scientific Reports* 6: 33733.

257 Zhang, Z., Zhang, Q., Shi, J. et al. (2016). A self-forming composite electrolyte for solid-state sodium battery with ultralong cycle life. *Advanced Energy Materials* 7 (4): 1601196.

258 Lei, D., He, Y.B., Huang, H. et al. (2019). Cross-linked beta alumina nanowires with compact gel polymer electrolyte coating for ultra-stable sodium metal battery. *Nature Communications* 10 (1): 4244.

259 Eshetu, G.G., Elia, G.A., Armand, M. et al. (2020). Electrolytes and interphases in sodium-based rechargeable batteries: recent advances and perspectives. *Advanced Energy Materials* 10 (20): 2000093.

260 Hwang, J., Matsumoto, K., and Hagiwara, R. (2018). $Na_3V_2(PO_4)_3$/C positive electrodes with high energy and power densities for sodium secondary batteries with ionic liquid electrolytes that operate across wide temperature ranges. *Advanced Sustainable Systems* 2 (5): 1700171.

261 Chen, M., Hua, W., Xiao, J. et al. (2020). Development and investigation of a NASICON-type high-voltage cathode material for high-power sodium-ion batteries. *Angewandte Chemie-International Edition* 59 (6): 2449–2456.

262 Liu, T., Zhang, Y., Jiang, Z. et al. (2019). Exploring competitive features of stationary sodium ion batteries for electrochemical energy storage. *Energy & Environmental Science* 12 (5): 1512–1533.

263 Zhang, G., Ge, S., Xu, T. et al. (2016). Rapid self-heating and internal temperature sensing of lithium-ion batteries at low temperatures. *Electrochimica Acta* 218: 149–155.

264 Wang, C.Y., Zhang, G., Ge, S. et al. (2016). Lithium-ion battery structure that self-heats at low temperatures. *Nature* 529 (7587): 515–518.

265 Yang, C., Liu, X., Lin, Y. et al. (2023). Entropy-driven solvation toward low-temperature sodium-ion batteries with temperature-adaptive feature. *Advanced Materials* 35 (28): e2301817.

266 Vartanian, A. (2023). A battery electrolyte adapts to the cold. *Nature Reviews Materials* 8: 496.

267 Fang, Y., Xiao, L., Ai, X. et al. (2015). Hierarchical carbon framework wrapped $Na_3V_2(PO_4)_3$ as a superior high-rate and extended lifespan cathode for sodium-ion batteries. *Advanced Materials* 27 (39): 5895–5900.

268 Xu, Y., Wei, Q., Xu, C. et al. (2016). Layer-by-layer $Na_3V_2(PO_4)_3$ embedded in reduced graphene oxide as superior rate and ultralong-life sodium-ion battery cathode. *Advanced Energy Materials* 6 (14): 1600389.

269 Lin, X., Du, X., Tsui, P.S. et al. (2019). Exploring room- and low-temperature performance of hard carbon material in half and full Na-ion batteries. *Electrochimica Acta* 316: 60–68.

270 Ponrouch, A. and Palacín, M.R. (2015). On the high and low temperature performances of Na-ion battery materials: hard carbon as a case study. *Electrochemistry Communications* 54: 51–54.

271 Li, Z.P., Zhu, K.J., Liu, P. et al. (2021). 3D confinement strategy for dendrite-free sodium metal batteries. *Advanced Energy Materials* 12 (4): 2100359.

272 Lee, B.Y., Paek, E., Mitlin, D. et al. (2019). Sodium metal anodes: emerging solutions to dendrite growth. *Chemical Reviews* 119 (8): 5416–5460.

273 Ma, L., Cui, J., Yao, S. et al. (2020). Dendrite-free lithium metal and sodium metal batteries. *Energy Storage Materials* 27: 522–554.

274 Sun, B., Xiong, P., Maitra, U. et al. (2020). Design strategies to enable the efficient use of sodium metal anodes in high-energy batteries. *Advanced Materials* 32 (18): e1903891.

275 Lee, Y., Lee, J., Lee, J. et al. (2018). Fluoroethylene carbonate-based electrolyte with 1 M sodium bis(fluorosulfonyl)imide enables high-performance sodium metal electrodes. *ACS Applied Materials & Interfaces* 10 (17): 15270–15280.

276 Ruiz-Martínez, D., Kovacs, A., and Gómez, R. (2017). Development of novel inorganic electrolytes for room temperature rechargeable sodium metal batteries. *Energy & Environmental Science* 10 (9): 1936–1941.

277 Lee, J., Kim, J., Kim, S. et al. (2020). A review on recent approaches for designing the SEI layer on sodium metal anodes. *Materials Advances* 1 (9): 3143–3166.

278 Wang, S., Chen, Y., Jie, Y. et al. (2020). Stable sodium metal batteries via manipulation of electrolyte solvation structure. *Small Methods* 4 (5): 1900856.

279 Gao, L., Chen, J., Liu, Y. et al. (2018). Revealing the chemistry of an anode-passivating electrolyte salt for high rate and stable sodium metal batteries. *Journal of Materials Chemistry A* 6 (25): 12012–12017.

280 Luo, W., Lin, C.F., Zhao, O. et al. (2016). Ultrathin surface coating enables the stable sodium metal anode. *Advanced Energy Materials* 7 (2): 1601526.

281 Zhao, C., Liu, L., Qi, X. et al. (2018). Solid-state sodium batteries. *Advanced Energy Materials* 8 (17): 1703012.

282 Luo, W., Zhang, Y., Xu, S. et al. (2017). Encapsulation of metallic Na in an electrically conductive host with porous channels as a highly stable Na metal anode. *Nano Letters* 17 (6): 3792–3797.

283 Chi, S.S., Qi, X.G., Hu, Y.S. et al. (2018). 3D flexible carbon felt host for highly stable sodium metal anodes. *Advanced Energy Materials* 8 (15): 1702764.

284 Xiong, W.S., Jiang, Y., Xia, Y. et al. (2018). A robust 3D host for sodium metal anodes with excellent machinability and cycling stability. *Chemical Communications* 54 (68): 9406–9409.

285 Luo, J., Wang, C., Wang, H. et al. (2018). Pillared MXene with ultralarge interlayer spacing as a stable matrix for high performance sodium metal anodes. *Advanced Functional Materials* 29 (3): 1805946.

286 Ding, J., Zhang, H., Stephenson, T. et al. (2017). Exceptional energy and new insight with a sodium–selenium battery based on a carbon nanosheet cathode and a pseudographite anode. *Energy & Environmental Science* 10: 153–165.

287 Wang, Y.X., Yang, J., Lai, W. et al. (2016). Achieving high-performance room-temperature sodium-sulfur batteries with S@interconnected mesoporous carbon hollow nanospheres. *Journal of the American Chemical Society* 138 (51): 16576–16579.

288 Ye, H., Yin, Y.-X., Zhang, S.-F. et al. (2014). Advanced Se–C nanocomposites: a bifunctional electrode material for both Li–Se and Li-ion batteries. *Journal of Materials Chemistry A* 2 (33): 13293–13298.

289 Wang, Y.X., Zhang, B., Lai, W. et al. (2017). Room-temperature sodium-sulfur batteries: a comprehensive review on research progress and cell chemistry. *Advanced Energy Materials* 7 (24): 1602829.

290 Xin, S., Yu, L., You, Y. et al. (2016). The electrochemistry with lithium versus sodium of selenium confined to slit micropores in carbon. *Nano Letters* 16 (7): 4560–4568.

291 Yuan, B., Sun, X., Zeng, L. et al. (2018). A freestanding and long-life sodium-selenium cathode by encapsulation of selenium into microporous multichannel carbon nanofibers. *Small* 14 (9): 1703252.

292 Adelhelm, P., Hartmann, P., Bender, C.L. et al. (2015). From lithium to sodium: cell chemistry of room temperature sodium-air and sodium-sulfur batteries. *Beilstein Journal of Nanotechnology* 6: 1016–1055.

293 Li, Q., Liu, H., Yao, Z. et al. (2016). Electrochemistry of selenium with sodium and lithium: kinetics and reaction mechanism. *ACS Nano* 10 (9): 8788–8795.

294 Chen, Y.-M., Liang, W., Li, S. et al. (2016). A nitrogen doped carbonized metal–organic framework for high stability room temperature sodium–sulfur batteries. *Journal of Materials Chemistry A* 4 (32): 12471–12478.

295 Ding, J., Zhou, H., Zhang, H. et al. (2017). Selenium impregnated monolithic carbons as free-standing cathodes for high volumetric energy lithium and sodium metal batteries. *Advanced Energy Materials* 8 (8): 1701918.

296 Qiang, Z., Chen, Y.-M., Xia, Y. et al. (2017). Ultra-long cycle life, low-cost room temperature sodium-sulfur batteries enabled by highly doped (N,S) nanoporous carbons. *Nano Energy* 32: 59–66.

297 Ceylan Cengiz, E., Erdol, Z., Sakar, B. et al. (2017). Investigation of the effect of using Al_2O_3–nafion barrier on room-temperature Na–S batteries. *The Journal of Physical Chemistry C* 121 (28): 15120–15126.

298 Yu, X. and Manthiram, A. (2015). Ambient-temperature sodium-sulfur batteries with a sodiated nafion membrane and a carbon nanofiber-activated carbon composite electrode. *Advanced Energy Materials* 5 (12): 1500350.

299 Abouimrane, A., Dambournet, D., Chapman, K.W. et al. (2012). A new class of lithium and sodium rechargeable batteries based on selenium and selenium-sulfur as a positive electrode. *Journal of the American Chemical Society* 134 (10): 4505–4508.

300 Yao, Y., Zeng, L., Hu, S. et al. (2017). Binding $S_{0.6}$ $Se_{0.4}$ in 1D carbon nanofiber with C-S bonding for high-performance flexible Li-S batteries and Na-S batteries. *Small* 13 (19): 1603513.

301 Li, Z., Zhang, J., Lu, Y. et al. (2018). A pyrolyzed polyacrylonitrile/selenium disulfide composite cathode with remarkable lithium and sodium storage performances. *Science Advances* 4 (6): eaat1687.

302 Zhang, J., Fang, Y., Xiao, L. et al. (2017). Graphene-scaffolded $Na_3V_2(PO_4)_3$ microsphere cathode with high rate capability and cycling stability for sodium ion batteries. *ACS Applied Materials & Interfaces* 9 (8): 7177–7184.

5

All-Solid-State Batteries

Yixin Wu[1], *Yang Wang*[1], *Zhen Chen*[1], *and Minghua Chen*[2]

[1] *Harbin University of Science and Technology, School of Electrical and Electronic Engineering, No. 52 Xuefu Road, Harbin, 150080, China*
[2] *Harbin University of Science and Technology, School of Electrical and Electronic Engineering, Key Laboratory of Engineering Dielectric and Applications (Ministry of Education), No. 52 Xuefu Road, Harbin, 150080, China*

5.1 Introduction

Since their first commercialization in the 1990s, lithium-ion batteries (LIBs) have become an indispensable component of daily lives as high-performance energy storage devices. Over the past few decades, rapid progress has been made in the Li-ion battery technology, with applications ranging from portable electronic devices to electric vehicles (EVs) and grid energy storage [1, 2]. Although individual lithium-ion cells have sufficient reliability, their overall reliability significantly diminishes when they are employed collectively, such as in an EV power system consisting of thousands of cylindrical batteries. If a single cell experiences overheating, short circuits, or combustion, a hazardous chain reaction can be initiated that causes failures throughout the battery pack. This uncontrolled and rapid release of the stored energy generates substantial heat that is capable of causing combustion or even explosion. Consequently, these aspects necessitate higher energy density and safety demands for LIBs in EVs compared with their counterparts in portable electronic devices [3]. Frequent incidents of spontaneous combustion and explosions in EVs and energy storage power stations have raised significant public concerns regarding the inability of current state-of-the-art LIBs to meet stringent safety requirements [4–6].

At present, electrolytes based on carbonic acid esters and organic ethers are used in commercially available LIBs. These electrolytes have poor oxidation resistance, are flammable and prone to leaks, and react strongly with electrode materials, including lithium metal, silicon, carbon materials, and nickel-ternary-based oxides. Such properties result in significant safety concerns that not only hinder the further deployment of LIBs in large-scale applications but also restrict the use of high-energy-density electrode materials. Although the issue of range anxiety

Towards Next Generation Energy Storage Technologies: From Fundamentals to Commercial Applications, First Edition. Edited by Minghua Chen.

has been partially resolved through battery structure and shape design, conventional LIBs are gradually reaching their theoretical energy density limit, which presents significant challenges in meeting the increasing demand for high energy density [7–9].

To overcome this limitation, it is imperative to use materials with higher energy density and working voltage, such as high-nickel-ternary materials and lithium metal cathodes. In this context, using lithium metal anodes becomes highly advantageous due to their significantly higher theoretical specific capacity (3860 mAh g^{-1}) and reduced reduction potential (−3.04 V versus standard hydrogen electrode), in comparison with commercially available graphite [10, 11]. Substituting graphite with a Li metal anode is considered an effective method to increase the energy density by 40–50% at the cell level [12, 13]. Unfortunately, the initially commercialized lithium batteries developed by Moli Energy in the late 1980s had failed, leading to the callback of their cells from the market [14, 15]. This failure is primarily attributable to the inability of the liquid electrolyte and separators to effectively suppress the growth of lithium dendrites, thereby resulting in cell short circuits that posed thermal runaway risks and potential explosion hazards [16–18]. Among them, the high-nickel-ternary cathode has a higher working potential (up to 4.5 V) and specific capacity. However, the application of high-nickel-ternary materials also poses numerous challenges, including air sensitivity, electrolyte reactivity, cation mixing, lattice oxygen precipitation, transition metal ion migration, and performance degradation, as well as safety concerns arising from microcrack formation and expansion during cycling.

High-energy-density anodes and high-voltage cathodes are crucial components for enhancing the energy density of batteries while also imposing higher requirements on the electrolyte system. Compared with traditional electrolytes, solid-state electrolytes (SSEs) show superior characteristics such as limited flow and combustion, stronger mechanical properties that inhibit lithium dendrite growth, wider electrochemical stability windows that match with those of high-electric materials, internal battery stacking that simplifies packaging, and a broader operating temperature range. Therefore, solid-state batteries (SSBs) utilizing SSEs show excellent features of both high energy density and safety. Hence, the development of solid-state lithium batteries has become a general trend in order to meet the growing demand for comprehensive performance.

The unique advantages of SSEs can be summarized in three key points. First, their exceptional mechanical strength and superior electrochemical stability enhance the potential for the application of high-energy-density lithium metal anodes and high-potential cathodes, enabling the achievement of both safety and energy density simultaneously [19]. Second, the use of solid and gel electrolytes in batteries enables their widespread adoption. Unlike batteries that use liquid electrolytes, batteries with SSEs can be easily assembled by stacking the cells in series, in contrast to the complex external circuit required for the parallel stacking of liquid electrolytes. This method enhances both pack efficiency and volumetric energy density while reducing the design complexity associated with the battery management system. Moreover, due to the absence of liquid media, SSEs are easy to construct, thus with reduced costs [20]. Third, solid electrolytes are characterized by higher thermal stability and conductivity than their liquid counterparts, thereby minimizing heat generation in solid batteries and avoiding the requirement of cooling systems. As a result, the application of solid electrolytes leads to reduced accessory costs. Considering these benefits, the shift from liquid to solid electrolytes is inevitable across different applications.

5.2 Ion Transport Mechanism

A solid electrolyte is a type of ionic conductor. Ion migration in an electrolyte is driven by the chemical gradient ($\nabla \varphi_i$) and electrochemical gradient (∇c_i) in the electrochemical system. Its current density (j) can be expressed by the sum of all ionic fluxes:

$$j = F \Sigma_i z_i N_i \tag{5.1}$$

where F represents the Faraday constant and N_i represents the flux of substance i, whose charge number is z_i (assumed to be 1, ignored in subsequent calculations). Ion flux in an ideal electrolyte solution can be described by the Nernst–Planck equation, which expresses the flux of solute substances through diffusion, migration, and convection:

$$N_i = -D_i \nabla c_i - F \mu_i c_i \nabla \varphi_i + c_i u \tag{5.2}$$

where c_i represents the concentration of ion i, D_i represents the diffusion coefficient, μ_i represents the mobility, φ_i represents the electrolyte potential, and u represents the convection velocity of ion transport. When the Nernst–Planck Equation (5.2) is substituted into current density expression (5.1), the following equation can be obtained:

$$j = -F\Sigma_i D_i \nabla c_i - F^2 \nabla \varphi_i \Sigma_i \mu_i c_i + Fu\Sigma_i c_i \tag{5.3}$$

In solid electrolytes, the ion concentration gradient is small, and the diffusion and convection terms are negligible. Therefore, the conductivity (σ) of the electrolyte can be obtained by simplifying Eq. (5.4) as follows:

$$\sigma = -j/\nabla \varphi_i = F^2 \Sigma_i u_i c_i \tag{5.4}$$

The relationship between mobility and diffusion coefficient can be expressed as follows:

$$u_i = D_i/RT \tag{5.5}$$

As can be seen from Eq. (5.4), high ionic conductivity requires good ionic mobility and a sufficient number of dissociated ion pairs, both of which are indispensable. Therefore, when designing solid electrolytes with high ion conductivity, it is necessary to consider the strategies to promote the dissociation of lithium-ion pairs and reduce the resistance encountered in ion movement. In addition, although both cation and anion migration contribute to the total current, in most solid electrolytes, carriers that generate the transport current are primarily cations. Therefore, effective ionic conductivity generally refers to the migration rate of metal cations in a solid electrolyte.

5.2.1 Ion Transport of Inorganic Solid Electrolytes

Most inorganic solid-state electrolytes (ISSEs) have the periodic structure of coordination polyhedra, in which the migration and diffusion of lithium ions is carried out in the defects existing in the skeleton. These defects primarily include point defects, line defects, plane defects, volume defects, and electronic defects, among which point defects play an important role in the ion transport process of ISSEs. Typical point defects include the Schottky defect and Frenkel defect (Figure 5.1) [21].

In ISSEs, the transport mechanism of lithium ions chiefly includes the vacancy mechanism and gap mechanism. The vacancy mechanism is a large number of vacancies created by Schottky defects for lithium-ion migration. When a lithium ion migrates, a new vacancy is created in its original position, and this cycle is repeated, thus realizing the continuous transport of lithium ions. In the gap mechanism, ions diffuse through Frenkel defects, and lithium ions continuously displace adjacent effective sites so as to achieve diffusion between molecular frameworks. In addition, the lattice of the electrolyte may be distorted near the boundary and the interface, and lithium ions may be rearranged in accordance with the potential difference, generating additional interfacial charge transport paths [22].

5.2.2 Ion Transport in Solid Polymer Electrolytes

The ion conduction mechanism in of solid polymer electrolytes (SPEs) has been extensively investigated over the course of several decades. The proposed mechanism involves the hopping transfer of Li ions between sites within a single polymer chain or across different segments of the polymer (Figure 5.2a) [25]. Upon dissolution of Li salts in the polymer matrix, dissociated Li ions can coordinate with polar groups present in the polymer, such as —O—, —S—, and —C=N. Subsequently,

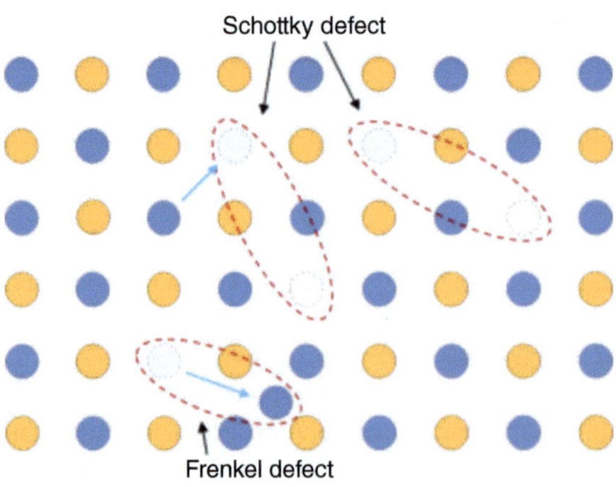

Figure 5.1 Schematic diagram of defects. Source: Reproduced with of permission [21]. © 2017/Elsevier.

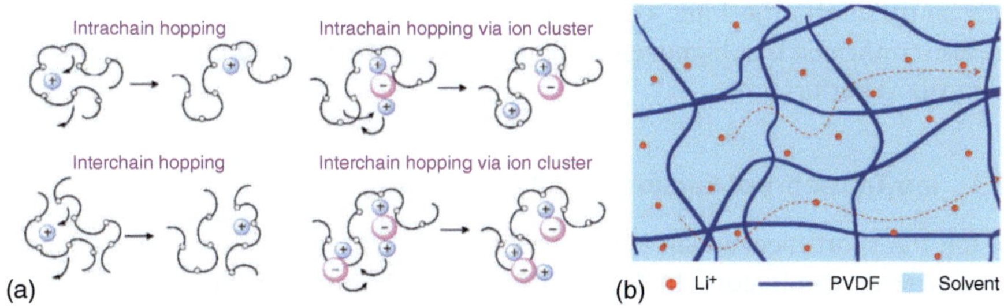

Figure 5.2 (a) Schematic of predominant Li-ion transport mechanisms in a coordinating polymer solid electrolyte. Source: Reproduced with permission from Xue et al. [23]. © 2015/Royal Society of Chemistry. Ion migration pathway in non-coordinating polymer PVDF-based (b) gel electrolytes. Source: Reproduced from Wu et al. [24]. © 2021/Elsevier.

under the influence of an external electrical field, Li ions show mobility by jumping between these coordinated sites. Consequently, ionic conductivity is directly influenced by the mobility of Li ions, which is determined by both the degree of dissociation of Li salts and their interaction with the polymer chain [26]. When the effect of segmental motion of polymer chains is sufficiently small to be negligible, the Arrhenius model could be used to explain the ion migration process which depends on the hopping mechanism (Eq. (5.6)):

$$\sigma = \sigma_0 \exp\left(\frac{-E_a}{kT}\right) \tag{5.6}$$

where σ_0, E_a, and k refer to the pre-exponential factor, activation energy, and Boltzmann constant, respectively. Ionic conductivity is inversely proportional to activation energy, where the latter is defined as the energy barrier between adjacent available sites [27].

Above glass transition temperature (T_g), ion transport primarily occurs in the amorphous region, while ionic conductivity is negligible in the crystal region. Lithium ions can move within the free volume of the polymer matrix, migrate from one coordination site to a new coordination site, or jump from one chain to another in a segment motion of the polymer.

Ion migration in homogeneous media such as conventional liquid electrolytes, ionic liquids, and amorphous polymers can be described using the Vogel–Tammann–Fulcher model [28]:

$$\sigma = \sigma_0 T^{-\frac{1}{2}} \exp\left(\frac{-E_a}{T - T_0}\right) \tag{5.7}$$

where E_a is the pseudo-activation energy associated with the polymer segmental motion and T_0 is the reference temperature, which normally falls 10–50 K below the experimental (kinetic) glass transition, T_g.

A gel polymer electrolyte (GPE) facilitates the rapid transport of lithium ions, similar to liquid electrolytes, while the polymer framework acts as the carrier for retaining the liquid electrolyte (Figure 5.2b). The liquid component also functions as a plasticizer, augmenting the mobility of polymer molecules. Solvent concentration significantly affects the ionic conductivity of gel electrolytes. Numerous studies have also confirmed that higher electrolyte uptake is often correlated with improved ion transport performance.

5.2.3 Ion Transport of Complex Solid Electrolytes

The ion transport mechanism is more complicated in polymer electrolytes doped with fast ionic conductors. In 1998, Croce et al. [29] studied the effect of nanoparticles on the ionic conductivity of polymer electrolytes based on polyethylene oxide (PEO), thus initiating the rapid progress in the application of nanoparticles in the design and preparation of composite solid electrolytes. Fillers in composite solid electrolytes include inert and active types, both of which can improve the ionic conductivity of polymer electrolytes [30].

Fan et al. [31] presumed that inert particles could improve the ionic conductivity of composite electrolytes, primarily due to the following two mechanisms. First, filler particles reduce the crystallinity of the polymer matrix, and the Lewis acid–base interaction between the particle surface and the polymer chain also inhibits the crystallinity of the polymer chain kinetically, thus enhancing the mobility of the polymer chain segment and increasing the mobility of lithium ions. Second, the Lewis acid–base interaction between filler particles and lithium salts promotes the dissociation of lithium salts and increases the concentration of mobile lithium ions. In addition, the interface layer formed between filler particles and the polymer can also be a path for fast ion transport. Compared with inert particles, active particles are more effective in enhancing ionic conductivity. In composite solid electrolytes, not only active particles play the role of inert particles, i.e. the Lewis acid–base interaction creates the lithium-ion transport path in the interface layer between the polymer and the particles, but also being lithium-ion conductors they can promote the transport of lithium ions. Therefore, research on active fillers has become a hot topic.

Zheng et al. [32] prepared a composite electrolyte using cubic-phase $Li_7La_3Zr_2O_{12}$ (LLZO) garnet and PEO and assembled a lithium symmetric battery. They monitored the replacement of ^7Li in the composite electrolyte by ^6Li in the metal electrode during the battery cycle using isotope labeling and nuclear magnetic resonance (NMR) spectroscopy to identify the transport path of lithium ions in the LLZO/PEO composite electrolyte. The experimental results showed that the LLZO ceramic phase is more conducive to the passage of lithium ions than the PEO/LLZO interface or PEO matrix phase. Chen et al. [33] prepared a series of PEO/garnet composite solid electrolytes using a solvent-free hot-pressing method. They found that as $Li_{6.4}La_3Zr_{1.4}Ta_{0.6}O_{12}$ (LLZTO) content increases, the composite solid electrolyte gradually transitions from a ceramic-in-form ceramic to a polymer-in-ceramic form. At the same time, the lithium conduction channel in the electrolyte is also transformed from the single-polymer lithium conduction channel to the polymer and LLZTO dual-phase lithium conduction channel (Figure 5.3).

Gupta et al. [34] investigated three possible transport paths for lithium ions in LLZTO- composite polymer electrolytes (CPE): (i) through the polymer matrix; (ii) through the polymer and inorganic phase interface; and (iii) through the

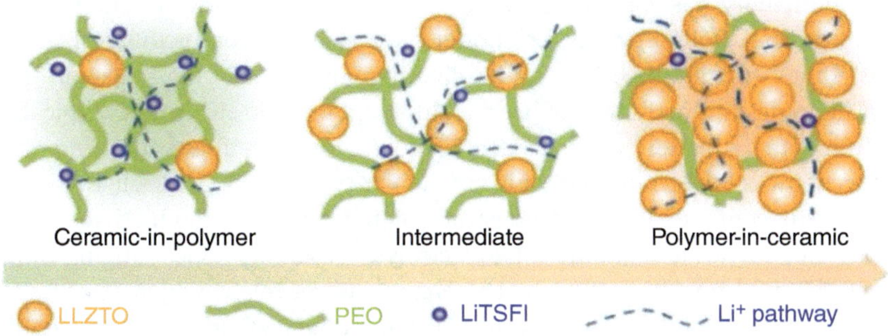

Figure 5.3 Transition of lithium-conducting channels in composite solid electrolytes with increasing LLZTO content. Source: Reproduced with permission from Chen et al. [33]. © 2017/Elsevier.

inorganic filler phase. They originally presumed that the ionic conductivity of the polymer was low and that the content of inorganic components in CPE was not high, so they assumed that path (ii) should be the main transport path of lithium ions. However, experimental results showed that the transmission impedance of path (ii) is higher than that of path (i), so path (i) is the primary ion transmission path in CPE. They believed that if the ion transport rate of path (ii) can be enhanced, the ionic conductivity of CPE can be greatly improved. The high interfacial impedance between PEO and LLZTO is the major factor affecting the conductivity of the CPE ion, while the impurities on the LLZTO surface and the difference in the concentration of lithium ions between PEO and LLZTO are the primary reasons for high interfacial impedance.

As discussed earlier, lithium-ion transport in CPE can take equally different routes through the polymer phase, inorganic phase, and inorganic/polymer interface. Due to the complex ion transport mechanism in the complex solid electrolyte, more experiments and simulations are needed to reveal the specific transport behavior. It is certain that the interface between the polymer chain and the filler plays a crucial role in creating a fast ion transport channel, and both electrolyte structure and the uniform dispersion of the filler are essential factors affecting the conductivity of CPE ions [35].

5.3 Key Performance Parameters

5.3.1 Ionic Conductivity

Ionic conductivity is a crucial parameter for evaluating the ion transport ability in electrolyte systems, and it plays a pivotal role in determining the electrochemical performance of SSBs. Ionic conductivity can be expressed as follows:

$$\sigma = \sum n_i q_i \mu_i \tag{5.8}$$

where σ is the ionic conductivity at a certain temperature (T), n_i is the number of carriers, q_i is the charge number of the carrier, and μ_i is the carrier mobility.

Ionic conductivity has a direct impact on the overall resistance of the battery system. In general, solid electrolytes should have an ionic conductivity of more than 10^{-4}S cm^{-1} to guarantee the regular charging and discharging behavior of a battery. The ionic conductivity of conventional liquid organic electrolytes typically ranges from approximately 10^{-3} to 10^{-2} S cm^{-1} at room temperature, whereas most SPEs show a lower range of ionic conductivity, generally between 10^{-6} and 10^{-5} S cm^{-1}. In contrast, ISSEs demonstrate higher levels of conductivity, with oxide solid electrolytes and sulfide solid electrolytes reaching values ranging from 10^{-5} to 10^{-4} S cm^{-1} and from 10^{-3} to 10^{-2} S cm^{-1}, respectively, which essentially satisfies the operational requirements for SSBs. To enhance the ionic conductivity of SPEs and improve the carrier transfer rate, it is essential to increase the dissociation capacity of lithium salts, increase the number of carriers, and decrease the crystallinity of the polymer matrix. Furthermore, the polymer electrolyte should be an electrical insulator for efficient ion transport and effective reduction in self-discharge.

5.3.2 Li$^+$ Transference Number

The Li$^+$ transference number, denoted as $t_{Li}{}^+$, is a fundamental factor for evaluating the proportion of charge transported by Li ions to the total charge passed in the electrolyte:

$$t_{Li}{}^+ = \mu_{Li}{}^+ / \left(\mu_{Li}{}^+ + \mu_{anion}\right) \tag{5.9}$$

where $\mu_{Li}{}^+$ and μ_{anion} are the mobility of Li$^+$ and anions, respectively.

Considering that ISSEs transport Li$^+$ exclusively through the lattice and gaps within the crystal structure and anions are unable to permeate, they can be regarded as single conductors for Li$^+$ without considering their $t_{Li}{}^+$. Therefore, $t_{Li}{}^+$ is commonly utilized as an assessment criterion for liquid electrolytes, polymer electrolytes, and composite electrolyte systems. Since anions do not take part in reversible electrochemical reactions, the charge transported by their motion does not add to the output charge of the battery. In addition to high ionic conductivity, a high Li$^+$ transference number is also critical for the operation of solid batteries, particularly at high current densities.

5.3.3 Electrochemical Stability Window

The compatibility of the electrolyte with positive and negative electrodes significantly affects battery performance. Due to thermodynamic limitations, irreversible reduction or oxidation of the electrolyte may occur at the negative or positive electrode, respectively, leading to a decreased coulombic efficiency and hindrance in charge–discharge processes. Therefore, expanding the electrochemical stability window of the electrolyte can lead to enhanced performance in terms of coulombic efficiency and capacity retention.

Generally, the oxidation potential of electrolytes should be above 4.5 V and stable for lithium metal. Oxide ceramic electrolytes usually have high oxidation resistance and are not oxidized by the positive electrode, but some of them are unstable for lithium; for example, sodium superionic conductor (NASICON)-type materials containing Ti^{4+} and Ge^{4+} are easily reduced by lithium, resulting in interface deterioration. Oxide ceramic electrolytes typically have exceptional oxidation resistance and remain unoxidized by the positive electrode. However, some ceramics like NASICON-type materials containing Ti^{4+} and Ge^{4+} are unstable for lithium and experience easy reduction, leading to interface deterioration. The electrochemical stability window of sulfide electrolytes is generally narrow. Therefore, additional interface treatment is necessary to ensure their long-term stability. Halide solid electrolytes generally have good oxidation resistance but are easily reduced by lithium. As for polymer electrolytes, in common PEO-based SPEs, ether linkages are oxidized when the working potential exceeds 4 V versus Li/Li^+. Polyester- and nitrile-based SPEs suffer from poor stability against electrochemical reduction when paired with a lithium metal anode [36]. Therefore, it is imperative to improve the electrochemical stability window of polymer electrolytes.

5.3.4 Mechanical Property

The mechanical property is a crucial parameter for solid electrolytes, as it directly influences the contact between the electrolyte and electrodes and the inhibition of the growth of lithium dendrites. Oxide ceramics have outstanding hardness and modulus, resulting in their hard and brittle properties and poor interfacial contact with electrodes. Moderate hardness and ductility are more conducive to the pressing of ceramic sheets and achieving superior interface contact. In polymer electrolytes, it is the mechanical property that determines the feasibility of producing a freestanding membrane. Using CPE as the separator eliminates the need for an additional separator, thereby mitigating issues related to incompatibility and ion transport resistance associated with using extra separators. Moreover, studies have reported that high mechanical strength effectively suppresses Li dendrite growth. As a consequence, high mechanical strength is desirable for CPEs. Usually, the stress–strain test is conducted to characterize the mechanical strength of CPEs. Stress continuously increases with an increasing tensile strength applied until the CPE breaks with a sharp loss of stress. Generally, CPEs show lower tensile stress but better strain than those of commercial separators, which is beneficial for applications in flexible batteries.

5.3.5 Thermal Property

Ceramic electrolytes are relatively stable at ambient temperatures, and their thermal properties are often used to describe polymer characteristics. Parameters such as melting temperature (T_m), glass transition temperature (T_g), and decomposition temperature (T_d) are important factors determining the performance of SPEs and CPEs. T_g corresponds to the transition of a polymer from a rubbery state to a glassy state; i.e. the segmental motions of polymer chains begin to occur above T_g. Therefore, a lower T_g of polymer electrolyte is generally correlated with higher ionic conductivity. Melting point (T_m) indicates the transition of a crystalline polymer from a long-range ordered state to a disordered state. The change in T_m is related to the molecular weight and crystallinity of the polymer. The degree of crystallinity is calculated by the ratio of the T_m peak area to that of the crystalline polymer. It provides useful information for the evaluation of CPEs besides assessing ionic conductivity. Currently, thermogravimetry (TG) and differential scanning calorimetry (DSC) are the popular techniques used to analyze these properties. The TG curve generally provides information on the evaporation or decomposition temperature and further helps determine the content of certain components in electrolytes. Wu et al. [37] investigated the thermal properties of polymer electrolytes using TG. Both the polymer and Celgard membranes, which had been pre-absorbed with the liquid electrolyte, showed weight loss starting at 65 °C, indicating the evaporation of the liquid phase. The gentler slope observed for the polymer membrane suggests a slower rate of liquid loss, thereby confirming its advantage in minimizing liquid-phase leakage. Additional thermal properties can be investigated from the DSC curve.

5.3.6 Electrolyte Uptake

The electrolyte uptake is a special parameter for GPEs. As Li-ion conduction in liquid is much faster than that in polymers, a higher electrolyte uptake usually means a higher Li-ion conductivity. To accommodate more electrolytes, the polymer matrix should have appropriate porosity. Pore size and distribution also play important roles. Good affinity between polymers and liquid electrolytes is beneficial for increasing the electrolyte uptake [38]. The liquid trapped in the polymer matrix can minimize the leakage and evaporation of the liquid and enhance battery safety. Measuring electrolyte uptake is straightforward. A polymer electrolyte is soaked in a liquid electrolyte for sufficient time and taken out, and then the residual electrolyte on the surface is removed [39]. The weight change of the polymer electrolyte after soaking presents the electrolyte uptake, as shown in Eq. (5.10):

$$\eta = (W_t - W_0)/W_0 \times 100\% \tag{5.10}$$

where W_0 and W_t are the weight of the polymer electrolyte before and after soaking, respectively. In some cases, electrolyte uptake can be as high as 500–700%. However, high electrolyte uptake usually deteriorates the mechanical strength of the film; thus, it is meaningful to make the compromise to obtain a more balanced electrolyte uptake [40].

5.4 Classification of Solid Electrolytes

Solid electrolytes are generally classified into ISSEs and organic polymer solid electrolytes. ISSEs primarily include oxides, sulfides, and halides. Oxide solid electrolytes primarily include NASICON type (such as $LiGe_2(PO_4)_3$), garnet type (such as LLZO), and perovskite type (such as $LiLaTiO_3$). Sulfide solid electrolytes are mostly in a glassy state (such as $20Li_2S$-$80P_2S_5$), glassy ceramic state

Table 5.1 Advantages and disadvantages of solid electrolytes (SEs).

SEs	Advantages	Disadvantages
Oxide	Wide electrochemical window	High interfacial impedance and high sintering temperature
Sulfide	High ionic conductivity and low interfacial impedance	Poor air stability and narrow electrochemical window
Halide	High oxidation stability and simple synthesis process	Poor reduction stability and hygroscopicity
Polymer	Processibility and flexibility	Low ionic conductivity and poor thermal stability

(such as $Li_7P_3S_{11}$), or crystalline state (such as $Li_{10}GeP_2S_{12}$). Halide electrolytes have a type of transition metal element cations with a high valence state introduced into a lithium halide LiX (X = Br, Cl, F) to form Li_a-M-X_b compounds. SPEs primarily include polyether-based electrolytes, PVDF-based electrolytes, polyester-based electrolytes, etc., presenting different characteristics and unique properties. The advantages and disadvantages of solid electrolytes in various systems are shown in Table 5.1.

Solid oxide electrolytes have high ion conductivity and good stability, but with high particle rigidity and poor contact. They show high ion conductivity only after high-temperature sintering into blocks and eliminating grain impedance. Sulfide solid electrolytes have the highest ionic conductivity, which can be comparable to that of liquid electrolytes, and the particles are relatively soft, which can show high ionic conductivity with simple powder cold pressing. However, most sulfide solid electrolytes can react with moist air, and their chemical stability is poor, making their preparation difficult. The ionic conductivity of halide solid electrolytes can reach $10^{-3}\ S\ cm^{-1}$ at room temperature, and the theoretical ionic conductivity can reach $10^{-2}\ S\ cm^{-1}$, but they are easily incompatible with lithium anodes. Polymer electrolytes have good interfacial compatibility and machinability, but compared with ISSEs, they have lower ionic conductivity and a narrower temperature range.

5.4.1 Inorganic Solid Electrolytes

In general, ISSEs are ionic conductors that depend on the concentration and distribution of defects [41]. ISSEs with high ionic conductivity should meet the following criterion: (i) the number of sites available for mobile ions to occupy should be much greater than that of all mobile ions; (ii) the migration energy barrier between adjacent available sites should be lower than that which allows ions to jump easily from one site to another; (iii) the available sites should be connected to each other, forming a continuous diffusion path; (iv) ISSEs should have a 3D solid framework with channels that preferably allow "molten" migrating ions without structure destruction; and (v) the anion should possess high polarizability to accommodate migrating ions through covalent bonds [42].

5.4.1.1 Oxide Solid Electrolytes

LISICON-type Electroltyes The crystalline structure of thio-lithium superionic conductor (thio-/LISICON) compounds is consistent with that of γ-Li_3PO_4, with an orthorhombic unit cell and *Pnma* space group where all cations are in a tetrahedral coordination [43]. Since the 1980s, a significant amount of research effort has been focused on LISICONs. As the first typical LISICON electrolyte, $Li_{14}Zn(GeO_4)_4$ was synthesized by H. Hong in 1978, consisting of Li_4GeO_4 and Zn_2GeO_4 forming intersolution, with the $[GeO_4]^{4-}$ anion group as the skeleton matrix. A part of Li^+ ions is involved in the structure in the form of an equilibrium charge, and three Li^+ ions fill the void in the skeleton structure, which constitutes ion transport performance. This architecture is similar to that of γ-Li_3PO_4, which has a PO_4 tetrahedron as a skeleton. The ionic conductivity at room temperature is only 10^{-6}~10^{-7} S cm^{-1}. Replacing the P(Ge) element with a low-priced element can increase the concentration of Li^+ in the structure and improve the ionic conductivity [44]. Subsequently, LISICON-type SSEs such as Li_4GeO_4–$LiVO_4$ systems, $Li_{2+2x}Zn_{1-x}GeO_4$, Li_2O–V_2O_5–SiO_2 systems, and Li_4SiO_4–Li_3PO_4 systems with higher ionic conductivity have been reported [45, 46].

In the crystal structure of *Pnma*, Li^+ is filled in the gaps of the PO_4 tetrahedron, and Li^+ can only be transported by the gap mechanism in this structure [47]. Generally, electrolytes with a γ-Li_3PO_4 structure can be called LISICON solid electrolytes. The $Li_{3+x}X_xY_{1-x}O_4$ solid solution formed by Li_4XO_4 and Li_3YO_4 (X = Si, Ge, Ti; Y = P, As, V, Cr) shows the structure of γ-Li_3PO_4. The solid solution formed by $LiSiO_4$–Li_3PO_4 has good chemical and electrochemical stability, and the room temperature conductivity of the intermediate component $Li_{3.5}Si_{0.5}P_{0.5}O_4$ is 3×10^{-6} S cm^{-1}, which is the highest among Li_4SiO_4–$LiPO_4$ series solid solutions. LISICON electrolytes are less reported because of their relatively low lithium-ion conductivity (10^{-7} S cm^{-1} at room temperature) and poor stability in contact with lithium and air.

In the following, we will discuss the origin of NASICON-type glass-ceramics electrolyte. The superionic conducting glass ceramics based on a NASICON structure are primarily glass ceramics containing lithium or sodium. Fu et al. [48, 49], based on previous studies, systematically studied the crystallizable titanium phosphate system for the first time and successfully synthesized NASICON-type glass-ceramic electrolyte. Among them, the conductive-phase $Li_{1+x}Al_xTi_{2-x}(PO_4)_3$ showed a conductivity of 1.3×10^{-3} S cm^{-1} at room temperature, with good Li_3N conductive performance. When Ti^{4+} (ionic radius 0.0605 nm) in the structure is replaced by Al^{3+} (ionic radius 0.0535 nm) in the lower state, in order to maintain charge balance, more Li^+ will be introduced and the volume of the cell will decrease. The cell parameters first shrink along the direction of the *c*-axis [50], forming a more favorable crystal structure for ion migration. The efficiency of ion conduction is improved. Similar doping effects also appear in the NASICON system introduced by Ga^{3+}, Sc^{3+}, and Y^{3+} [51, 52]. The NASICON solid solution allows a wide range of compositions. Therefore, the microcrystalline glass method has been applied to many precursor glass systems. Such materials can be classified as lithium-ion- and sodium-ion-conductive types. Compared with the electrolyte of lithium-ion-conductive microcrystalline glass, sodium-ion conductivity of the conductive system is relatively low, the temperature (300 °C) under the condition of the electrical conductivity of the corresponding glass being $10^{-3} - 10^{-4}$ S cm^{-1}. In view of its advantages in resource storage and raw material cost, this type of sodium-ion-conductive glass ceramics has good research value, and its performance can be improved by doping appropriate ions or optimizing preparation conditions.

Regarding the structure and transport mechanism of NASICON solid electrolytes, glass ceramics are primarily composed of conductive crystal phases, impurity phases, and residual glass matrixes. Based on their special structure, they can combine the advantages of glass and ceramics and achieve improved comprehensive performance. Therefore, the research on glass-ceramic solid electrolytes, especially the NASICON system, has aroused widespread attention among researchers. Based on its structure and composition, the precursor glass matrix corresponding to the NASICON glass-ceramic electrolyte is primarily a phosphate system or a silicate system. The phosphate system, due to its low melting and softening temperature, component flexibility, high thermal expansion coefficient, and excellent ionic conductivity, has been compared with the existing glass ceramics based on silicate and borosilicate. They have potential advantages. In recent years, there have been many studies on the application potential of crystalline phosphate systems as SSEs. Abo-Mosallam et al. [53] proposed doping Al_2O_3 with Fe_2O_3 in a lithium aluminum phosphate (LAP) phosphate system to study the crystallization characteristics of $12Li_2O$-$(18-x)$ Al_2O_3-xFe_2O_3-$70P_2O_5$ (mol%) (x = 0, 9, 18) components. Because the radius of Fe^{3+} is higher than that of Al^{3+}, the doping of Fe^{3+} to Al^{3+} can form a channel more conducive to the migration of Al^{3+}, resulting in improved conductivity. Luo et al. [54] prepared glass ceramics of a xFe_2O_3-$50Li_2O$-$10B_2O_3$-$(40-x)$ P_2O_5 (mol%) ($x = 0, 1, 2, 3, 4, 5$) system via two-step heat treatment. They found that the highest conductivity is achieved when $x = 4$, i.e. when the doping amount of Fe_2O_3 is 4 mol%. Moreover, the conductivity of the sample after crystallization is about 10 times higher than that of the corresponding glass, because the precipitation of an appropriate amount of highly conductive $Li_9Fe_3P_8O_{29}$ is conducive to improving the ion conduction efficiency. These results provide a good reference for phosphate glass ceramics as solid electrolyte materials.

NASICON solid electrolytes have high electrical conductivity and an isometric three-dimensional network structure, which are generally comparable to those of liquid electrolytes at higher temperatures [55]. NASICON-based glass ceramics are one of the most striking systems. The microcrystalline glass composite obtained from the precipitation of conductive NASICON phase as the matrix glass has high selectivity for network components and strong shape plasticity and can avoid the shortcomings of high porosity and morphology in ceramic systems. In recent years, researchers have developed and evaluated a wide range of microcrystalline glass-based NASICON-type solid electrolyte materials with excellent electrochemical, mechanical, and thermal properties. Studies on glass systems with high electrical conductivity have also been extensively carried out in the past decades [56], but in general, their electrical conductivity is lower than that of corresponding conductive crystalline phases. In a conductor glass system containing lithium ions or sodium ions, conductivity can only reach 10^{-3} S cm^{-1} at 300 °C, while in corresponding microcrystalline glass systems, conductivity with the same order of magnitude can be achieved at room temperature through crystallization heat treatment [57]. As mentioned earlier, some glasses can be converted into ceramic superionic electrolytes under an appropriate temperature regime. The obtained microcrystalline glass will have high conductivity, with no complex interparticle transfer interface in the glass state.

With respect to the crystal structure of NASICON-type conductors, Goodenough and Hong [58] reported in 1976 that they have excellent ion conduction properties, with the general formula $Na_{1+x}Zr_2Si_xP_{3-x}O_{12}$ ($0 \leq x \leq 3$). The frame structure in $NaZr_2(PO_4)_3$ where P^{5+} is partially replaced by Si^{4+} is composed of ordered skeleton rows, providing the channels required for sodium-ion migration in a three-dimensional space. $Na_{1+x}Zr_2Si_xP_{3-x}O_{12}$ ($0 \leq x \leq 3$) compounds generally have a rhombic hexahedral structure with a spatial group of $R\bar{3}c$ (Figure 5.4a). However, in the range of $1.8 \leq x \leq 2.2$, the crystal structure changes to the monoclinic system with space group *C2/c* (as shown in Figure 5.4b) [59, 60]. It is worth noting that Na is usually in Na1 (coordinate (0, 0, 0)) in the rhombic hexahedron structure. In a monoclinic system, the sodium ion is located in three positions: Na1 (coordinate (0.25, 0.25, 0.50)), Na2 (coordinate (0.50, 0.95, 0.25)), and Na3 (coordinate (0.83, 0.10, 0.70)). However, due to the small size of lithium ions, their free migration in a three-dimensional network structure in a quasi-system is difficult, resulting in low ionic conductivity. Aono et al. [61] studied the phase and ionic conductivity characteristics of a $Li_{1+x}M_xTi_{2-x}(PO_4)_3$ system in 1986 and found that the optimal conductivity group was divided into $Li_{1.3}Al_{0.3}Ti_{1.7}(PO_4)_3$ (M = Al, Sc). The maximum conductivity below 298 K can be up to 7×10^{-4} S cm^{-1}, suggesting that the body of the rhombic hexahedral structure is advantageous to lithium-ion conduction. However, for some systems containing large tetravalent ions, such as LiM_2

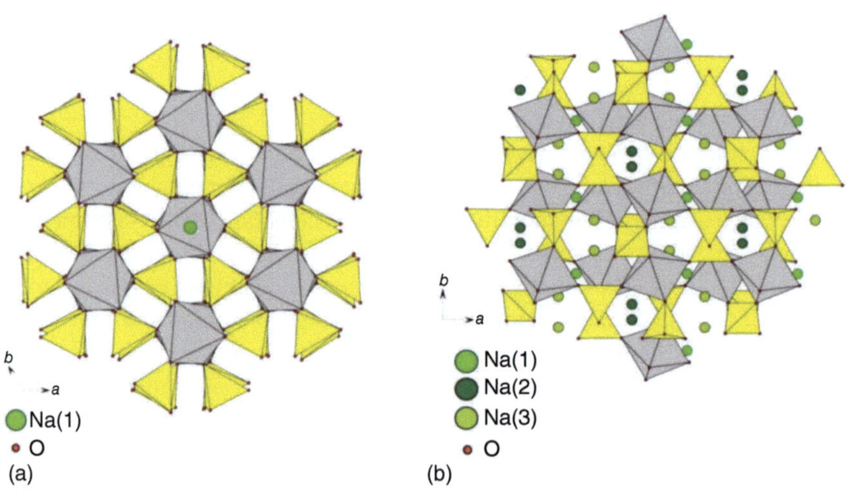

Figure 5.4 Projection profiles in the a–b plane: (a) rhombohedral $R\bar{3}c$ structure of $NaZr_2(PO_4)_3$ and (b) monoclinic *C2/c* structure of $Na_3Zr_2(SiO_4)_2(PO_4)$. Source: Reproduced with permission from Goodenough et al. [58]. © 1976/Elsevier.

$^{IV}(PO_4)_3$ (M = Zr, Hf, Sn), it is easy to distort into a low-symmetry triclinic system (C1) at low temperatures [62], and this low-symmetry structure is not conducive to the rapid conduction of lithium ions. Due to differences in components and preparation methods, sodium superionic structure can also be in a rhombic or trisomorphic system [63].

Currently, the most studied NASICON materials are $LiTi_2(PO_4)_3$-phase structures with $R\bar{3}c$ spatial groups. As shown in Figure 5.5a, the crystal structure of $LiTi_2(PO_4)_3$ consists of $[TiO_6]$ octahedrons and $[PO_4]$ tetrahedrons, which are interchangeably joined by co-angles to form a rigid skeleton $[Ti_2(PO_4)_3]$, providing a lithium-ion migration channel. In the NASICON structure, O occupies the Wyckoff site at (36f), P and Ti are located at the Wyckoff site at (18e) and (12c), respectively, and Li is primarily located at the (6b) site. The lithium ion is located between the two $[TiO_6]$ octahedrons only along the *c*-axis (Figure 5.5b) [64]. In most of the studies, the position where lithium ions are preferentially distributed and surrounded by six oxygen ions is called the M1 site. In addition, there are M2 sites, which are surrounded by ten oxygen ions and distributed symmetrically around three axes [65, 66]. It is worth noting that each M1 site is surrounded by six M2 sites at the same time; i.e. the number of M2 sites is three times that of M1 sites. In the parent compound $LiTi_2(PO_4)_3$, the M1 site is completely occupied, while the M2 site is usually empty. In the triclinic system, structural deformation drives the lithium ion to the more stable M12 site, which is intermediate between the M1 and M2 sites with a four-valent oxygen coordination [21, 67]. These sites are arranged alternately along ion migration channels, and

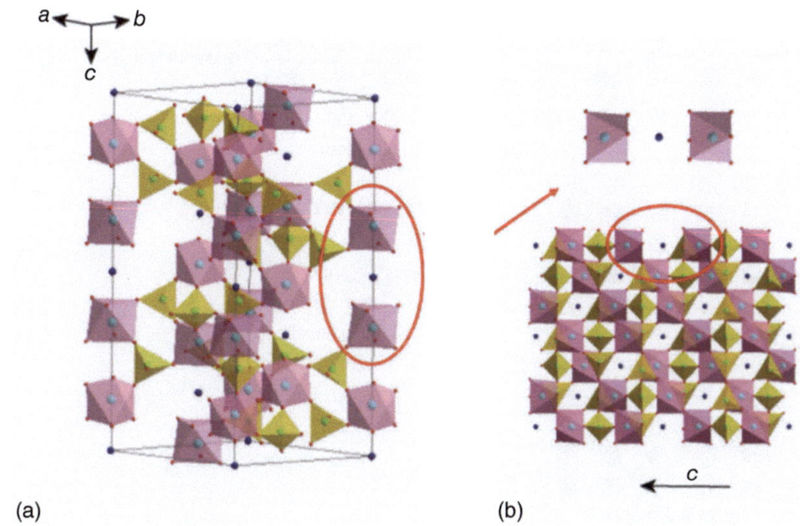

Figure 5.5 (a) Rhombohedral (R3c) structure of $LiTi_2(PO_4)_3$. Oxygen is shown in red. The purple octahedra and yellow tetrahedra correspond to $[TiO_6]$ and $[PO_4]$ units, respectively. Lithium ions are shown in blue at the M1 site with the Wyckoff site 6b. (b,c) Distribution of Li^+ at the M1 site along the *c*-axis. Source: Reproduced with permission Chunchun et al. [64]. © 2021/Elsevier.

ion migration in this structure is primarily achieved by jumping between the $[TiO_6]$ octahedron and the $[PO_4]$ tetrahedron over the obstacles of the network skeleton.

NASICON solid electrolyte consists of a network skeleton formed by anion groups and mobile lithium ions. The migration of lithium ions in the structure is realized through the movement of point defects, as shown in Figure 5.6a; i.e. the interaction between lithium ions and the fixed skeleton has a crucial effect on the migration mode of ions (including its migration channels and dimensions). Point defects in materials are primarily classified into Frankel defects (simultaneous existence

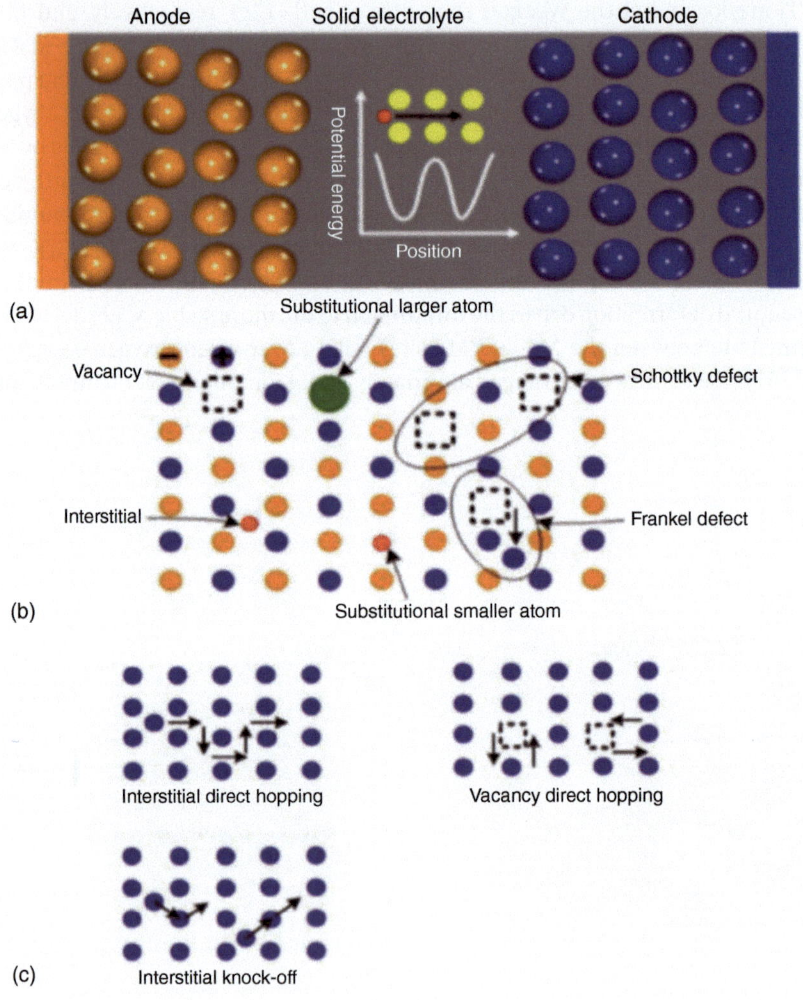

Figure 5.6 Schematic representations of (a) an all-solid-state lithium-ion battery, (b) typical point defects, and (c) mobile ion diffusion types. Source: Zhang et al. [21]. © 2017/Elsevier.

of vacancies and gap particles) and Schottky defects (proportional occurrence of positive and negative ion vacancies), i.e. intrinsic defects, whose characteristics are shown in Figure 5.6b. Under external forces, defects can be transferred directionally, thus realizing various transmission processes of lithium ions in materials. As shown in Figure 5.6c, the movement of thermal defects under the action of an external electric field realizes ionic conduction of crystals [21]. The Nernst–Einstein equation is shown in Eq. (5.11):

$$\sigma = \frac{nz^2e^2}{KT}D. \tag{5.11}$$

According to Eq. (5.11), the ionic conductivity of the crystal primarily depends on the number of thermal defects in the crystal (n) and the drift rate and diffusion coefficient of defects under the action of the electric field (D). In short, the conductivity of the material can be changed by controlling the concentration of defects and the influence of the ion migration barrier on the diffusion coefficient.

It is well known that interstitial location is closely related to ion diffusion in $Li_{1.3}Al_{0.3}Ti_{1.7}(PO_4)_3$ (LATP) materials. The conductivity of pure $LiTi_2(PO_4)_3$ (LTP) is relatively low ($10^{-5} - 10^{-8}$ S cm^{-1}), making it difficult to meet the standard of an ideal electrolyte. As shown in Figure 5.7a [68], the substitution of doped Ti^{4+} ions can significantly improve ionic conductivity. Specifically, partial substitution of Ti^{4+} with Al^{3+} can significantly reduce the length of the c-axis and further affect the a-axis through the gradual contraction of the AlO_6 octahedron along the c-axis, thus improving ion transport channels and achieving higher ionic conductivity than the parent matrix. The introduction of heterovalent ions, on the one hand, increases carrier concentration and, on the other hand, introduces ion migration channels with lower activation energy, which effectively improves ion conduction efficiency. In the doping system, a new site for Li+ ions is introduced to maintain charge balance. Different locations of lithium ions have been proposed in the literature, but in general, there will be a new site different from the M1 and M2 sites, a stable site with a lower activation energy for Li^+ ions to conduct. Neutron diffraction analysis showed that the M2′ site (Li2′) exists in the NASICON $Li_{1+x}M_xTi_{2-x}(PO_4)_3$-component structural model. Pérez-Estébanez et al. [71] discovered additional lithium ions at the M2′ site (36f, (0.47, 0.31, 0.25)), which is near the M2 site but is displaced from the center of the polyhedron along the b-axis. Arbi et al. [69] and Rettenwander et al. [50] explained that the lithium ion preferentially occupies the M1 site in the $LiTi_2(PO_4)_3$ sample, while occupying the M1 and M3 sites in an aluminum-doped sample (Figure 5.7b). The increase in the lithium content is beneficial for the distribution of M3 sites (0.07, 0.34, 0.07) along the ternary axis, thus reducing the electrostatic repulsion between Li1 and Li3 ions and being more conducive to ion migration. The ^7Li, ^{27}Al, and ^{31}P multiple magnetic angle rotation NMR (MAS-NMR) spectra of each component showed that the insertion of lithium resulted in the reduction of Ti^{4+} to Ti^{3+}, while Li^+ ions were located in the M2 cavity, which is the position of the trioxide surface. In $Li_3Al_xTi_{2-x}(PO_4)_3$ samples, the M3/M3′ of the four coordination sites near the M2 position of Li are (0.030, 0.319, 0.045) and (0.055, 0.373, 0.117), respectively.

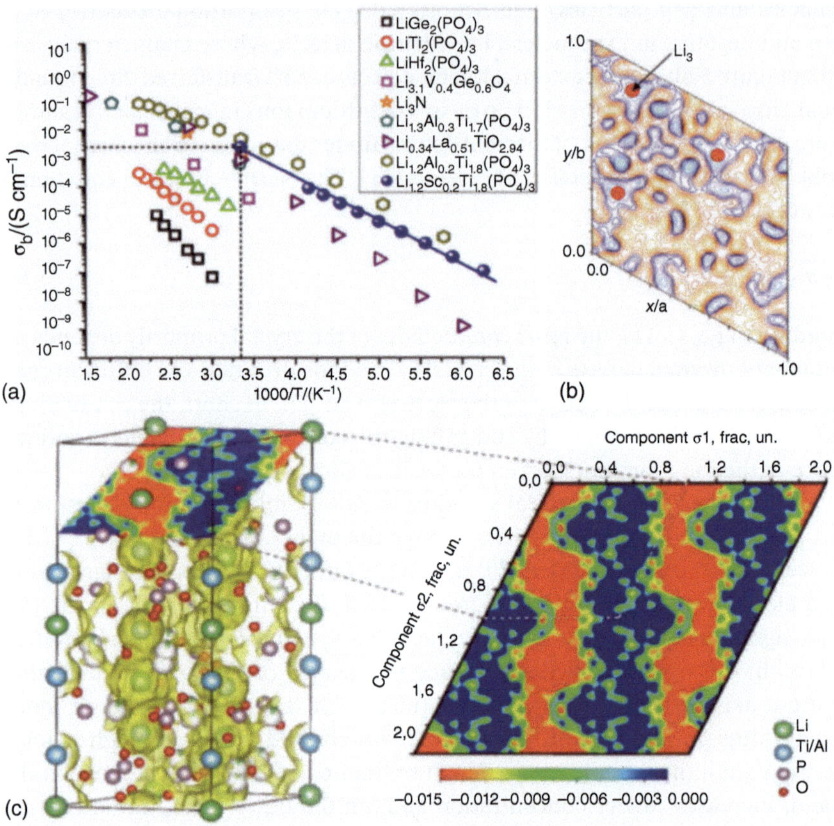

Figure 5.7 (a) Ion conductivities of NASICON-structured Li-ion solid electrolytes. Source: Reproduced with permission from Kahlaoui et al. [68]. © 2017/American Chemical Society. (b) Fourier map differences calculated at $z/c = 0$ from the ND patterns recorded at 5 K, obtained from Rietveld refinement data of $Li_{1+x}Al_xTi_{2-x}(PO_4)_3$. Source: Reproduced with permission from Arbi et al. [69]. © 2013/American Chemical Society. (c) MEM-reconstructed negative nuclear density maps and (0 1-4) sections of the bond–valence mismatch and negative nuclear densities in $Li_{1.3}Al_{0.3}Ti_{1.7}(PO_4)_3$. Source: Reproduced with permission from Monchak et al. [70]. © 2016/American Chemical Society.

Activation energy is a crucial factor affecting ionic conductivity and ion conduction rate. Monchak et al. [70] systematically studied the crystal structure of $Li_{1.3}Al_{0.3}Ti_{1.7}(PO_4)_3$ ionic conductors using a combination of neutron- and synchrotron-based high-resolution powder diffraction, as shown in Figure 5.7c. After reconstruction using the maximum entropy method (MEM), the lithium diffusion paths obtained by the differential bond valence theory and the negative kernel density map are compared, which shows the zigzag shape path of lithium diffusion in LATP, indicating that the migration of Li in LATP is primarily between Li1(6b) and Li3(36f) sites. The potential energy distribution curve along the Li1-Li3-Li3-Li1 chain corresponds to an activation barrier of 0.33 eV due to the electrostatic action around lithium ions. This value is consistent with the volume activation energy of LATP-based materials reported in the literature, which varies between 0.15 and 0.30 eV, replacing 0.41 eV of LTP materials.

The impedance of a NASICON lithium-ion electrolyte material is primarily attributable to the volume phase impedance and grain boundary impedance. In order to reduce the impedance and improve the conductivity, on the one hand, the synthesis technology is improved and, on the other hand, the electrolyte material is doped. The results show that the ionic conductivity of electrolytes can be enhanced by the following three aspects of doping modification: (i) changing the skeleton structure to adapt to the transmission of lithium ions; (ii) changing the concentration of lithium ions to make their migration in the anion skeleton easier; and (iii) increasing the density of electrolyte materials. For the NASICON lithium-ion electrolyte $Li[A_2P_3O_{12}]$, the A position can be doped by other elements and the P position can be doped or substituted to improve the conductivity of the electrolyte. In addition, some researchers have also achieved good results by co-doping two sites.

Doping Modification at Position A Site NASICON lithium-ion electrolyte A has usually Ti^{4+}, Zr^{4+}, Ge^{4+}, Sn^{4+}, Hf^{4+}, and other tetravalent cations, according to different elements in the A site, composed of a variety of electrolyte systems with different properties. However, ionic conductivity of such undoped electrolytes is low at room temperature, and cations such as Al^{3+}, Cr^{3+}, Sc^{3+}, Fe^{3+}, Y^{3+}, and Mg^{2+} are usually modified by doping in the position A site. Due to the excellent electrochemical performance of Al doping and the cheap availability of materials, studies on LATP, $Li_{1+x}Al_xZr_{2-x}(PO_4)_3$, and $Li_{1+x}Al_xGe_{2-x}(PO_4)_3$ (LAGP) electrolyte systems are widespread. The NASICON-type inorganic lithium-ion electrolyte material commonly used in the market is LATP, which has high total ionic conductivity ($10^{-4}\,S\,cm^{-1}$). It has been reported that the maximum total ionic conductivity in LATP at room temperature is $1.21 \times 10^{-3}\,S\,cm^{-1}$ [72]. However, when lithium metal is used as the negative electrode, Ti^{4+} is easily reduced to Ti^{3+}, which limits the application of LATP. In $LiZr_2(PO_4)_3$ Li-ion conductors, the stability of Zr^{4+} with lithium metal is considered an advantage in the assembly of Li metal batteries. However, the total conductivity of this type of electrolyte is low, only $10^{-8}\,S\,cm^{-1}$ at room temperature, so improving the total conductivity is the biggest challenge in this type of conductor [73]. LAGP has a high total ionic conductivity ($10^{-4}\,S\,cm^{-1}$) at room temperature, and its structure is relatively stable, which is expected to be used as solid electrolyte materials for LIBs.

Ortiz et al. [74] doped Fe^{3+} and Ca^{2+} in $LiTi_2(PO_4)_3$ and showed that partial substitution of Ti^{4+} with Fe^{3+} and Ca^{2+} not only increased rhombohedral cell parameters and reduced grain boundary resistance but also increased the Li^+ content, resulting in significantly improved electrical conductivity. Perez-Estebanez et al. [71] found that in $LiTi_2(PO_4)_3$ electrolytes, the largest contribution to the resistance was from grain boundary resistance. Even with minimal amounts of Cr^{3+}, Al^{3+}, or Fe^{3+}, Li^+ migration between the M1 and M2 sites was enhanced, significantly reducing grain boundary resistance. At room temperature, the total conductivity of three doped elements is of the following order: $Al^{3+} > Fe^{3+} > Cr^{3+}$.

Kothari et al. added trace Sc^{3+} into the LATP system via the high-temperature solid-phase method and showed that the electrostatic repulsion of Sc^{3+} on Li^+ could promote its migration from the M1 site to the M2 site, thus improving electrical conductivity. In the $LiZr_2(PO_4)_3$ electrolyte system, Xie et al. [75] synthesized $Li_{1.2}Zr_{1.9}Ca_{0.1}(PO_4)_3$ using the high-temperature solid-phase method and found that the electrolyte doped with Ca^{2+} can maintain the rhomboid phase at room temperature with a high total ionic conductivity ($4.9\times10^{-5}\,S\,cm^{-1}$). Li et al. [76] doped Y^{3+} and Er^{3+} in a $LiZr_2(PO_4)_3$ system, respectively, and successfully prepared solid electrolytes that can exist stably in the ring-cubic phase at room temperature via the high-temperature solid-phase method, which greatly improved ionic conductivity. In a $LiGe_2(PO_4)_3$ electrolyte system, Xu et al. [77] found that the total ionic conductivity increased by two orders of magnitude by Cr^{3+} doping, reaching $5.7\times10^{-4}\,S\,cm^{-1}$. Illbeigi et al. [78] obtained glass-ceramic mixture $Li_{1.5}Al_{0.4}Cr_{0.1}Ge_{1.5}(PO_4)_3$ using the melt-quenching technology and subsequent heat treatment and reported a total ionic conductivity of up to $6.65\times10^{-3}\,S\,cm^{-1}$ at room temperature and an activation energy of 0.29 eV. For NASICON-type lithium-ion electrolytes, appropriate doping modification can greatly improve ionic conductivity, and there are numerous elements that can be doped to modify the A position, which makes it a hot topic of research.

At present, there is little research on doping modification of the P position in NASICON lithium-ion electrolytes, but the substitution of phosphate groups can directly change the skeleton structure of electrolyte materials, thus changing the performance of electrolytes. As early as 1999, Best et al. [79] found that substituting anionic groups can simultaneously affect the electrolyte's bulk and grain boundary resistance, thus affecting the conductivity of the electrolyte. Geng et al. [80] partially replaced PO_4^{3-} in $Li_3SC_2(PO_4)_3$ with VO_4^{3-} and reported an increase of two orders of magnitude in the electrolyte's total ionic conductivity at room temperature, reaching $8.41\times10^{-6}\,S\,cm^{-1}$. Fu et al. [81] added SiO_2 into LATP and $Li_{1.5}Cr_{0.5}Ti_{1.5}(PO_4)_3$ (LCTP), respectively, and synthesized glass-ceramic electrolyte using the melt-quenching technology. The total conductivity at room temperature is as high as 10^{-3} and $10^{-2}\,S\,cm^{-1}$, comparable to that of some liquid electrolytes. These studies show that doping or replacing P sites with suitable elements may also be an effective means to improve conductivity.

Given the obvious influence of the doping of A and P sites on the electrochemical performance of NASICON lithium-ion conductors, some researchers have studied the co-doping of two sites. Xu et al. [56] studied the double-doped electrolyte system $Li_{1+2x+2y}Al_xZn_yTi_{2-x-y}Si_xP_{3-x}O_{12}$ and showed that the proper introduction of Zn^{2+} was beneficial to Li^+ migration. They reported that when $x = 0.1$ and $y = 0.5$, the electrolyte had the highest total ionic conductivity of $7.2\times10^{-5}\,S\,cm^{-1}$ at room temperature and a conductance activation energy of only $20\,kJ\,mol^{-1}$ (0.2 eV). In addition, there are numerous clays and minerals in nature that often contain elements such as Al, Ca, Fe, Mg, and Si, and the introduction of these materials may improve the electrochemical performance of NASICON lithium-ion electrolytes. Lu et al. [82] doped 10% SBA-15 molecular sieve powder into a LAGP electrolyte and reported a

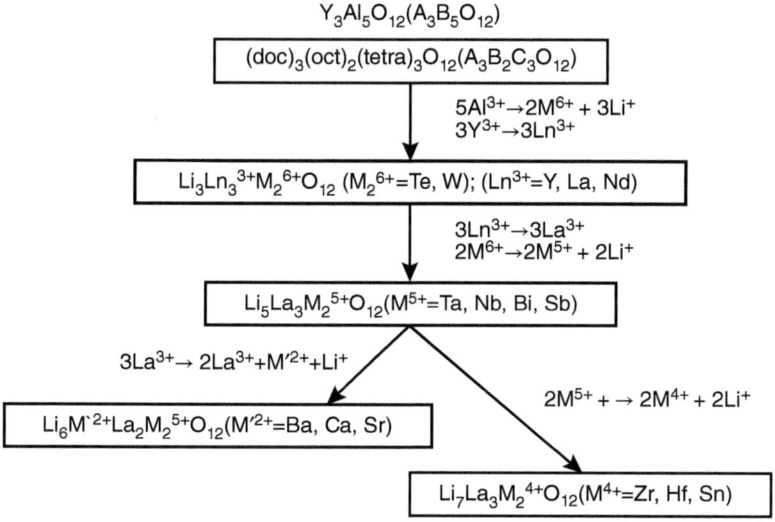

Figure 5.8 Schematic diagram of evolution of garnet type solid electrolyte family.

smaller particle size of the electrolyte than that of LAGP and an total ionic conductivity improved by one order of magnitude.

Garnet-type Electrolyte Figure 5.8 shows the evolution of the garnet solid electrolyte family [83]. When the A site is occupied by Li^+, it serves as a pathway for the transport of lithium ions, thereby exhibiting lithium ion conduction ability. In 2003, Thangadurai et al. discovered a new type of lithium-rich garnet solid electrolyte $Li_5La_3M_2O_{12}$ (M = Ta, Nb), whose ionic conductivity is about 10^{-5} S cm^{-1}, which is suitable for high-voltage cathode materials [84]. In 2005, Thangadurai and Weppner et al. [85] found that the ionic conductivity of the $Li_6BaLa_2Ta_2O_{12}$ solid electrolyte is about 4.0×10^{-14} S cm^{-1} and the interface contact with the electrode material is stable. In 2007, Thangadurai and Murugan et al. [86] obtained the cubic-phase $Li_7La_3Zr_2O_{12}$ (LLZO), whose ionic conductivity at room temperature is about 4×10^{-4} S cm^{-1} with high stability. Awaka et al. [87] obtained tetragonal-phase $Li_7La_3Zr_2O_{12}$ at the sintering temperature of 980 °C. The conductivity of this material was lower than that of cubic-phase LLZO, and the conductivity of the lower body phase at room temperature was only 1.63×10^{-6} S cm^{-1}. A large number of studies have shown that the formation of tetragonal-phase LLZO can be inhibited using Al_2O_3 crucibles during the synthesis of LLZO materials or actively adding a small amount of Al_2O_3 during the sintering process to format a stable cubic LLZO products. Thompson et al. [88] compared tetragonal and cubic LLZO materials and proposed that 0.4~0.5 Li holes per molecule is a necessary condition to stabilize cubic materials.

Crystal Structure The traditional garnet electrolyte is a type of orthosilicate, with the general structure $A_3B_2SiO_{12}$ (A = Ca, Mg; B = Al, Cr, Fe). The material structure is composed of an AO_8 dodecahedron, BO_6 octahedron, and SiO_4 tetrahedron, with a face-centered cubic structure and *Ia*3-*d* space group. By substituting Si in

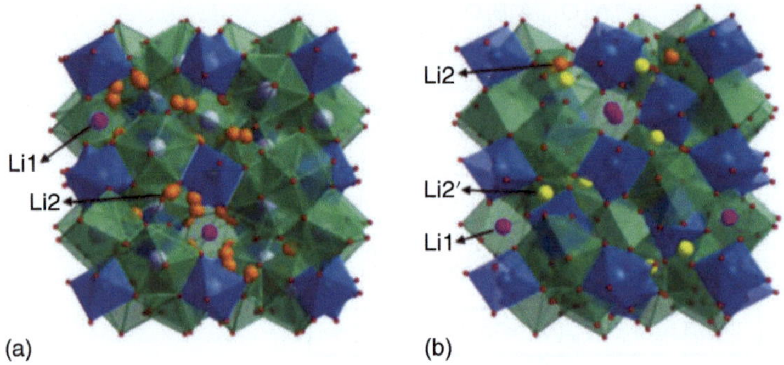

Figure 5.9 Two crystal structures of LLZO. (a) Cubic. (b) Tetragonal. Source: Reproduced with permission from Wang et al. [90]. © 2020/American Chemical Society.

$A_3B_2Si_3O_{12}$, a garnet oxide solid electrolyte was obtained, with the chemical formula $A_3B_5O_{12}$ (A = Mg, Ca, Ln, La, Y; B = Fe, Ga, Al, Ge, Mn, Ni, V) [89].

Garnet-type electrolytes can be classified into four subtypes: Li3, Li5, Li6, and Li7. LLZO is a Li7-type $Li_7La_3X_2O_{12}$ (X = Zr, Sn, Ta). Compared with the other three types (i.e. Li3-type $Li_3Ln_3Te_2O_{12}$ (Ln = Y, Pr, Nd, Sm–Lu), Li5-type $Li_5La_3M_2O_{12}$ (M = Nb, Ta), and Li6-type $Li_6ALa_2M_2O_{12}$ (A = Mg, Ca, Sr, Ba)), the Li7-type LLZO has 1 order of magnitude higher ionic conductivity (25 °C, 3×10^{-4} S cm^{-1} versus 10^{-5} S cm^{-1}). LLZO has two crystal forms: tetragonal phase and cubic phase (Figure 5.9) [90]. Both these phases have the same skeleton structure, and the difference between them lies in the distribution of lithium ions in the structure. The tetragonal-phase space group is *I*41/*acd* (No.142) with lattice constants $a = b = 13.134$ Å and $c = 12.663$ Å. In the tetragonal LLZO crystal structure, ions occupy tetrahedral gap Li1(8a), octahedral gap Li2(16f), and octahedral eccentric gap Li3(32g). In the tetragonal-phase LLZO, the three gap positions are all occupied by Li^+, thus showing a high degree of orderliness. Nevertheless, the ionic conductivity of LLZO with this highly symmetric structure is approximately one to two orders of magnitude lower than that of the cubic phase LLZO with an asymmetric structure.

The space group of the cubic-phase LLZO is la-3d(No.230) with lattice constant $a = 12.9682$ Å. The ion part fills the Li1(24d), Li2(48g), and Li3(96h) locations, where 24d is the tetrahedral gap location, 48g is the octahedral gap location, and 96h is the two eccentric locations of the octahedron. Because the charges repel each other, the adjacent 96h positions cannot be occupied simultaneously. According to the symmetry, there are 24 Li1 positions, 48 Li2 positions, and 96 Li3 positions in a single LLZO cell. There are eight $Li_7La_3Zr_2O_{12}$ in a single cell, namely 56 lithium ions, which are randomly distributed in the gap between tetrahedron and octahedron, resulting in various randomly distributed potassium ion spatial arrangement structures.

Due to the random distribution of Li in the LLZO structure of the cubic phase and a large number of Li vacancies, the Li^+ migration capacity of the cubic phase is higher than that of the tetragonal phase. The distribution of Li^+ in the LLZO structure (ordered arrangement, random distribution) is an important factor leading to the significant difference between the cubic phase and the tetragonal phase of

LLZO. The ionic conductivity of the cubic-phase LLZO (10^{-4} S cm^{-1}) is 2 orders of magnitude higher than that of the tetragonal phase (10^{-6} S cm^{-1}). Meier et al. [91] used first-principles calculations to obtain the migration modes of Li^+ in cubic and tetragonal phases LLZO. Their results showed that the LLZO ion migration mode is synchronous migration with an activation energy of 0.4 eV. However, in the cubic phase, the Li distribution mode is partially occupied, and there are many random Li vacancies in the crystal structure village, showing a mixture of co-migration and independent migration to the vacancy, and the activation energy is only 0.1–0.3 eV. Due to the difference in the transition ability of the tetragonal-phase and cubic-phase LLZO, the cubic-phase LLZO with a higher ionic conductivity is preferred to be synthesized in the preparation process.

The element doping process is simple, less expensive, and easy to operate and is one of the most effective and common methods to improve the ionic conductivity of garnet solid electrolyte. By selecting other elements with different ionic valence states and radii to partially dope LLZO with Li, La, Zr, and O sites, more lithium vacancies are introduced to make lithium arrangement more disordered, and finally, the disordered cubic phase is obtained, thus improving ionic conductivity. At present, the lithium vacancies required to stabilize the cubic phase range from 0.125 to 0.500 mol per unit. The density function theory is used to calculate the Li-doped elements, such as Al^{3+}, Ga^{3+}, and Fe^{3+}. To maintain charge balance in the lattice, replacing Li^+ with Al^{3+} will generate two lithium vacancies simultaneously, which improves ionic conductivity to some extent. However, the doped Al^{3+} will enter the Li^+ channel, blocking its transmission path and reducing the ionic conductivity. Therefore, the doping amount should be controlled reasonably. Neutron diffraction studies show that the lithium content in the cubic-phase garnet crystal is up to 7.5, and when the lithium content is 6.4 ± 0.1, the conductivity of Li^+ is the highest. Guo et al. found that when Al and Ga are used to dope Li sites, the optimal doping amount is $0.25 \sim 0.3$, and the ionic conductivity of the synthesized pure c-LLZO at room temperature exceeds 10^{-4} S cm^{-1} [92]. At present, La-doped elements include Ca^{2+}, Ba^{2+}, Mg^{2+}, Sr^{2+}, etc. Yang et al. successfully synthesized Mg-doped cubic-phase LLZO at 1000 °C, with an ionic conductivity up to 5.56×10^{-5} S cm^{-1} at room temperature and a sintering temperature about 230 °C lower than that of undoped Mg. In addition, the transmittance of LLZO doped with Mg to visible light at $500 \sim 800$ nm can reach $10\% \sim 12\%$, which provides a reference for the future application of transparent solid electrolytes [93]. However, La site doping can change the size of Li^+ migration channel to some extent, thus changing the migration path of Li^+ and affecting the transmission rate of Li^+, so there are few studies on La site doping at present. In addition, element doping can also be carried out at the Zr site, such as Mo^{6+}, Nb^{5+}, Ga^{3+}, and Ta^{5+}. The valence state and radius of doping ions have a great influence on the performance of LLZO. Doping of some high-valence metal ions can increase the concentration of lithium vacancy in the crystal structure. Li^+ is more likely to migrate and can also stabilize the cubic-phase structure. Huang et al. [94] applied Sr^{2+} and Mo^{6+} co-doping to successfully prepare $Li_{6.65}La_{2.95}Sr_{0.05}Zr_{1.8}Mo_{0.2}O_{12}$ at 1100 °C, with an ionic conductivity of up to 6.43×10^{-4} S cm^{-1} and a relative density of 95%. Among them, Mo^{6+} doping can effectively guarantee the formation of the cubic phase. Sr^+ doping can improve the ionic conductivity and relative density of samples. The F element has a higher electronegativity than O and a stronger ionic bond with La and Zr, which is more conducive to the conduction and migration of Li^+. Pure c-LLZO can be obtained by proper fluoride doping, and the inner grain boundary of the electrolyte can be eliminated to obtain a dense microstructure and improve ionic conductivity [95].

5.4.1.2 Sulfide Solid Electrolytes

Compared with oxygen ions, sulfur ions are less electronegative and less bound to lithium ions. At the same time, the large radius of sulfur ions makes the transmission channel of lithium ions in the crystal structure wider, which is conducive to the movement of lithium ions. Thus, sulfide solid electrolytes have the highest ionic conductivity of the three electrolytes. Therefore, sulfide solid electrolytes have become a research hotspot for all kinds of enterprises and institutions represented by Toyota. The evolution of sulfide solid electrolytes is shown in Figure 5.10 [96], which indicates the invention and ionic conductivity values of several important sulfide solid electrolytes.

"Sulfide solid electrolyte" is a general term for a class of solid electrolytes with different crystal structures. In the 1980s and 1990s, glassy sulfide solid electrolytes represented by Li_2S-P_2S_5 were developed. After years of development, although their ionic conductivity could be increased to 6.9×10^{-4} S cm^{-1} by doping, they could not meet the demand of use. To further improve the conductivity, a thio-LISICON structure Li_2S-P_2S_5 element sulfide solid electrolyte material was proposed in 2001, which started the development of high-conductivity sulfide solid electrolytes. In 2005, the $Li_7P_3S_{11}$ sulfide solid electrolyte material with a higher room temperature ionic conductivity (3.2×10^{-3} S cm^{-1}) was discovered, which laid a solid foundation for the practical application of sulfide solid electrolytes. Driven by the development of solid electrolytes with the high ionic conductivity Li_7P_3S11 structure, researchers have subsequently developed sulfide solid electrolyte materials such as $Li_{10}GeP_2S_{12}$ terpolymer solid electrolytes and Li_6PS_5X (X = Cl, Br, I) solid electrolytes with high ionic conductivity. Moreover, the sulfide solid electrolyte with the highest ionic conductivity is $Li_{9.54}Si_{1.74}P_{1.44}S_{11.7}Cl_{0.3}$, which up to 2.5×10^{-2} S cm^{-1}, which indicates that a great potential for their application.

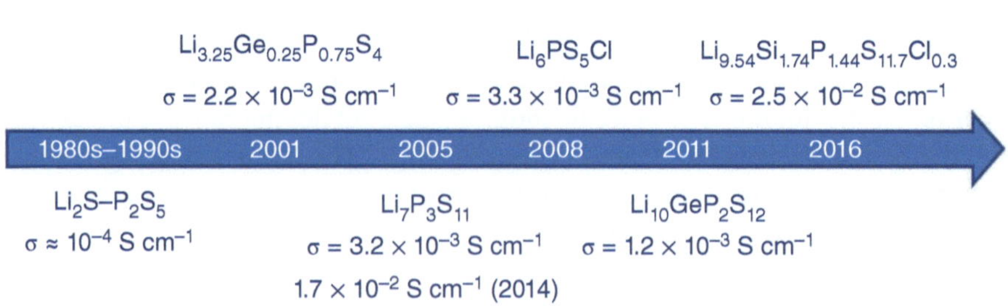

Figure 5.10 Evolution of sulfide solid electrolytes. Source: Reproduced with permission from Zhang and Wei [96]. © 2022/Elsevier.

Sulfide electrolytes can be classified into three categories according to their crystalline states: glass (amorphous), crystal, and glass-ceramic. Glass electrolytes are favored for their isotropic ion conduction, zero grain boundary resistance, and low preparation cost. The ionic conductivity of glass electrolytes can reach about $10^{-4}\,S\,cm^{-1}$ at room temperature. The ionic conductivity of glass-ceramic electrolytes obtained from the crystallization of glassy electrolytes is about $10^{-3}\ S\,cm^{-1}$, and some glass-ceramic electrolytes show an ionic conductivity as high as $10^{-2}\,S\,cm^{-1}$. This is because glassy electrolytes have a softening and reduced grain boundary impedance, and precipitation of some crystals is also conducive to lithium-ion conduction. Based on the crystal structure of electrolytes, sulfide electrolytes can be classified into two types: fast ion conductors (thio-LISICON) as well as their variants, and Li-argyrodites type. In addition, according to the number of constituent elements, they can be classified as binary system of Li_2S-MS_x (M = P, Si, and Ge) and ternary system of Li_2S-P_2S_5-MeS_x (Me = Si, Ge, and Sn) (Figure 5.11).

The glassy sulfide solid electrolyte was obtained by mechanical ball milling or rapid cooling after high-temperature melting, and there was no obvious peak in X-ray diffraction (XRD) characterization. Glass-ceramic sulfide solid electrolytes are usually obtained by one-step sintering at low temperatures after ball milling. They belong to the metastable phase that is mixed with glassy and crystalline states and have a few peaks characterized by XRD. Crystalline sulfide solid electrolytes are generally sintered at high temperatures and have a clear crystal structure and XRD peaks. Studies have shown that glassy solid electrolytes are primarily composed of four kinds of microcrystals, namely orthophosphates, pyrophosphates, met-thiophosphates, and hypophosphates (Figure 5.12) [97], and the mechanism of their ion conduction is not very clear.

Crystalline sulfide solid electrolytes are usually obtained via high-temperature sintering after high-energy ball milling. Some studies have also used high-energy ball milling, sintering after grinding, and liquid-phase methods to prepare crystalline sulfide solid electrolytes. Crystalline sulfide solid electrolytes can be classified into

Figure 5.11 Classification of sulfide electrolytes.

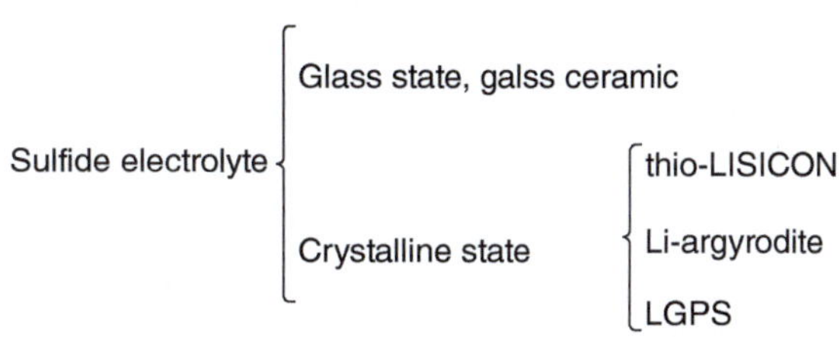

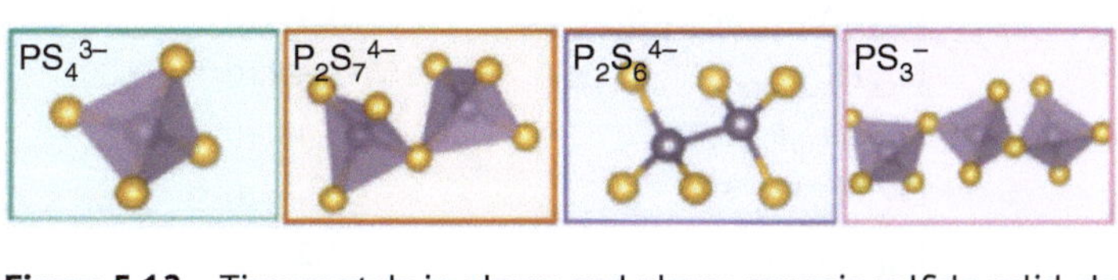

Figure 5.12 Tiny crystals in glassy and glassy ceramic sulfide solid electrolytes. Source: Reproduced with permission from Dietrich et al. [97]. © 2017/Royal Society of Chemistry.

thio-LISICON type, LGPS type, and Li-argyrodite type. These three types of electrolytes have specific crystal structures and lithium-ion transport channels, and their structural composition and ion migration mechanism are relatively clear.

At present, the types and properties of major sulfide solid electrolytes are classified and summarized according to the crystal structure, as shown in Table 5.2. Among them, crystalline sulfide solid electrolytes have the highest ionic conductivity. Among cold-pressed powders, the Li-argyrodite solid electrolyte $Li_{6.6}Si_{0.6}Sb_{0.4}S_5I$ has the highest ionic conductivity, which was discovered by Canadian scholars in 2019, and its powder ionic conductivity is 14.8 mS cm^{-1}. In 2016, Japanese scholars discovered a LGPS-type solid electrolyte $Li_{9.54}Si_{1.74}P_{1.44}S_{11.7}Cl_{0.3}$, which exhibited an ionic conductivity of 25 mS cm^{-1}. This value is notably higher than that of most solid electrolytes currently available [98, 99].

$Li_{10}GeP_2S_{12}$, with an ionic conductivity of 1.2×10^{-2} S cm^{-1} at room temperature, has an ionic conductivity higher than that of organic electrolytes used in commercial batteries. XRD shows that the phase structure of $Li_{10}GeP_2S_{12}$ is different from that of thio-LISICON previously reported. High ionic conductivity is achieved by fast ion conduction through a zigzag transmission path formed by the LiS_4 tetrahedron at the 16h and 8f sites along the *c*-axis [100].

The ionic conductivity of $Li_{10}GeP_2S_{12}$ is comparable to that of organic electrolytes. Sulfide solid electrolytes are usually classified into glassy state, glassy ceramic state, and crystalline state based on the crystal structure. Among them, $Li_{3.25}Ge_{0.25}P_{0.7}S_4$

Table 5.2 Ionic conductivity of several typical sulfide solid electrolytes.

Sulfide solid electrolyte	Chemical composition	Ionic conductivity (S cm^{-1}) (powder/block a)
Glassy state	$Li_2S\text{-}SiS_2$	5.3×10^{-4}
	$Li_2S\text{-}P_2S_5\text{-}LiI$	6.5×10^{-4a}
	$Li_2S\text{-}SiS_2\text{-}LiI$	1.8×10^{-3}
	Li_4SnS_4 LiI	2.1×10^{-4a}
Glass ceramics	$Li_7P_3S_{11}$	1.7×10^{-2a}
	$80Li_2S\text{-}20P_2S_5$	1.7×10^{-4}
	95 ($0.8Li_2S$ $0.2P_2S_5$) 5LiI	2.7×10^{-3a}
	$9Li_2S$ $3P_2S_5$ Ni_3S_2	2.0×10^{-3}
Crystalline state	$Li_{3.25}Ge_{0.25}P_{0.75}S_4$	2.2×10^{-3a}
	$Li_{3.833}As_{0.166}Sn_{0.833}S_4$	1.4×10^{-3}
	$Li_{5.5}PS_{4.5}C_{11.5}$	9.6×10^{-3}
	$Li_{6.6}Ge_{0.6}P_{0.4}S_5I$	5.4×10^{-3}
	$Li_{6.6}Si_{0.6}Sb_{0.4}S_5I$	2.4×10^{-2a}
	$Li_{10}GeP_2S_{12}$	1.2×10^{-2a}
	$Li_{9.54}Si_{1.74}P_{1.44}S_{11.7}Cl_{0.3}$	2.5×10^{-2a}

belongs to thio-LISICON sulfide solid electrolytes, Li_6PS_5X (X = Cl, Br, I) belongs to Li-argyrodite solid electrolytes, and $Li_{10}GeP_2S_{12}$ belongs to LGPS solid electrolytes.

Although $Li_{10}GeP_2S_{12}$ has high ionic conductivity, the high cost of Ge elements and their instability to metals limit their large-scale application in all-solid-state lithium batteries. As a result, many low-cost elements are used to replace Ge elements. Ong et al. [48] calculated the ionic conductivity of solid electrolytes in this system and confirmed that replacing germanium with relatively less expensive tin or silicon does not affect the conductivity. Meanwhile, Bron et al. [49] successfully synthesized $Li_{10}SnP_2S_{12}$ with a grain ion conductivity of 7 mS cm^{-1} and a total conductivity of 4 mS cm^{-1}. Structural characterization showed that $Li_{10}SnP_2S_{12}$ and $LiGeP_2S_{12}$ have similar structures, the spatial group is *P42/mc*, and ionic conductivity is high.

The high ionic conductivity and low cost of $Li_{10}SnP_2S_2$ make it an attractive sulfide solid electrolyte for use in all-solid-state batteries. Subsequently, White et al. [54] used Si instead of Ge to report a $LiSiPS_{12}$ similar to $Li_{10}GeP_2S_{12}$. Aluminum is also used in place of Ge to prepare $Li_{10}AlP_2S_{12}$ with a structure similar to that of $Li_{10}GePS_{12}$ [56]. Although its ionic conductivity of 0.82 mS cm^{-1} is much lower than that of $LiGePS_{12}$, $Li_{10}AlP_2S_{12}$ shows a wider electrochemical window and excellent electrochemical stability.

The Li_6PS_5X (X = Cl, Br, I) electrolyte is a sulfide solid electrolyte with a crystal structure evolved from Ag_8GeS_6 with high Ag^+ conductivity. Replacing Ag with Li and P with Ge, Li_7PS_6 silver germanium ore-type lithium sulfur compounds with a higher ionic conductivity can be achieved. However, Li_7PS_6 is a cubic-phase material at high temperatures and turns into an orthorhombic-phase one at low temperatures, with poor temperature stability. In 2008, Deiseroth et al. [101] partially substituted S^{2-} ions with halogen anions to obtain the general formula Li_6PS_5X (X = Cl, Br, I). Theoretical calculations showed that the incorporation of halogen ions can stabilize the cubic structure at room temperature and improve the ionic conductivity. The ionic conductivity of Li_6PS_5CI and Li_6PS_5Br prepared using the solid-phase method is 1.8×10^{-3} S cm^{-1} and 2.58×10^{-3} S cm^{-1} [102], respectively, with a lower ionic conductivity for Li_6PS_5I [103]. In addition, the Li_6PS_5X electrolyte has a wide electrochemical window, good stability against lithium, and low preparation cost, so it has a broad space in the preparation of all-solid-state lithium batteries.

The crystal structure of Li_6PS_5I is shown in Figure 5.13 [104]. In its crystal structure, the tetrahedral structure is surrounded to form a "cage" in which the anions are placed, and the space group of the cubic cell is $F\overline{4}3m$. The halide anion I^- at site 4a forms a face-centered cubic structure, the isolated PS_4 tetrahedron (P at site 4) is located at the octahedral position point, the free S (site 4d) is located in the middle of the two tetrahedrons, and Li^+ is randomly distributed across the tetrahedral gap, occupying the 48h and 24g sites. For Li_6PS_5I, the S and 4a I atoms at site 4d are ordered due to the size mismatch between S and the larger I atom, while for Li_6PS_5C1, the S and 4a I atoms at site 4d are completely disordered. The crystal structure of Li_6PS_5Br has both ordered and disordered structures. A total of 84% of S and 16% of Br are located at site 4d, and 60% of Br and 40% of S are located at site 4a. The ordered arrangement of the S and 4a I atoms at site 4d of the Li_6PS_5I

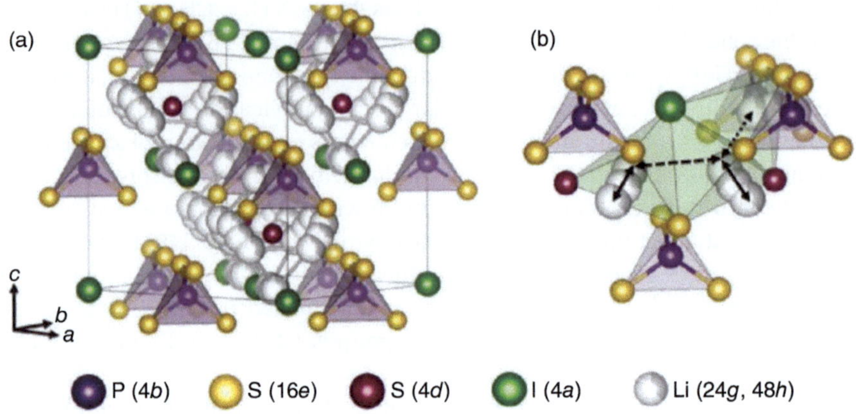

Figure 5.13 (a) Crystal structures of Li_6PS_5I. (b) The rate-determining step of the inter-cage 48h–48h jump in Li^+ migration, as shown by the LiS_3I polyhedral. Source: Reproduced with permission from Kraft et al. [104]. © 2018/American Chemical Society.

crystal structure results in a higher activation barrier of ion migration and lower ionic conductivity of the Li_6PS_5I electrolyte compared with other electrolytes in the silver and germanite system [105].

Cationic Substitution Li^+ diffusion in solid electrolyte materials is primarily carried out through ion migration channels. In these channels, Li^+ diffuses from one site to an adjacent vacancy, completing Li^+ conduction. The whole diffusion process is greatly affected by the crystal structure, especially by the size of ion migration channels. In addition, the higher the concentration of lithium ions, the faster the diffusion rate and the higher the ionic conductivity. Based on this, various strategies to improve the room temperature conductivity of sulfide electrolytes have been designed and proposed, among which polyvalent cation substitution in the crystal structure is an effective strategy. In order to ensure the valence balance of the main body of the electrolytic substance, more lithium ions are sometimes introduced to compensate for the introduction of low-price cations, so that the concentration of migrating ions increases, thus achieving higher conductivity.

Ge et al. [106] prepared a 2 mol% Ni_2P-doped $Li_7P_3S_{11}$ electrolyte, with an ionic conductivity of 2.22×10^{-3} S cm^{-1}, 1.6 times that of the original electric conductivity. Wu et al. [107] prepared $70Li_2S$-(30−x) P_2S_5-$x$$SeS_2$ (x = 0, 0.3, 0.5, 1, 3, 5) series glass-ceramic electrolytes by ball milling. The highest conductivity of $70Li_2S$-$29P_2S_5$-$1SeS_2$ is 5.28×10^{-3} S cm^{-1} at 20 °C. Yamauchi et al. [108] demonstrated that the addition of $LiBH_4$ to the (100−x) ($0.75Li_2S{\cdot}0.25P_2S_5$) −$x$$LiBH_4$ glass electrolyte not only improves the conductivity, but also increases its electrochemical stability window to 5 V versus Li/Li^+. Kraft et al. [109] systematically investigated the effect of the heterovalent ion substitution on $Li_{6+x}P_{1-x}Ge_xS_5I$. With the increase in the Ge content, the disorder of the S element and I element sites will be increased, and the energy barrier of ion migration will be significantly reduced. After annealing, the conductivity of electrolyte ions is as high as 18.4 ± 2.7 mS cm^{-1}.

Anion Substitution Doping an appropriate amount of oxide can simultaneously improve the conductivity of electrolyte ions and the stability of lithium metal negative electrodes. Tao et al. [110] found that the conductivity of the $75Li_2S$-$25P_2S_5$ electrolyte doped with 1 mol% P_2O_5 can reach 8×10^{-4} S cm^{-1}, which is 56% higher than that of the undoped electrolyte. In addition, the $1P_2O_5$-$75Li_2S$-$24P_2S_5$ electrolyte has good electrochemical stability with lithium metal. Huang et al. [111] also came to a similar conclusion that the conductivity of the $Li_7P_3S_{11}$ electrolyte doped with Li_3PO_4 (1.87×10^{-3} S cm^{-1}) is higher than that of the undoped $Li_7P_3S_{11}$ electrolyte (1.07×10^{-3} S cm^{-1}). For ternary sulfide electrolytes, Kim et al. [112] prepared a new $Li_{10}SiP_2S_{12-x}O_x$ solid electrolyte by replacing a portion of S with O. When $x = 0.7$, the maximum ionic conductivity of the O-substituted sample can reach 3.1×10^{-3} S cm^{-1}.

The addition of halides can also improve the ionic conductivity of the Li_2S-P_2S_5 electrolyte. Ujiie et al. [113] compared the doping effects of different halides and showed that the introduction of lithium bromide can effectively improve the conductivity of $Li_7P_3S_{11}$ glass ceramics to 6.5×10^{-3} S cm^{-1}. They prepared (100–x) ($0.7Li_2S_{0.3}P_2S_{5-x}$LiI) glass and glass-ceramic electrolytes using the mechanical ball milling method. The composition range of (100–x) ($0.7Li_2S_{0.3}P_2S_{5-x}$LiI) and glass-ceramic electrolytes was $0\leq x$ (mol%)≤ 20. The conductivity of glass-phase electrolytes increased with an increase in the LiI content, and the conductivity was the highest when $x = 20$ (5.6×10^{-4} S cm^{-1}). More importantly, the electrochemical window for 80 ($0.7Li_2S_{0.3}P_2S_5$)-20LiI glassy electrolytes is as wide as 10 V (versus Li/Li$^+$) based on cyclic voltammetry. In glass-ceramic solid electrolytes, the conductivity of the electrolyte decreases sharply with an increase in the LiI content.

Regarding the Li_6PS_5X electrolyte, the type and content of halide anions in the electrolyte structure have a great influence on ionic conductivity. Halide anions can affect the crystal structure and ion transport activation barrier of the electrolyte, thus affecting the ionic conductivity of the electrolyte. The Li_6PS_5X (X = Cl, Br, I) electrolyte system has the highest ionic conductivity [114]. Adeli et al. reported that an increase in the Cl^-/S^{2-} ratio in the $Li_{6-x}PS_{5-x}Cl1_{+x}$ electrolyte system has a significant effect on the Li^+ diffusivity in the lattice. After annealing and sintering, the room temperature conductivity of $Li_{5.5}PS_{4.5}Cl_{1.5}$ can be up to 12.0 ± 0.2 mS cm^{-1}. Under the same preparation conditions, its conductivity is almost four times that of Li_6PS_5Cl. The replacement of S^{2-} by Cl^- weakens the interaction between lithium ions and the surrounding skeleton anion, which enhances the diffusivity of lithium ions. Meanwhile, the disorder of Cl and S elements and the number of lithium-ion vacancies increase in the crystal structure, further improving the transport rate of lithium ions [115].

In addition to single-element doping, multi-element substitution, especially doping of different kinds of double-positive ions with synergistic effects, has been proven to be an effective method to improve the electrochemical performance of electrolytes. Yang et al. [116] studied the effects of multi-element doping on ion channel width and electrolyte activation energy and synthesized a Sn-/Se-co-doped $Li_{10}GeP_2S_{12}$ electrolyte using the solid-phase method. Compared with the single element doping, the ionic conductivity of Sn-/Se-double-doped $5Li_2S$-P_2S_5-0.6

GeS_2-0.4$SnSe_2$ ($Li_{10}Ge_{0.6}Sn_{0.4}P_2S_{11.2}Se_{0.8}$) can reach 2.75×10^{-3} mS cm^{-1} at room temperature and activation energy can reach 16 kJ mol^{-1}. The activation energy is the lowest reported one for lithium-ion conductors at that time. Kanno et al. [117] found that changing the Sn/Si ratio and (Sn^{4+} and Si^{4+})/P^{5+} ratio can adjust the size of the lithium-ion conduction channel, which is conducive to the optimization of conductivity. The optimized ionic conductivity is 1.1×10^{-2} S cm^{-1}, which is close to that of the original $Li_{10}GeP_2S_{12}$ electrolyte, and the electrolyte does not contain Ge, so the lower cost is more suitable for practical applications.

5.4.1.3 Halide Solid Electrolytes

Halide solid electrolytes (lithium metal halide solid-state electrolytes) have an ionic conductivity of up to 10^{-3} S cm^{-1} at room temperature and a theoretical ionic conductivity of 10^{-2} S cm^{-1}. Halide electrolytes in LIBs are derived from the introduction of high-valence transition metal cations into Li-halide LiX (X = Br, Cl, F) to adjust the concentration of Li^+ and vacancy, thus forming Li_a-M-X_b compounds. By regulating the component characteristics of different halide anions, the anionic skeleton in such compounds can generally achieve a large polarization. In addition, the interaction with lithium ions is weak. Theoretical simulation results show that halides have higher redox potential than other solid electrolytes (chloride > 4 V versus Li/Li^+, fluoride > 6 V versus Li/Li^+), with better compatibility with high-voltage cathode materials, allowing stable cycling under high voltage windows.

The chemical formula for common metal halide electrolytes is Li_a-M-X_b, where M is a metal element and X is a halogen element. The structure of metal halide electrolytes depends on the ionic radius and arrangement of ions. The radii of halide ions (Cl^- 167 pm, Br^- 182 pm, I^- 202 pm, F^- 122 pm) are much higher than those of Li^+ (73–90 pm) and other metal ions. According to ion binding, the coordination number of the space lattice where the ion resides is determined by the ratio of ionic radii. In most metal fluoride electrolytes, the ionic radius ratio of metal elements to halogens is 0.732, resulting in the formation of the octa-coordination cubic lattice structure MX_8 similar to CsCl. All chlorides, bromides, and iodides, as well as some fluorides, have cation radius ratios between 0.414 and 0.732, resulting in the formation of a hexagonal octahedron of MX_6. Unlike common sulfides, which require the formation of bcc lattices to achieve fast Li^+ transport channels with high lithium-ion conductivity, ionic bonding is weak in metal halide electrolytes, and Li+ ions are mobile in most anionic lattice structures. The crystal structure classification of some halide electrolytes is shown in Figure 5.14 [118]. According to the substitution amount of Li ions by transition metal elements, halide electrolytes commonly studied at present generally fall into three categories: (i) Li_a-M-Cl_6, (ii) Li_a-M-Cl_4, and (iii) Li_a-M-Cl_8 halides (M = transition metal elements). Li_a-M-Cl_6 and part of Li_2-M-Cl_4 are halide electrolytes with a conductivity of up to 10^{-3} S cm^{-1}. In addition, early studies have primarily focused on Li_a-M-Cl_8 electrolytes. For example, Li_6FeCl_8 and other high-temperature halide electrolytes have low conductivity and temperature instability, as well as other challenges, which limits their practical applications.

The structure of halide electrolytes is chiefly determined by the coordination number corresponding to the atomic radius ratio of metal elements and halogen atoms.

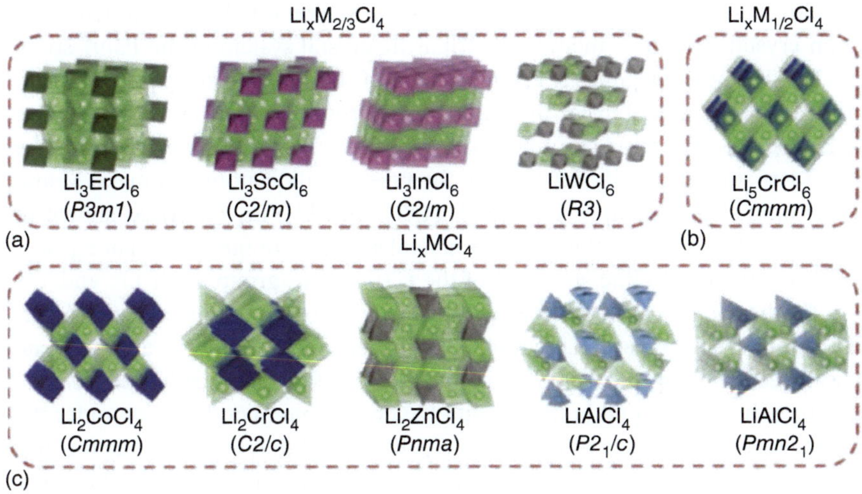

Figure 5.14 The typical crystal structure of $Li_xM_{2/3}Cl_4$ (Li_xMCl_6), $Li_xM_{1/2}Cl_4$, and Li_xMCl_4 halide solid electrolytes based on simulation results. Source: Reproduced with permission from Kim et al. [118]. © 2021/American Chemical Society.

Therefore, Al and Ga in group IIIA trivalent metal halides have difficulty reacting with lithium salts to form a high-coordination mixed structure due to their low ionic radius. Generally, halides of Ga and other metal elements can only form $LiAlX_4$ [119], $LiGaX_3$ [120], $LiGaX_4$, and other structures. Except for Li_3InCl_6, Li_a-M-X_6 halide electrolytes are mostly composed of group IIIB metal ions. Lithium halide electrolytes have attracted extensive attention in recent years since Asano et al. [121] have successfully prepared Li_3YCl_6 (LYC) and Li_3BrCl_6 (LYB) with a room temperature ionic conductivity of 5.1×10^{-4} S cm^{-1} and 1.7×10^{-3} S cm^{-1} using ball milling and sintering processes. Liang et al. [122] studied the Li_xScCl_{3+x} electrolyte and found that when $x = 3$, it had the highest room temperature lithium-ion conductivity (3.02×10^{-3} S cm^{-1}), and when x was lower than 3, the lithium-ion conductivity of the electrolyte would decrease significantly. Li_a-M-X_6 halide electrolytes generally have high ionic conductivity at room temperature ($>1 \times 10^{-3}$ S cm^{-1}). Currently, studies on halide electrolytes primarily focus on this type of material. The Li-ion conduction characteristics of Li_a-M-X_6 electrolytes are chiefly affected by the occupation of the octahedral site by transition metal ions and Li^+ ions. Sun et al. [123] reported a series of highly crystalline Li_3InCl_6 halide electrolytes with high positive electrode stability, whose theoretical ionic conductivity can reach up to 6.4×10^{-3} S cm^{-1}. In practice, the room temperature lithium-ion conductivity of Li_3InCl_6 electrolytes prepared via ball milling sintering can reach 1.49×10^{-3} S cm^{-1}. Recently, Li_2ZrCl_6 electrolytes based on Zr^{4+} have been widely investigated for their low-cost advantages. Although the room temperature ionic conductivity of the Li_2ZrCl_6 electrolyte itself is low, it can be effectively increased to 10^{-3} S cm^{-1} by heterovalently substituting or doping halide electrolytes to generate different vacancies and gaps [124]. Examples include $Li_{2+x}Zr_{1-x}In_xCl_6$ [125] and $Li_{2+x}Zr_{1-x}M_xCl_6$ [126].

According to the laws of ion bonding of Li_a-M-X_6 metal halide electrolytes, the common crystal structures include the tripartite crystal system of the $P\bar{3}m1$ space group (hcp-T), the orthogonality crystal system of the *Pnma* space group (hcp-O), and the cubic crystal system of the $C2/m$ space group (ccp) [127]. Figure 5.14b,c shows the common hcp and ccp anion structure skeletons, respectively. The change of crystal structure primarily depends on the type of halide atom and metal M and the ionic radius. For example, in Li_3YCl_6 (LYC) and Li_3YBr_6 (LYB), as shown in Figure 5.15a,b [128], LYC is the tripartite crystal structure of the $P\bar{3}m1$ space group, while LYB is the monocline structure of the $C2/m$ space group. Both Li^+ and Y^{3+} occupy the hexagonal octahedral void formed by halogen anions, and the incorporation of trivalent metal ions will form a trivalent metal ion layer and generate Li^+ vacancies to generate ion conduction channels. After modification of the crystal structure, it is found that both LYB and LYC have better conductivity under the $C2/m$ structure. In the study of $Li_3Y_{1-x}In_xCl_6$, it is found that by increasing the proportion of In^{3+}, the structure of $Li_3Y_{1-x}In_xCl_6$ can be transformed from the hcp-T structure to the hcp-O structure and finally to the $C2/m$ structure, and its electrochemical performance as an electrolyte can be greatly improved. $Li_{3-x}Yb_{1-x}Hf_xCl_6$ has similar characteristics [129]. At the same time, the structural parameters can also be adjusted by sintering and other processes, such as low-crystallinity $P\bar{3}m1$ space group $Li_{2.25}Zr_{0.75}In_{0.25}Cl_6$ obtained by ball milling and high-crystallinity phase of $C2/m$ space group obtained by low-temperature sintering.

Li_a-M-Cl_4 electrolytes are primarily divided into positive spinel phases composed of bivalent metal ions M, such as Li_2MnCl_4 and Li_2MnCl_4, and halide electrolytes formed by trivalent and other valence metal ions M, such as $LiYbF_4$ and $LiAlF_4$. The ionic conductivity of these halide electrolytes synthesized in the early stage is low, and some of them cannot exist stably at room temperature, so there is little research on Li_a-M-Cl_4 electrolytes. In the spinel structure, halide anions are tightly packed to form a ccp skeleton and form Li_2MX_4 solid electrolytes such as Li_2ZnCl_4, where M^{2+} occupies the tetrahedral void and Li^+ occupies the octahedral void. In the inverse spinel structure, M^{2+} is distributed in the octahedral void formed by the halogen

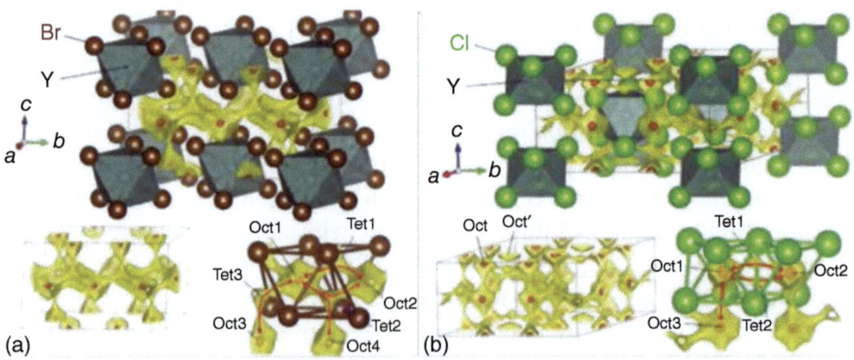

Figure 5.15 (a) $P\bar{3}m1$ structure of Li_3YCl_6 and (b) $C2/m$ structure of Li_3YBr_6. Source: Reproduced with permission from Asano et al. [128]. © 2018/John Wiley & Sons.

anion, and Li^+ is distributed in both tetrahedral and octahedral spaces. When metal ions of higher valence state are used or the substitution amount of transition metal elements is further increased, the defective inverse spinel structure will be formed, because in the inverse spinel structure, Li^+ can occupy both octahedral and tetrahedral point sites. It is generally believed that the diffusion of Li^+ depends on the direct diffusion of common-sided point sites between the octahedral and tetrahedral point sites. Hence, it is presumed that positive spinel halides, such as Li_2ZnCl_4, which occupy only octahedral point sites but not tetrahedral point sites, have low conductivity due to the limited Li^+ ion transport. In defective anti-spinel structures, the introduction of high-valence metal cations will generate more vacancies to maintain electrical neutrality, which is conducive to the transmission of Li^+ ions. Therefore, this type of defective spinel generally has higher electrical conductivity than positive and anti-spinel structures, such as $Li_{2-2x}Fe_{1+x}Cl_4$. Olivine structure is commonly seen in the sintered products of Li_2ZnX_4. After sintering of Li_2ZnX_4 with a positive spinel structure, a hcp structure with X^- as a skeleton (generally *Pnma* group) can be obtained [130]. Nazar et al. [131] successfully prepared a $Li_2SC_{2/3}Cl_4$ electrolyte by replacing primary divalent ions with trivalent ions and improved the conductivity of spinel halides at room temperature to 1.5×10^{-3} S cm^{-1} for the first time. The room temperature conductivity of the electrolytes was further increased to 2.0×10^{-3} S cm^{-1} by In^{3+} doping. In this type of typical defective spinel halides, the whole crystal structure of the standard spinel structure Li_2MgCl_4 can be regarded as the co-octahedral composition of $(Mg_1/Li_2)Cl_6$. A part of Li occupies the tetrahedral gap position formed by halogen atoms, and the rest of Li occupies the octahedral gap position together with Mg. However, after the introduction of trivalent metal ions, halides showed a defective spinel structure. Due to the introduction of trivalent ions, the uptake of octahedral and tetrahedral gaps is lower, so lithium ions diffuse more rapidly and have higher conductivity. Meanwhile, with the increase in the halogen transition metal element content, the crystal structure will also change. The spinel-phase $Li_2In_{1/3}Sc_{1/3}Cl_4$ is transformed into monoclinic $Li_3In_{2/3}Sc_{1/3}Cl_6$ [132].

Li_a-M-X_8 halide electrolytes are primarily Suzuki lattices, in which MX_6 structures exist in the form of octahedrons and are separated from each other, and all octahedral voids are occupied by Li^+. In such electrolytes, ionic conductivity is generally not high due to the low substitution of metal M and few vacancies for Li^+ transmission. For example, the conductivity of Li_6CoCl_8 at a high temperature (400 °C) is 9.3×10^{-2} S cm^{-1}, which is lower than 10^{-5} S cm^{-1} at room temperature [133]. Neutron diffraction experiments on the Li_6VCl_8 electrolyte showed that the vacancy in the structure of the materials will significantly improve the migration rate of Li^+. Therefore, when applied in electrolytes, such materials need to increase the substitution amount of transition metal elements further to obtain higher conductivity. Specific materials and Li^+ transmission problems need to be further studied.

The transport pathway of lithium ions is a critical factor in the study of solid electrolytes. In solid electrolytes, the lattice sites on the lattice position will generally block the transition and transport pathway of Li^+ ions. Asano et al. [128] presumed that Li^+ ions in metal halide electrolytes with the ccp structure are connected through a 3D isotropic network. Since Li^+ occupies tetrahedral and octahedral sites simultaneously, it can jump from the tetrahedron to other sites of the adjacent common-sided octahedron (LYB Oct1-[Tet1 or Tet2]Oct2 and Oct1-Tet3-Oct3). However, in hcp, Li^+ ions are anisotropic diffused, with one-dimensional *c*-axis fast diffusion channels and Li^+ jumps in octahedral voids. These one-dimensional fast *c*-axis channels are connected by tetrahedral point sites with slower diffusivity in the ab plane, forming an anisotropic three-dimensional diffusion network (LYC Oct1-[Tet1 or Tet2]-oct2 and Oct1 – Oct3 along *c*-axis). In the case of spinel halides, it is commonly accepted that lithium ions conduct coplanar in the octahedral and tetrahedral gaps, namely Li1-Li2-Li3. Therefore, when Li^+ occupies tetrahedral and octahedral point sites at the same time, i.e. when the structure is more inclined to defective spinel structure, higher ionic conductivity can be obtained.

The vacancy mechanism of Li^+ in solid electrolytes is the main conduction mechanism. Wang et al. [134] reported that compared with the substitution and gap mechanisms, the vacancy mechanism requires a lower barrier, so vacancy concentration is the key factor affecting the migration of lithium ions. Fixed vacancy can improve the ionic conductivity in the material, and when doping LiX with trivalent metal ions, three Li^+ will be removed and two additional vacancies will be formed. In addition, Liu et al. [135] found that the mixed arrangement of cations with different atomic radii leads to the energy difference between lattices, which will affect the ion migration barrier in chloride. In the point sites with more dispersed Li^+ ion arrangement and lower concentration, the migration barrier will be very low even if Li^+ is largely replaced by other metal ions.

There are other crystallographic and electronic factors that affect the mechanics of Li^+ transport, such as the induction of ion movement by the polarization process, atomic vibration, different crystallinity and grain boundaries, and impurities. Wakamura et al. [136] proposed that an anion lattice more susceptible to polarization in the transport of Li^+ ions is conducive to ion transport. Earlier studies have proposed that the high-coordination amorphous intermediate phase of lithium can improve the conductivity. However, the current commonly used sintering process leads to a change of crystallinity and has different effects on the conductivity of different electrolytes. Most sulfide solid electrolytes with an ionic conductivity higher than $10^{-2}\ S\ cm^{-1}$ are sintered to reduce the influence of grain boundary energy. However, Kwak et al. [126] found that after sintering Li_2ZrCl_6, the conductivity of halide electrolytes declines sharply and the Li^+ transport barrier increases, which may be because sintering has no obvious effect on the grain boundary energy of halide electrolytes, and the more stable structure after sintering leads to the reduction of Li^+ carrier. The transmission channel is narrowed and a stronger coulomb repulsion is generated between Li^+ and Zr^{4+}. Jung et al. [137] obtained $Li_{2+x}Zr_{1-x}In_xCl_6$ by In^{3+} doping and low-temperature annealing, while maintaining the improvement in Li^+ conductivity. They presumed that in $C2/m$ structure halides obtained by sintering, In^{3+} gradually replaces the point site of Zr^{4+} in the process of In^{3+} doping, promoting the distribution of Li^+. The concentration of Li^+ in the diffusion path is increased. Meanwhile, the anisotropic lattice expansion caused by the covalent bond of different metal halogen atoms promotes the migration of Li^+. The crystalline $C2/m$ type

$Li_{2+x}Zr_{1-x}In_xCl_6$ halides obtained by sintering are also significantly different from the low-crystallinity $Li_{2+x}Zr_{1-x}In_xCl_6$ halide electrolytes obtained by ball milling.

5.4.2 Polymer Electrolytes and Composite Electrolytes

Replacing conventional liquid electrolytes with polymer electrolytes has been recognized as a promising approach to address the safety concerns associated with Li-based batteries. Generally, polymer electrolytes for Li-based batteries can be categorized into two major types: (i) solvent-free SPEs, also known as all-solid-state polymer electrolytes; and (ii) GPEs or quasi-solid polymer electrolytes (QPEs), which contain fewer liquid components. However, the inherent properties of polymers often fail to meet the requirements of lithium batteries. To enhance the electrochemical performance of electrolytes, researchers have incorporated various inorganic fillers, including inert fillers and active fillers (fast ion conductors), into the polymer matrix, resulting in CPEs [138].

5.4.2.1 Solid Polymer and Composite Electrolytes

Among all the polymer matrix of SPEs, PEO is the first identified and the most extensively studied material. The concept of polymer electrolytes dates back to the 1970s, with the breakthrough discovery by Wright and Armand et al. [139, 140], which demonstrated that alkali metal salts dissolve in PEO to form ionic conductive complexes. In addition, Feuillade et al. introduced organic plasticizers into the polymer–salt binary system to develop GPEs for Li-based batteries in 1975 [141]. Another important breakthrough happened in 1988. Skaarup et al. achieved significant progress by introducing fast ionic conductor particles, Li_3N (which had a high room temperature ionic conductivity of $\sim 10^{-3}\,S\,cm^{-1}$), as fillers into the PEO–$LiCF_3SO_3$ matrix [142]. As a result, the obtained CPE showed much higher ionic conductivity than the blank PEO-based polymer electrolyte. More interestingly, Wieczorek et al. found that the addition of inert fillers such as Al_2O_3 can also improve the ionic conductivity of PEO-based polymer electrolytes by at least one order of magnitude, which is primarily attributable to the increase in the amorphous phase with high-ion mobility [143]. Since these pioneering works, polymer electrolytes have undergone rapid progress because of their prominent advantages such as low flammability, easy processability, higher tolerance to vibration, shock, mechanical deformation compared with liquid electrolytes, and better electrode–electrolyte interfacial contact as well as compatibility than ISSEs.

In PEO-based SPEs, a commonly accepted mechanism is that Li^+ coordinates to ether bonds on the polymer host and is transported by intrachain/interchain jumps, or with polymer chain segment motion (Figure 5.16a). This is corroborated by the X-ray-determined structure of the crystalline $P(EO)_6$-$LiAsF_6$ complex, where Li ions are solvated by PEO sheaths. Under an electric field, these cations tend to hop from one coordinating site (usually composed of more than three electron donor groups) to another. Such ion hopping is facilitated by either a segmental motion of the polymer chains or an ion-cluster-assisting function, in which temporary reassociation with the counterion occurs before being resolvated by electron donor groups in the

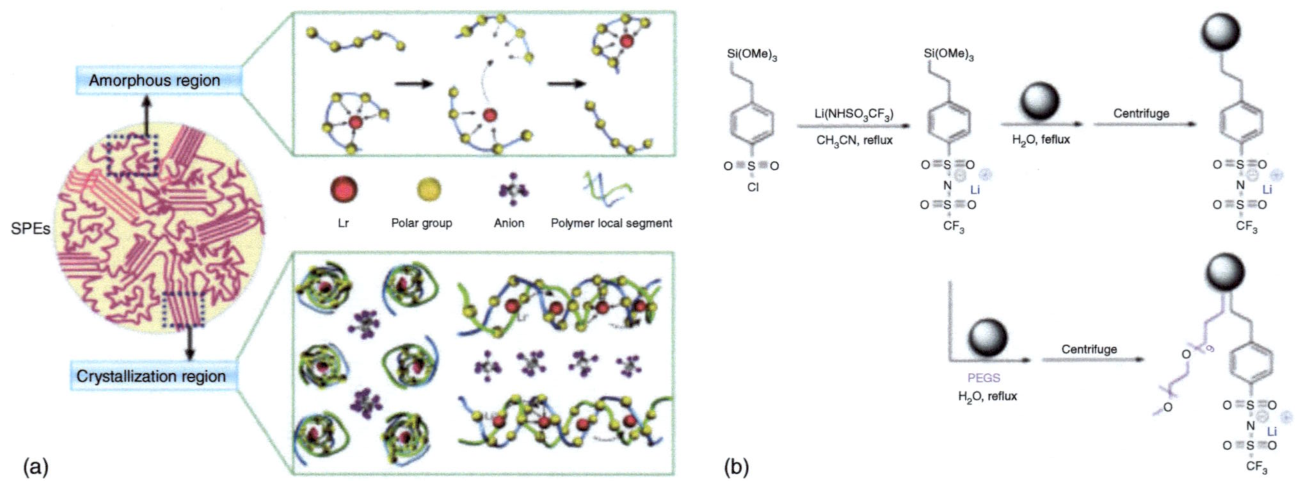

Figure 5.16 Polyethylene oxide-based solvent-free polymer electrolytes. (a) Schematic of Li-ion transport mechanism in PEO-based SPEs. Source: Reproduced from Chen et al. [144]/Royal Society of Chemistry. (b) Synthesis route to obtain grafted and co-grafted inorganic nanoparticles. Source: Reproduced from Zhang et al. [14]. © 2018/John Wiley & Sons.

polymer [144]. It is thus generally presumed that the ionic conduction in PEO-based SPEs primarily occurs in the amorphous part of the PEO matrix, while the crystalline part delivers limited ion motion. Therefore, the ionic conductivity of PEO-based SPEs highly depends on the nature of the polymer matrix in terms of its ability to solvate Li salts and the ratio of the crystalline phase to the amorphous phase. Since the PEO chain is primarily crystalline below 65 °C, PEO-based SPEs generally present low ionic conductivities with an order of 10^{-8}–10^{-7} S cm^{-1} at room temperature. Such limited ionic conductivity greatly restricts the application of PEO-based SPEs. Consequently, modification strategies are widely applied to improve the ionic conductivity of PEO-based SPEs, including polymer structural modification, optimizing the structure and proportion of the Li salts, and introducing organic plasticizers or inorganic fillers.

This was finally solved by introducing perfluorocarbon moieties as electron-withdrawing groups to the anionic centers (e.g. sulfonate ($-SO_3^-$), carboxylate ($-CO_2^-$), and sulfonyl imide ($-SO_2N^{(-)}SO_2^-$) on the polymer matrix). Bouchet et al. synthesized a SPE for lithium-based batteries, incorporating a central block composed of linear PEO and a block based on poly(styrene trifluoromethane sulfonyl imide of lithium) P(STFSILi) (Figure 5.16b). The SPE obtained showed a high ionic conductivity of 1.3×10^{-5} S cm^{-1} and lithium-ion transference number (>0.85) at 60 °C, similar to a single-ion conductor due to the block architecture anchoring the anion. The Li|SPE|$LiFePO_4$ battery also showed stable cycling at 60 °C [145].

Another viable strategy to achieve single- or almost-single-ion-conducting properties is via grafting or co-grafting an anion onto nanoparticles. Lago et al. functionalized Al_2O_3 and SiO_2 nanoparticles with lithium-4-[2-(trimethoxysilyl) ethyl]benzene-1-sulfonyl [(trifluoromethyl)-sulfonyl]amide or a combination of this Li salt with poly(ethylene glycol) (PEG) strands (Figure 5.16c) [146]. When combined with PEO, the hybrid SPEs comprised of grafted nanoparticles showed an ionic conductivity of 1.9×10^{-4} S cm^{-1} at 70 °C and a stable cycling lifespan in Li||$LiFePO_4$ batteries.

Non-ionic organic solvents with strong polarity are typically used in traditional liquid electrolytes, as an environment with a high dielectric constant promotes lithium dissociation. This principle also applies to the selection of polymer electrolytes, particularly in liquid-deficient all-solid-state systems. The polarity of the polymer significantly influences the electrostatic force between ions [147]. Compared with PEO's permittivity of 5, poly(vinylidene difluoride) (PVDF) possesses a higher relative permittivity of ~8.3, which weakens the interaction between lithium salt ion pairs and enhances lithium-ion dissociation. Furthermore, PEO materials have inherent issues such as poor stability, inadequate mechanical properties, and a narrow voltage window. In contrast, PVDF materials show desirable physicochemical properties for solid electrolytes, including a wide electrochemical stability window, superior mechanical strength, and thermal stability.

The first PVDF electrolytes date back to the 1980s. Because of its attractive properties, PVDF has been chosen as a polymer host for lithium battery applications. Tsunemi et al. investigated the effects of different plasticizers and lithium salts on electrolyte properties [148]. Then, Tarascon et al. produced the first LIB with a fluorinated polymer (PVDF-HFP) as a battery separator in 1996 [149]. Until the twentieth century, studies on PVDF-based electrolytes gradually increased, among which gel electrolytes were the primary ones. Goncalves et al. prepared SPEs based on PVDF-HFP copolymer and lithium bis((trifluoromethyl)sulfonyl)azanide (LiTFSI) by solvent casting without other fillers and evaluated the effects of the LiTFSI content on physical properties such as ionic conductivity [150]. This work proved that PVDF material as a solid electrolyte affects the dissociation of lithium salts. Nevertheless, the pure polymer with the lithium salt system makes it difficult to meet the needs of electrolytes for ion transfer, and the performance of the battery made of these electrolytes is obviously inferior compared with that of organic liquid electrolytes.

However, due to its higher crystallinity and lack of groups that coordinate with lithium ions, PVDF does not perform well in SPEs, while it is more popular in GPEs. In 2019, Cewen Nan et al. [151] found that the ionic conductivity of the PVDF-based solid electrolyte prepared using LiFSI instead of the commonly used lithium salt LiTFSI was higher by about an order of magnitude (Figure 5.17a). According to their study, a nanoscale interfacial layer capable of forming multicomponents was obtained using a simple solution casting method, effectively inhibiting lithium dendrites and enhancing the cycling stability of lithium metal batteries (Figure 5.17b–d). Nevertheless, controversy arose due to the presence of approximately 10 wt% DMF solvent molecules in the electrolyte, which deviates from the conventional understanding of dry solvent-free all-solid electrolytes. Further investigation revealed that these solvent molecules in PVDF-based electrolytes show distinct bonding behavior when complexed with lithium ions as $[Li(DMF)_x]^+$ carriers (Figure 5.18e), resulting in a polymer–solvent molecule–lithium salt ternary electrolyte system closely resembling an all-solid-state electrolyte transport mechanism [153, 154].

Despite this, the dielectric constant of PVDF is lower compared to high-dielectric-constant solvents such as EC and PC (ε_r: 30–90). In recent years, He Yanbing et al. found that polyvinylidene-trifluoro-vinyl chloride (P(VDF-TrFE-CTFE)) with relaxor ferroelectric (RFE) behavior, selected as the matrix to prepare a polymer solid electrolyte, can achieve higher lithium-ion conductivity in their research

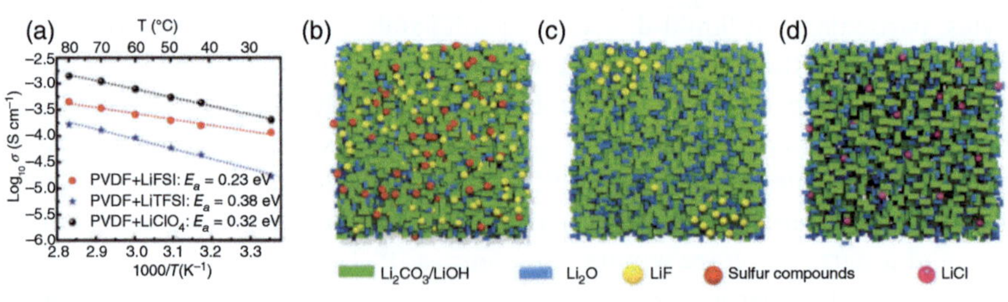

Figure 5.17 (a) Arrhenius plots of PVDF–LiFSI, PVDF–LiTFSI, and PVDF–$LiClO_4$ electrolytes. Morphological characterizations of the interface layer. Schematic of the component distribution at (b) PVDF–LiFSI|Li, (c) PVDF–LiTFSI|Li, and (d) PVDF–$LiClO_4$|Li interfaces. Source: Reproduced with permission from Zhang et al. [151]. © 2018/John Wiley & Sons.

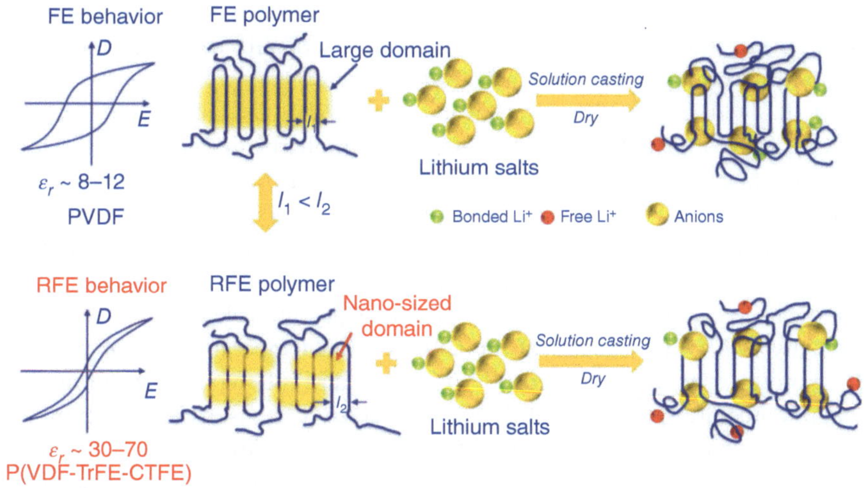

Figure 5.18 Schematic of the dissociation of lithium salts by a FE polymer (e.g. PVDF) with a lower ε_r and an RFE polymer [e.g. P(VDF-TrFE)-based terpolymer] with a higher ε_r. Source: Reproduced with permission from Huang et al. [152]. © 2021/Royal Society of Chemistry.

on polymer solid electrolytes (Figure 5.18). Since the dielectric constant of the polymer is greatly increased (ε_r: 30–70), the dissociation of lithium salts can be significantly promoted and the ionic conductivity of the P(VDF-TrFE-CTFE) solid electrolyte at room temperature increases from $1.77 \times 10^{-5}\,\mathrm{S\,cm^{-1}}$ of PVDF to $3.10 \times 10^{-4}\,\mathrm{S\,cm^{-1}}$. Thus, high-performance operation of SSBs at room temperature can be achieved [152].

Polyacrylonitrile (PAN) contains a polar electron-absorbing nitrile group, C≡N, which has excellent electrochemical stability, a wide electrochemical window, good thermal stability, and considerable mechanical strength [155]. In particular, the high antioxidant potential of PAN-based electrolytes allows them to be adapted to high-voltage cathode materials to achieve high energy density. In the PAN/$LiClO_4$ system, lithium ions can bind to up to four CN groups in the PAN chain in a comb-like manner. However, the ionic conductivity of the dried PAN/$LiClO_4$ SPE is still as low as $10^{-7}\,\mathrm{S\,cm^{-1}}$ at room temperature, which is comparable to the ionic conductivity of PEO-based electrolytes. Although LiTFSI has been reported to reduce the crystallinity of PAN, in contrast to the mechanism of PEO-based solid electrolytes, the ionic conductivity of PAN-based electrolytes is almost independent of the mobility of the PAN polymer chains [156]. However, the two strong electron-withdrawing groups on the nitrogen atom in LiTFSI keep the imide anion stable, and the large charge dissipation region of the imide anion promotes ion dissociation. The ionic conductivity of dry PAN-LiTFSI SPEs has been reported to be as high as $10^{-6}\,\mathrm{S\,cm^{-1}}$ [157]. Wang et al. [158] showed that the addition of plasticizers is an effective way to disrupt the interaction between Li^+ and PAN and increase the ionic conductivity of PAN-based solid electrolytes. In addition to plasticizers, the introduction of inorganic fillers into PAN is another strategy to increase ionic conductivity. Wang et al. [159] also showed that ceramic fillers with modified surfaces can act as a source of charge carriers, thereby increasing the ionic conductivity according to the Lewis acid–base theory.

Liu et al. [160] reported a PAN/$LiClO_4$ CPE with $Li_{0.33}La_{0.557}TiO_3$ (LLTO) nanowires and compared it with the effect of nanoparticles synthesized using a sol-gel method on ionic conductivity enhancement at the same filler concentration (15 wt%) (Figure 5.19a). As shown in Figure 5.19b, results revealed that the ionic conductivities of the CPE with 15 wt% LLTO nanowires at various temperatures were significantly higher than those of the CPE with 15 wt% LLTO nanoparticles. In comparison with the filler-free polymer electrolyte, the ionic conductivity of 2.4×10^{-4} S cm^{-1} of the CPE at room temperature was increased by three orders of magnitude. Recently, Yi Cui et al. [161] reported an ultrathin and high-temperature-resistant SPE. They observed that using electrospinning technology in combination with the PEO/LiTFSI osmosis treatment, PAN-based SPE films with a thickness of 5 μm can be easily prepared, which have high mechanical strength, good interfacial stability with lithium metal, and excellent thermal stability (Figure 5.19c). Compared with ordinary PEO/LiTFSI electrolytes, its modulus is increased by five orders of magnitude, and it can form a stable interface layer containing LiF and Li_3N to inhibit the growth of lithium dendrites and prevent short circuits. In addition, the electrospun PAN film also showed good thermal stability, maintaining excellent cyclic stability up to 150 °C.

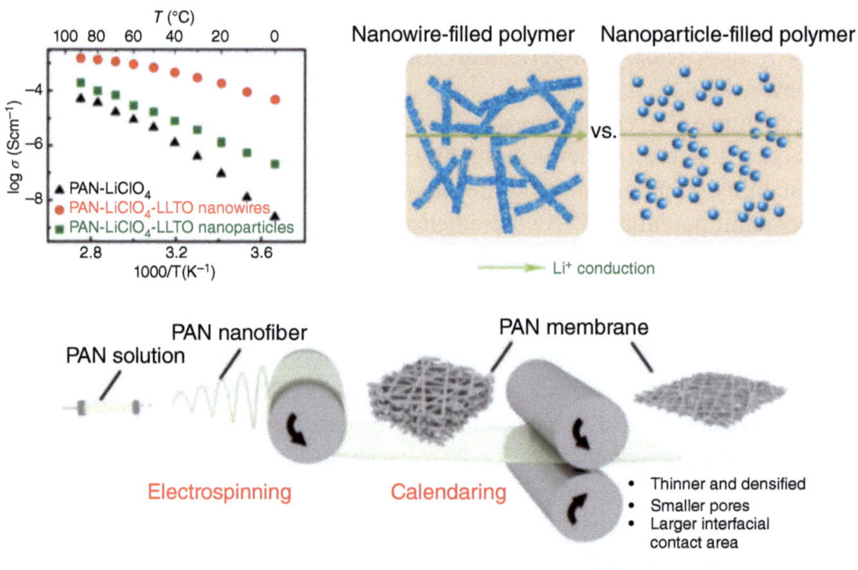

Figure 5.19 (a) Comparison of possible Li ion conduction pathways in nanowire-filled and nanoparticle-filled CPEs. (b) Ionic conductivities of the filler-free PAN/$LiClO_4$ SPE and the corresponding CPEs containing LLTO nanowires and nanoparticles as a function of temperature. Source: Reproduced with permission from Liu et al. [160]. © 2015/American Chemical Society. (c) Schematic illustration of the preparation of the PAN membrane. Source: Reproduced with permission from Ma et al. [161]. © 2021/John Wiley & Sons.

However, PAN has low solubility in most volatile solvents such as methanol, acetone, and THF. A common method to obtain homogeneous polymer solutions is to dissolve PAN in high-boiling organic solvents such as *N*,*N*-dimethylformamide (DMF) and dimethyl sulfoxide (DMSO). However, this requires the removal of large amounts of energy, which is detrimental to the large-scale production of large PAN-based SPEs [162]. In addition, the unstable interface between PAN and the lithium metal anode needs to be addressed.

The low dielectric constant of PEO leads to insufficient dissociation of lithium salts, which results in the agglomeration of ions in the SSE and reduction in ionic conductivity. Aliphatic or cyclic carbonate esters [—O—(C=O)—O—] have high dielectric constants, which can ensure adequate dissociation of lithium salts and are often used as solvents or additives in liquid electrolytes [163]. It has been demonstrated that the introduction of such structures into the ester polymer backbone or functional side chains can improve their ionic conductivity and electrochemical stability [164].

Poly(ethylene carbonate) (PEC), poly(cyclopropylene carbonate) (PTMC), poly(propylene carbonate) (PPC), poly(vinylidene carbonate) (PVCA), etc., are promising polycarbonate-based polymer solid electrolytes. Kimura et al. [165] reported a PEC-based SSE assembled with polyimide (PI) as the support and a PEC-LiFSI composite SSE with an ionic conductivity of 1.6×10^{-5} S cm^{-1} at room temperature and a Li^+ mobility number of as high as 0.5. Chai et al. [166] reported a PVCA-based SPE prepared via an in situ polymerization of the vinylene carbonate (VC) monomer. The PVCA-based SPE presented an ionic conductivity of 9.82×10^{-5} S cm^{-1} at 50 °C and an electrochemical stability window of up to 4.5 V versus Li/Li^+, which enabled excellent cycling performance in high-voltage Li|PVCA-based SPE|Li_xCoO_2 batteries (Figure 5.20). Zhang et al. [167] prepared a cellulose-supported PPC-based SPE for LIBs that showed adequate mechanical strength, high ionic conductivity (3×10^{-4} S cm^{-1} at 20 °C), high electrochemical stability (4.6 V versus Li/Li^+), and outstanding interfacial compatibility with Li metal (the Li|SPE interface can maintain stable values of resistance during a 25-day storage). Its assembled Li|CCPP-SPE|$LiFePO_4$ battery has a cycle life of more than

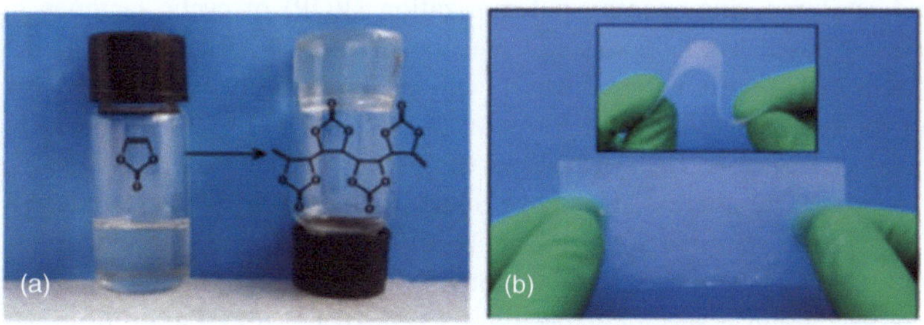

Figure 5.20 (a) A typical image of in situ polymerization of VC into PVCA after heating at 60 °C for 24 hours. (b) Digital images of cellulose/PVCA-LiDFOB composite solid polymer electrolyte. Source: Chai et al. [166]. ©2016/John Wiley & Sons/CC BY 4.0.

1000 cycles and high rate performance and safety performance, which has a high potential as an all-solid-state electrolyte.

In 2017, Zhang et al. [168] used $Li_{6.5}La_3Zr_{2.5}Ti_{0.5}O_{12}$(LLZTO) as an inorganic filler to further optimize the performance of a PPC electrolyte, with an improved ion conductivity, mechanical strength, and Li^+ migration number to varying degrees. They realized an all-solid-state battery that can withstand high pressure and normal temperature operations. It is an excellent idea to use solid polymers with commonly used electrolyte solvents and electrolyte additives for solid electrolytes, because these polymers have the advantages of high dielectric constant, high ambient ionic conductivity, and good interface compatibility with both positive and negative electrodes, but the processing technology for all-solid-state electrolytes needs further exploration.

Aliphatic polyesters featuring low glass transition temperatures and high dielectric constants are also used as polymer hosts for SPEs. The most extensively researched structures include polycaprolactones and polylactic acids. Polyesters typically serve as the primary component of SPE blends or block copolymers, wherein another polymer is introduced to improve the mechanical strength of the polyester-based SPE. Mindemark et al. [169] used a random copolymer of trimethylene carbonate (TMC) and caprolactone (CL) as the SPE for Li-based batteries. The low degree of crystallinity in this SPE benefited the existence of large amorphous regions, which facilitated ionic conduction. For an optimized ratio of TMC: CL (60, 40), a high ionic conductivity of the order of 10^{-5} S cm^{-1} was achieved at 60 °C. While the conductivity of this polyester may appear lower than that of conventional PEO-based SPEs, its pseudo-activation energy is significantly lower than that of PEO. This allows an experimental battery to function at room temperature, albeit with a low current density and small areal capacity. The most noteworthy benefit of polyester-based electrolytes is their high electrochemical stability against oxidation, making them ideal for coupling with high-voltage cathodes in Li-based batteries. Furthermore, it is established that individual esters undergo a reaction with Li metal because of the acidity of the α-hydrogen to the carbonyl group. However, there is a lack of research on the reactivity of these electrolytes with Li metal. Xue et al. [170] used the one-pot combination of reversible addition-fragmentation chain transfer (RAFT) polymerization and organic-acid-catalyzed ring-opening polymerization (ROP) of cyclic lactones to design polyester-based linear blocks and graft copolymers with controlled structures (Figure 5.21a). The ionic conductivity can be significantly enhanced by the poly(2-hydroxyethyl methacrylate)-graft-poly(ε-caprolactone) (PHEMA-g-PCL) graft copolymer (Figure 5.21b) synthesized using the one-pot method. At 30 °C, the ionic conductivity increased by two magnitudes, from 2.70×10^{-5} S cm^{-1} to 4.17×10^{-5} S cm^{-1}, compared with the linear polyester polymer electrolyte (Figure 5.21c). The assembled Li||LFP battery maintained a high coulombic efficiency of 99.7% and a specific capacity of 136 mAh g^{-1} (0.1 C) after 70 cycles (Figure 5.21d).

Polysiloxanes, also known as silicones and inorganic comb polymers, are characterized by a flexible backbone, chemical stability, thermal stability, and low toxicity.

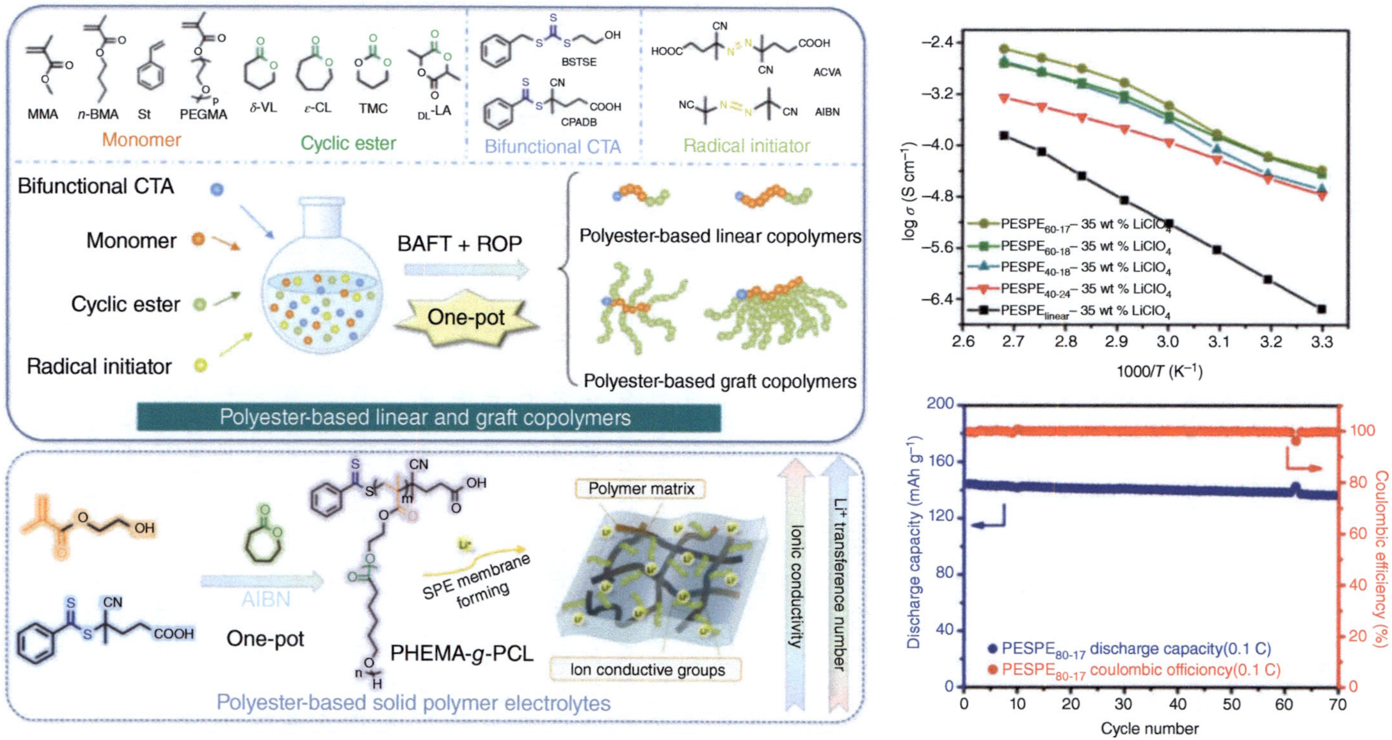

Figure 5.21 (a) Synthesis of linear block and graft polyester-based copolymers. (b) Schematic of the preparation of brush-shaped PESPEs. (c) Temperature dependence of ionic conductivity for PESPEs with 35 wt% of lithium salts. (d) Cycle performances of Li||LFP battery at 0.1 C, 30 °C. Source: Reproduced with permission from Guo et al. [170]. © 2021/Chinese Chemical Society.

They can form a variety of substrates by combining with silicon in a backbone of alternating silicon and oxygen atoms [171]. Polysiloxanes are widely regarded as promising SSEs due to their low glass transition temperature (T_g), high room temperature ionic conductivity, high thermal stability, and high chemical stability, but the low mechanical strength of polysiloxanes prevents them from being used as practical SSEs. Therefore, polysiloxane-based solid electrolytes combining both high ionic conductivity and high mechanical strength have been extensively studied in recent years. Zhang et al. [172] used oligoethylene glycol instead of polysiloxane as the ion-conducting support and designed a novel crosslinked polysiloxane-based solid electrolyte with oligoethylene glycol units as the internal plasticized chain segments, with an ionic conductivity as high as 1.33×10^{-4} S cm^{-1} at room temperature. This is attributable to the flexible siloxane backbone and the rapid oscillation of the short-chain amorphous PEO region, which provides a fast transport channel for Li^+. Subsequently, they [173] investigated a series of polysiloxane-based crosslinked networked polymer electrolytes containing oligo-poly(ethylene glycol), $(OCH_2CH_2)n$, etc., as internal free chain segments, and the highest ionic conductivity of 1.62×10^{-4} S cm^{-1} was achieved using $LiN(CF_3SO_2)_2$ as the lithium salt. They found that ionic conductivity depends on the crosslinking density, the low repetition of the poly(ethylene glycol) chain segments, and the length of the crosslinked skeleton. This combination of crosslinked backbone and free chain segments provides an important method to improve the ionic conductivity and mechanical strength of polymer electrolytes.

Plastic crystals possess positional order but orientational disorder in a certain temperature range, originating from the rotational motions of molecules or ions. Materials thus far known to exhibit the behavior of plastic crystals include inorganic salts, such as Li_2SO_4 and a-Na_3PO_4, as well as molecular species like succinonitrile (SN) and sebaconitrile, alongside organic ionic plastic crystals. Compared with traditional rigid crystals, plastic crystals show greater diffusivity and plasticity. Consequently, solid electrolytes based on plastic crystals containing dissolved Li salts generally have significantly higher conductivities. In recent years, nitrile materials have been widely used as solid electrolytes due to their high thermal stability and oxidation resistance. Alarco et al. [174] reported for the first time that a variety of salts can be dissolved in the plastic-crystalline phase of SN, and when the plastic-crystalline phase of SN is used as a solid electrolyte, an ionic conductivity as high as 3 mS cm^{-1} can be achieved at room temperature. The high polarity of the plastic-crystal phase makes it capable of dissociating various types of lithium salts, and its disordered structure makes it more plastic and diffusible than common rigid crystals. However, the addition of lithium salts will reduce the melting point of the plastic crystal, so that it shows a liquid-like behavior at room temperature and low mechanical strength, and it cannot form a self-supporting solid electrolyte layer. Therefore, SN is mostly added to mechanical reinforcement as a functional additive to prepare composite solid electrolytes with high ionic conductivity and high mechanical strength [175].

In order to further improve the mechanical strength of plastic-crystal-based solid electrolyte, Zhou et al. [176] formed a crosslinked network by in situ polymerization

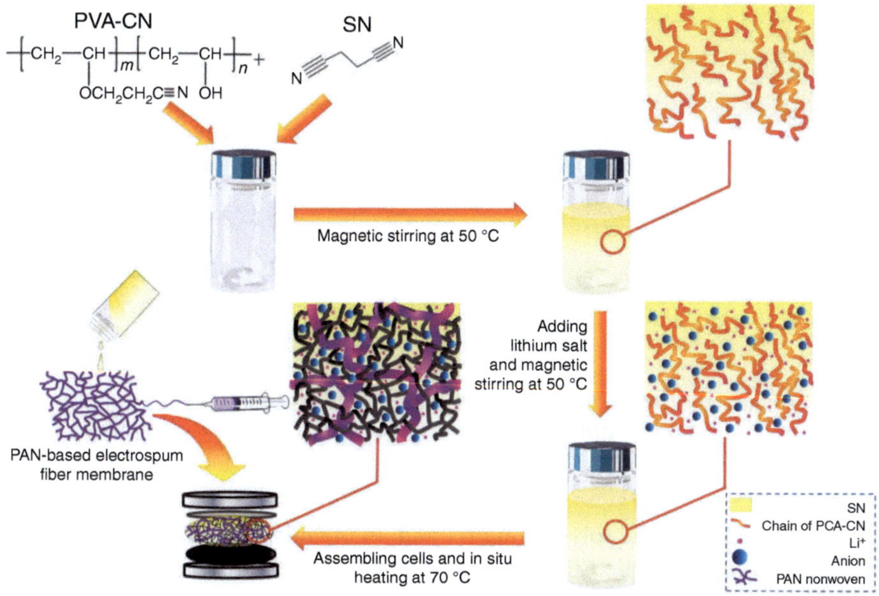

Figure 5.22 Schematic of the in situ synthesis route of SSEs based on nitrile materials. Source: Reproduced with permission from Zhou et al. [176]. © 2015/John Wiley & Sons.

of cyanoethyl polyvinyl alcohol (PVA-CN) in the solid electrolyte of SN and prepared a multilevel all-solid plastic-crystal-based solid electrolyte in situ using the PAN film obtained by electrospinning as the supporting skeleton. The solid electrolyte has a high room temperature ionic conductivity (3.02×10^{-3} S cm^{-1}), high Li^+ migration number (0.57), and high mechanical strength (15.31 MPa), which can effectively inhibit the growth of lithium dendrites, but the interfacial incompatibility between nitrile and lithium metal is not well resolved (Figure 5.22).

5.4.2.2 Gel Polymer Electrolytes

Although the pure polymer solid electrolyte has the advantage of high flexibility, its ionic conductivity is low and needs to be further improved. When an appropriate amount of a liquid component is added to the polymer electrolyte, the liquid component is dispersed in the polymer network as a filler. Although this system contains liquid components with no fluidity, it is called a gel electrolyte. In order to alleviate interface problems, SSBs are usually prepared by adding a liquid electrolyte to them. According to the content of the liquid electrolyte in batteries, they can be classified into gel-state, semi-solid-state, quasi-solid-state, and all-solid-state batteries (Figure 5.23). Some scholars also refer to SSBs containing part of liquid as semi-solid batteries. A semi-solid battery does not fundamentally solve the safety problem, so it is only a transitional product. With the gradual decrease in the liquid content, an all-solid-state battery is the final solution. However, at this stage, the GPE is the most promising electrolyte that can be easily adapted to the existing battery technology. The liquid component that swells the polymer matrix eliminates leakage problems and ensures the use of a GPE in LMBs.

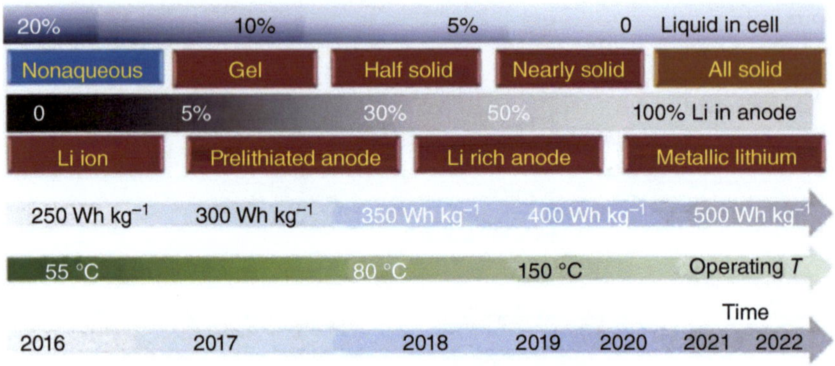

Figure 5.23 A roadmap from traditional lithium-ion batteries to solid-state lithium batteries. Source: Reproduced with permission from Li [177]. © 2016/Editorial office of Energy Storage Science and Technology.

A GPE, composed of a polymer matrix and an organic liquid component (typically referred to as organic LE), combines the advantages of both liquid and solid electrolytes. The polymer matrix used in GPEs exhibits no significant deviation from the aforementioned polymers, while high-mechanical-strength polymers like PVDF offer distinct advantages. The incorporation of the liquid component plays a crucial role in enhancing ionic conductivity and interfacial stability, while the polymer matrix offers sufficient mechanical strength and shape flexibility. Moreover, due to the exceptional flexibility and elasticity inherent in the polymer structure, conventional diaphragms are capable of withstanding volumetric effects from electrodes as well as short circuits caused by dendrite punctures during cell operation. Consequently, the rational design and fabrication of advanced GPEs represent a viable approach toward achieving safe and durable lithium metal batteries. Typically, a GPE consists of a polymeric backbone, lithium salts, and organic solvents. The essential prerequisites for a qualified polymer matrix include fast segmental movement, the ability to dissociate lithium salts, a low glass transition temperature, and good electrochemical/thermal stability.

GPEs play an important role in capturing electrochemical by-products from S or LMO cathodes and stabilizing cathode-electrolyte interphase (CEI) in high-voltage cathodes such as NMC and LNMO. Due to the size exclusion effect, GPEs with a porous structure have a physical adsorption effect on lithium polysulfide. However, physical blocking is relatively weak, and therefore, the chemical capture of by-products has become a popular research topic. GPEs with abundant polar groups are capable of strong chemical interactions with polysulfide and Mn ions, which can prevent the migration of by-products to the anode side. At the same time, the introduction of catalysts in GPEs can promote the conversion kinetics of polysulfides in LSBs. In addition, GPEs can stabilize CEI by two principles. First, they allow for higher electrochemical stability, thus preventing side reactions between the electrode and the electrolyte at high pressures. Second, they can form a stable interface with the electrode or participate in stabilizing the formation of CEI.

GPEs offer overwhelming advantages in protecting the cathode. Although various studies have shown that specific polar groups contribute to the trapping of by-products, a deep understanding of by-product–GPE interactions is lacking. Meanwhile, research has primarily focused on the development of GPE functionality in specific battery systems. Little attention has been paid to their mechanical properties, which are important for practical use. Therefore, a rational design of the polymer structure should achieve both excellent electrochemical/mechanical properties and functionality to solve cathode problems.

5.4.3 Performance Requirements and Design Strategies

A solid electrolyte is the core component of all-solid-state lithium batteries. For all-solid-state lithium batteries, especially those that can adapt to future EVs and large-scale energy storage applications, SSEs should meet the following requirements.

(1) High room temperature conductivity ($>10^{-4}$ S cm^{-1}).
(2) Electronic insulation (Li^+ migration number is approximately 1).
(3) Electrochemical window width (higher than 4.5 V relative to Li^+/Li).
(4) Good compatibility with electrode materials.
(5) Good thermal stability, moisture resistance, and excellent mechanical properties.
(6) Easy access to raw materials, low cost, and simple synthesis method.

To improve the ionic conductivity of solid electrolytes, researchers worldwide are trying to develop a variety of new solid electrolytes.

The crystal structure of ISSEs is an important factor in determining their ionic conductivity. In recent years, researchers have proposed many effective strategies for optimizing the crystal structure of ISSEs, including improved preparation methods, such as high-temperature solid-phase reaction, melt quenching, co-precipitation, and mechanochemical and gas-phase reaction. Doping different compounds or introducing different element ions, such as oxides, sulfides, halides, and cations with large ionic radius, has also been studied [178].

To investigate the relationship between the composition structure of superionic conductors and the electrical conductivity of lithium ions, Liang et al. [64] prepared a series of $Li_{3-3x}M_{1+x}Cl_6$($-0.14 < x \leq 0.2$, M = Tb, Dy, Ho, Y, Er, Tm) chloride superionic conductors. They found that the spatial group of the electrolyte structure changes from the tripartite phase (*P*3*m*1) to the orthogonal phase (*Pnma*) with an increase in the *x* value. Both *P*-3*m*1 and *Pnma* space groups are composed of hcp structures of Cl^- anions, but their cations (including Li^+ and M^{3+}) are arranged differently. In different Li-M-Cl systems, the activation energy barrier of lithium-ion diffusion is reduced and the diffusion rate of lithium ions is increased during the transition to *Pnma* phase. The room temperature ionic conductivity of new orthotropic materials such as $Li_{2.73}Ho_{1.09}Cl_6$ is more than four times that of the tripartite material Li_3HoCl_6 (0.3×10^{-3} S cm^{-1}). The ionic conductivity of a superionic conductor can be improved by changing the structure of the space group of ISSEs.

The ionic conductivity of typical sulfide electrolytes (Li_6PS_5Cl) can be improved by heteroatom substitution at Li, P, and S positions. In addition, the content of chlorine also has a significant influence on the crystal structure of Li_6PS_5Cl. Zeng et al. [179] explored the Cl^- content in a $Li_{7-x}PS_{6-x}Cl_x$ system, its microstructure, Li|SSE interface evolution, and the influence of lithium dendrite resistance. The experimental results showed that for medium and high degrees of chlorine content, Cl^- ions are not completely located in the lattice of argyrodite, but tend to form LiCl existing on the surface of the grain and interconnect to form the LiCl nanoshell framework. In the process of electrochemical cycle, Cl^- of the LiCl nanoshell will migrate to the lithium metal anode and electrolyte interface (Li|SSE) to form high-quality LiCl since the restrictions of the interface layer effectively suppress the interface between the cathode electrolyte electrochemical reaction and lithium dendrite growth and improve the lithium-ion transport.

Mo et al. [180] reported a lead-free organic–inorganic halide perovskite $MASr_{0.8}Li_{0.4}Cl_3$ (MA = methyl ammonium, formula CH_3NH_3) electrolyte, which has a highly symmetrical crystal structure and good composition adjustability. The density functional theory calculation showed that the hybrid perovskite framework can allow rapid lithium-ion migration without damaging the crystal structure. They also systematically studied the influence of different lithium contents on the migration rate of lithium ions in $MASr_{1-x}Li_{2x}Cl_3$ ($x = 0.1, 0.2, 0.3, 0.4$) and found that the ionic conductivity of $MASr_{0.8}Li_{0.4}Cl_3$ was the highest ($7.0\times10^{-6}\ S\ cm^{-1}$) when the lithium content was $x = 0.2$. In other words, the transition energy barrier of ions can be reduced by balancing lithium ions and Sr vacancies in $MASr_{0.8}Li_{0.4}Cl_3$ so as to realize the rapid transmission of lithium ions.

The preparation of garnet solid electrolytes usually requires sintering temperatures higher than 1100 °C to achieve densification and room temperature ionic conductivity higher than $0.1\ mS\ cm^{-1}$. Currently, Al_2O_3, Li_3BO_3, Li_4SiO_4, and other sintering additives are commonly used to reduce the sintering temperature of garnet electrolytes. Chen et al. [181] adopted the atomic layer deposition technology to uniformly introduce a layer of nano-amorphous alumina coating on the surface of ceramic particles as a sintering agent. This method of coating the sintering agent can improve the distribution uniformity and sintering activity of the sintering agent and successfully reduce the sintering temperature of the garnet electrolyte to below 1000 °C. The garnet electrolyte showed a higher room temperature ionic conductivity ($0.13\ mS\ cm^{-1}$) after sintering at 980 °C.

Polymer chain motion, dissociation energy between lithium ions and polymer chains, and the transport path of lithium ions are the key factors affecting the conductivity of SPE ions. For example, in PEO, there are three main strategies to improve its ionic conductivity: (i) a plasticizer is used to promote the dissociation of the lithium salt, reduce the crystallinity of the polymer, and improve the motility of the polymer chain; (ii) new lithium salt additives can be used to inhibit anion transport, enhance lithium-ion migration, and thus improve effective ion conductivity; (iii) the internal structure of the polymer intrinsic molecules can be improved and SPEs can be developed by optimizing the ionic conduction framework.

Xu et al. [182] proposed a PEO-based solid electrolyte design scheme. Using butyl dinitrile as a plasticizer, they constructed a homogeneous SPE with continuous lithium-ion transport channels by adjusting the molar ratio of butyl dinitrile to PEO. By adjusting the molar ratio of butylene dinitrile and PEO, the PEO-based solid electrolyte can be made homogeneous, the crystallization of PEO is inhibited, and the affinity between EO and the lithium ion is reduced. In other words, a fast lithium-ion transport channel is formed in the uniform SPE, and the ionic conductivity is increased by 100 times.

Arrese-Igor et al. [183] developed a bilayer polymer electrolyte composed of PPC and PEO. They measured the number of lithium-ion migrations using Bruce–Vincent and pulse-gradient NMR (PFG-NMR) methods. Experimental results showed that after the addition of poly-4-styrene sulfonyl (trifluoromethyl sulfonyl) imide lithium salt (LiPSTFSI), anionic ion (TFSI) and polystyrene chain bonding can greatly improve lithium-ion migration, inhibit the accumulation of anionic ions on the negative surface, and reduce the occurrence of side reactions, so as to show good lithium-ion transport performance.

Wen et al. [184] simultaneously induced 1,3-dioxy-naphthene (DOL) and tri-hydroxymethylpropane triglycide ether (TTE) to perform cationic ring-opening crosslinking polymerization in ultrathin vertical porous polymer film (LP) and successfully prepared an ultrathin crosslinked SPE (poly(DOLTTE)-LP). The obtained three-dimensional network crosslinking structure can greatly reduce the crystallinity between the local molecular chains and the number of readily oxidized hydroxyl groups at the end of the polymer solid electrolyte, resulting in a higher ionic conductivity at room temperature.

The properties of inorganic fillers are the key factors affecting the ionic conductivity of complex solid electrolytes. The interaction and binding mode between the inorganic filler and the polymer matrix are hot research topics in recent years. At present, there are three design ideas for solid electrolytes: (i) adding an inert or active filler to the polymer matrix; (ii) constructing an organic/inorganic double-layer or multilayer structure; and (iii) filling the three-dimensional inorganic skeleton with organic-phase components.

For example, Wang et al. [185] introduced lithium iodide (LiI) into a PEO-based electrolyte as a bifunctional additive to improve its ionic conductivity and construct a stable Li/PEO interface. Experimental results showed that there will be strong electrostatic interaction between the iodide ion and the hydrogen atom in the PEO chain, forming a large number of I—H bonds, making the PEO chain crosslinked, reducing the crystallinity of PEO, and thus improving the transmission rate of lithium ions. Zhang et al. [186] prepared a PAN/LLZTO fiber network material using electrostatic spinning technology and combined it with a PEO polymer as a filler to prepare a three-dimensional fiber-network-reinforced PEO solid electrolyte. Among them, the fiber/ceramic network material with a 3D skeleton structure can be evenly distributed in the PEO polymer matrix, significantly reducing the crystallinity of the PEO polymer and thus successfully building a continuous lithium-ion transport channel with high ionic conductivity ($1.76 \times 10^{-4}\,S\,cm^{-1}$, 30 °C). Wang et al. [187] used NaCl as a template to prepare a LATP framework with a three-dimensional interconnecting porous structure and then introduced PEO and LiTFSI into porous ceramic pores to construct a composite electrolyte (CPE-3D). This LATP-reinforced CPE shows high ionic conductivity and mechanical strength. The ionic conductivity of CPE-3D is $7.47 \times 10^{-4}\,S\,cm^{-1}$ at 60 °C, which is significantly higher than that of PEO ($2.2 \times 10^{-4}\,S\,cm^{-1}$), due to the formation of long-range continuous lithium-ion transport channels between the porous LATP framework and the PEO electrolyte.

5.5 Practical Problems and Critical Challenges

However, SSBs face several technical challenges and hurdles to commercialization. One of the major problems is the high cost and the limitation of production scale. At present, SSBs are expensive to manufacture and require mass production to achieve cost reduction. In addition, there are some technical problems in the process of SSBs, such as the interface problem between the electrode and the solid electrolyte, and the stability and reliability of the solid electrolyte.

5.5.1 Slow Li^+ Transportation

A common problem with SSEs is their relatively low ionic conductivity. Compared with electrolytes that can reach an ionic conductivity of $10^{-2}\,S\,cm^{-1}$ at room temperature, current state-of-the-art oxide and polymer SSEs typically only reach $10^{-4}\,S\,cm^{-1}$. The problem of this low ionic conductivity severely limits the performance of SSBs and their application scenarios.

For SPEs, the problem of low ionic conductivity is attributable to several factors:

(1) *Poor polymer chain mobility*: Solid-state polymeric electrolytes are typically composed of polymeric materials with a flexible chain structure. However, under solid-state conditions, the movement of the polymer chains is restricted, limiting the rapid transport of ions through the electrolyte material. Some polymer electrolyte materials have a high degree of crystallinity in the solid state and ion transport in the crystalline region is more difficult, resulting in lower ionic conductivity.

(2) *Interaction of lithium ions with the polymer*: The interaction between the functional groups on the polymer chains and the ions can also slow down the rate of ion transport. Ions in solid-state polymer electrolytes should interact with the polymer chains in order to be transported. If this interaction is too weak, it is difficult for the lithium ions to dissociate, resulting in low ionic conductivity. However, if such interactions are too strong, it can result in a strong bond between the ions and the polymer, e.g. the ether–oxygen bond in PEO is too strong for Li^+ coordination, which facilitates the dissociation of the lithium salt but hinders ion transport, resulting in a low lithium-ion mobility number.

(3) *Pore structure*: SPEs are usually prepared with a certain pore structure associated with them and are solvent free. The pore structure of all-solid electrolytes is often detrimental to ion transport and may also reduce the actual contact area with the electrode similarly in ceramic electrolytes, but the preparation of high-density solid electrolytes is still quite difficult.
(4) *Low temperature effect*: At low temperatures, the movement of the polymer chains is further restricted, so SPEs will have poorer ion transport performance at low temperatures. For high-activation-energy electrolyte systems, the temperature effect will be even more severe.

In addition, ISSEs suffer from some similar problems:

(1) *Crystal structure*: ISSEs usually consist of inorganic salts with a specific crystal structure. Ions in the crystal structure tend to be transported along specific pathways, which may be restricted, resulting in low ion transport rates.
(2) *Interaction of ions with crystal defects*: Ions in ISSEs should also interact with crystal defects in order to be transported. Crystal defects can provide additional vacancies and channels that facilitate ion transport. However, the interaction of ions with crystal defects may not be strong enough, limiting the efficiency of ion transport.
(3) *Crystal growth*: The preparation of ISSEs usually involves a crystal growth process. However, during crystal growth, the integrity and crystallinity of the lattice can be compromised, which in turn affects the efficiency of ion transport. Defects and imperfections in the crystal growth process can lead to interruptions and distortions in the ion transport pathway, and ion conductivity is usually lower at grain boundaries, making ion transport across grain boundaries more difficult than in crystals, which in turn limits the ion conductivity of solid electrolytes.
(4) *Space charge layer*: When the transition metal oxide cathode comes into contact with the sulfide electrolyte, lithium ions are driven by electric field forces to migrate from the sulfide electrolyte into the oxide cathode material until the potential at both ends of the interface reaches equilibrium, as lithium ions have a higher potential in the oxide than in the sulfide. However, when equilibrium is reached, a region of low lithium-ion concentration is formed at the interface between the sulfide electrolyte and the oxide cathode material, similar to a PN junction in an electron conductor, and this region is known as the space charge layer. Because of the low lithium-ion concentration in the space charge layer, the ionic conductivity in this region is low, resulting in a high migration potential for ions, causing a large increase in impedance.

5.5.2 Chemical and Electrochemical Instabilities

The interfacial instability of solid electrolytes at high or low voltages will lead to interfacial side reactions and electrolyte degradation, resulting in battery degradation or even failure. Therefore, the rational use of solid electrolytes within the electrochemical stability window of the electrolyte is essential to improve battery cycle life. Many theoretically calculated electrochemical stability windows for solid electrolytes are lower than those measured in experiments. This discrepancy is because first-principles calculations do not consider kinetic reaction factors in the cell, such as the chemical inertness of the electrode and passivation layer and the size of the effective contact area.

5.5.2.1 Chemical and Electrochemical Instability of SPEs

(i) Degradation of polymers: as lithium is extremely reductive, some polymers may react with lithium, resulting in interfacial degradation; e.g. PAN reacts with lithium, resulting in a high impedance interfacial layer, and PPC degrades when in contact with lithium. (ii) Side reactions from solvents: some SPEs are prepared using organic solvents or by adding solvents or plasticizers to improve the solubility and conductivity of lithium ions. However, during long-term use of SSBs, organic solvents may escape from the electrolyte or react with the electrode, resulting in reduced performance due to solvent loss from the electrolyte and increased interfacial impedance due to side reactions at the electrode/electrolyte interface. (iii) Typical polymeric solid electrolytes such as PEO and other polyether-based polymers have poor oxidative stability of the ether bonds (~3.8 V versus Li/Li^+) and are susceptible to oxidative decomposition when matched to ternary cathodes, leading to the degradation of electrolyte performance, low coulombic efficiency, and rapid loss of cell performance.

5.5.2.2 Chemical and Electrochemical Instability of Solid Oxide Electrolytes

(i) Lithium in solid oxide electrolytes is prone to volatilization during high-temperature calcination, leading to a decrease in ionic conductivity. LLZO ceramics react with moisture and CO_2 in air, leading to the formation of Li_2CO_3 on the surface, and lithium carbonate has poor infiltration of lithium metal, while the low ionic conductivity can seriously affect the performance of ceramic electrolytes. (ii) Lithium metal is extremely reductive, and Ti^{4+}-containing oxide electrolytes such as LLTO and LATP can be reduced to form titanium ions or lower-valence titanium metals. In situ transmission electron microscopy results show that the reaction of LAGP with lithium metal results in amorphization and volume expansion, leading to cracks in the solid electrolyte. Interfacial reactions between the solid oxide electrolyte and the electrode can degrade the electrochemical properties of the electrolyte and result in a high interfacial impedance, limiting ion transport.

5.5.2.3 Chemical Instability and Electrochemical Instability of Solid Sulfide Electrolytes

(i) The main reason for the poor air stability of sulfide electrolytes is that they are highly sensitive to moisture and can easily react with it to produce H_2S gas, especially for sulfide electrolytes containing P elements, as the bond energy of the P—S bond is much lower than that of the P—O bond, making the P-S structure easily oxidized or desulfurized. When a sulfide solid electrolyte material reacts irreversibly with oxygen and water vapor in the air, the material undergoes structural changes resulting in a reduction in ionic conductivity. (ii) The electrochemical stability window of solid sulfide electrolytes is narrow. At present, both theoretical calculations and experimental results indicate that the electrochemical stability window of solid sulfide electrolytes is highly limited, being about 1.6–2.3 V versus Li^+/Li [188, 189]. Therefore, almost all commonly used cathode materials are unable to form thermodynamically stable interfaces with solid sulfide electrolytes. In fact, despite the much higher mechanical strength of solid electrolytes compared with lithium metal, lithium dendrite growth was found to be easier in solid sulfide electrolytes than in liquid electrolytes. The voids and grain boundaries within the solid sulfide electrolyte and the instability of the lithium cathode/solid electrolyte interface are generally considered the two primary causes of lithium dendrite formation in sulfide electrolytes. Due to its narrow electrochemical stability window, the solid sulfide electrolyte is also unstable toward the lithium metal cathode and is reduced to an interfacial layer when in contact with the lithium metal.

5.5.3 Ineffective Facial Contact

The most obvious difference between a "solid–solid" and a "solid–liquid" contact is that the latter is a "soft" contact in the form of infiltration, whereas the former is a "hard" contact that is difficult to fully adhere. This directly contributes to the problem of reduced lithium-ion channels and stress build-up in all SSBs. If the solid electrolyte has poor contact with the electrode, the effective contact area is reduced, resulting in blocked ion transport between the electrolyte and the electrode. This can lead to an increase in the internal resistance of the cell, limiting the current density of the device and consequently reducing the power output capability of the cell. It can also lead to a reduction in cell capacity and a reduction in the charge/discharge rate and a reduction in cycle life, severely limiting the range of applications and performance of the cell. As current passes through the electrolyte, the contact resistance between the electrolyte and the electrode creates a potential difference, and the increased interfacial impedance caused by poor contact between the solid electrolyte and the electrode will result in more potential landing at the interface, exacerbating the polarization phenomenon, which will cause the battery to generate more heat during operation, affecting the efficiency and life of the battery. In addition, poor contact can result in uneven current distribution across the contact area. This will result in high current density in some areas and low current density in others. This unevenness can lead to localized overcharge and over-discharge of the cathode material or the formation of lithium dendrites, resulting in damage to the cell material or even cell failure. As poor contact increases resistance and energy loss, the current can cause localized heating in the contact area. This can lead to increased temperatures, further degrading the quality of the contact and even triggering the risk of device failure or thermal runaway, reducing the life and reliability of the SSB.

For example, in the process of delithiation or lithiation of the cathode active material, its cell parameters will inevitably shrink or expand, resulting in volume deformation of the material particles [190]. Especially in SSBs, the solid electrolyte is in rigid contact with the electrode particles, which is more sensitive to the volume change of the electrode material. During cycling, the contact between the electrode particles and between the electrode particles and the electrolyte is easily degraded, or the mechanical properties of the electrolyte fail due to the accumulation of stress, which in turn leads to the degradation of the electrochemical performance of the battery. The inevitable volume changes in the electrode material during cell cycling will lead to stress concentrations at the electrolyte/anode material particle interface and may even lead to the development of cracks. In LIBs, most cathode electrode materials undergo phase changes, lattice expansion/contraction, and structural changes during the lithium embedding/delithiation process, resulting in volume changes in the material after charging and discharging. This leads to a reduction in capacity.

For the lithium metal on the negative side, the cycling process is also associated with a huge change in volume; 1 mAh cm^{-2} of surface capacity corresponds to about 5 μm of lithium metal, while the surface capacity of commercial LIB electrodes is at least 3 mAh cm^{-2} or more. The shedding of lithium metal leaves voids at the interface, and continued shedding leads to the accumulation of interfacial holes, reduced contact area, and higher impedance, resulting in open circuit failure, while the high current density generated locally can also lead to dendrite growth and battery failure.

Therefore, good interfacial mechanical stability is also key to achieving high performance in SSBs.

5.5.4 Solid-State Composite Cathodes

Conventional cathodes are usually composed of active substances (AM), adhesives, and carbon additives. The presence of a liquid electrolyte provides good interfacial contact between the electrode/electrolyte and the inside of the electrode. However, in SSBs, almost all contact points are solid, and the interface polarization is serious due to the volume expansion of electrode materials during the cycle (Figure 5.24). In order to ensure ion conduction in the thick cathode of SSBs, it is necessary to introduce an ion conduction medium to construct a composite cathode. Common ion transport media are ionic liquids, small molecule plasticizers (butanedionitrile, etc.), ISSEs, etc.

For all-solid-state batteries, a sufficient amount of SSE should be introduced to ensure effective ion conduction in the cathode, thus improving the utilization rate of AM. The introduction of SSE does bring about a reduction in energy density compared to conventional commercial cathodes. The structure of the composite cathode also affects the transfer of Li^+ under high current density in practical applications, which has attracted wide attention. With respect to the optimization of SSE and AM, it is necessary to construct an effective ion and electron transport network inside the cathode to improve the energy density. For solid-state lithium batteries, SSEs are added to the composite positive electrode in order to establish an effective ion transfer network, but their inherent electronic insulation damages the overall electronic conductivity. Therefore, the cathode structure should be carefully designed to balance the conductivity of ions and electrons.

The first common strategy when formulating a composite cathode is to use a cathode electrolyte made from SSB components. Blue Solutions' only commercial

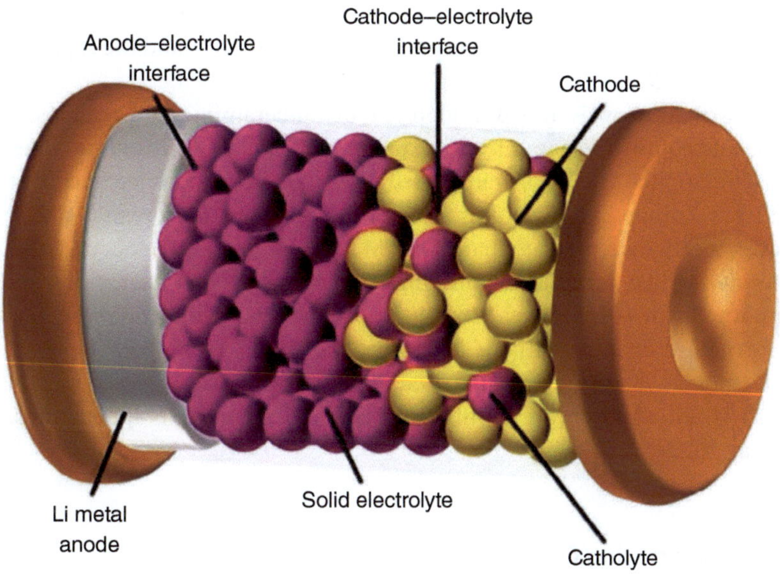

Figure 5.24 Schematic of the structure of an all-solid-state battery. Source: Reproduced with permission from Ujiie et al. [113]. © 2018/American Chemical Society.

SSB uses a cathode electrolyte with the same composition as the PEO: LiTFSI electrolyte. Passerini and coworkers [191] reported a Li/SPE/LFP SSB with a volume retention rate of ≈100% after 100 cycles at 0.1 C. The formula for the composite cathode they used is as follows: 43% LFP, 50% SPE (PEO LiTFSI-PYR14LiTFSI) as an ionic conductor and mechanical binder, and 7% conductive carbon. The PEO-LiTFSI-LLZO polymer-ceramic composite has also been used as a cathode electrolyte for lithium/PEO-LiTFSI-LLZO/LFP, reporting a peak specific discharge capacity of 137 mAh g^{-1} at 0.2 C. It is worth noting that the initial specific discharge capacity is only 94 mAh g^{-1}, and it takes 21 charge and discharge cycles to reach the peak specific discharge capacity [192]. This behavior was also observed in some other SSBs, suggesting that it takes a long time for cells to achieve complete interlayer homogenization and complete activation of AM particles.

A second common strategy in composite cathode formulations is to use a semicrystalline polymer electrolyte as the cathode electrolyte, even though SSE consists of a polymer-ceramic composite. This is because these polymers are more effective as adhesives than ceramics [193–195]. Zhang and coworkers [196] reported a high-performance Li/PEO-LLZTO-SN/LFP SSB with the following cathode composition: 70% LFP, 10% Super P, 15% PEO-LiTFSI, and 5% SN. The SSB they assembled achieved an initial specific discharge capacity of 151.1 mAh g^{-1} at 0.5 C and 60 °C and maintained 98% of the initial capacity after 200 cycles. In addition, butanedionitrile, a common polymer plasticizer, has also been shown to be an effective ionic conductor in composite cathodes, but it cannot be used as a binder alone because it does not provide mechanical stability. Therefore, it is sometimes coupled with conventional PVDF binders [197, 198]. The Li/SN-LLZNO/LFP cell with 35.7% SN in the cathode formula after 100 cycles at 0.05 C and room temperature, the specific discharge capacity reaches 149.8 mAh g^{-1} and the capacity retention rate reaches 99%.

A less common strategy is the development of all-ceramic oxide composite cathodes, typically coupled to ceramic SSE. The main advantage of these cathodes over the abovementioned cathodes is increased safety, as they are more capable of preventing thermal runaway by avoiding the use of any combustible organic matter [199]. However, recent data indicate that high AM loads are difficult to obtain (load range 50–58 wt%), which complicates the implementation of high-energy-density batteries with these types of cathodes. In addition, controlling interfacial resistance in the microstructure of these cathodes becomes tricky due to the poor bonding properties of ceramics. Kim et al. summarized different composite cathode formulations and preparation techniques for lithium garnet electrolyte SSB applications, as shown in Figure 5.25 [200].

5.5.5 High Manufacturing Costs

Currently, the cost of manufacturing SSBs is quite high, with the cost of the solid electrolyte accounting for a significant proportion of the cost; for example, the synthesis of LLZO requires mechanical chemical milling (e.g. ball milling) and long periods of high-temperature sintering (>1000 °C), while the sintering step often requires an additional sacrifice of LLZO as a "mother powder" to prevent lithium loss, further increasing the cost. Sulfide electrolytes are highly unstable in air and must be operated in an argon-filled environment, adding significantly to the complexity and cost of the manufacturing process. This results in the preparation and subsequent assembly of solid sulfide full cells, which must be operated in an argon-protected environment, making production costs high and severely limiting the pace of their industrialization. Techno-economic analysis suggests that the solid electrolyte should represent <35% of the total manufacturing cost. Unfortunately, current solid electrolyte processing is estimated to account for nearly 70% of the cost of manufacturing SSBs. The cost of processing one kilogram of LLZO, LGPS, and Li_6PS_5Cl is approximately US$ 2,000, US$ 69,500, and US$ 36,000, respectively, compared to that of solid polymers such as PEO and lithium salts (US$ 700 kg^{-1} and US$ 1980 kg^{-1}, respectively) [201].

In addition, laboratory-scale and benchmark studies typically use far too much lithium metal, with thicknesses of hundreds of microns, which is not practical. Lithium foil processing requires energy-intensive purification processes and an inert (argon) working environment, and lithium is difficult to process on a roll-to-roll basis due to its viscosity and low mechanical strength, so the price of lithium foil increases dramatically with decreasing thickness, from around US$ 310 kg^{-1} for large lithium metal to US$ 3953 kg^{-1} for 100 μm lithium foil, considering that commercial SSBs require a controlled excess of lithium to achieve high energy density. The requirement for lithium foil thickness on the negative side could be 20 μm, which would cost around US$ 12,327 kg^{-1} [202].

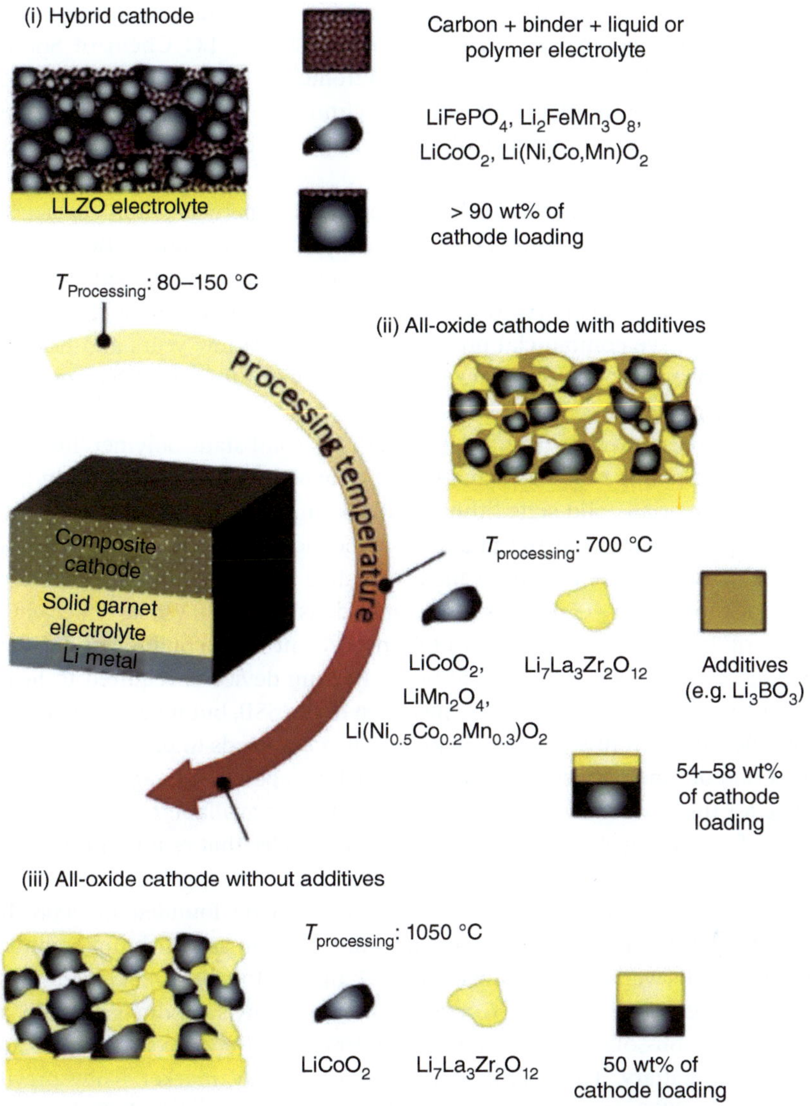

Figure 5.25 An overview of composite cathode processing for each design option. Source: Reproduced with permission from Kim and Rupp [200]. © 2020/Royal Society of Chemistry.

5.6 Practical Advances in Electric Vehicles and Other Areas

The industrialization of SSBs is still in its early stages, and large-scale commercial production has yet to be fully realized. Nevertheless, numerous companies and research institutions have made significant strides in the development of SSB technology, investing substantial amounts of money and resources into research, development, and promotion. Some large multinational companies and start-ups have carried out research and development work on SSBs and have made some breakthroughs. These companies include Toyota of Japan, LG Chem of South Korea, and QuantumScape of the USA. Some automobile manufacturers are also actively involved in the development of SSB technology, hoping to improve the driving range and safety performance of EVs through the commercial application of SSBs.

Although SSBs face some technical challenges and commercialization obstacles, the industrialization of SSBs is still a concern and is considered one of the important development directions of next-generation battery technology. With further advances in technology and efforts to address the aforementioned challenges, SSBs are expected to achieve commercial production in the next few years and lead to higher-performance and safer and more reliable battery solutions for EVs, portable devices, and energy storage.

In 2011, Bollore, a French firm, commercialized solid-state polymer lithium metal batteries for public rental EVs and public transport. It is the first case in the world to use polymer all-solid-state lithium batteries in EVs, and it is also the only all-solid-state power battery at the practical application level. The battery uses a $LiFePO_4$ positive electrode, a lithium metal negative electrode, and a SPE, with a cycle life of more than 4000 times, an energy density of 230 Wh kg^{-1} per cell, and a range of 250 km. However, due to the insufficient ionic conductivity of the electrolyte at room temperature, an additional heating device is required to heat the battery pack to 50–80 °C to meet the operation of the SSB, but no cooling device is required. Because the all-solid-state battery of this car needs to be preheated for a certain period of time before use in order to reach its operating temperature, it is not suitable for the mass market that demands immediate availability. Instead, it is better suited for passenger cars or other industrial vehicles that require prolonged operation.

ProLogium is a Taiwan-based SSB technology company founded in 2006. It focuses on the development and commercialization of solid-state lithium ceramic batteries. Their SSBs were mass-produced in 2017, mainly for consumer electronics and wearable electronics markets. Currently, the company's products contain a small amount of non-solid components (about 10%), but they are pretty close to an all-solid-state battery. Then, ProLogium entered the EV market. They previously said that in 2023, the world's first solid-state lithium ceramic battery mass production line was launched for trial production, from 0.5 GWh, toward the full annual capacity of 3 GWh (equivalent to 30,000 units of 100 kWh EVs) target mass production.

Qingtao Energy Development Co., Ltd., was founded in 2014 by Nan Cewen, an academician of the Chinese Academy of Sciences. In March 2021, Qingtao completed the F round of financing led by the Shanghai Science and Technology Fund, and the company's valuation has exceeded 10 billion yuan. According to the disclosure of Qingtao, the company has built a complete industrial ecological chain of "new energy materials – solid lithium batteries – automation equipment – comprehensive utilization of lithium battery resources – scientific research achievements incubation – industrial investment." In terms of technical route,

Qing Tao primarily uses nanoscale lithium lanthanum titanium oxide and lithium lanthanum zirconium oxide as solid electrolyte materials. In terms of production capacity, Qingtao built the first solid-state lithium battery production line in China in 2018, with a production capacity of 0.1 GWh, and the products are applied in the field of special safety. In July 2020, Qingtao completed and put into operation the first-phase solid-state lithium battery project with an annual output of 1 GWh in Yichun City, Jiangxi Province, with an investment of 550 million yuan, and plans to add 9 GWh of capacity in the second phase with an investment of 4.95 billion yuan. The first generation of mass-produced power solid-state lithium battery products developed by Qingtao passed the national motor vehicle product inspection in 2019 and carried out cooperation and communication with many automakers such as BAIC, SAIC, GAC, and United. United and BAIC demonstrated a pure electric prototype equipped with the Qingtao SSB system and the BAIC Blue Valley SSB version in July 2020.

Beijing WeLion New Energy Technology Co., Ltd., was established in 2016, relying on the Institute of Physics of the Chinese Academy of Sciences (CAS), and was co-founded by Chen Liquan, an academician of the Chinese Academy of Engineering; Li Hong, a researcher at the Institute of Physics of CAS; and Yu Huigen, former chief engineer of BAIC New Energy. They used the oxide SSB technology. According to public information, the company has completed the design and development of high-nickel-ternary cathode hybrid SSBs with more than 300 Wh kg^{-1}, which have been sent to a number of domestic and international OEMs for sample testing and have successfully passed abuse tests such as pinprick, extrusion, overcharge, and short circuit, with a cycle life of more than 1200 times. On June 30, 2023, Beijing WeLion New Energy 360 Wh kg^{-1} LIB cells were delivered to NIO Inc. The battery uses a semisolid electrolyte, silicon–carbon composite anode material, and ultrahigh nickel cathode material, with a single energy density of 360 Wh kg^{-1}, a range of up to 1000 km, and less than 10% liquid electrolyte content, which is a semisolid battery.

Toyota, one of the world's largest automakers, began investing in SSBs in 2008. Toyota and Panasonic partnered in 2020 to establish Prime Planet Energy & Solutions, a company dedicated to the development and commercial application of the SSB technology, which Toyota plans to incorporate into its long-term electrification strategy using the SSB technology in its future EVs. In 2020, Toyota unveiled its concept car with an all-solid-state battery but did not reveal any details on technical parameters or manufacturing costs. At the same time, Toyota has updated the timetable for the all-solid-state battery to "come on board," announcing that it will be available in battery electrical vehicles (BEVs) and challenging to put it into practical use in 2027–2028. According to the Nikkei Index and Tokyo-based research firm Patent Result, Toyota will have 1331 known patents on SSBs by 2022, including battery structures, materials, and manufacturing processes. On July 5, 2023, they claimed to have made a major breakthrough in SSB technology, capable of halving the weight, size, and cost of batteries, which could provide a huge boost to the development of EVs. Keiji Kaita, president of Toyota's Carbon Neutral R&D Centre, said: "Whether it's a liquid or solid-state battery, our goal is to completely change the current situation where batteries are too big, too heavy and too expensive, and in terms of the potential for this, we will aim to halve these factors." He noted that Toyota had developed ways to improve battery durability and believes it can now build SSBs with a range of 1200 km and a recharge time of 10 minutes or less.

Liquid lithium batteries have almost reached the energy density limit, and there remains the risk of thermal runaway. SSBs have the potential to significantly increase battery safety and energy density, making them a critical factor in the next round of power battery technology competition. During the past two decades, solid electrolytes and batteries have made remarkable progress, resulting in numerous new materials, structures, designs, and concepts emerging. Simultaneously, with the speedy growth of sectors like EVs and energy storage, materials associated with SSBs are ripe for extensive development opportunities. Leading automobile and battery companies, investment institutions, and worldwide scientific research institutions are actively investing financial and technological resources in fostering the industrialization of SSBs. From the current viewpoint, SSBs are still in their early stages of industrialization, and it will take a minimum of 3–10 years to achieve the commercialization and widespread use of SSBs, especially all-solid-state batteries.

References

1 Quartarone, E. and Mustarelli, P. (2011). Electrolytes for solid-state lithium rechargeable batteries: recent advances and perspectives. *Chemical Society Reviews* 40 (5): 2525–2540.

2 Chen, Z., Kim, G.-T., Kim, J.-K. et al. (2021). Highly stable quasi-solid-state lithium metal batteries: reinforced $Li_{1.3}Al_{0.3}Ti_{1.7}(PO_4)_3$/Li interface by a protection interlayer. *Advanced Energy Materials* 11 (30): 2101339.

3 Li, Y., Zhang, J., Chen, Q. et al. (2021). Emerging of heterostructure materials in energy storage: a review. *Advanced Materials* 33 (27): 2100855.

4 Sun, P., Bisschop, R., Niu, H. et al. (2020). A review of battery fires in electric vehicles. *Fire Technology* 56 (4): 1361–1410.

5 Feng, X., Ouyang, M., Liu, X. et al. (2018). Thermal runaway mechanism of lithium ion battery for electric vehicles: a review. *Energy Storage Materials* 10: 246–267.

6 Roth, E.P. and Orendorff, C.J. (2012). How electrolytes influence battery safety. *Interface Magazine* 21 (2): 45–49.

7 Niu, C., Lee, H., Chen, S. et al. (2019). High-energy lithium metal pouch cells with limited anode swelling and long stable cycles. *Nature Energy* 4 (7): 551–559.

8 Chen, Z., Gao, X., Kim, J.-K. et al. (2021). Quasi-solid-state lithium metal batteries using the $LiNi_{0.8}Co_{0.1}Mn_{0.1}O_2$–$Li_{1+x}Al_xTi_{2-x}(PO_4)_3$ composite positive electrode. *ACS Applied Materials & Interfaces* 13 (45): 53810–53817.

9 Yu, Z., Cui, Y., and Bao, Z. (2020). Design principles of artificial solid electrolyte interphases for lithium-metal anodes. *Cell Reports Physical Science* 1 (7): 100119.

10 Asenbauer, J., Eisenmann, T., Kuenzel, M. et al. (2020). The success story of graphite as a lithium-ion anode material – fundamentals, remaining challenges, and recent developments including silicon (oxide) composites. *Sustainable Energy & Fuels* 4 (11): 5387–5416.

11 Chen, Z., Kim, G.-T., Wang, Z. et al. (2019). 4-V flexible all-solid-state lithium polymer batteries. *Nano Energy* 64: 103986.

12 Lin, D., Liu, Y., and Cui, Y. (2017). Reviving the lithium metal anode for high-energy batteries. *Nature Nanotechnology* 12 (3): 194–206.

13 Weber, R., Genovese, M., Louli, A.J. et al. (2019). Long cycle life and dendrite-free lithium morphology in anode-free lithium pouch cells enabled by a dual-salt liquid electrolyte. *Nature Energy* 4 (8): 683–689.

14 Zhang, X.-Q., Cheng, X.-B., and Zhang, Q. (2018). Advances in interfaces between Li metal anode and electrolyte. *Advanced Materials Interfaces* 5 (2): 1701097.

15 Cheng, X.-B., Zhang, R., Zhao, C.-Z. et al. (2017). Toward safe lithium metal anode in rechargeable batteries: a review. *Chemical Reviews* 117 (15): 10403–10473.

16 Wang, F., Li, Y., Xia, X. et al. (2021). Metal–CO_2 electrochemistry: from CO_2 recycling to energy storage. *Advanced Energy Materials* 11 (25): 2100667.

17 Wang, Q., Jiang, L., Yu, Y. et al. (2019). Progress of enhancing the safety of lithium ion battery from the electrolyte aspect. *Nano Energy* 55: 93–114.

18 Zhao, X., Liang, X., Li, Y. et al. (2021). Challenges and design strategies for high performance aqueous zinc ion batteries. *Energy Storage Materials* 42: 533–569.

19 Lin, D., Liu, Y., and Cui, Y. (2017). Reviving the lithium metal anode for high-energy batteries. *Nature Nanotechnology* 12 (3): 194.

20 Or, T., Gourley, S.W.D., Kaliyappan, K. et al. (2020). Recycling of mixed cathode lithium-ion batteries for electric vehicles: current status and future outlook. *Carbon Energy* 2 (1): 6–43.

21 Zhang, B., Tan, R., Yang, L. et al. (2018). Mechanisms and properties of ion-transport in inorganic solid electrolytes. *Energy Storage Materials* 10: 139–159.

22 Zhao, Q., Stalin, S., Zhao, C.-Z. et al. (2020). Designing solid-state electrolytes for safe, energy-dense batteries. *Nature Reviews Materials* 5 (3): 229–252.

23 Xue, Z., He, D., and Xie, X. (2015). Poly(ethylene oxide)-based electrolytes for lithium-ion batteries. *Journal of Materials Chemistry A* 3 (38): 19218–19253.

24 Wu, Y., Li, Y., Wang, Y. et al. (2022). Advances and prospects of PVDF based polymer electrolytes. *Journal of Energy Chemistry* 64: 62–84.

25 Wu, F., Zhang, K., Liu, Y. et al. (2020). Polymer electrolytes and interfaces toward solid-state batteries: recent advances and prospects. *Energy Storage Materials* 33: 26–54.

26 Tang, S., Guo, W., and Fu, Y. (2021). Advances in composite polymer electrolytes for lithium batteries and beyond. *Advanced Energy Materials* 11 (2): 2000802.

27 Dirican, M., Yan, C., Zhu, P. et al. (2019). Composite solid electrolytes for all-solid-state lithium batteries. *Materials Science and Engineering: R: Reports* 136 (APR.): 27–46.

28 Aziz, S.B., Woo, T.J., Kadir, M.F.Z. et al. (2018). A conceptual review on polymer electrolytes and ion transport models. *Journal of Science: Advanced Materials and Devices* 3 (1): 1–17.

29 Croce, F., Appetecchi, G., Persi, L. et al. (1998). Nanocomposite polymer electrolytes for lithium batteries. *Nature* 394 (6692): 456–458.

30 Meng, N., Zhu, X., and Lian, F. (2022). Particles in composite polymer electrolyte for solid-state lithium batteries: a review. *Particuology* 60: 14–36.

31 Fan, L.-Z., He, H., and Nan, C.-W. (2021). Tailoring inorganic–polymer composites for the mass production of solid-state batteries. *Nature Reviews Materials* 6 (11): 1003–1019.

32 Zheng, J., Tang, M., and Hu, Y.Y. (2016). Lithium ion pathway within $Li_7La_3Zr_2O_{12}$-polyethylene oxide composite electrolytes. *Angewandte Chemie International Edition* 55 (40): 12538–12542.

33 Chen, L., Li, Y., Li, S.-P. et al. (2018). PEO/garnet composite electrolytes for solid-state lithium batteries: from "ceramic-in-polymer" to "polymer-in-ceramic". *Nano Energy* 46: 176–184.

34 Gupta, A. and Sakamoto, J. (2019). Controlling ionic transport through the PEO-LiTFSI/LLZTO interface. *The Electrochemical Society Interface* 28 (2): 63.

35 Ji, X., Zhang, Y., Cao, M. et al. (2022). Advanced inorganic/polymer hybrid electrolytes for all-solid-state lithium batteries. *Journal of Advanced Ceramics* 11 (6): 835–861.

36 Lou, S., Zhang, F., Fu, C. et al. (2021). Interface issues and challenges in all-solid-state batteries: lithium, sodium, and beyond. *Advanced Materials* 33 (6): 2000721.

37 Zhu, Y., Wang, F., Liu, L. et al. (2013). Composite of a nonwoven fabric with poly(vinylidene fluoride) as a gel membrane of high safety for lithium ion battery. *Energy & Environmental Science* 6 (2): 618–624.

38 Bi, H., Sui, G., and Yang, X. (2014). Studies on polymer nanofibre membranes with optimized core–shell structure as outstanding performance skeleton materials in gel polymer electrolytes. *Journal of Power Sources* 267: 309–315.

39 Prasanth, R., Shubha, N., Hng, H.H. et al. (2014). Effect of poly(ethylene oxide) on ionic conductivity and electrochemical properties of poly(vinylidenefluoride) based polymer gel electrolytes prepared by electrospinning for lithium ion batteries. *Journal of Power Sources* 245: 283–291.

40 Zhu, M., Wu, J., Wang, Y. et al. (2019). Recent advances in gel polymer electrolyte for high-performance lithium batteries. *Journal of Energy Chemistry* 37: 126–142.

41 Manthiram, A., Yu, X., and Wang, S. (2017). Lithium battery chemistries enabled by solid-state electrolytes. *Nature Reviews Materials* 2 (4): 1–16.

42 Gao, Z., Sun, H., Fu, L. et al. (2018). Promises, challenges, and recent progress of inorganic solid-state electrolytes for all-solid-state lithium batteries. *Advanced Materials* 30 (17): 1705702.

43 Bachman, J.C., Muy, S., Grimaud, A. et al. (2016). Inorganic solid-state electrolytes for lithium batteries: mechanisms and properties governing ion conduction. *Chemical Reviews* 116 (1): 140–162.

44 Knauth, P. (2009). Inorganic solid Li ion conductors: an overview. *Solid State Ionics* 180 (14–16): 911–916.

45 Deng, Y., Eames, C., Chotard, J.-N. et al. (2015). Structural and mechanistic insights into fast lithium-ion conduction in Li_4SiO_4–Li_3PO_4 solid electrolytes. *Journal of the American Chemical Society* 137 (28): 9136–9145.

46 Kanno, R., Hata, T., Kawamoto, Y. et al. (2000). Synthesis of a new lithium ionic conductor, thio-LISICON–lithium germanium sulfide system. *Solid State Ionics* 130 (1-2): 97–104.

47 Theo, H. (2005). *International Tables for Crystallography*, Volume A: Space-Group Symmetry. Dordrecht: Springer.

48 Fu, J. (1997). Superionic conductivity of glass-ceramics in the system Li_2O-Al_2O_3-TiO_2-P_2O_5. *Solid State Ionics* 96 (3-4): 195–200.

49 Fu, J. (1998). Fast Li^+ ion conduction in Li_2O–(Al_2O_3 $Ga_{22}O_3$)–TiO_2–P_2O_5 glass–ceramics. *Journal of Materials Science* 33 (6): 1549–1553.

50 Rettenwander, D., Welzl, A., Pristat, S. et al. (2016). A microcontact impedance study on NASICON-type $Li_{1+x}Al_xTi_{2-x}(PO_4)_3$ ($0 \leqslant x \leqslant 0.5$) single crystals. *Journal of Materials Chemistry A* 4 (4): 1506–1513.

51 Vizgalov, V.A., Nestler, T., Trusov, L.A. et al. (2018). Enhancing lithium-ion conductivity in NASICON glass-ceramics by adding yttria. *CrystEngComm* 20 (10): 1375–1382.

52 Silva, d.'A.A., I., Nieto-Muñoz, A.M., Rodrigues, A.C.M. et al. (2020). Structure and lithium-ion mobility in $Li_{1.5}M_{0.5}Ge_{1.5}(PO_4)_3$ (M= Ga, Sc, Y) NASICON glass-ceramics. *Journal of the American Ceramic Society* 103 (7): 4002–4012.

53 Abo-Mosallam, H.A. and Farag, M. (2020). Preparation, crystallization features and electro-magnetic properties of phosphate based glass-ceramics as solid electrolyte for lithium-ion batteries. *Journal of the Australian Ceramic Society* 56: 353–361.

54 Luo, Z., Qin, C., Wu, Y. et al. (2020). Structure and properties of Fe_2O_3-doped $50Li_2O$-$10B_2O_3$-$40P_2O_5$ glass and glass-ceramic electrolytes. *Solid State Ionics* 345: 115177.

55 Ren, Y., Chen, K., Chen, R. et al. (2015). Oxide electrolytes for lithium batteries. *Journal of the American Ceramic Society* 98 (12): 3603–3623.

56 Chandra, A., Bhatt, A., and Chandra, A. (2013). Ion conduction in superionic glassy electrolytes: an overview. *Journal of Materials Science and Technology* 29 (3): 193–208.

57 Thokchom, J.S. and Kumar, B. (2010). The effects of crystallization parameters on the ionic conductivity of a lithium aluminum germanium phosphate glass–ceramic. *Journal of Power Sources* 195 (9): 2870–2876.

58 Goodenough, J.B., Hong, H.-P., and Kafalas, J. (1976). Fast Na^+-ion transport in skeleton structures. *Materials Research Bulletin* 11 (2): 203–220.

59 Anantharamulu, N.A., Koteswara Rao, K., Rambabu, G. et al. (2011). A wide-ranging review on Nasicon type materials. *Journal of Materials Science* 46: 2821–2837.

60 Guin, M. and Tietz, F. (2015). Survey of the transport properties of sodium superionic conductor materials for use in sodium batteries. *Journal of Power Sources* 273: 1056–1064.

61 Aono, H., Sugimoto, E., Sadaoka, Y. et al. (1990). Ionic conductivity of solid electrolytes based on lithium titanium phosphate. *Journal of the Electrochemical Society* 137 (4): 1023.

62 Iglesias, J.E. and Pecharromán, C. (1998). Room temperature triclinic modification of NASICON-type $LiZr_2(PO_4)_3$. *Solid State Ionics* 112 (3-4): 309–318.

63 Wang, Y., Song, S., Xu, C. et al. (2019). Development of solid-state electrolytes for sodium-ion battery–a short review. *Nano Materials Science* 1 (2): 91–100.

64 Chunchun, Q., Zhiwei, L., Haozhang, L. et al. (2021). Research status and prospect of NASICON-type glass-ceramic electrolytes. *Journal of Ceramics* 42 (1): 1–20.

65 Francisco, B.E., Stoldt, C.R., and M'Peko, J.-C. (2015). Energetics of ion transport in NASICON-type electrolytes. *The Journal of Physical Chemistry C* 119 (29): 16432–16442.

66 Arbi, K., Mandal, S., Rojo, J. et al. (2002). Dependence of ionic conductivity on composition of fast ionic conductors $Li_{1+x}Ti_{2-x}Al_x(PO_4)_3$, $0 \leqslant x \leqslant 0.7$. A parallel NMR and electric impedance study. *Chemistry of Materials* 14 (3): 1091–1097.

67 Lang, B., Ziebarth, B., and Elsässer, C. (2015). Lithium ion conduction in $LiTi_2(PO_4)_3$ and related compounds based on the NASICON structure: a first-principles study. *Chemistry of Materials* 27 (14): 5040–5048.

68 Kahlaoui, R., Arbi, K., Sobrados, I. et al. (2017). Cation miscibility and lithium mobility in NASICON $Li_{1+x}Ti_{2-x}Sc_x(PO_4)_3$ ($0 \leqslant x \leqslant 0.5$) series: a combined NMR and impedance study. *Inorganic Chemistry* 56 (3): 1216–1224.

69 Arbi, K., Hoelzel, M., Kuhn, A. et al. (2013). Structural factors that enhance lithium mobility in fast-ion $Li_{1+x}Ti_{2-x}Al_x(PO_4)_3$($0 \leqslant x \leqslant 0.4$) conductors investigated by neutron diffraction in the temperature range 100–500 K. *Inorganic Chemistry* 52 (16): 9290–9296.

70 Monchak, M., Hupfer, T., Senyshyn, A. et al. (2016). Lithium diffusion pathway in $Li_{1.3}Al_{0.3}Ti_{1.7}(PO_4)_3$ (LATP) superionic conductor. *Inorganic Chemistry* 55 (6): 2941–2945.

71 Pérez-Estébanez, M., Isasi-Marín, J., Többens, D. et al. (2014). A systematic study of Nasicon-type $Li_{1+x}M_xTi_{2-x}(PO_4)_3$ (M: Cr, Al, Fe) by neutron diffraction and impedance spectroscopy. *Solid State Ionics* 266: 1–8.

72 Ma, F., Zhao, E., Zhu, S. et al. (2016). Preparation and evaluation of high lithium ion conductivity $Li_{1.3}Al_{0.3}Ti_{1.7}(PO_4)_3$ solid electrolyte obtained using a new solution method. *Solid State Ionics* 295: 7–12.

73 Kuwano, J., Sato, N., Kato, M. et al. (1994). Ionic conductivity of $LiM_2(PO_4)_3$ (M=Ti, Zr, Hf) and related compositions. *Solid State Ionics* 70–71: 332–336.

74 Ortiz, G.F., López, M.C., Lavela, P. et al. (2014). Improved lithium-ion transport in NASICON-type lithium titanium phosphate by calcium and iron doping. *Solid State Ionics* 262: 573–577.

75 Xie, H., Goodenough, J.B., and Li, Y. (2011). $Li_{1.2}Zr_{1.9}Ca_{0.1}(PO_4)_3$, a room-temperature Li-ion solid electrolyte. *Journal of Power Sources* 196 (18): 7760–7762.

76 Li, Y., Liu, M., Liu, K. et al. (2013). High Li^+ conduction in NASICON-type $Li_{1+x}Y_xZr_{2-x}(PO_4)_3$ at room temperature. *Journal of Power Sources* 240: 50–53.

77 Xu, X.-X., Wen, Z.-Y., Gu, Z.-H. et al. (2004). High lithium conductivity in $Li_{1.3}Cr_{0.3}Ge_{1.7}(PO_4)_3$ glass-ceramics. *Materials Letters* 58 (27): 3428–3431.

78 Illbeigi, M., Fazlali, A., Kazazi, M. et al. (2016). Effect of simultaneous addition of aluminum and chromium on the lithium ionic conductivity of $LiGe_2(PO_4)_3$ NASICON-type glass–ceramics. *Solid State Ionics* 289: 180–187.

79 Best, A., Newman, P.J., Macfarlane, D.R. et al. (1999). Characterisation and impedance spectroscopy of substituted $Li_{1.3}Al_{0.3}Ti_{1.7}(PO_4)_{3-x}(ZO_4)_x$ (Z= V, Nb) ceramics. *Solid State Ionics* 126 (1-2): 191–196.

80 Geng, S., Yang, Y., Zhang, Y. et al. (2015). Effect of $VO_4{}^{3-}$ substitution for $PO_4{}^{3-}$ on electrical conductivity in the Nasicon $Li_3Sc_2(PO_4)_3$ compound. *Electrochimica Acta* 176: 327–333.

81 Fu, J. (1997). Fast Li^+ ion conduction in Li_2O-Al_2O_3-TiO_2-SiO_2-P_2O_2 glass-ceramics. *Journal of the American Ceramic Society* 80 (7): 1901–1903.

82 Lu, X. and Xiao, H. (2016). Improving ionic conductivity of polycrystalline lithium ion conductors by interacting with mesoporous materials. *Materials Letters* 177: 50–53.

83 Mengqin, T.A.O., Zhenfei, C.A.I., Huimin, W.U. et al. (2022). Research progress of $Li_7La_3Zr_2O_{12}$ solid electrolyte based on first-principles calculation. *Jorunal of Functional Materials* 53 (8): 8067–8077.

84 Thangadurai, V., Kaack, H., and Weppner, W.J.F. (2003). Novel fast lithium ion conduction in garnet-type $Li_5La_3M_2O_{12}$ (M = Nb, Ta). *Journal of the American Ceramic Society* 86 (3): 437–440.

85 Thangadurai, V. and Weppner, W. (2005). Investigations on electrical conductivity and chemical compatibility between fast lithium ion conducting garnet-like $Li_6BaLa_2Ta_2O_{12}$ and lithium battery cathodes. *Journal of Power Sources* 142 (1): 339–344.

86 Murugan, R., Thangadurai, V., and Weppner, W. (2007). Fast lithium ion conduction in garnet-Type $Li_7La_3Zr_2O_{12}$. *Angewandte Chemie International Edition* 46 (41): 7778–7781.

87 Awaka, J., Kijima, N., Hayakawa, H. et al. (2009). Synthesis and structure analysis of tetragonal $Li_7La_3Zr_2O_{12}$ with the garnet-related type structure. *Journal of Solid State Chemistry* 182 (8): 2046–2052.

88 Thompson, T., Wolfenstine, J., Allen, J.L. et al. (2014). Tetragonal vs. cubic phase stability in Al–free Ta doped $Li_7La_3Zr_2O_{12}$(LLZO). *Journal of Materials Chemistry A* 2 (33): 13431–13436.

89 Wang, Y., Chen, Z., Jiang, K., et al. (2024). Accelerating the Development of LLZO in Solid-State Batteries Toward Commercialization: A Comprehensive Review. *Small* n/a: 2402035.

90 Wang, C., Fu, K., Kammampata, S.P. et al. (2020). Garnet-type solid-state electrolytes: materials, interfaces, and batteries. *Chemical Reviews* 120 (10): 4257–4300.

91 Meier, K., Laino, T., and Curioni, A. (2014). Solid-state electrolytes: revealing the mechanisms of Li-ion conduction in tetragonal and cubic LLZO by first-principles calculations. *The Journal of Physical Chemistry C* 118 (13): 6668–6679.

92 Wu, J.-F., Chen, E.-Y., Yu, Y. et al. (2017). Gallium-doped $Li_7La_3Zr_2O_{12}$ garnet-type electrolytes with high lithium-ion conductivity. *ACS Applied Materials & Interfaces* 9 (2): 1542–1552.

93 Yang, X., Kong, D., Chen, Z. et al. (2018). Low-temperature fabrication for transparency Mg doping $Li_7La_3Zr_2O_{12}$ solid state electrolyte. *Journal of Materials Science: Materials in Electronics* 29 (2): 1523–1529.

94 Zhou, X., Huang, L., Elkedim, O. et al. (2022). Sr^{2+} and Mo^{6+} co-doped $Li_7La_3Zr_2O_{12}$ with superior ionic conductivity. *Journal of Alloys and Compounds* 891: 161906.

95 Ma, X. and Xu, Y. (2022). Efficient anion fluoride-doping strategy to enhance the performance in garnet-type solid electrolyte $Li_7La_3Zr_2O_{12}$. *ACS Applied Materials & Interfaces* 14 (2): 2939–2948.

96 Zhang, Z. and Wei, B. (2022). Research and application of sulfide all-solid-state battery. *Marine Electric & Electronic Engineering* 010: 042.

97 Dietrich, C., Weber, D.A., Sedlmaier, S.J. et al. (2017). Lithium ion conductivity in Li_2S–P_2S_5 glasses–building units and local structure evolution during the crystallization of superionic conductors Li_3PS_4, $Li_7P_3S_{11}$ and $Li_4P_2S_7$. *Journal of Materials Chemistry A* 5 (34): 18111–18119.

98 Park, K.H., Bai, Q., Kim, D.H. et al. (2018). Design strategies, practical considerations, and new solution processes of sulfide solid electrolytes for all-solid-state batteries. *Advanced Energy Materials* 8 (18): 1800035.

99 Kato, Y., Hori, S., Saito, T. et al. (2016). High-power all-solid-state batteries using sulfide superionic conductors. *Nature Energy* 1 (4): 1–7.

100 Kamaya, N., Homma, K., Yamakawa, Y. et al. (2011). A lithium superionic conductor. *Nature Materials* 10 (9): 682–686.

101 Deiseroth, H.J., Kong, S.T., Eckert, H. et al. (2008). Li_6PS_5X: a class of crystalline Li-rich solids with an unusually high Li^+ mobility. *Angewandte Chemie International Edition* 47 (4): 755–758.

102 de Klerk, N.J.J., Rosłoń, I., and Wagemaker, M. (2016). Diffusion mechanism of Li argyrodite solid electrolytes for Li-ion batteries and prediction of optimized halogen doping: the effect of Li vacancies, halogens, and halogen disorder. *Chemistry of Materials* 28 (21): 7955–7963.

103 Rao, R.P. and Adams, S. (2011). Studies of lithium argyrodite solid electrolytes for all-solid-state batteries. *Physica Status Solidi* 208 (8): 1804–1807.

104 Kraft, M.A., Ohno, S., Zinkevich, T. et al. (2018). Inducing high ionic conductivity in the lithium superionic argyrodites $Li_{6+x}P_{1-x}Ge_xS_5I$ for all-solid-state batteries. *Journal of the American Chemical Society* 140 (47): 16330–16339.

105 Bachman, J.C., Muy, S., Grimaud, A. et al. (2016). Inorganic solid-state electrolytes for lithium batteries: mechanisms and properties governing ion conduction. *Chemical Reviews* 116 (1): 140–162.

106 Ge, Q., Zhou, L., Lian, Y.-M. et al. (2018). Metal-phosphide-doped $Li_7P_3S_{11}$ glass-ceramic electrolyte with high ionic conductivity for all-solid-state lithium-sulfur batteries. *Electrochemistry Communications* 97: 100–104.

107 Wu, Z., Xie, Z., Yoshida, A. et al. (2020). Novel SeS_2 doped Li_2S-P_2S_5 solid electrolyte with high ionic conductivity for all-solid-state lithium sulfur batteries. *Chemical Engineering Journal* 380: 122419.

108 Yamauchi, A., Sakuda, A., Hayashi, A. et al. (2013). Preparation and ionic conductivities of (100−x)($0.75Li_2S{\cdot}0.25P_2S_5$)·x$LiBH_4$ glass electrolytes. *Journal of Power Sources* 244: 707–710.

109 Kraft, M.A., Ohno, S., Zinkevich, T. et al. (2018). Inducing high ionic conductivity in the lithium superionic argyrodites $Li_{6+x}P_{1-x}Ge_xS_5I$ for all-solid-state batteries. *Journal of the American Chemical Society* 140 (47): 16330–16339.

110 Tao, Y., Chen, B.S., and Liu, Z.D. et al, Lithium superionic conducting oxysulfide solid electrolyte with excellent stability against lithium metal for all-solid-state cells. *Journal of the Electrochemical Society* 163: A96.

111 Huang, B., Yao, X., Huang, Z. et al. (2015). Li_3PO_4-doped $Li_7P_3S_{11}$ glass-ceramic electrolytes with enhanced lithium ion conductivities and application in all-solid-state batteries. *Journal of Power Sources* 284 (jun.15): 206–211.

112 Kim, K.H. and Martin, S.W. (2019, 2019). Structures and properties of oxygen-substituted $Li_{10}SiP_2S_{12-x}O_x$ solid-state electrolytes. *Chemistry of Materials* 31: 3984–3991.

113 Ujiie, S., Inagaki, T., Hayashi, A. et al. (2014). Conductivity of 70Li2S·30P2S5 glasses and glass–ceramics added with lithium halides. *Solid State Ionics* 263: 57–61.

114 Zhou, L., Park, K.H., Sun, X. et al. (2019). Solvent-engineered design of argyrodite Li_6PS_5X (X = Cl, Br, I) solid electrolytes with high ionic conductivity. *ACS Energy Letters* 4 (1): 265–270.

115 Adeli, P., David Park, J., HoKochetkov, K. et al. (2019). Boosting solid-state diffusivity and conductivity in lithium superionic argyrodites by halide substitution. *Angewandte Chemie* 58 (26).

116 Yang, K., Dong, J., Long, Z. et al. (2016). Dual doping: an effective method to enhance the electrochemical properties of $Li_{10}GeP_2S_{12}$-based solid electrolytes. *Journal of the American Ceramic Society* 98 (12): 3831–3835.

117 Hori, Satoshi, Suzuki et al. (2017). Superionic conductors: $Li_{10+\delta}[Sn_ySi_{1-y}]_{1+\delta}P_{2-\delta}\ S_{12}$ with a $Li_{10}GeP_2S_{12}$-type structure in the Li_3PS_4-Li_4SnS_4-Li_4SiS_4 quasi-ternary system. *Chemistry of Materials: A Publication of the American Chemistry Society* 29 (14): 5858–5864.

118 Kim, K., Park, D., Jung, H.-G. et al. (2021). Material design strategy for halide solid electrolytes Li_3MX_6 (X = Cl, Br, and I) for all-solid-state high-voltage li-ion batteries. *Chemistry of Materials* 33 (10): 3669–3677.

119 Flores-González, N., Minafra, N., Dewald, G. et al. (2021). Mechanochemical synthesis and structure of lithium tetrahaloaluminates, $LiAlX_4$ (X= Cl, Br, I): a family of Li-ion conducting ternary halides. *ACS Materials Letters* 3 (5): 652–657.

120 Hönle, W., Miller, G., and Simon, A. (1988). Preparation, crystal structures, and electronic properties of $LiGaCl_3$ and $LiGaI_3$. *Journal of Solid State Chemistry* 75 (1): 147–155.

121 Asano, T., Sakai, A., Ouchi, S. et al. (2018). Solid halide electrolytes with high lithium-ion conductivity for application in 4 V class bulk-type all-solid-state batteries. *Advanced Materials* 30 (44): 1803075.

122 Liang, J., Li, X., Wang, S. et al. (2020). Site-occupation-tuned superionic Li_xScCl_{3+x} halide solid electrolytes for all-solid-state batteries. *Journal of the American Chemical Society* 142 (15): 7012–7022.

123 Li, X., Liang, J., Chen, N. et al. (2019). Back cover: water-mediated synthesis of a superionic halide solid electrolyte. *Angewandte Chemie International Edition* 58 (46): 16698–16698.

124 Wang, K., Ren, Q., Gu, Z. et al. (2021). A cost-effective and humidity-tolerant chloride solid electrolyte for lithium batteries. *Nature Communications* 12 (1): 4410.

125 Chen, S., Yu, C., Chen, S. et al. (2022). Enabling ultrafast lithium-ion conductivity of Li_2ZrCl_6 by indium doping. *Chinese Chemical Letters* 33 (10): 4635–4639.

126 Kwak, H., Han, D., Lyoo, J. et al. (2021). New cost-effective halide solid electrolytes for all-solid-state batteries: mechanochemically prepared Fe^{3+}-substituted Li_2ZrCl_6. *Advanced Energy Materials* 11 (12): 2003190.

127 Liang, J., Li, X., Adair, K.R. et al. (2021). Metal halide superionic conductors for all-solid-state batteries. *Accounts of Chemical Research* 54 (4): 1023–1033.

128 Asano, T., Sakai, A., Ouchi, S. et al. (2018). Solid halide electrolytes with high lithium-ion conductivity for application in 4 V class bulk-type all-solid-state batteries. *Advanced Materials* 30 (44): 1803075.

129 Park, J., Han, D., Kwak, H. et al. (2021). Heat treatment protocol for modulating ionic conductivity via structural evolution of $Li_{3-x}Yb_{1-x}M_xCl_6$ (M= Hf^{4+}, Zr^{4+}) new halide superionic conductors for all-solid-state batteries. *Chemical Engineering Journal* 425: 130630.

130 Kanno, R., Takeda, Y., Takada, K. et al. (1983). Phase diagram and ionic conductivity of the lithium chloride-iron (II) chloride system. *Solid State Ionics* 9: 153–156.

131 Zhou, L., Kwok, C.Y., Shyamsunder, A. et al. (2020). A new halospinel superionic conductor for high-voltage all solid state lithium batteries. *Energy & Environmental Science* 13 (7): 2056–2063.

132 Zhou, L., Zuo, T.-T., Kwok, C.Y. et al. (2022). High areal capacity, long cycle life 4 V ceramic all-solid-state Li-ion batteries enabled by chloride solid electrolytes. *Nature Energy* 7 (1): 83–93.

133 Kanno, R., Takeda, Y., Takahashi, A. et al. (1987). New double chloride in the LiCl $CoCl_2$ system: I. Preparation, crystal structure, phase transformation, and ionic conductivity of Li_6CoCl_8. *Journal of Solid State Chemistry* 71 (1): 189–195.

134 Wang, S., Bai, Q., Nolan, A.M. et al. (2019). Lithium chlorides and bromides as promising solid-state chemistries for fast ion conductors with good electrochemical stability. *Angewandte Chemie International Edition* 58 (24): 8039–8043.

135 Liu, Y., Wang, S., Nolan, A.M. et al. (2020). Tailoring the cation lattice for chloride lithium-ion conductors. *Advanced Energy Materials* 10 (40): 2002356.

136 Wakamura, K. (1997). Roles of phonon amplitude and low-energy optical phonons on superionic conduction. *Physical Review B* 56 (18): 11593.

137 Kwak, H., Han, D., Son, J.P. et al. (2022). Li+ conduction in aliovalent-substituted monoclinic Li_2ZrCl_6 for all-solid-state batteries: $Li_{2+x}Zr_{1-x}M_xCl_6$ (M = In, Sc). *Chemical Engineering Journal* 437: 135413.

138 Wang, Y., Chen, Z., Wu, Y. et al. (2023). PVDF-HFP/PAN/PDA@LLZTO Composite Solid Electrolyte Enabling Reinforced Safety and Outstanding Low-Temperature Performance for Quasi-Solid-State Lithium Metal Batteries. *ACS Applied Materials & Interfaces* 15 (17): 21526–21536.

139 Fenton, D.E., Parker, J.M., and Wright, P.V. (1973). Complexes of alkali metal ions with poly(ethylene oxide). *Polymer* 14 (11): 589–589.

140 Armand, M., Chabagno, J., and Duclot, M. (1978). *Second International Meeting on Solid Electrolytes*, 20–22. Scotland: St Andrews.

141 Feuillade, G. and Perche, P. (1975). Ion-conductive macromolecular gels and membranes for solid lithium cells. *Journal of Applied Electrochemistry* 5 (1): 63–69.

142 Skaarup, S., West, K., and Zachau-Christiansen, B. (1988). Mixed phase solid electrolytes. *Solid State Ionics* 28-30: 975–978.

143 Wieczorek, W., Florjanczyk, Z., and Stevens, J.R. (1995). Composite polyether based solid electrolytes. *Electrochimica Acta* 40 (13): 2251–2258.

144 Chen, R., Qu, W., Guo, X. et al. (2016). The pursuit of solid-state electrolytes for lithium batteries: from comprehensive insight to emerging horizons. *Materials Horizons* 3 (6): 487–516.

145 Bouchet, R., Maria, S., Meziane, R. et al. (2013). Single-ion BAB triblock copolymers as highly efficient electrolytes for lithium-metal batteries. *Nature Materials* 12 (5): 452–457.

146 Lago, N., Garcia-Calvo, O., Lopez del Amo, J.M. et al. (2015). All-solid-state lithium-ion batteries with grafted ceramic nanoparticles dispersed in solid polymer electrolytes. *ChemSusChem* 8 (18): 3039–3043.

147 Zhou, D., Shanmukaraj, D., Tkacheva, A. et al. (2019). Polymer electrolytes for lithium-based batteries: advances and prospects. *Chem* 5 (9): 2326–2352.

148 Tsunemi, O.K. (1983). Conduction of lithium ions in polyvinylidene fluoride and its derivatives—I. *Electrochimica Acta* 28 (5): 591–595.

149 Watanabe, M., Kanba, M., Matsuda, H. et al. (1981). High lithium ionic conductivity of polymeric solid electrolytes. *Die Makromolekulare Chemie. Rapid Communications* 2 (12): 741–744.

150 Gonçalves, R., Miranda, D., Almeida, A.M. et al. (2019). Solid polymer electrolytes based on lithium bis(trifluoromethanesulfonyl)imide/poly(vinylidene fluoride -co-hexafluoropropylene) for safer rechargeable lithium-ion batteries. *Sustainable Materials and Technologies* 21: e00104.

151 Zhang, X., Wang, S., Xue, C. et al. (2019). Self-suppression of lithium dendrite in all-solid-state lithium metal batteries with poly(vinylidene difluoride)-based solid electrolytes. *Advanced Materials* 31 (11): 1806082.

152 Huang, Y.-F., Gu, T., Rui, G. et al. (2021). A relaxor ferroelectric polymer with an ultrahigh dielectric constant largely promotes the dissociation of lithium salts to achieve high ionic conductivity. *Energy & Environmental Science* 14 (11): 6021–6029.

153 Zhang, X., Wang, S., Xue, C. et al. (2020). Response to comment on "self-suppression of lithium dendrite in all-solid-state lithium metal batteries with poly(vinylidene difluoride)-based solid electrolytes". *Advanced Materials* 32 (14): 2000026.

154 Liu, Q., Yang, G., Li, X. et al. (2022). Polymer electrolytes based on interactions between [solvent-Li+] complex and solvent-modified polymer. *Energy Storage Materials* 51: 443–452.

155 Hu, P., Chai, J., Duan, Y. et al. (2016). Progress in nitrile-based polymer electrolytes for high performance lithium batteries. *Journal of Materials Chemistry A* 4 (26): 10070–10083.

156 Li, J., Huang, X., and Chen, L. (2000). X-ray diffraction and vibrational spectroscopic studies on PAN-LiTFSI polymer electrolytes. *Journal of the Electrochemical Society* 147 (7): 2653.

157 Wu, B., Wang, L., Li, Z. et al. (2016). Performance of "polymer-in-salt" electrolyte PAN-LiTFSI enhanced by graphene oxide filler. *Journal of the Electrochemical Society* 163 (10): A2248.

158 Wang, Z., Huang, B., Xue, R. et al. (1999). Spectroscopic investigation of interactions among components and ion transport mechanism in polyacrylonitrile based electrolytes. *Solid State Ionics* 121 (1): 141–156.

159 Wang, Z., Hu, Y., and Chen, L. (2005). Some studies on electrolytes for lithium ion batteries. *Journal of Power Sources* 146 (1): 51–57.

160 Liu, W., Liu, N., Sun, J. et al. (2015). Ionic conductivity enhancement of polymer electrolytes with ceramic nanowire fillers. *Nano Letters* 15 (4): 2740–2745.

161 Ma, Y., Wan, J., Yang, Y. et al. (2022). Scalable, ultrathin, and high-temperature-resistant solid polymer electrolytes for energy-dense lithium metal batteries. *Advanced Energy Materials* 12 (15): 2103720.

162 Jia, W., Li, Z., Wu, Z. et al. (2018). Graphene oxide as a filler to improve the performance of PAN-$LiClO_4$ flexible solid polymer electrolyte. *Solid State Ionics* 315: 7–13.

163 Zhang, J., Yang, J., Dong, T. et al. (2018). Aliphatic polycarbonate-based solid-state polymer electrolytes for advanced lithium batteries: advances and perspective. *Small* 14 (36): 1800821.

164 Xu, K. (2014). Electrolytes and interphases in Li-Ion batteries and beyond. *Chemical Reviews* 114 (23): 11503–11618.

165 Kimura, K., Yajima, M., and Tominaga, Y. (2016). A highly-concentrated poly(ethylene carbonate)-based electrolyte for all-solid-state Li battery working at room temperature. *Electrochemistry Communications* 66: 46–48.

166 Chai, J., Liu, Z., Ma, J. et al. (2017). In situ generation of poly (vinylene carbonate) based solid electrolyte with interfacial stability for $LiCoO_2$ lithium batteries. *Advanced Science* 4 (2): 1600377.

167 Zhang, J., Zhao, J., Yue, L. et al. (2015). Safety-reinforced poly(propylene carbonate)-based all-solid-state polymer electrolyte for ambient-temperature solid polymer lithium batteries. *Advanced Energy Materials* 5 (24): 1501082.

168 Zhang, J., Zang, X., Wen, H. et al. (2017). High-voltage and free-standing poly(propylene carbonate)/$Li_{6.75}La_3Zr_{1.75}Ta_{0.25}O_{12}$ composite solid electrolyte for wide temperature range and flexible solid lithium ion battery. *Journal of Materials Chemistry A* 5 (10): 4940–4948.

169 Mindemark, J., Törmä, E., Sun, B. et al. (2015). Copolymers of trimethylene carbonate and ε-caprolactone as electrolytes for lithium-ion batteries. *Polymer* 63: 91–98.

170 Guo, K., Li, S., Chen, G. et al. (2021). One-pot synthesis of polyester-based linear and graft copolymers for solid polymer electrolytes. *CCS Chemistry* 4 (9): 3134–3149.

171 Kang, Y., Lee, J., Suh, D.H. et al. (2005). A new polysiloxane based cross-linker for solid polymer electrolyte. *Journal of Power Sources* 146 (1): 391–396.

172 Zhang, Z., Sherlock, D., West, R. et al. (2003). Cross-linked network polymer electrolytes based on a polysiloxane backbone with oligo(oxyethylene) side chains: synthesis and conductivity. *Macromolecules* 36 (24): 9176–9180.

173 Zhang, Z.C., Jin, J.J., Bautista, F. et al. (2004). Ion conductive characteristics of cross-linked network polysiloxane-based solid polymer electrolytes. *Solid State Ionics* 170 (3): 233–238.

174 Alarco, P.-J., Abu-Lebdeh, Y., Abouimrane, A. et al. (2004). The plastic-crystalline phase of succinonitrile as a universal matrix for solid-state ionic conductors. *Nature Materials* 3 (7): 476–481.

175 Fan, L.-Z. and Maier, J. (2006). Composite effects in poly(ethylene oxide)–succinonitrile based all-solid electrolytes. *Electrochemistry Communications* 8 (11): 1753–1756.

176 Zhou, D., He, Y.-B., Liu, R. et al. (2015). In situ synthesis of a hierarchical all-solid-state electrolyte based on nitrile materials for high-performance lithium-ion batteries. *Advanced Energy Materials* 5 (15): 1500353.

177 Li Hong, X.X. (2016). R&D vision and strategies on solid lithium batteries. *Energy Storage Science and Technology* 5 (5): 607–614.

178 Kwak, H., Wang, S., Park, J. et al. (2022). Emerging halide superionic conductors for all-solid-state batteries: design, synthesis, and practical applications. *ACS Energy Letters* 7 (5): 1776–1805.

179 Zeng, D., Yao, J., Zhang, L. et al. (2022). Promoting favorable interfacial properties in lithium-based batteries using chlorine-rich sulfide inorganic solid-state electrolytes. *Nature Communications* 13 (1): 1909.

180 Mo, H., Yin, Y.-C., Luo, J.-D. et al. (2022). Lead-free solid-state organic–inorganic halide perovskite electrolyte for lithium-ion conduction. *ACS Applied Materials & Interfaces* 14 (15): 17479–17485.

181 Chen, S., Hu, X., Bao, W. et al. (2021). Low-sintering-temperature garnet oxides by conformal sintering-aid coating. *Cell Reports Physical Science* 2 (9): 100569.

182 Xu, S., Sun, Z., Sun, C. et al. (2020). Homogeneous and fast ion conduction of PEO-based solid-state electrolyte at low temperature. *Advanced Functional Materials* 30 (51): 2007172.

183 Arrese-Igor, M., Martinez-Ibañez, M., Pavlenko, E. et al. (2022). Toward high-voltage solid-state Li-metal batteries with double-layer polymer electrolytes. *ACS Energy Letters* 7 (4): 1473–1480.

184 Wen, S., Luo, C., Wang, Q. et al. (2022). Integrated design of ultrathin crosslinked network polymer electrolytes for flexible and stable all-solid-state lithium batteries. *Energy Storage Materials* 47: 453–461.

185 Wang, H., Hou, T., Cheng, H. et al. (2022). Bifunctional LiI additive for poly(ethylene oxide) electrolyte with high ionic conductivity and stable interfacial chemistry. *Journal of Energy Chemistry* 71: 218–224.

186 Zhang, Z., YingHuang; Zhang, G. et al. (2021). Three–dimensional fiber network reinforced polymer electrolyte for dendrite–free all–solid–state lithium metal batteries. *Energy Storage Materials* 41: 631–641.

187 Wang, G., Liu, H., Liang, Y. et al. (2022). Composite polymer electrolyte with three-dimensional ion transport channels constructed by NaCl template for solid-state lithium metal batteries. *Energy Storage Materials* 45: 1212–1219.

188 Zhu, Y., He, X., and Mo, Y. (2015). Origin of outstanding stability in the lithium solid electrolyte materials: insights from thermodynamic analyses based on first-principles calculations. *ACS Applied Materials & Interfaces* 7 (42): 23685–23693.

189 Han, F., Zhu, Y., He, X. et al. (2016). Electrochemical stability of $Li_{10}GeP_2S_{12}$ and $Li_7La_3Zr_2O_{12}$ solid electrolytes. *Advanced Energy Materials* 6 (8): 1501590.

190 Koerver, R., Aygün, I., Leichtweiß, T. et al. (2017). Capacity fade in solid-state batteries: interphase formation and chemomechanical processes in nickel-rich layered oxide cathodes and lithium thiophosphate solid electrolytes. *Chemistry of Materials* 29 (13): 5574–5582.

191 Wetjen, M., Kim, G.-T., Joost, M. et al. (2014). Thermal and electrochemical properties of PEO-LiTFSI-Pyr_{14}TFSI-based composite cathodes, incorporating 4 V-class cathode active materials. *Journal of Power Sources* 246: 846–857.

192 Pitillas Martinez, A.I., Aguesse, F., Otaegui, L. et al. (2019). The cathode composition, a key player in the success of Li-metal solid-state batteries. *The Journal of Physical Chemistry C* 123 (6): 3270–3278.

193 Chen, R.-J., Zhang, Y.-B., Liu, T. et al. (2017). Addressing the interface issues in all-solid-state bulk-type lithium ion battery via an all-composite approach. *ACS Applied Materials & Interfaces* 9 (11): 9654–9661.

194 Wan, Z., Lei, D., Yang, W. et al. (2019). Low resistance–integrated all-solid-state battery achieved by $Li_7La_3Zr_2O_{12}$ nanowire upgrading polyethylene oxide (PEO) composite electrolyte and PEO cathode binder. *Advanced Functional Materials* 29 (1): 1805301.

195 Chen, F., Zha, W., Yang, D. et al. (2018). All-solid-state lithium battery fitted with polymer electrolyte enhanced by solid plasticizer and conductive ceramic filler. *Journal of the Electrochemical Society* 165 (14): A3558.

196 Zha, W., Chen, F., Yang, D. et al. (2018). High-performance $Li_{6.4}La_3Zr_{1.4}Ta_{0.6}O_{12}$/poly (ethylene oxide)/succinonitrile composite electrolyte for solid-state lithium batteries. *Journal of Power Sources* 397: 87–94.

197 Zhang, B., Chen, L., Hu, J. et al. (2019). Solid-state lithium metal batteries enabled with high loading composite cathode materials and ceramic-based composite electrolytes. *Journal of Power Sources* 442: 227230.

198 Hu, J., He, P., Zhang, B. et al. (2020). Porous film host-derived 3D composite polymer electrolyte for high-voltage solid state lithium batteries. *Energy Storage Materials* 26 (283–289): 2020.

199 Liu, K., Liu, Y., Lin, D. et al. (2018). Materials for lithium-ion battery safety. *Science Advances* 4 (6): eaas9820.

200 Kim, K.J. and Rupp, J.L. (2020). All ceramic cathode composite design and manufacturing towards low interfacial resistance for garnet-based solid-state lithium batteries. *Energy & Environmental Science* 13 (12): 4930–4945.

201 Zaman, W. and Hatzell, K.B. (2022). Processing and manufacturing of next generation lithium-based all solid-state batteries. *Current Opinion in Solid State and Materials Science* 26 (4): 101003.

202 Chen, H., Yang, Y., Boyle, D.T. et al. (2021). Free-standing ultrathin lithium metal–graphene oxide host foils with controllable thickness for lithium batteries. *Nature Energy* 6 (8): 790–798.

6

Lithium–Sulfur Battery

Saisai Qiu[1] *and Minghua Chen*[2]

[1] *Harbin University of Science and Technology, School of Electrical and Electronic Engineering, Department of Electrical Engineering, No. 52, Xuefu Road, Nangang District, Harbin, 150080, China*
[2] *Harbin University of Science and Technology, School of Electrical and Electronic Engineering, Department of New Energy Materials and Devices, Key Laboratory of Engineering Dielectric and Applications (Ministry of Education), No. 52, Xuefu Road, Nangang District, Harbin, 150080, China*

The growing reliance on and demand for high-energy-density power sources has stimulated the development of sophisticated energy storage systems. Currently, lithium-ion batteries (LIBs) dominate the portable energy storage market with their high reliability and proven battery pack technology. Although commercialized LIBs based on insertion chemistry are approaching the theoretical value of energy density, they may not be sufficient for high-energy-density devices such as electric vehicles, smart grids, and drones, leading to the rapid development of higher-energy-density energy storage systems, including silicon-based anode and lithium anode LIBs, lithium–sulfur (Li-S), and Li–air batteries. Among them, Li-S batteries are one of the most promising high-capacity energy storage devices for commercialization due to their low cost, high sulfur content, high theoretical specific capacity ($1675\,mAh\,g^{-1}$), and high energy density ($2600\,Wh\,kg^{-1}$), endowing them with a very high potential for achieving a battery energy density of $500\,Wh\,kg^{-1}$ or higher. In Li-S batteries, the cathode utilizes naturally abundant, cost-effective, nontoxic, and eco-friendly sulfur as its active material. Generally, Li-S batteries are composed of a lithium metal anode, an organic electrolyte, a separator, and a cathode containing sulfur. The sulfur cathode typically comprises sulfur particles, a carbon additive serving as a conductive agent, and a polymer binder. The separator functions as a barrier separating the anode and cathode to prevent internal short circuits. Lithium metal exhibits a specific capacity of $3860\,mAh\,g^{-1}$ in battery applications, which makes it an excellent choice for the anode. These two electrode materials are the cathode and anode materials, respectively, with the highest unit mass capacity. On the one hand, unlike commercial graphite anodes, directly using lithium metal as the anode does not require the addition of conductive carbon and binder and a large amount of electrolyte that wets the anode, which reduces the additional weight of the battery. In addition, the energy storage mechanism of the inner Li-S battery is based on the conversion reaction

Towards Next Generation Energy Storage Technologies: From Fundamentals to Commercial Applications, First Edition. Edited by Minghua Chen.

between sulfur, lithium polysulfides (LiPSs), and lithium sulfide (Li_2S). It does not need to be limited to maintaining the integrity of the crystal structure of the electrode material like LIBs. Each sulfur atom can combine with two lithium atoms and simultaneously fix two electrons, which greatly increases the charge storage capacity of the battery and can achieve a significant increase in the energy density of the battery.

Although the application potential of the Li-S battery technology is enormous, its comprehensive performance indicators, such as energy density, power density, and cycling stability (coulombic efficiency and lifespan), are still lower than commercial battery standards. During the charging and discharging process of Li-S batteries, the sulfur cathode will go through a complex process involving multiphase electrochemical reactions: during the discharge, the electrons obtained from the solid-phase elemental sulfur will be reduced to liquid-phase long-chain LiPSs, liquid-phase short-chain LiPSs, solid-phase Li_2S_2, and solid-phase Li_2S, whereas during the charging process, the reverse is true. During this cycle, the utilization rate and reaction depth of the sulfur active material in the positive electrode determine the energy density of Li-S batteries. The electrochemical reaction rate of the sulfur in the positive electrode directly affects the power density of the Li-S battery, including the diffusion of LiPSs in the electrolyte, the mutual conversion of sulfur species, and the reaction activity of the electrode. The kinetics of the conversion reaction of LiPSs is significantly related to the actual working voltage of the battery, which directly determines the output power of the battery. In addition, the cycling stability of Li-S batteries is determined by the degree of reversibility of the electrochemical reaction of sulfur. When the battery is discharged, the dissolved LiPSs have a "shuttle effect" and are consumed in large quantities. Furthermore, the ultimately formed discharge products undergo uneven deposition on the electrode, forming insulating large particles, thus making it difficult for the reactive active substance to be reused.

Due to the inherent disadvantages of Li-S batteries, their actual energy density is much lower than the theoretical energy density. These disadvantages include the insulating properties of sulfur and solid-state discharge products (Li_2S_2 and Li_2S), the volume expansion of the cathode after the reduction of S_8 to Li_2S, the shuttling effect of soluble LiPSs (Li_2S_x, $x \geq 4$), and the presence of lithium dendrites and dead cells on the lithium anode. In the past ten years, the Li-S battery technology has made great progress and has attracted extensive attention from academia and industry. Since the inception of this technology, great efforts have been made to address the abovementioned challenges of Li-S batteries. Figure 6.1 details the major milestones in the development of lithium batteries to date [1]. The concept of the Li-S battery originated in the 1960s, introduced by Herbet and Ulam. Early investigations were centered on achieving reversible operation of the battery. Following a period of inactivity, substantial advancements in cathode design occurred in the twenty-first century, markedly enhancing the cycling stability of Li-S batteries and sparking a renewed technological revolution [2]. The research boom of lithium batteries appeared again (purple). In 2002, Wang et al. demonstrated the application of a composite S/porous carbon cathode. It has a good sulfur utilization rate and

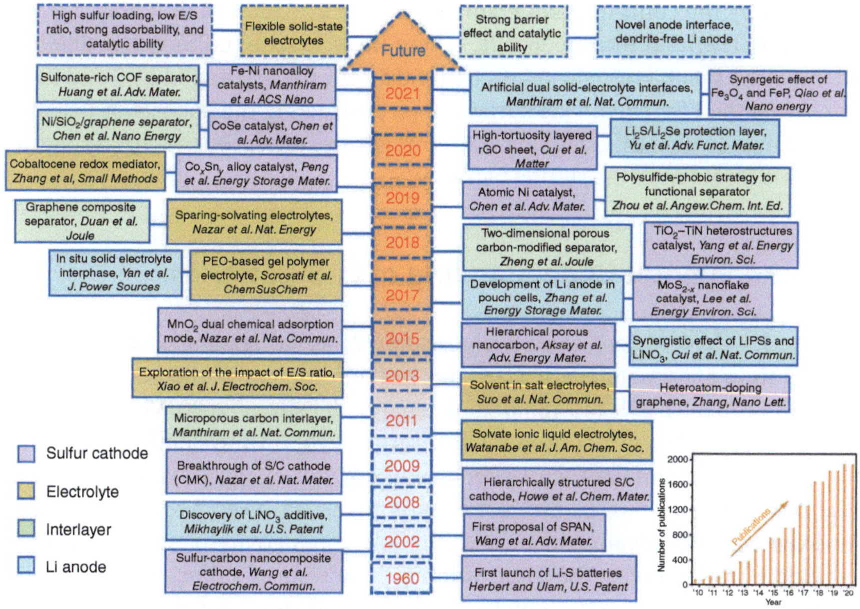

Figure 6.1 Overview of the development of Li-S batteries from the first proposal to the present. Source: Reproduced with permission from Huang et al. [1]. © 2020/John Wiley & Sons./CC BY 4.0.

better recyclability [2]. Crucially, Nazar and colleagues presented an innovative composite S/mesoporous carbon cathode (S/C, carbon: CMK-3) in 2009, realizing a substantial capacity of 1320 mAh g^{-1} and prolonged cycle life. This landmark achievement catalyzed the dynamic evolution of Li-S battery cathodes. Addressing the same challenge, battery performance can be enhanced through novel electrolyte configurations (depicted in brown) and alterations in the separator structure (illustrated in light green). Furthermore, advancements in solid electrolyte interface (SEI) engineering for stable anodes, such as the identification of the liquid electrolyte additive $LiNO_3$, have increased the coulombic efficiency of Li-S batteries, bringing it closer to practical utility. In essence, the historical progression of Li-S batteries is intricately intertwined with the evolution of strategies to inhibit the shuttling of LiPSs. As long as the LiPS shuttling effect persists, meeting the practical demands for the cycle performance of the battery remains challenging. The shuttle behavior of LiPSs results in diminished coulombic efficiency, rapid capacity deterioration, and corrosion of lithium metal throughout cycling. Moreover, the severity of these challenges increases with increased sulfur loading, particularly in practical pouch cells.

Considering the behavior of LiPSs, when solid sulfur incorporates Li^+ ions from a liquid electrolyte, it gets converted into soluble LiPSs with extended chains. If host materials inadequately capture these lengthy LiPSs for subsequent redox reactions, they diffuse into the electrolyte [3]. Subsequently, as LiPSs dissolve, the electrolyte's viscosity increases steadily, significantly increasing the transfer resistance of Li^+

ions. Driven by the concentration gradient, soluble extended-chain LiPSs traverse the separator, reaching the anode side, where these lengthy LiPSs either undergo reactions with Li metal to yield short-chain LiPSs or establish a solid Li_2S and Li_2S_2 layer, coating the Li metal anode surface, thus obstructing the passage of Li+ ions [4]. Hence, it is evident that soluble LiPSs exert adverse effects on the corresponding components throughout the shuttle pathway.

6.1 Fundamental Understanding of Li-S Batteries

Li-S batteries have been extensively explored as a leading contender for next-generation energy storage systems due to their impressive theoretical specific capacity of 1675 mAh g^{-1} and remarkable energy density of 2600 Wh kg^{-1}. These values far exceed those of conventional insertion-type LIB systems. Meanwhile, sulfur is naturally abundant, cost-effective, and eco-friendly; thus, the potential for the commercialization of Li-S batteries is high. However, Li-S batteries also suffer from some disadvantages such as the low conductivity of sulfur and Li_2S, the shuttle effect of LiPSs, the large volume expansion of sulfur electrodes, and unstable lithium metal anodes, which can hinder their commercial application.

6.1.1 Reaction Mechanism

A typical Li-S battery primarily encompasses a sulfur cathode, lithium anode, organic electrolyte, and separator. Elemental sulfur reacts with lithium and forms Li_2S during discharging; then, Li_2S decomposes into Li and S during charging (Figure 6.2). The overall electrochemical charge transfer reaction at the cathode can be represented by Eq. (6.1):

$$S_8 + 16e^- + 16Li^+ \rightarrow 8Li_2S \tag{6.1}$$

While Li_2S represents the end product of discharging, the redox reactions between sulfur and Li_2S encompass multiple steps, involving intricate alterations in composition and structure. The discharge process can be categorized into four distinct stages. Li metal is oxidized to produce Li^+ ions and electrons at the anode during the discharge process. The Li^+ ions then move to the cathode via the electrolyte, while the electrons travel through the external circuit to generate electricity. At the cathode side, sulfur reacts with the Li^+ ions and electrons and is reduced in a stepwise manner. Figure 6.1 shows a typical two-platform discharge curve corresponding to the oxidation process of Li-S batteries.

Stage I: The reduction of solid sulfur to highly soluble long-chain Li_2S_8, corresponding to a solid–liquid two-phase reaction is presented in Eq. (6.2):

$$S_8 + 2e^- + 2Li^+ \rightarrow Li_2S_8 \tag{6.2}$$

Due to the continuous dissolution of Li_2S_8 into the electrolyte, the cathode becomes porous and undergoes volume contraction. Therefore, a small ramp plateau appears during discharge at a relatively high voltage of approximately 2.3 V.

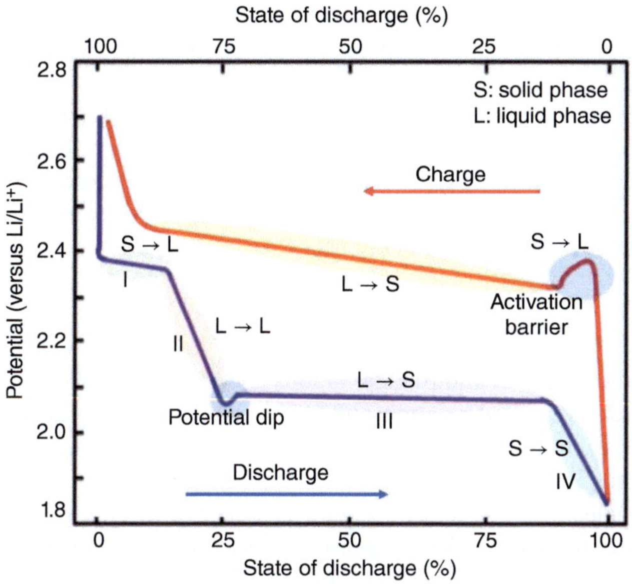

Figure 6.2 Schematic of operating principles of Li-S battery. Source: Reproduced with permission from Fang et al. [2]. © 2017, John Wiley & Sons.

Stage II: Li_2S_8 is further reduced to Li_2S_6 and Li_2S_4, both of which are soluble in organic electrolytes according to Eqs. (6.3) and (6.4). At the end of this discharge process, the high electrolyte viscosity caused by the dissolution of LiPSs leads to a high overpotential. Stages I and II contribute to one-fourth (419 mAh g^{-1}) of the overall theoretical specific capacity, corresponding to the acceptance of 0.5 electrons per sulfur atom.

$$3Li_2S_8 + 2e^- + 2Li^+ \rightarrow 4Li_2S_6 \tag{6.3}$$

$$2Li_2S_6 + 2e^- + 2Li^+ \rightarrow 3Li_2S_4 \tag{6.4}$$

Stage III: Soluble Li_2S_4 is reduced to insoluble Li_2S_2 and Li_2S. This stage involves liquid–solid conversions shown in Eqs. (6.5) and (6.6), resulting in gradually decreased redox kinetics:

$$Li_2S_4 + 2e^- + 2Li^+ \rightarrow 2Li_2S_2 \tag{6.5}$$

$$Li_2S_4 + 6e^- + 6Li^+ \rightarrow 4Li_2S \tag{6.6}$$

These reduction reactions occur simultaneously in competition with each other, resulting in a long lower voltage plateau of about 2.1 V.

Stage IV: In the last stage, further reduction of Li_2S_2 to Li_2S occurs, which suffers from high polarization and slow kinetics due to the solid–solid transformation reaction, according to Eq. (6.7):

$$Li_2S_2 + 2e^- + 2Li^+ \rightarrow 2Li_2S \tag{6.7}$$

Stages III and IV contribute to three-fourth (1256 mAh g^{-1}) of the overall theoretical specific capacity, corresponding to the acceptance of an additional 1.5 electrons per sulfur atom.

In the subsequent charging phase, solid Li_2S and Li_2S_2 undergo reversible conversion into diverse soluble LiPSs, subsequently oxidizing to solid S_8. This corresponds to a prolonged, single charging plateau at approximately 2.4 V. This charging process exhibits sluggish kinetics, given that the reverse transition needs to overcome the added activation energy due to the aggregation of Li_2S.

6.1.2 Challenges

Conductivity influences the transport and transfer of ions and electrons and determines the kinetics of reactions. Rapid electron and ion transport is highly beneficial for electrochemical reactions, but the extremely poor electrical conductivity of sulfur and the discharge product Li_2S makes the electrochemical reaction more sluggish. The transport of electrons and ions during the electrochemical reaction is limited by the insulating nature of sulfur, which greatly reduces the utilization of the active material and even causes the so-called dead sulfur. Even worse, precipitated Li_2S/Li_2S_2 is not easily delithiated and decomposed, which is prone to covering the cathode to generate an inert layer. These are detrimental to the commercial application of Li-S batteries.

Notably, long-chain LiPSs (Li_2S_x, $4 \leq x \leq 8$) dissolve readily into the electrolyte. Due to the concentration gradient, the mobility of LiPSs allows its irreversible dissolution into the electrolyte and diffusion to the anode, where they are subsequently chemically reduced by Li metal (rather than electrochemically) to form short-chain LiPSs or even solid Li_2S_2 and Li_2S coverings on the anode. The subsequent diffusion of short-chain LiPSs back to the cathode and the subsequent charging phase result in the generation of long-chain LiPSs. Therefore, this round-trip transport of LiPSs within the Li^+ battery is identified as the notorious "shuttle effect." This shuttle phenomenon has adverse implications for the performance of Li^+ batteries. The most apparent consequence is the depletion of active sulfur. When soluble LiPSs lose electrical contact with the collector, they fail to participate in subsequent electrochemical reactions, leading to substantial sulfur losses. Additionally, the insoluble Li_2S_2/Li_2S passivation layer accumulating on the surface of the Li anode cannot be reoxidized to long-chain LiPSs and S_8. This not only results in an irreversible loss of active material but also impedes Li^+ diffusion and transport due to its poor electrical conductivity, ultimately causing rapid capacity decay and limiting the cycle life. Moreover, as the electrolyte's viscosity increases due to the dissolution of LiPSs, the charge transfer resistance increases. A significant voltage drop occurs at the transition's outset, aligning with the nucleation potential of solid Li_2S_2/Li_2S. The decomposition process also necessitates overcoming the additional activation energy.

The huge volume variation is another challenging issue for sulfur cathodes. The volume change of sulfur electrodes during charging and discharging is up to 80% due to the large density difference between sulfur (2.07 g cm^{-3}) and Li_2S

($1.66\,g\,cm^{-3}$). This continuous volume variation during the lithiation process might result in cathode breakage. In this case, the isolated/detached sulfur cannot be effectively utilized. Overall, these challenges will inevitably give rise to low sulfur utilization and rapid capacity attenuation.

Dendritic metallic lithium structures, known as lithium dendrites, grow during the reduction of lithium ions in the charging process of LIBs. The growth of these dendrites introduces instability at the interfaces between the electrode and electrolyte throughout LIB cycling, disrupting the resulting SEI film. These lithium dendrites may penetrate the separator, causing internal short circuits in LIBs. This in turn triggers thermal runaway, potentially leading to combustion and explosion. When soluble LiPSs traverse to the anode surface via the separator and interact with Li metal, they initiate the formation of inert Li_2S and Li_2S_2. This process results in the depletion of active sulfur species, irreversible Li consumption, and compromised coulombic efficiency. Additionally, the subsequently generated inert and heterogeneous SEI layer on the anode surface obstructs Li^+ diffusion. This obstruction induces nonuniform current density distribution, promoting uncontrollable, rapid growth of Li dendrites. Consequently, this cascade of events leads to suboptimal cycling performance and poses safety hazards.

6.2 Sulfur Cathode

A Li-S battery is a type of rechargeable battery that has the potential to offer higher energy density than traditional LIBs, and the sulfur cathode is a critical component of a Li-S battery. In a Li-S battery, the sulfur cathode undergoes a unique electrochemical reaction with lithium to store and release energy. During charging, lithium ions from the anode move through the electrolyte and deposit on the sulfur cathode. Then, sulfur undergoes a series of chemical reactions to form Li_2S and store energy. This is a conversion reaction, where sulfur molecules are transformed into a different compound. When the battery is discharged, the lithium ions return to the cathode from the anode through the electrolyte. The lithium ions react with Li_2S on the cathode, converting it back into elemental sulfur. This process releases the stored energy as electric current.

It is important to note that the sulfur cathode experiences significant volume changes during charging and discharging cycles. The expansion and contraction of the sulfur material can lead to mechanical stress and degradation of the electrode, which is one of the challenges in the development of Li-S batteries. Meanwhile, several challenges and research areas are associated with sulfur cathodes in Li-S batteries:

Sulfur utilization: Sulfur cathodes often suffer from poor sulfur utilization, meaning that not all of the sulfur in the electrode participates in the desired chemical reactions. This limits the energy density and overall performance of the battery.

Cathode cycling stability: The expansion and contraction of sulfur during charging and discharging cycles can cause structural degradation of the cathode material over time. This leads to reduced capacity and reduced battery life.

Electrolyte compatibility: Li-S batteries require specific electrolytes that can accommodate the chemical reactions between sulfur and lithium. Developing stable electrolytes that do not degrade during cycling is a challenge.

Cathode design and architecture: Researchers have been exploring various cathode designs, including nanostructured materials, sulfur composites, and protective coatings, to improve the overall stability and performance of sulfur cathodes.

Safety concerns: Sulfur cathodes can lead to the formation of LiPSs, which can migrate through the electrolyte and cause issues such as the "shuttle effect." This can lead to decreased battery performance and safety concerns.

The development of Li-S batteries is an active area of research and innovation in the field of energy storage. Researchers have been working on addressing the challenges associated with sulfur cathodes to unlock the full potential of Li-S batteries for applications requiring high energy density, such as electric vehicles and renewable energy storage.

6.2.1 Sulfur Cathode

6.2.1.1 S_8 Cathode

The poor conductivity of the sulfur cathode and the dissolution of polysulfides limit the further applications of Li-S batteries due to subsequent effects such as the poor electron and ion conductivity of the active material and the shuttle effect of LiPSs. Thus, it is necessary to encapsulate the active sulfur material in a host material with a high specific surface area and more electrons or ions. Usually, the following encapsulation methods of active materials are used: (i) ball milling method, (ii) melt diffusion method, (iii) dissolution recrystallization method, and (iv) chemical deposition method. In order to address the abovementioned issues, in early studies, sulfur was encapsulated in various carbon-based materials, including porous carbon, hollow carbon spheres, carbon nanotubes, graphene, and graphene oxide, to synthesize carbon–sulfur (C-S) nanocomposites. The introduction of carbon-based materials used as sulfur hosts was found to effectively improve the electron and ion transfer capacity, tackle the problem of volume expansion, and make use of the sparse and porous characteristics of carbon-based materials to diffuse the adsorption of LiPSs, effectively improving the conversion of positive porous carbon materials into microporous carbon, mesoporous carbon, and macroporous carbon depending on the pore size. In Li-S batteries, microporous carbon can be used as an ideal container for active materials; mesoporous carbon can reduce sulfur loss and improve ion and electrolyte transport channels; and macroporous carbon is usually an interwoven network of carbon nanotubes or carbon nanofibers, which can inhibit the migration of LiPSs and improve the electrolyte infiltration capacity. Chien-Hsun Yu et al. demonstrated the effect of various pores by generating non-porous, microporous, micro-mesoporous, and macroporous C-S composite cathodes through a drop-in method used for nanocomposites synthesized by sulfur melting method and electrolyte and active material mixtures on electrode substrates. In their experiments, conductive microporous carbon with abundant micropores led to improved electrochemical utilization of highly loaded sulfur cathodes [5]. However, some Li-S batteries with C-S cathodes have low coulombic efficiency, rapid capacity decay, and severe anode decomposition with a limited lifetime.

In 2011, Demir-cakan et al. first applied metal-organic frameworks (MOFs) to Li-S battery cathode materials and mesoporous chromium-based MOFs (MIL-100(Cr)) as the cathode backbone and found that the utilization and capacity retention of Li-S cathode materials were significantly improved. Since then, the application of MOFs has been gradually increased [6]. MOFs connect metal nodes and organic ligands through strong coordination bonds and have attracted extensive research interest due to their unique properties [7]. Since the mid-1990s, more than 20,000 different MOFs have been reported. MOFs have ultrahigh porosity, huge specific surface area, and flexible and tunable structure, but their conductivity is poor. Thus, MOFs are usually treated by annealing and other methods, and the materials obtained after treatment are called MOF derivatives. MOF derivatives can not only inherit the characteristics of their MOF precursors, but also acquire new morphology and structure due to adjustments to certain parameters. More importantly, the annealing treatment of MOFs results in in situ carbon substrates, forming conductive networks and increased electron contacts, which can further improve the electrical conductivity of MOF derivatives.

Based on these properties of MOFs, researchers have used a combination of carbon-based materials and MOF materials. Wang et al. synthesized CC@CoP/C by growing Co MOF nanosheet arrays on flexible carbon cloth (CC) and annealing them in argon and hydrogen atmosphere. The synthesized MOF nanosheet arrays, when covered with carbon cloth, showed good electrical conductivity with an accommodative design, combining the advantages of carbon-based materials with those of MOFs, and this combination has also been widely used [8].

6.2.1.2 Li_2S Cathode

The discharge product of Li-S batteries is Li_2S, which, as a fully lithium state of sulfur, replaces elemental sulfur, and Li_2S as a cathode enables the use of safer lithium-free anodes (e.g. graphite, silicon, and alloys) and provides a solution to the problem of large volume changes present in sulfur cathodes [9], offering a good prospect for lithium-free metal anodes. Since the charging and discharging process of Li-S batteries is based on the conversion between Li_2S and S, the use of Li_2S as a cathode material also does not completely address the problems inherent to Li-S and, like sulfur cathodes, requires the mixing or encapsulation of Li_2S with electron- or ion-conducting substrates. Li_2S has a high melting point of 938 °C and can withstand high encapsulation temperatures, whereas the encapsulation temperature of pure sulfur is below 200 °C [10]. Therefore, Li_2S has a surplus of S composites for preparation.

The particle size of Li_2S affects the overpotential for decomposition, so different methods have been developed to transform Li_2S particles into nanoscale or amorphous Li_2S particles. They can be categorized into physical methods and chemical methods. The ball milling method is a simple physical method used to reduce the particle size. It can be used to grind large-sized Li_2S into a size smaller than 400 nm, and it can also effectively mix Li_2S particles with carbon matrix to prepare the Li_2S cathode [11]. However, even after a long ball milling process, some Li_2S particles can be in the submicron scale, with poor electrochemical performance. Another physical method to reduce the particle size of Li_2S is recrystallization. First, large-sized Li_2S is dissolved in an organic solvent such as absolute ethanol and then recrystallized by solvent evaporation. This method is more effective in obtaining fine Li_2S particles.

In addition to these physical methods, chemical methods can also be used to achieve small-sized Li_2S particles. Li_2S particles ranging in size from 30 to 500 nm can be prepared by disproportionation of Li_2S_3 or Li_2S_6 solutions. These solutions are prepared by the stoichiometric reaction of sulfur and trace amounts of Li_2S and heating them above 200 °C to decompose. However, the cost and environmental impact of solvents, including 1,2-dimethoxyethane (DME) and tetrahydrofuran (THF), may limit their application [10].

6.2.1.3 SPAN Cathode

Constructing lithium polyacrylonitrile (Li-SPAN) batteries using sulfurized polyacrylonitrile (SPAN) as a sulfur cathode is considered a promising direction to realize high-performance Li-S batteries. The SPAN cathode was first reported by Wang and colleagues in 2002 [12]. They synthesized SPAN by simple annealing of a mixture of sulfur and polyacrylonitrile (PAN) under an inert atmosphere. The SPAN cathode is made of additional conductive carbon and polymer binder. Compared with the basic C-S cathode, the conductivity of SPAN is higher because pyrolyzed PAN-conjugated chains provide sufficient migration channels for electrons. More importantly, the active sulfur atoms within SPAN are dispersed and covalently bonded to the conductive framework without elemental sulfur, thus providing very limited soluble LiPSs in the electrolyte during cycling. Therefore, the SPAN cathode exhibits a unique solid–solid switching charge/discharge behavior different from the solid–liquid–solid pathway of conventional cathodes. The electrochemical reaction pathway without soluble LiPSs promotes the utilization of active sulfur, inhibits the corrosion of the Li metal anode during cycling, improves the coulombic efficiency, and improves the cycling stability of the battery [13]. Because of these advantages, this technique has attracted much research attention.

6.2.2 Typical Strategy

Developing an efficient and stable sulfur cathode is crucial for improving the performance and practicality of Li-S batteries. Researchers employ various strategies to enhance the performance of sulfur cathodes and address challenges associated with them (Figure 6.3) [14].

6.2.2.1 Physical Confinement

To address the challenges associated with the volume expansion and contraction of sulfur during charge and discharge cycles, physical confinement of sulfur cathodes in Li-S batteries can be used. This expansion and contraction can lead to structural degradation, loss of electrical contact, and decreased cycling stability. Confining

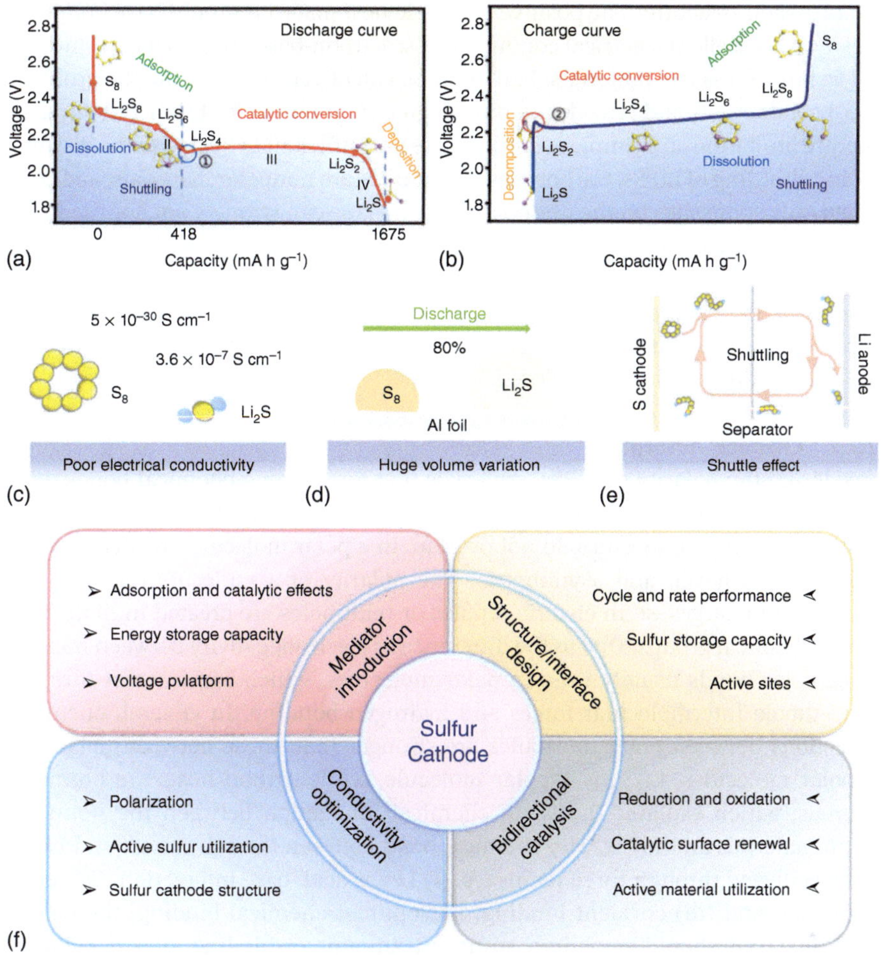

Figure 6.3 Schematic diagram of (a) discharge and (b) charge voltage curves and corresponding redox reaction products of Li-S batteries. Schematic diagram of main challenges for sulfur cathodes: (c) poor electrical conductivity of sulfur and Li_2S; (d) large variation of sulfur electrode volume; and (e) shuttle effect of LiPSs. (f) Key factors for sulfur electrodes.

sulfur within a specific structure or framework helps mitigate these issues. This confinement strategy aims to improve the cycling stability, capacity retention, and overall performance of Li-S batteries by addressing the challenges posed by the volume changes of sulfur. By containing and controlling sulfur's expansion and contraction, the practicality and longevity of Li-S battery systems can be enhanced.

As a common strategy for anchoring LiPSs, physical confinement aims to confine LiPSs in the sulfur cathode space and to design the spatial structure of the cathode host material. Ideal host materials usually have high specific surface area and effective physical confinement space and can not only accommodate more active substances but also effectively prevent the migration of LiPSs. The structure of the positive electrode host material is mainly divided into two types: porous structure and core–shell structure. The positive electrode host material should have both rich porosity and excellent electrical conductivity, so carbon-based materials are initially used as hosts for active materials. Participation can effectively alleviate the problem of poor conductivity and provide a certain degree of limitation for LiPSs, but physical structure limitation and simple adsorption are not sufficient to deal with the dissolution and shuttling of LiPSs. Carbon-based materials are nonpolar materials, and conjugated only some sites on the nonpolar carbon plane can strongly anchor the Li_2S_n species; thus, the interaction between the carbon-based material and Li_2S_n species is weak. Nonpolar carbon-based materials cannot effectively anchor polar polysulfides during the charge/discharge reaction, especially in long-cycle processes, and the problem of the shuttle effect has not been resolved. Therefore, polar materials with adsorption capacity are explored.

6.2.2.2 Chemical Adsorption

Considering the polarity of LiPSs, materials that form strong chemical bonds with LiPSs are better suited as sulfur hosts. In chemistry, the centers of positive and negative charges in molecules do not overlap. In a polar molecule, the distribution of charges is uneven and asymmetric. The polarity of a molecule refers to the separation of charges, so in electric dipoles or multipoles are created in molecules or their chemical groups. Due to the difference in electronegativity between bonded atoms, polar bonds usually exist in polar molecules, which can interact through dipole–dipole intermolecular forces and hydrogen bonding. In general, chemical interactions between polar molecules are stronger than those between polar and nonpolar molecules. Li_2S_n is a polar molecule, while carbon hosts are nonpolar materials, which explains the weak chemical interaction between the nonpolar carbon materials and LiPSs. The chemical bond between the sulfur host and LiPSs can be achieved through three pathways: (i) Lewis acid–base interaction, (ii) redox interaction, and (iii) covalent binding. For optimal chemical binding, the design of the material should provide a sufficient amount of binding sites capable of chemically interacting with polysulfides. Therefore, structures with a large specific surface area and pores are considered ideal sulfur host materials. Researchers have used various materials as cathode hosts to confine LiPSs, such as MOFs (V-MOF [15]), polar metal oxides (MnO_2 [16], TiO_2, Nb_2O_5), metal sulfides (CoS_2), and metal carbides. Materials like MXenes have been used as positive hosts for Li-S batteries.

In summary, the electrochemical kinetics of the interface is primarily determined by the binding ability of the material to LiPSs and the charge transfer on the boundary. Carbon, as a typical nonpolar material, cannot combine with polar LiPSs, resulting in an intermediate reaction during the charging and discharging process of Li-S batteries. Polar insulators with the polarity to adsorb LiPSs can alleviate a series of problems of intermediates, but their low electrical conductivity hinders the direct surface transformation on polar hosts. In contrast, polar materials with high conductivity can both possess polar adsorption capacity and provide charge transfer capacity on the surface, which can be used to enhance the electrochemical kinetics and thus are the best choice for positive hosts.

6.2.2.3 Catalytic Conversion

Based on the gradual and slow conversion mechanism of LiPSs in Li-S batteries, to avoid the dissolution of soluble LiPSs and alleviate the shuttle effect, accelerating the conversion process of LiPSs is necessary. Based on traditional catalysis, the electrocatalytic effect is introduced into Li-S batteries. In this approach, the initial adsorption method and catalytic method are used as independent modulation methods. In recent years, it has been discovered that adsorption and catalysis are correlated in Li-S battery cathode materials. The combined effect of adsorption and catalysis also accelerates the conversion of long-chain LiPSs in the electrolyte. After the sulfur cathode is introduced into the electrocatalyst, the electrocatalyst first adsorbs the middle LiPSs, which is the prelude to electrocatalysis. The rate of charge transfer is closely related to the intrinsic conductivity and catalytic properties of the material. It is worth noting that catalysts in Li-S batteries are different from traditional chemical polar adsorbents. The following are the major differences between catalytic materials and polar adsorbents: a catalytic material not only captures LiPSs on its surface but also enhances the kinetics of the redox reaction of adsorbed LiPSs through the migration of ions and electrons, whereas a polar adsorbent only captures LiPSs on its surface and does not further enhance the kinetics of the redox reaction of adsorbed LiPSs [17]. The slow conversion of LiPSs on polar adsorbents leads to the saturated state of LiPSs and even results in the production of a large amount of dead sulfur. However, the rapid conversion of LiPSs on a catalytic material leads to the unsaturated state of LiPSs, so further adsorption of LiPSs is feasible.

During the reverse charging process, the kinetics of the oxidation process of LiPSs is slow, which limits the application of intermediates and requires higher activation energy. In order to suppress the shuttling effect and improve sulfur utilization, accelerating the conversion kinetics between LiPSs and Li_2S on the substrate through the catalytic effect is more beneficial than the simple adsorption of LiPSs on the host. On the catalytic surface, LiPSs undergo a smooth "adsorption–diffusion–conversion" process due to moderate interactions, facilitating electron transfer to sulfur species, accompanied by abundant active sites. Materials like this that can not only capture LiPSs but also accelerate the conversion process between LiPSs and Li_2S are catalytic materials [18].

The following conditions are proposed for ideal catalytic materials for Li-S batteries: (i) good electrical conductivity and rapid electron and ion transport; (ii) strong binding ability with LiPSs and strong anchoring ability to LiPSs, which can avoid the loss of intermediates and alleviate the shuttle effect. Too strong absorption ability will also cause surface vulcanization or slow diffusion process of LiPSs; (iii) full exposure of active sites to accelerate the redox reaction; and (iv) low Li_2S formation and decomposition barrier. The Li_2S redox reaction is the major step leading to low sulfur utilization, poor rate performance, and fast capacity fading in Li-S battery systems. With the advances in the field of catalytic materials in Li-S batteries, many functional materials have been developed. For example, carbon materials containing exposed sites such as N, O, and S can adsorb and enrich LiPSs at the electrochemical interface through the formation of chemical bonds, and some transition metal compounds, including metal sulfides, nitrides, and oxides, due to their strong interaction with LiPSs, can promote the transformation of long-chain intermediates into short-chain intermediates. However, these materials also have some limitations such as low catalytic ability of oxides and weak affinity of sulfides. Developing Li-S cathode host materials with good physical confinement, chemical adsorption, and effective catalysis is the aim of research on Li-S batteries.

In summary, the sulfur cathode is a critical component of Li-S batteries, offering a high theoretical energy density. Addressing challenges related to polysulfide dissolution and volume expansion is key to unlocking the full potential of Li-S batteries for various applications, including electric vehicles and renewable energy storage. While Li-S batteries hold great promise, challenges related to their sulfur cathode stability, cycling life, and safety need to be overcome for their widespread commercial use. Ongoing research is focused on developing novel cathode materials, understanding the reaction mechanisms of sulfur, and optimizing cell components to achieve practical and durable Li-S battery systems.

6.3 Electrolyte

The mechanism of Li-S batteries is based on the reaction between lithium and sulfur compounds. In Li-S batteries, the electrolyte is a crucial component that enables the flow of ions between the electrodes. Liquid electrolytes are the most commonly used ones in Li-S batteries due to their high ionic conductivity, low resistance, and good compatibility with electrode materials. They are composed of a lithium salt dissolved in an organic solvent that may also contain additives to improve their performance, such as conductive carbon additives, solvents with a high boiling point, and stabilizing agents. However, liquid electrolytes are flammable and volatile and pose safety issues, which has led to the development of solid-state electrolytes as a safer alternative for applications in Li-S batteries.

Solid electrolytes are a promising alternative to liquid electrolytes in Li-S batteries due to their improved safety, higher energy density, and longer cycle life, which endows them with the potential for applications in the development of next-generation Li-S batteries. They are typically composed of a lithium salt and a solid polymer or ceramic material. However, the development of solid electrolytes is constrained by challenges such as their low ionic conductivity, poor mechanical properties, and high cost. Researchers have been actively working on developing new solid electrolyte materials and improving their properties to enable the commercialization of solid-state Li-S batteries. In the following section, the fundamental understanding of and recent advances in liquid and solid electrolytes are presented.

6.3.1 Liquid Electrolyte

Organic liquid electrolytes have high ionic conductivity and low resistance, which makes them good candidates for high-performance Li-S batteries.

6.3.1.1 Solvent

When selecting a suitable solvent for a battery electrolyte, several factors need to be considered, such as the electrochemical voltage window, ionic conductivity, transference number, temperature window, dielectric constant, viscosity, inertness to other cell components, flammability, toxicity, and cost. Organic solvents such as ether, sulfone, or carbonate solvents are commonly used in Li-S batteries due to their good properties and compatibility with different cathode materials.

Ether Solvent Among well-known solvents, 1,3-dioxolane (DOL) and DME are commonly used as ether solvents in Li-S batteries. These solvents can be used individually or in combination to form a suitable electrolyte for Li-S batteries. DOL has a lower solubility for LiPSs than DME, which can lead to lower levels of LiPS shuttling between the cathode and anode and improved cycle life. DOL also forms a more stable SEI on the lithium anode, which can improve the safety and stability of the battery. On the other hand, DME has a higher solubility for LiPSs than DOL, which can improve the sulfur utilization and cycling performance of the battery. DME also has higher lithium transport kinetics, which can lead to faster ion transport through the electrolyte and improved battery performance. Therefore, DOL and DME are often used in combination in Li-S batteries to take advantage of the properties of both compounds. The ratio of DOL to DME can be optimized to achieve the desired discharge capacity and average discharge voltage and to balance the solubility of LiPSs and lithium transport kinetics.

Ternary solvents can be used to enhance ether-based electrolytes in Li-S batteries. A ternary solvent blend comprising THF, toluene, and DOL in a 1 : 1 : 8 ratio is an example. This composition demonstrated increased sulfur utilization and superior conductivity compared with their binary counterparts. Another ternary solvent comprises dimethyl disulfide (DMDS), a liquid organosulfur compound, as a cosolvent with DOL/DME. DMDS undergoes a reaction with sulfur, generating soluble methyl-terminated LiPSs, which are subsequently reduced to lithium organosulfides during discharge. However, in this approach, $LiNO_3$ is continuously required to shield the lithium anode and enhance the charge efficiency due to potential lithium anode corrosion risks from soluble organopolysulfides and DMDS, which can induce shuttle effects and self-discharge issues.

An alternative cosolvent, dimethyl trisulfide (DMTS), integrated into a DOL/DME mixture, shows increased capacity and excellent cycling performance due to its enhanced reactivity with Li_2S and intrinsic theoretical capacity compared with DMDS. In addition, a diisopropyl ether (DIPE)-based electrolyte, in which 50% of DOL/DME is substituted with DIPE incorporating 5.0% ethylene carbonate (EC) is incorporated as the cathode film-forming agent, shows a prolonged lifespan in Li-S batteries featuring SPAN cathodes compared with the standard EC/diethyl carbonate (DEC) electrolyte. DIPE, with appropriate Li salt solubility, ensures necessary ionic conductivity, while LiPSs face challenges in dissolving within DIPE.

Using a fluorinated ether, specifically 1,1,2,2-tetrafluoroethyl-2,2,3,3-tetrafluoropropyl ether (TTE), is an effective cosolvent for Li-S batteries. TTE, characterized by reduced electron-donating power due to fluorine substitution, significantly reduces the solubility of LiPSs. Moreover, it seamlessly integrates with the solvating component, actively participating in redox reactions across both electrodes to formulate a stable SEI layer. This SEI layer serves as a protective barrier, curbing parasitic reactions, enhancing the utilization of active materials, and improving coulombic efficiency. Establishing a stable SEI layer on both the sulfur cathode and lithium anode remains crucial for the long-term stability of Li-S batteries. Furthermore, the synergistic use of TTE and $LiNO_3$ proves particularly effective in safeguarding both electrodes, fostering stable SEI layer formation, reducing the dissolution of LiPSs, and promoting efficient conversion kinetics from polysulfides to Li_2S/Li_2S_2. This approach results in enhanced cycling stability and increased capacity retention in the battery.

Sulfone Solvent Sulfone solvents, such as sulfolane, dimethyl sulfoxide (DMSO), and dimethyl sulfone, have been investigated as organic liquid electrolytes for Li-S batteries due to their high chemical stability, high boiling points, and low vapor pressure. These solvents have high donor numbers, which enable them to solvate lithium ions effectively, while their high acceptor numbers help reduce the solubility of LiPSs, which can improve the cycling stability and coulombic efficiency of the battery. Furthermore, sulfone solvents can effectively suppress the shuttle effect by forming a stable SEI layer on the surface of the electrodes. The SEI layer can prevent the dissolution of LiPSs, which reduces the capacity deterioration of the battery during cycling. In addition, sulfone solvents can promote the conversion of LiPSs to Li_2S, which enhances the rate capability and capacity of the battery. However, sulfone solvents have a high viscosity and low ionic conductivity, which can limit the performance of the battery. To overcome this limitation, researchers have investigated the use of cosolvents, such as DME and tetraethylene glycol dimethyl ether (TEGDME), to improve the ionic conductivity of the electrolyte while maintaining the high stability and low solubility of LiPSs.

Carbonate Solvent Carbonate solvents, such as EC, propylene carbonate (PC), and dimethyl carbonate, present commendable attributes, including a high dielectric constant, increased ionic conductivity, and effective Li^+ solvation. However, their propensity to react strongly with lithium metal poses challenges, giving rise to the development of a formidable and obstructive SEI layer. This in turn curtails the overall performance and cycling stability of the battery. Furthermore, the unsuitability of carbonate solvents for Li-S batteries stems from the nucleophilic attack reaction between LiPSs and the carbonyl group present in carbonate molecules. Despite these challenges, limited investigations have been conducted on Li-S batteries that use carbonate-based electrolytes, particularly in conjunction with innovative sulfur cathode systems featuring diminutive sulfur molecules at low sulfur loadings.

Phosphate Solvent Phosphate solvents, such as triethyl phosphate and tripropyl phosphate, have high donor numbers, which enable them to solvate Li^+ effectively, while their high acceptor numbers help reduce the solubility of LiPSs, which can improve the cycling stability and coulombic efficiency of the battery. Compared with carbonate and sulfone solvents, phosphate solvents have a lower viscosity and a higher flash point, which makes them safer to handle and transport. However, phosphate solvents have a lower ionic conductivity than carbonate solvents, which can limit the performance of the battery. Given this, researchers have investigated the use of cosolvents, such as DME and TEGDME, to improve the ionic conductivity of the electrolyte while maintaining the stability and low solubility of LiPSs.

Nitrile Solvent Nitrile solvents, such as acetonitrile, propionitrile, and benzonitrile, have been shown to have a low solubility for LiPSs, which can improve the cycling stability and coulombic efficiency of the battery. However, nitrile solvents have a higher viscosity and a lower boiling point compared with carbonate solvents, which can limit their performance at high temperatures. To overcome this limitation, researchers have investigated the use of cosolvents, such as DMSO and THF, to improve the ionic conductivity and reduce the viscosity of the electrolyte. Nitrile solvents have also been investigated as a component of solid-state electrolytes for Li-S batteries. Solid-state electrolytes have the potential to improve the safety and energy density of Li-S batteries by replacing the flammable and volatile liquid electrolytes with a solid-state material. In addition, nitrile solvents have been shown to form a stable interface with solid-state electrolytes and enhance the Li^+ transport through the solid-state material.

6.3.1.2 Salt

Studies show that lithium hexafluorophosphate ($LiPF_6$) can be used as a Li salt in commercial rechargeable Li-S batteries, showcasing exceptional overall performance across various factors (temperature range, passivation, conductivity, etc.). However, its optimal use in Li-S batteries might not be achieved unless combined with a carbonate solvent in scenarios without LiPSs. Consequently, sulfonyl groups are prevalent in Li salts of Li-S batteries, with triflate being the simplest anion in this category. The application of several Li salts, such as bis(fluorosulfonyl)imide (FSI), bis(trifluoromethanesulfonyl)imide (TFSI), and bis(perfluoroethanesulfonyl)imide (BETI or sometimes PFSI), have been demonstrated in the literature. Comparative studies involving LiTFSI, Li oxalyldifluoroborate (LiODFB), Li bis(oxalato) borate (LiBOB), $LiBF_4$, $LiPF_6$, and $LiClO_4$ in ether solvents using high-performance liquid chromatography revealed that none of these materials are particularly suitable for Li-S batteries. This lack of suitability arises from either the dissolved LiPSs causing destruction to the unstable SEI layer on the Li anode or the instability of Li salts with LiPS anions. The advantages and disadvantages of common Li salts for Li-S batteries are detailed in Table 6.1.

6.3.1.3 Additive

The significance of functional additives in the electrolyte of Li-S batteries is widely acknowledged, which play an important role similar to that of solvents and salts. Using functional additives is a cost-effective and efficient strategy for enhancing cell performance. Unlike LIBs, Li-S batteries utilize a limited number of additives

Table 6.1 Advantages and disadvantages of common Li salts for Li-S batteries.

Li salt	
Advantages	**Disadvantages**
Excellent cycling stability, practical capacity, coulombic efficiency, and rate capability	High viscosity and slow reaction with LiPSs
Excellent coulombic efficiency and cycling stability in high temperatures	Weak coordination by other solvent molecules
Good temperature stability and improved chemical properties	Slow reaction with LiPSs
Well-defined physicochemical properties in carbonate-based solvents	Inducing the polymerization of DOL
Wide operating temperatures and good film-forming ability in carbonate-based solvents	Reaction with LiPSs and cannot be used along with DOL
Excellent coulombic efficiency and shuttle suppression	Low ionic conductivity in ether-based electrolyte

because of the increased demands imposed by unique challenges in this battery system, encompassing issues such as the dissolution and shuttle of LiPSs, growth of Li dendrites on the Li metal anode, electrolyte instability, and safety concerns.

Among the widely used electrolyte additives, $LiNO_3$ has garnered extensive attention and thorough investigation, which plays an essential role in the latest advancements in Li-S battery technology. Being a highly effective additive, $LiNO_3$ finds widespread applications in Li-S batteries to alleviate the shuttle effect of LiPSs, thereby enhancing the cycling performance of these batteries. Reduction of $LiNO_3$ can be used to safeguard the lithium metal anode in Li-S batteries, which may lead to the creation of nitrogen-containing species, such as $R\text{-}NO_2$ and/or nitrite (NO_2^-) groups, Li_2O, and $LiNO_2$. Additionally, $LiNO_3$ facilitates the oxidation of LiPSs shuttled into Li_2SO_3 and Li_2SO_4, in turn being reduced to $LiNO_2$. As the reactions progress, the accumulation of Li_2SO_3 and Li_2SO_4 steadily increases, forming a stable layer on the surface of the lithium anode. This layer effectively hinders the rapid loss of lithium metal and electrolyte, as well as the formation of lithium dendrites, which are well-known factors responsible for the suboptimal cycling performance of Li-S batteries.

Nevertheless, studies show that $LiNO_3$ exhibits dual effects on both the Li anode and sulfur cathode within the Li-S battery. On the Li anode, $LiNO_3$ accelerates the development of a stable passivation film, a known factor that significantly mitigates the redox shuttle of LiPSs. This passivation film continuously expands as $LiNO_3$ is consumed. On the cathode, $LiNO_3$ may undergo irreversible reduction on the carbon surface at a potential below 1.6 V, and the resulting reduction products detrimentally affect the reversibility and capacity of the Li-S battery. Increasing the discharge cutoff voltage helps prevent the irreversible reduction on the cathode, necessitating a minimum discharge potential higher than 1.8 V versus Li, as suggested by Rosenman. In addition, $LiNO_3$ can react with Li_2S_n, forming Li_xSO_y species that negatively influence the irreversible oxidation of sulfur's active material. Furthermore, when binders with oxygen-containing functional groups are used, the cycling stability is compromised due to the decomposition of $LiNO_3$. Thus, careful consideration of binders with oxygen-containing functional groups is essential to prevent rapid capacity decay. Therefore, although $LiNO_3$ serves as an effective additive for suppressing the shuttle effect, it presents drawbacks in the context of Li-S battery applications. The thickness of the passivation layer remains uncontrollable, depleting with each deposition, and its efficacy diminishes rapidly over successive cycles.

LiPSs find widespread application as an electrolyte supplement in Li-S batteries and contribute to the creation of inorganic layers within the SEI in a LiPS-$LiNO_3$-infused ether-based electrolyte. In an electrolyte comprising 0.02 M Li_2S_5 (0.10 M sulfur) and 5.0 wt% $LiNO_3$, an in situ-formed stable inorganic layer safeguards the Li metal anode. Conversely, when the concentration of LiPSs exceeds 0.50 M sulfur in the organic electrolyte, in situ SEI formation becomes inadequate, leading to gradual Li metal etching.

Besides the formidable lithium anode challenge, another crucial concern in Li-S batteries is the mass loss of the cathode's active material caused by the high solubility of intermediate LiPSs in organic solvents. Introducing stabilizing additives for the S cathode proves beneficial in enhancing the electrochemical reaction kinetics and postponing the dissolution of LiPSs. A polymerized protective layer forming at high potentials on the SEI of the S cathode effectively inhibits the dissolution of LiPSs. Pyrrole serves as an electrolyte additive, trapping LiPSs in Li-S batteries by forming a surface protective layer of polypyrrole through pyrrole oxidation on the sulfur cathode. The inclusion of toluene in an electrolyte containing 1 M $LiCF_3SO_3$ in TEGDME yields a stable, thin, porous, and conductive electrode–electrolyte interface on the sulfur cathode, further preventing LiPSs from being transported away from the sulfur electrode.

Using combustible materials, including highly reactive lithium metal, lithium nitrate, carbon materials with enhanced specific surfaces, sulfur, and ether-based electrolytes with a low boiling point, leads to substantial safety concerns, posing significant challenges to the practical implementation of Li-S batteries. These safety challenges can be addressed by incorporating additives into the electrolyte, which offers a straightforward and efficient approach to enhancing the safety of Li-S batteries. Research has shown that various groups, such as alkyl phosphates, fluorinated alkyl phosphates, ionic liquids, and phosphazenes, can be used as electrolyte additives with flame-inhibiting properties, contributing to the enhanced safety of Li-S batteries. Similarly, triphenyl phosphite (TPPi) functions as a flame retardant, improving the safety of Li-S batteries. A conventional electrolyte containing 5 wt% TPPi not only improves thermal stability but also enhances electrochemical performance. TPPi initiates the generation of free radicals, such as $PO^{\cdot}$, which can actively capture other free radicals generated during electrolyte combustion (typically $H^{\cdot}$ and $^{\cdot}OH$) to retard the reaction. In addition, dimethyl methylphosphonate (DMMP) has been explored as a flame retardant in carbonate electrolytes. Optimally adjusting the DMMP content (7–11 wt%) significantly suppresses flammability, enhances thermal stability in Li-S batteries, and modifies the interfacial film, thereby improving Li-ion diffusion coefficients and expediting electrochemical reactions.

6.3.2 Solid Electrolytes

The electrolyte serves as the conduit for ion transport and is typically constituted of organic solvent and lithium salt. This combination facilitates the flow of conductive ions between positive and negative electrodes, ensuring high voltage and specific energy in Li-S batteries. Over the past few decades, different types of liquid electrolytes, including organic-based, ionic liquid-based, and aqueous-based liquid electrolytes, have predominantly been used in Li-S batteries. Nonetheless, some properties of these liquid electrolytes, such as diffusion, migration, solubility, and precipitation, contribute to irreversible capacity loss in Li-S batteries. For instance, the shuttle effect arising from the dissolution and diffusion of soluble LiPSs in these liquid electrolytes leads to rapid capacity degradation and suboptimal cycle performance. Hence, optimized electrolyte systems need to be developed for Li-S batteries. Solid-state electrolytes, which can prevent the dissolution and diffusion of LiPSs to some extent, present a viable alternative to conventional liquid electrolytes. Since liquid electrolytes effectively prevent the dissolution of LiPSs, the shuttle effect can also be avoided – a trend that has gained considerable attention in recent years. As shown in Figure 6.4, sulfur undergoes direct reduction to Li_2S in a single step without forming polysulfide intermediates. Furthermore, solid electrolytes suppress the penetration of lithium dendrites and the leakage of liquid electrolytes. Although the inhibition of polysulfide ion shuttling by solid-state electrolytes enhances the specific capacity of Li-S batteries, these electrolytes have drawbacks such as

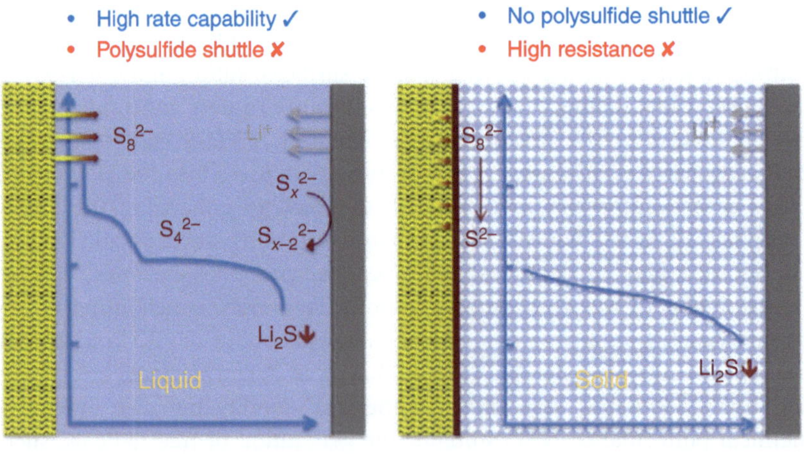

Figure 6.4 Schematic of Li-S batteries with liquid electrolytes and solid-state electrolytes. Source: Reproduced with permission from Li et al. [19]. © 2020, John Wiley & Sons.

low ionic conductivity and increased interfacial resistance at room temperature, resulting in compromised rate performance and cyclability. Thus, researchers have explored the application of different types of solid electrolytes in Li-S batteries, such as inorganic solid electrolytes (including oxide, sulfide, and hydride variants), solid polymer electrolytes (comprising polyethylene oxide (PEO) and other polymer-based options), and composite solid electrolytes (encompassing polymer/inorganic filler composites and other composite varieties) [20].

6.3.2.1 Inorganic Solid Electrolytes

Inorganic solid electrolytes show a Li^+ transfer number close to 1, and they inhibit polysulfide dissolution and diffusion or prevent polysulfide formation during all-solid-state Li-S battery cycling. Inorganic solid electrolytes used in Li-S batteries can be categorized into three groups: solid oxide electrolytes (e.g. LISICON, NASICON, and garnet), solid sulfide electrolytes (comprising silicon sulfide and $Li_2S–P_2S_5$ glass ceramics), and hydrides (specifically $LiBH_4$ type). Due to inadequate interfacial contact with electrodes, oxide-based solid electrolytes are seldom used independently; instead, they are often combined with liquid electrolytes or polymers to mitigate interfacial resistance. In such configurations, dense oxide-based solid electrolytes physically obstruct dissolved polysulfides. Nevertheless, solid-state batteries using solid oxide electrolytes present limitations in operating at low current densities due to insufficient solid–solid contacts leading to high interfacial resistance. Researchers have addressed this challenge using $Li_{1+x+y}Al_xTi_{2-x}Si_yP_{3-y}O_{12}$ (LATP) as a separator for liquid-electrolyte-based Li-S batteries to minimize electrical resistance and suppress polysulfide ion shuttling. For instance, LATP thin films serve as separators for Li/Li_2S batteries. In Li-S batteries, LATP acts as a permeable barrier for Li ions, preventing polysulfide anion migration. This configuration confines polysulfides to the cathode side, eliminating the shuttling effect and resulting in a battery with high coulombic efficiency, excellent cycling stability, and zero self-discharge. Recently, sulfide solid electrolytes, which present high ionic conductivity up to 10^{-2} S cm^{-1}, have gained attention in all-solid-state batteries. In sulfide-based batteries, polysulfide generation during discharge does not occur, thus overcoming the polysulfide shuttling issue. However, unstable electrochemical windows and poor electrolyte–electrode contact hinder their practical applications. In thiolisicon, for example $Li_{11-x}M_{2-x}P_{1+x}S_{12}$ (M = Ge, Sn, and Si), S elements in the LISICON framework are replaced with O elements. Kamaya et al. reported the lithium superionic conductor $Li_{10}GeP_2S_{12}$ (LGPS) with a 3D framework structure, which exhibits extremely high Li-ion conductivity (1.2×10^{-2} S cm^{-1} at 27 °C), surpassing that of liquid electrolytes. Despite its high conductivity, studies reveal unstable contact between $Li_{10}GeP_2S_{12}$ and Li metal due to Ge center reactivity.

Furthermore, a different type of thiolisicon solid-state electrolyte, $Li_{3.25}Ge_{0.25}P_{0.75}S_4$, finds application in all-solid-state Li-S batteries. However, due to the scarcity and cost of Ge, Yi and colleagues synthesized the solid-state electrolyte $Li_{10}SnP_2S_{12}$ (LSPS) through a more straightforward and efficient method, which shows high ionic conductivity (3.2×10^{-3} S cm^{-1} at RT). Charge–discharge curves revealed only one discharge and one charge plateau, indicating the direct reduction of S_8 to Li_2S without forming polysulfides. The elimination of polysulfide shuttling resulted in a high initial discharge capacity of 1601.7 mAh g^{-1} at 0.025 °C, maintaining 1180.3 mAh g^{-1} after 50 cycles with a 100% coulombic efficiency.

Li_2S–P_2S_5 glass–ceramic, another sulfide solid electrolyte with an amorphous structure, presents high ionic conductivity (10^{-5} – 10^{-3} S cm^{-1} at RT). Various Li_2S–P_2S_5 glass–ceramic electrolytes, prepared by mechanical milling, have been used in all-solid-state Li-S batteries. Different ratios of Li_2S and P_2S_5 yield various Li_2S–P_2S_5 electrolytes, including $80Li_2S$–$20P_2S_5$ ($Li_{4.0}PS_{4.5}$), $78Li_2S$–P_2S_5, $75Li_2S$–P_2S_5 (Li_3PS_4), $70Li_2S$–P_2S_5 ($Li_7P_3S_{11}$), and $60Li_2S$–$40P_2S_5$ ($Li_{1.5}PS_{3.3}$). Introducing additional materials into the structure is a viable strategy to enhance the ionic conductivity of Li_2S–P_2S_5 glass–ceramic electrolytes. In 2017, a high-quality MoS_2-doped Li_2S–P_2S_5 electrolyte ($Li_7P_{2.9}S_{10.85}Mo_{0.01}$) with high ionic conductivity (4.8×10^{-3} S cm^{-1} at room temperature) was reported. Mo doping also improves the compatibility of solid electrolytes with Li metal. However, addressing the high electrical resistance of thick solid electrolytes is crucial. Hence, Wang et al. used a Kevlar nonwoven scaffold as the mechanical support for cathode-supported all-solid-state LIBs, preparing a thin Li_3PS_4 solid electrolyte layer (approximately 100 μm thick). Since no polysulfide intermediates form in an all-solid-state system, the polysulfide shuttle is eliminated. At 25 °C and 0.05 C, LIB exhibits an initial discharge capacity of 949.9 mAh g^{-1}, a coulombic efficiency of 88.8%, and a stable specific capacity for 100 cycles at 0.2 C.

6.3.2.2 Solid Polymer Electrolytes

Solid polymer electrolytes, characterized by macromolecular or supramolecular structures, are typically prepared by embedding lithium salts within polymer matrices. In contrast to the previously mentioned dense inorganic solid electrolytes, solid polymer electrolytes, due to their lower density, offer increased gravimetric energy density, enhanced design flexibility, and improved electrode contact. Nevertheless, they consistently demonstrate suboptimal ionic conductivity at room temperature. A more significant concern arises from the potential dissolution and diffusion of polysulfides in the polymer matrix, leading to the shuttling effect. In this section, solid polymer electrolytes are categorized into two groups: PEO-based solid polymer electrolytes and alternative polymer-based solid polymer electrolytes.

Among various polymer matrices, PEO emerges as a promising candidate for solid polymer electrolytes due to its excellent flame retardancy, leak resistance, flexibility, low cost, and lightweight characteristics. However, its relatively low conductivity at room temperature poses a significant constraint to practical application in Li-S batteries. PEO exists in a semi-molten state at the working temperature and exhibits electrical conductivity above 10^{-4} S cm^{-1}. In PEO-based solid-state Li-S batteries, LiPSs show good solubility in the PEO matrix, which is counterproductive to preventing polysulfide shuttling. Typically, PEO-based solid polymer electrolytes are prepared by dissolving lithium salts into the PEO matrix, such as $LiN(SO_2CF_3)_2$ (LiTFSI), $Li[N(SO_2F)_2]$ (LiFSI), $Li[N(SO_2F)(SO_2CF_3)]$

(LiFTFSI), $LiC(CN)_3$ (LiTCM), and $LiBF_4$. LiTFSI and LiFSI are among the most widely used lithium salts. Despite the occurrence of both dissolution and diffusion of polysulfides in the polymer electrolyte, the FSI^- of LiFSI aids in forming a stable polysulfide-resistant SEI film on the lithium anode surface. LiFTFSI, featuring both $-SO_2CF_3$ and $-SO_2F$ functional groups, combines the advantages of $TFSI^-$ and FSI^- in all-solid-state Li-S batteries. However, fluorine in these salt anions acts as an SEI component on the Li anode surface, resulting in the production of LiF with relatively poor ionic conductivity ($\approx 10^{-31}$ $S\,cm^{-1}$). Consequently, this leads to reduced reaction kinetics and suboptimal rate performance of the battery. In contrast to the widely used LiTFSI and LiFSI, fluorine-free LiTCM-based electrolytes show superior ionic conductivity, thermal stability, and anodic stability in all-solid-state Li-S batteries. LiTCM induces a thin and highly conductive SEI layer on the Li anode, thus preventing polysulfide shuttling and Li dendrite formation.

Due to the limited ionic conductivity of PEO at room temperature, alternative solid polymer electrolyte supports should be explored. Besides PEO, a range of polymer matrices has been used in solid polymer electrolytes, including PAN, polymethylmethacrylate (PMMA), polyvinylidene fluoride (PVDF), and polyvinylidene fluoride-*co*-hexafluoropropylene (PVDF-HFP). PAN, an organic polymer renowned for its excellent antioxidant properties, has high thermal, mechanical, and electrochemical stability, which makes it a promising polymer matrix in solid polymer electrolytes. Sim et al. devised a PAN-based polymer electrolyte by dissolving 60% PAN and 40% LiTFSI in *N*,*N*-dimethylformamide, achieving a significant ionic conductivity of 2.54×10^{-4} $S\,cm^{-1}$ at 80 °C.

PMMA, another polymer host with a polar carbonyl group (—C=O—), has the capability to form coordination bonds with lithium salts, facilitating rapid ion transport in solid polymer electrolytes. Among various lithium salts, $LiClO_4$ is frequently incorporated into PMMA, e.g. formulations such as 49% PMMA-grafted natural rubber (MG_{49})/$LiClO_4$, PMMA/$LiClO_4$/dimethylacetamide, and PMMA/$LiClO_4$/PC. Rajendran et al. successfully prepared polymer electrolyte membranes composed of PMMA, PVDF, and $LiClO_4$. The ionic conductivity of the [0.25PMMA/0.75PVDF]-$LiClO_4$ electrolyte stands at $3.14 \times 10\,S\,cm^{-1}$ at room temperature, a value influenced by the salt's solubility and ions' mobility within the amorphous PMMA phase.

The high dielectric constant and the presence of strong electron-withdrawing functional groups (—C—F—) in PVDF make it an appealing polymer matrix for solid polymer electrolytes in Li-S batteries. A solid polymer electrolyte of PVDF-LiX type (X = FSI^-, $TFSI^-$, ClO_4^-) is synthesized. Analyzing the Arrhenius plots of these PVDF-LiX electrolytes shows that the ionic conductivity is of the following order: PVDF-$LiClO_4$ > PVDF-LiFSI > PVDF-LiTFSI. Notably, these values surpass those of PEO- and PAN-based solid polymer electrolytes. To further enhance ionic conductivity, HFP is often introduced into PVDF chains, reducing crystallinity and forming a PVDF–HFP polymer matrix.

Apart from the extensively used PAN, PMMA, PVDF, and PVDF-HFP, researchers have explored alternative polymer supports for solid polymer electrolytes. For instance, Lin et al. developed a solid polymer electrolyte with food-grade starch as the host, which showed an impressive ionic conductivity of 3.39×10^{-4} S cm^{-1} at 25 °C.

Solid electrolytes have attracted considerable attention as potential components for advanced Li-S batteries, offering advantages over liquid electrolytes such as enhanced safety, stability, and the potential for mitigating the shuttle effect. In summary, the potential of solid electrolytes for addressing key challenges associated with conventional Li-S batteries, including safety enhancement and shuttle effect suppression, underscores their significance in the ongoing research and development endeavors in the energy storage field.

6.4 Anode

The anode in a Li-S battery plays a critical role in the overall battery performance and electrochemical reactions. Its primary function is to host lithium ions during discharge and release them during charge cycles. The anode in a Li-S battery is typically made of a material that can intercalate lithium ions. During discharge, these lithium ions move from the anode to the cathode through the electrolyte, releasing energy. During charge, they move from the cathode back to the anode. The reversible intercalation of lithium ions at the anode is crucial for the overall cycling performance of the battery [21].

Dendrites are needle-like structures that can form on lithium metal anodes during cycling. They can pierce the separator and lead to short circuits, potentially causing safety issues. Mitigating dendrite growth is a significant challenge in Li-S batteries. Anodes in Li-S batteries, especially lithium metal anodes, form a SEI layer during the initial cycles. The properties of the SEI layer determine the battery's performance, cycle life, and safety. The anode material should be compatible with the sulfur cathode and electrolyte to avoid side reactions and ensure stable performance. Li-Si alloys have been explored as anodes due to their high lithium storage capacity. However, they also have drawbacks such as volume expansion during lithiation, which can cause mechanical stress and electrode degradation. Composite anodes made of carbon materials with high lithium capacity (such as hard carbons) have been investigated to improve the overall energy density and cycling stability of Li-S batteries. Researchers have been investigating methods to design electrolyte additives and coatings that can stabilize the SEI layer, thereby improving the stability and cycling performance of the anode. 3D-structured anodes, such as porous carbon scaffolds, can accommodate volume changes and enhance lithium ion diffusion, thus addressing issues such as dendrite growth. Achieving good compatibility between the anode and sulfur cathode is essential for overall battery performance. An appropriate selection of the electrolyte and an understanding of electrochemical reactions at both electrodes are crucial.

Unlike lithium metal batteries based on intercalation-type cathode materials, soluble LiPSs formed in Li-S batteries during charge and discharge have a non-negligible effect on lithium anodes (Figure 6.5). First of all, during the charging

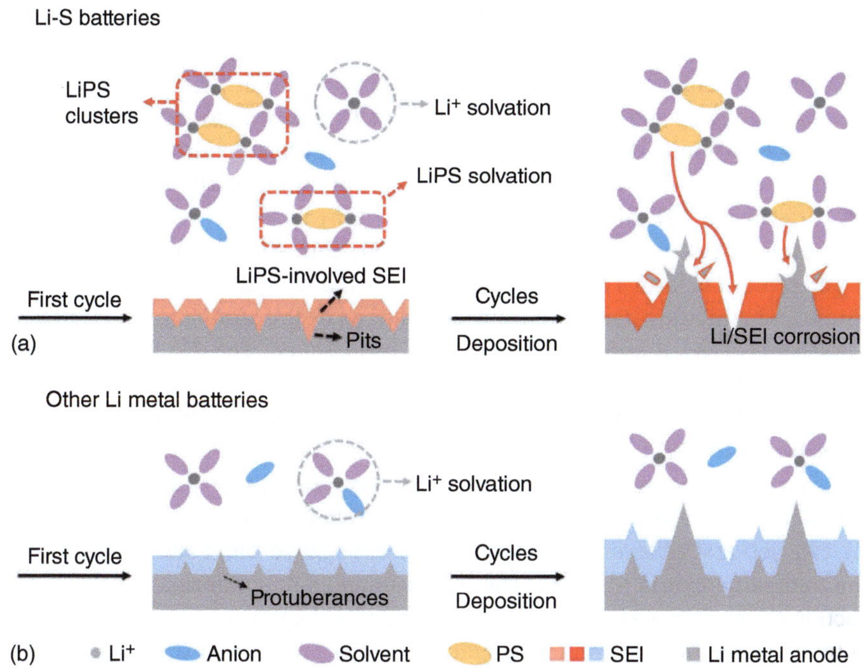

Figure 6.5 A comparison of a working Li metal anode in Li-S batteries and other Li metal batteries. The evolution of the Li metal anode in (a) Li-S batteries with the soluble LiPS intermediates and (b) other Li metal batteries using intercalation-type cathodes without LiPSs. Source: Reproduced with permission from Hou et al. [22]. © 2020, Elsevier.

and discharging process, LiPSs in the electrolyte undergo a dynamic change, which greatly affects the composition and physicochemical properties of the electrolyte. Secondly, after LiPSs diffuse to the negative electrode, they permeate through the SEI film or react with the exposed lithium, resulting in increased electrode corrosion and decreased coulombic efficiency. The Li_2S formed by this reaction changes the composition and structure of the SEI film, affecting the diffusion of Li^+, thereby changing the behavior of lithium deposition/dissolution. Therefore, metal lithium anodes in Li-S batteries have the disadvantages of not only dendrite growth and "dead lithium" that are typical for traditional metal lithium anodes but also loosening of the lithium metal surface and SEI due to the diffusion and corrosion of LiPSs. In addition, membrane rupture is a more serious issue. Especially under high sulfur loading and low electrolyte dosage conditions, the concentration of LiPSs in the electrolyte can reach more than 6 M [S], which greatly increases the viscosity of the electrolyte, leading to an increase in Li+ transport impedance, thus accelerating the growth of dendrites and even leading to micro-short circuits inside the Li-S battery. This shows that the existence of LiPSs and the complex transformation process significantly increase the difficulty in protecting lithium metal anodes.

Studies have shown that the SEI film is a key factor affecting the Li^+ flux and Li deposition/dissolution behavior. However, it has weak mechanical strength, exhibits spatial heterogeneity, and is extremely prone to cracking and damage under the volume fluctuation of cycles and corrosion of LiPSs. Therefore, constructing an SEI film with uniform Li^+ transport channels, low interfacial resistance, high mechanical strength, and LiPS shielding is an effective way to maintain stable and reversible lithium deposition/dissolution and inhibit dendrite growth. By adjusting the electrolyte components (such as solvent, lithium salt, and additives), a high-performance SEI film can be formed in situ by spontaneous reaction with metallic lithium. As an electrolyte additive, $LiNO_3$ can react synergistically with LiPSs and lithium to form a dense SEI film, thereby inducing the orderly deposition of lithium and inhibiting the "shuttle effect." Inspired by this, similar nitrate derivatives (such as $La(NO_3)_2$ and KNO_3) have been investigated. The SEI film is brittle and continuously generates cracks during the cycle, exposing lithium and forming a new SEI film. NO_3^- is a sacrificial additive, and its continuous consumption leads to the formation of the SEI film. It is difficult to maintain stability in long-term cycling. Introducing organic sulfur components can increase the elasticity and stability of SEI membranes. Electrolyte additives such as 3,5-bis(trifluoromethyl)thiophene (BTB) can self-assemble on the surface of the negative electrode to form an organic/inorganic hybrid SEI film. The planar main chain conformation and p–p interaction of aromatic-based substances can improve the flexibility of the SEI film. In addition, in the SEI film, the Ph-S group in LiPSs can prevent the corrosion of LiPSs through electrostatic shielding, reduce the consumption of BTB, and improve the stability of the SEI film.

Replacing the self-formed SEI film with a high-strength adjustable artificial coating can result in better protection for the lithium anode [23]. Electron beam reactive evaporation by nitrogen plasma-assisted deposition can establish a uniform, dense lithium phosphorus oxynitride (LiPON) protective layer on the surface of lithium metal, which exhibits high ionic conductivity, chemical stability, and mechanical strength that are comparable with that of lithium and sulfur. Using this approach, the reversibility of the deposition/dissolution of the lithium anode is effectively maintained in the pouch battery, with the number of stable cycles exceeding 150 cycles. However, there are a series of problems in the artificial coating strategy after matching the high-capacity sulfur cathode. Due to the unsupported nature of metallic lithium anodes, high-capacity lithium deposition/dissolution causes severe deformation of the anode, which may lead to cracking of the coating and even loss of close contact with the electrode, which is often impossible to repair. However, during the charging process, the thickness of the sulfur positive electrode shrinks, causing the metal lithium negative electrode to further advance toward the sulfur positive electrode, which not only intensifies the deformation of the electrode, but also easily causes lithium dendrites to penetrate the separator under pressure. It can be seen that the "magnification effect" brought about by the increase in sulfur loading not only deteriorates the performance of the sulfur cathode, but also poses a new challenge to the protection of the anode.

Confining Li within the self-supporting framework can significantly alleviate the volume fluctuations of repeated deposition/dissolution processes, reduce the local current density, homogenize the Li^+ flux, and stabilize the SEI film. Studies have shown that porous copper foil (3D Cu) obtained using the deposition–reduction method is an ideal lithium carrier. 3D Cu has a stable submicron-scale framework and rich pore structure, which significantly increases the specific surface area of the negative electrode and helps reduce the local current density on the electrode surface. In addition, the protruding tips on the copper skeleton can be used as nucleation centers to reduce the nucleation energy barrier of lithium on 3D Cu, increase the nucleation rate, and facilitate the orderly deposition of lithium. In order to further reduce the weight ratio of the carrier skeleton, the lightweight 3D carbon skeleton is a better choice, but there is a gap between carbon materials and lithium. The weak affinity limits its prospects as an anode carrier. Therefore, it is necessary to introduce lithiophilic heterogeneous nucleation centers on the carbon material matrix to improve the dispersion of lithium in the carbon framework. Studies on the binary phase diagram of lithium alloys and lithium deposition experiments have shown that there are obvious differences in the nucleation and growth modes of lithium on different metal substrates. Among these substrates, lithium has a higher nucleation energy barrier on the metals with low lithium solubility (Cu, Ni, and Sn), while the nucleation energy barrier on the surface of metals with higher solubility (Au, Ag, Zn, and Mg) is lower. Gold has been shown to have the best lithium affinity, and the nucleation overpotential of lithium on a gold surface is zero. This innovative work provides guidelines for the selection of metal-based lithium-friendly centers. In addition to transition metal modification, heteroatom doping on the carbon matrix is also an effective means to improve the lithium affinity of the framework, among which graphitized boron, edge graphitized boron, B-2C-O type boron, pyridine nitrogen, carboxyl group, and S- and P-doped structures all show good lithium affinity.

The larger specific surface area of the self-supporting lithiophilic support also increases the electrolyte consumption and, combined with the construction of an artificial SEI membrane, reduces the electrolyte consumption of the self-supporting lithium anode. Wei et al. investigated 3D lithium carbon composite material as the negative electrode and added diphenyl disulfide and LiNCb in the electrolyte to co-deposit thiophene ions and inorganic ions on the lithium surface. Lithium salt self-assembled into SEI film. Under the synergistic effect of the fibrous structure buffering the volume change and the self-assembled SEI membrane suppressing side reactions, the full battery assembled with the 3D lithium carbon composite material anode and the high-load (15 mg cm^{-2}) sulfur cathode realized a high energy density of 22.3 mWh cm^{-2}, 100 cycles after its capacity retention rate of 96.7%, and a coulombic efficiency of over 89.0%. Increasing the sulfur loading of the positive electrode inevitably increases the burden on the lithium negative electrode, and the growth of lithium dendrites and the accumulation of "dead lithium" are found to deteriorate, resulting in a decrease in the cycle performance and working life of the electrode.

Introducing a stable SEI film and a self-supporting lithium carrier is an effective strategy to maintain the stability of the electrode structure and improve the uniform deposition of lithium. Especially when matched with a high-capacity sulfur cathode, the self-supporting matrix with strong lithium affinity can homogenize the transfer of Li^{+}, buffer the severe volume fluctuations during cycling, reduce the pressure change in the battery, and maintain the SEI membrane. Stability is one of the important factors in improving the electrochemical performance of high-energy-density Li-S batteries.

Drawing from the aforementioned discourse, it becomes evident that every constituent within the battery employs unique methods to counteract the shuttling effect. The strategies aimed at mitigating this effect in Li-S batteries can be classified into four categories. The sulfur cathode assumes a pivotal role, acting as the active site for sulfur storage and reaction. The crux of the polysulfur shuttle effect lies within the sulfur cathode. Devising a cathode with a well-thought-out design can impede the dissolution and diffusion of polysulfides at the source. However, comprehending the electrocatalytic mechanism remains elusive, and challenges persist in achieving cathodes with high sulfur content and substantial areal sulfur loading, along with being lightweight criteria.

Regarding electrolytes, while liquid and gel electrolytes, along with certain solid electrolytes, fall short of eliminating the shuttle effect, the utilization of optimal all-solid electrolytes combined with sulfide-solid electrolytes or composite solid electrolytes without polymer/liquid components facilitates a one-step solid–solid conversion reaction during discharge. Using this approach, a Li-S battery without the shuttle effect can be realized. Nevertheless, issues persist with solid electrolytes, including low ionic conductivity, increased interfacial resistance, and a restricted voltage window. The application of solid electrolytes with exceptional performance and stability in Li-S batteries depends on addressing these challenges.

The separator coating functions as a supplementary and reinforcing element for the sulfur host and also serves as a secondary collector. Introducing this secondary modifying layer directly hinders the penetration of polysulfides into the separator. Moreover, polysulfides absorbed onto the modified layer can be electrochemically reactivated to engage in subsequent electrochemical reactions, enhancing sulfur utilization. To achieve this effect, factors such as volume loading, mass loading, thickness, and wettability to the electrolyte should be thoroughly investigated. Serving as the ultimate barrier in the polysulfide switching process, anode surface engineering acts as the optional screening factor, ensuring the smooth transport of Li+ ions while isolating direct contact between polysulfides and Li metal. This inhibition mitigates the solubility and corrosion of Li metal by LiPSs, thereby reducing unnecessary sulfur loss. Simultaneously, the instability of the SEI layer, prone to breaking during cycling, should also be taken into account.

In general, each component in a lithium battery possesses unique advantages in countering the shuttle effect. Thus, enhancing each component contributes to a corresponding improvement in performance. While there is no optimal solution to the shuttle effect yet, it is evident that collaborative efforts are crucial for improving battery performance. To sum up, the anode in a Li-S battery functions as the host for lithium ions during discharge and charge cycles. Progress in the engineering of anode materials is vital for enhancing the safety, energy density, and cycling stability of Li-S batteries, with the aim of making them suitable for diverse applications, such as electric vehicles and renewable energy storage.

6.5 Li-S Pouch Cell Analysis

6.5.1 Limitations of Coin Cells in Li-S Research

As the name suggests, the shape of the coin cells is like a small button, with a large diameter and low thickness. They have the advantages of being lightweight and small in size and are widely used in daily life. On the other hand, coin cells are often the test materials for laboratories engaged in battery research and development as they provide a convenient platform for rapid testing of new materials on a small scale. However, obtaining reliable, reproducible data using coin cells is inherently difficult, particularly in the full-cell configuration.

Most studies on Li-S coin cells (electrode sizes less than $1.6\,cm^{-2}$) have used low-sulfur-loaded electrodes ($<2\,mAh\,cm^{-2}$), excess electrolyte ($E/S > 38\,ml\,g^{-1}$, where E/S stands for electrolyte volume/sulfur mass), and excess lithium anode ($N/P > 25{,}250\,\mu m$ lithium, where N/P stands for anode/cathode capacity ratio). A typical coin cell (sulfur load of $2\,mg\,cm^{-2}$, $E/S = 38\,ml\,g^{-1}$ and $250\,\mu m$ lithium) has an electrolyte percentage higher than 78% of the total battery weight and actual battery energy less than $45\,Wh\,kg^{-1}$. Compared with the $350\,Wh\,kg^{-1}$ that can be achieved by the pouch cell, it is very low.

Li-S coin cells adopt low areal sulfur loading, flooded electrolyte, and excessive anode usage, which are defined as coin-cell-level evaluation conditions. In contrast, high areal sulfur loading, low E/S ratio, and low N/P ratio are essential to construct high-energy-density Li-S pouch cells, which are recognized as pouch-cell-level evaluation conditions. Therefore, it is obvious that pouch-cell-evaluation of Li-S batteries has more significance for high-energy-density Li-S batteries.

6.5.2 The Main Difference Between Pouch Cell and Coin Cell

Coin cells and pouch cells are designed for distinct applications. Coin cells are primarily applied in miniature electron devices, such as electronic scale. Since coin cells can be easily prepared, they are extensively adopted in laboratory investigation and are all coated on a single side. However, pouch cells are commonly used in applications with a high energy density demand, including mobile phones, laptops, and electric vehicles. Compared with other battery forms (coins and columns), pouch cells are lighter in weight and better at maximizing specific energy and can also evaluate changes in battery volume for better control or compensation [3–5]. It is feasible to change pouch cells from single layer to multilayer with controllable active S amount. The more the number of layers, the larger the electrode area, so more sulfur carbon composite is required. But as sulfur sublimes, the electrode coating solvent evaporates, so maintaining the uniformity of the battery's two-sided coating is a challenge.

Figure 6.6 shows the major differences between a coin cell and a multilayer pouch cell. Coin cells are single-layered, and the electrode requires only a single-sided coating, while pouch cells are generally multilayered, with the need for double-coated electrodes and more active material. In addition, pouch cells have

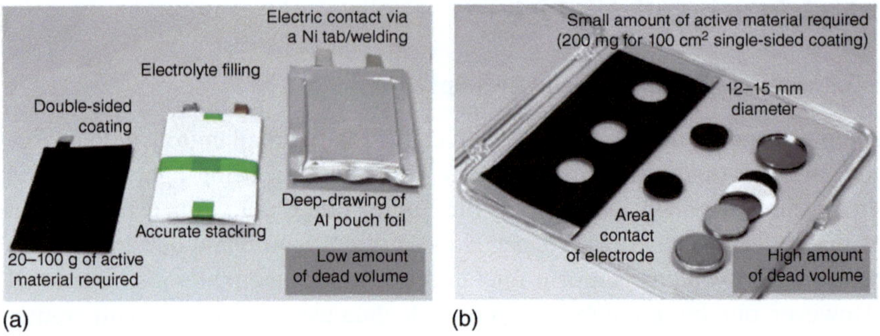

Figure 6.6 Photographs of a coin cell and a pouch cell and their major differences.

an aluminum–plastic casing, while coin cells have a metal casing; thus, pouch cells show a lower dead volume.

Multilayer pouch cells require several additional processing steps, thus increasing the complexity of battery assembly. Although recent papers have reported the high cycle life of lab-assembled pouch cells under high active material loads (>1000 cycles) and high specific capacity [5], the battery capacity declines rapidly in actual use, primarily due to electrolyte depletion and pulverization of the anode.

In conclusion, most lab-scale experiments in the Li-S literature do not allow a direct transfer to the practical pouch cell level, and as a consequence, important aspects of the relevant Li-S battery chemistry are overlooked. Multilayer pouch cells require several additional processing steps, thus adding complexity to the system. On the other hand, they also present significant advantages for reproducible performance evaluation such as reduction of dead volume, better control of electrolyte content, and external pressure. In contrast, the worse control of the abovementioned parameters is the major disadvantage of coin cell testing, which limits the acquisition of application-relevant data.

6.5.3 Key Challenges and Solutions for Pouch Cells

6.5.3.1 Cathode

The final discharge products of both sulfur and Li-S batteries are electronic insulators, and this property of positive active materials is one of the main reasons for the capacity decline of Li-S batteries after long cycles. In addition, active S materials undergo a volume change of about 80% during conversion to Li_2S, which reduces the physical contact between the active material and the conductive additives in the cathode composite. Poor cathode compaction and electrolyte expansion also lead to the loss of active material and electrolyte consumption. In addition, precipitation of sulfur substance causes compressive strain of slurry during the coating process, which needs to be resolved using cathode composite materials of different thicknesses.

Therefore, the strategy of increasing the sulfur load, utilization rate, and cathode compactness is the key factor for accelerating the commercialization of Li-S pouch cells. The major drawbacks of composite cathodes in Li-S batteries include low utilization rate of sulfur load, large volume variation during cycling, poor compaction density, polysulfide dissolution, and limited electronic conductivity. To enhance the cathode composite structure to improve electrochemical performance, the following strategies can be used: (i) decorating nanosulfur in a solid layered conductive substrate and (ii) using polar free radical compounds, functional adhesives, and external stacking pressure to improve compaction.

6.5.3.2 Electrolyte

An electrolyte not only provides the ion transport route between the two electrodes of a Li-S battery, but also acts as the reaction medium for sulfur conversion in the cathode. Electrolyte decomposition at the cathode–anode interface is the main reason for the degradation in Li-S battery performance, which leads to the low utilization rate of sulfur active substances in the battery, low rate capacity, and short cycle life. In Li-S batteries with excess electrolyte (high E/S ratio), the cathode material is completely wet, and the cycle life of the battery is relatively extended as it takes longer for the electrolyte to run out. However, a high E/S ratio results in a reduction in the weight and volume energy density of the battery. With the increase in sulfur load in a certain area of the battery, the influence of the E/S ratio on energy density by weight also increases. Therefore, in order to improve the specific capacity of Li-S pouch cells, it is important to improve their cycling ability under the condition of a lean electrolyte. At present, the primary methods used to reduce the electrolyte consumption of Li-S pouch cells are (i) adding electrolyte additives and (ii) improving the interface performance between the anode and electrolyte.

6.5.3.3 Separator

Traditional polyolefin separators, such as polypropylene (PP) and polyethylene (PE), are widely used in Li-S batteries due to their low cost, high mechanical strength, good thermal stability, and good electrolyte wettability. The separator in Li-S batteries chiefly plays the role of screening ions, regulating polysulfide shuttle, and preventing short circuits due to the growth of lithium dendrites. However, soluble polysulfides produced during sulfur conversion at the cathode can be dissolved in the electrolyte, which can travel through the separator to the anode. Once polysulfides reach the anode, they trigger a parasitic reaction that erodes the anode surface and leads to severe capacity loss. In addition, the thermal conductivity of PP and PE separators ($<0.40\,\mathrm{W\,m^{-1}\,K^{-1}}$) and thermal diffusivity($<2.60\,\mathrm{mm^2\,s^{-1}}$) are low, which makes it difficult for the heat generated inside the battery to dissipate, thus increasing the risk of heat runaway in Li-S pouch cells.

Thermal monitoring of the separator can improve the safety and stability of Li-S pouch cells. Carbon-based materials have relatively high thermal conductivity, which is favorable for the separator in the case of heat (charge and discharge process, or internal short circuit) rapid heat dissipation. By surface modification of carbon additives, the shrinkage rate of the PP/PE separators at high temperatures can be significantly reduced. Researchers have developed functional separators for improving heat accumulation and limiting polysulfide penetration. Figure 6.2 shows the strategies for modification of these separators. For Li-S pouch cells,

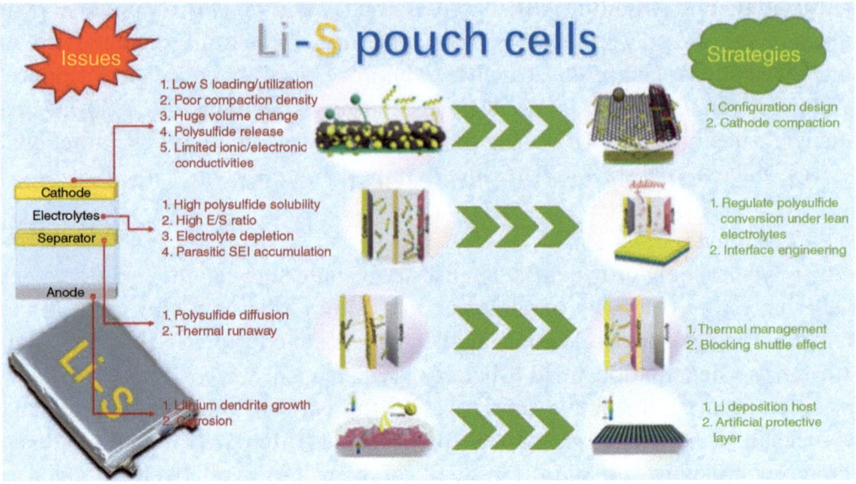

Figure 6.7 Overview of key issues and strategic solutions for the development of practical Li-S pouch cells.

separator modification measures can be classified fall two main categories: (i) thermal management and (ii) polysulfide restriction.

6.5.3.4 Anode

The major reason for the failure of button cells is polysulfide shuttle, whereas that for pouch cells is the lithium metal anode [18]. At high current densities, lithium deposition and volume expansion during plating can disrupt interfacial contact and lead to cell polarization. These factors may affect the performance of pouch batteries over long periods of time. Anode excess is common in Li-S batteries; however, polysulfide formation and shuttle effect can form thick SEI and lead to anode corrosion. Lithium metal anodes expand and contract during cycling, resulting in the decomposition of old SEI and the formation of new SEI, thereby exposing uncorroded lithium to cracks. The loss of lithium metal and the high resistance of the SEI layer result in a reduced Coulomb effect and deteriorated cycling performance. Figure 6.7 shows a summary of some of the strategies focusing on homogenizing lithium ion deposition, reducing the formation of "dead lithium," and inhibiting the corrosion of lithium metal anodes using lithium deposition hosts and artificial protective layers on the anode surface.

6.5.4 Application Prospect Analysis

Li-S batteries are considered the most promising next-generation energy storage systems due to their ultrahigh energy density and low-cost sulfur feedstock. However, their low coulombic efficiency, fast capacity decay, and safety hazards due to the shuttle effect hinder their commercialization. During the operation of Li-S batteries, as long as the shuttle behavior of LiPSs exists, the performance of the battery is difficult to meet the actual demand. Therefore, a deeper and more comprehensive understanding of the shuttle effect of LiPSs is urgently needed, which can also provide a reference for the suppression of the shuttle effect in future high-performance Li-S batteries. Although various strategies to suppress the shuttle effect have made significant progress and shown good results in improving the electrochemical performance of Li-S batteries, there are many issues and difficulties that need to be addressed for large-scale commercial applications in the near future.

With continuous research, the understanding of the failure mechanism of Li-S batteries has become deeper, and some effective protection strategies have been proposed. However, most of the studies have focused on coin cells or some in situ cells, which do not reflect the real-world operating conditions of Li-S batteries, especially under harsh conditions. Currently, it is challenging to study the failure mechanisms and regulation strategies of Li-S batteries because of the huge gap between lab-scale coin cells for academic research and equipment-grade pouch batteries used in practical applications. In order to bridge this gap, more and more attention has been paid to high energy density (>300 Wh kg^{-1}) at the pouch cell level. For example, Cheng et al. performed a failure analysis of Li-S pouch cells (300 Wh kg^{-1}) at a low E/S ratio of 3 μl mg^{-1} using a thin lithium anode of 50 μm [24]. They found that the failure mechanism of Li-S pouch cells is not as simple as that of a thin lithium anode. They concluded that the failure of the lithium anode was the main cause of rapid capacity degradation rather than the decomposition of the electrolyte. Liu et al. investigated the reaction non-homogeneity of a practical Li-S pouch battery with an energy density of 300 Wh kg^{-1} [25]. They proposed that the low mobility of the electrolyte is the major factor contributing to the inhomogeneous distribution of lithium ions, leading to a tendency for lithium ions to be deposited in electrolyte-rich regions, thus exacerbating lithium metal. Overall, the failure mechanisms of lithium batteries used in real-world situations are much more complex and challenging to study than those of lab-scale batteries. Therefore, it is crucial to find new strategies to study batteries in real-world conditions to reveal the failure mechanisms, which is a key factor in developing more reliable Li-S batteries.

In addition, the formation of lithium dendrites can puncture the diaphragm and cause internal short circuits, which is a major safety concern associated with Li-S batteries. Although modifying the lithium anode can help address this issue, it is also important to consider the safety of the cathode. Interactions between electrolytes and sulfur species, as well as redox reactions between lithium anodes and C-S cathodes, have been shown to cause self-heating and thermal runaway. To further elucidate the triggers of thermal runaway in Li-S batteries under real-world conditions, Jiang et al. systematically evaluated the thermal runaway characteristics of long-term Li-S pouch batteries (16 and 45 cycles) with and without electrolyte, showing that the reaction between higher-order polysulfides ($Li_2S_x \geq 6$) and Li is the most important trigger for thermal runaway in cycling Li-S pouch batteries.

To achieve high electrochemical performance and long cycle life of Li-S pouch cells, the challenges of polysulfide shuttle and lithium metal anode should be solved. There is a lot of work to be done on the cathode in terms of active material availability and sulfur load to achieve weight and volume energy density suitable for their commercial use. Good material structures can provide a stable carrier for the deposition of sulfur active materials and lithium metal. However, the design of substrate structures for sulfur cathodes and lithium metal anodes is in its infancy, and a unified approach is needed to evaluate design parameters for commercial batteries, such as the compacting density of the cathode, the expansion of the cell volume when injected with electrolyte, and the stacking pressure of the battery. At the same time, attention should be paid to parameters such as ionic conductivity, electrolyte absorption, and thermal safety of cathode/anode in the research of electrolytes, and the key step to realize the commercialization of Li-S pouch cells is by adding electrolytes, especially solid electrolytes. The design of the separator is also important because it directly limits the passage of the polysulfide.

Producing Li-S pouch cells with high energy density and long cycle life is a complex subject, which requires consideration of many factors and mechanisms. Factors such as sulfur utilization, cathode density, electrolyte usage, lithium usage, and safety need to be consistent with those intended for the commercial application of the battery. Addressing these issues will be a key step toward commercializing Li-S pouch cells.

Although various approaches to mitigating the polysulfide shuttling effect have made notable progress, exhibiting positive outcomes in enhancing the electrochemical performance of Li-S batteries, numerous challenges persist, hindering the realization of large-scale commercial applications in the foreseeable future. Consideration of all these issues and difficulties is imperative.

(1) *Comprehensive exploration of catalysis mechanism*: Despite the recognition of the catalytic impact of the medium on Li-S batteries, extensive experimental data affirming this effect exist. However, owing to the multistage and intricate reaction process, the precise mechanism remains elusive. Currently, there is a dearth of profound insights into the transfer and conversion of ions/electrons on the working surface/interface of electrocatalysts. The spatial distribution and nucleation behavior of Li_2S discharge products and S_8 oxidation products remain uncertain. A more direct description of catalysis in Li-S batteries can only be achieved with a better understanding of the adsorption catalysis mechanism at the molecular or atomic level. Fortunately, due to the continuous evolution of characterization techniques, the mystery of the catalytic mechanism has been gradually unraveled. The development and optimization of in situ characterization techniques, such as XAS, IR, XPS, and STEM during operation, facilitate the real-time monitoring of instantaneous changes and evolution processes within the battery during operation. This approach provides guidelines for designing advanced dielectric materials for Li-S batteries.

(2) *Advancement of polysulfide-free solid-state electrolytes*: It is an approach to effectively addressing the polysulfide shuttling predicament, which involves substituting liquid electrolytes with entirely solid counterparts. Notably, oxide-based and solid polymer electrolytes often amalgamate with liquid solvents or polymers, resulting in the generation of polysulfides.

    Sulfide-solid electrolytes or composite solid electrolytes are optimal choices for preventing polysulfide shuttling, as they do not have liquid components.

Nevertheless, they pose formidable challenges such as low ionic conductivity, increased interfacial resistance, and chemical and electrochemical instability, presenting obstacles for all-solid-state lithium batteries. Realization and ultimate resolution of these issues are imperative to facilitate the plausible application of all-solid-state Li batteries and alleviate the shuttle effect.

(3) *Investigation of complete cells and viable pouch cells*: In coin-type half-cell examinations using robust Li plates as anodes, the adverse influence of LiPSs on Li anodes is frequently disregarded at high N/P ratios (>150). Nevertheless, in fully charged Li-S batteries, particularly in practical pouch batteries, the catastrophic deterioration of the Li metal anode and the dearth of Li supply are crucial factors for substantial battery capacity loss and swift failure. Constrained Li mass loading in pouch cells with a low N/P ratio is inadequate to sustain continuous, irreversible Li consumption triggered by the reaction with LiPSs. Increased sulfur loading intensifies the concentration of LiPSs, exacerbating the shuttling effect and lithium corrosion. Consequently, while addressing the shuttle effect, safeguarding metal lithium and impeding dendrite growth in the battery should also be considered. Moreover, for the pragmatic implementation of Li-S batteries, emphasis should be placed not only on material exploration but also on electrode configurations and battery engineering.

(4) *Suitable for extensive practical deployment*: The energy density of Li-S batteries is highly influenced by high sulfur content, increased areal mass loading, and a low electrolyte/sulfur (E/S) ratio. To attain 300 Wh l^{-1}, a sulfur loading of at least 7 mg cm^{-2} is required. However, increasing the sulfur content generates more LiPSs, and the insufficient electrolyte volume presents various challenges to wettability and ion migration, placing increased demands on battery components. Apart from internal conditions, key parameters such as applied pressure, temperature ranges, storage conditions, and others necessitate further exploration in practical scenarios. Considering the potential safety risks, the thermal stability of Li-S batteries has been scarcely investigated and requires systematic exploration and attention. Furthermore, despite the nontoxic nature of sulfur, recovering lithium metal resources after large-scale commercial Li-S battery deployment remains challenging.

Ultimately, despite significant advances in the evolution of Li-S batteries, numerous challenges persist during the commercialization phase. It is anticipated that through advances in characterization technologies and a more comprehensive exploration of the reaction mechanism, a thorough understanding of Li-S batteries can be achieved. This in turn may facilitate the more effective prevention of the shuttle effect.

References

1 Huang, Y., Lin, L., Zhang, C. et al. (2022). Recent advances and strategies toward polysulfides shuttle inhibition for high-performance Li-S batteries. *Advancement of Science* 9 (12): 2106004.

2 Fang, R., Zhao, S., Sun, Z. et al. (2017). More reliable lithium-sulfur batteries: status, solutions and prospects. *Advanced Materials* 29 (48): 1606823.

3 Urbonaite, S., Poux, T., and Novák, P. (2015). Progress towards commercially viable Li-S battery cells. *Advanced Energy Materials* 5 (16): 1500118.

4 Jeong, Y.C., Kim, J.H., Nam, S. et al. (2018). Rational design of nanostructured functional interlayer/separator for advanced Li-S batteries. *Advanced Functional Materials* 28 (38): 1707411.

5 Yu, C.H., Yen, Y.J., and Chung, S.H. (2021). Nanoporosity of carbon-sulfur nanocomposites toward the lithium-sulfur battery electrochemistry. *Nanomaterials – Basel* 11 (6): 11061518.

6 Demir-Cakan, R., Morcrette, M., Nouar, F. et al. (2011). Cathode composites for Li-S batteries via the use of oxygenated porous architectures. *Journal of the American Chemical Society* 133 (40): 16154–16160.

7 Qiu, S., Zhang, J., Liang, X. et al. (2022). Tunable MOFs derivatives for stable and fast sulfur electrodes in Li-S batteries. *Chemical Engineering Journal* 450: 138287.

8 Wang, Z., Shen, J., Liu, J. et al. (2019). Self-supported and flexible sulfur cathode enabled via synergistic confinement for high-energy-density lithium-sulfur batteries. *Advanced Materials* 31 (33): 1902228.

9 Li, H., Li, Y., and Zhang, L. (2022). Designing principles of advanced sulfur cathodes toward practical lithium-sulfur batteries. *Sustainable Materials* 2 (1): 34–64.

10 Li, S., Leng, D., Li, W. et al. (2020). Recent progress in developing Li_2S cathodes for Li–S batteries. *Energy Storage Materials* 27: 279–296.

11 Chen, L., Liu, Y., Ashuri, M. et al. (2014). Li_2S encapsulated by nitrogen-doped carbon for lithium sulfur batteries. *Journal of Materials Chemistry A* 2 (42): 18026–18032.

12 Wang, J., Yang, J., Xie, J. et al. (2002). A novel conductive polymer-sulfur composite cathode material for rechargeable lithium batteries. *Advanced Materials* 14 (13–14): 963–965.

13 Chen, W.J., Li, B.Q., Zhao, C.X. et al. (2020). Electrolyte regulation towards stable lithium-metal anodes in lithium-sulfur batteries with sulfurized polyacrylonitrile cathodes. *Angewandte Chemie International Edition* 59 (27): 10732–10745.

14 Liang, X., Cai, J., Qiu, S. et al. (2022). Bidirectional catalyst design for lithium–sulfur batteries: phase regulation cooperates with N-doping. *Journal of Materials Chemistry A* 10 (44): 23780–23789.

15 Jing, E., Chen, L., Xu, S. et al. (2022). Dual redox catalysis of VN/nitrogen-doped graphene nanocomposites for high-performance lithium-sulfur batteries. *Journal of Energy Chemistry* 64: 574–582.

16 Lai, Y., Wang, P., Qin, F. et al. (2017). A carbon nanofiber@mesoporous δ-MnO_2 nanosheet-coated separator for high-performance lithium-sulfur batteries. *Energy Storage Materials* 9: 179–187.

17 Lim, W.G., Kim, S., Jo, C. et al. (2019). A comprehensive review of materials with catalytic effects in Li-S batteries: enhanced redox kinetics. *Angewandte Chemie International Edtion* 58 (52): 18746–18757.

18 Qiu, S., Liang, X., Niu, S. et al. (2022). Quantitative defect regulation of heterostructures for sulfur catalysis toward fast and long lifespan lithium-sulfur batteries. *Nano Research* 15 (9): 7925–7932.

19 Li, S., Zhang, W., Zheng, J. et al. (2020). Inhibition of polysulfide shuttles in Li–S batteries: modified separators and solid-state electrolytes. *Advanced Energy Materials* 11 (2): 2000779.

20 Qiu, S., Liang, X., Li, Y. et al. (2022). Recent advance on Co-based materials for polysulfide catalysis toward promoted lithium-sulfur batteries. *Nano Select* 3 (2): 298–319.

21 Chen, M., Zhou, X., Liang, X. et al. (2023). Constructing porous nickel-zinc alloy layer on nickel foam for dendritic-free lithium metal anode. *Electrochimica Acta* 460: 142615.

22 Hou, L.-P., Zhang, X.-Q., Li, B.-Q. et al. (2021). Challenges and promises of lithium metal anode by soluble polysulfides in practical lithium–sulfur batteries. *Materials Today* 45: 62–76.

23 Shan, X., Zhong, Y., Huang, L. et al. (2023). Carbon cloth supported lithiophilic carbon nanotubes@Ag skeleton for lithium anode via ultrafast Joule heating method. *Journal of Solid State Electrochemistry* 27 (6): 1391–1398.

24 Cheng, X.-B., Yan, C., Huang, J.-Q. et al. (2017). The gap between long lifespan Li-S coin and pouch cells: the importance of lithium metal anode protection. *Energy Storage Materials* 6: 18–25.

25 Shi, L., Bak, S.-M., Shadike, Z. et al. (2020). Reaction heterogeneity in practical high-energy lithium–sulfur pouch cells. *Energy & Environmental Science* 13 (10): 3620–3632.

7

Aqueous Multivalent Metal Ion Batteries: Fundamental Mechanism and Applications

Xingyu Zhao[1], *Chunyang Yang*[1], *and Minghua Chen*[2]

[1] *Harbin University of Science and Technology, School of Electrical and Electronic Engineering, Department of Electrical Theory and New Technology, 52 Xuefu Road, Harbin, 150080, China*
[2] *Harbin University of Science and Technology, School of Electrical and Electronic Engineering, Key Laboratory of Engineering Dielectric and Applications (Ministry of Education), Department of Electrical Theory and New Technology, 52 Xuefu Road, Harbin, 150080, China*

7.1 Introduction

The increasing energy consumption and the decreasing availability of nonrenewable fuel sources necessitate advances in the generation of alternative energy sources. Generation of clean and sustainable intermittent renewable energy for large-scale grid and transportation calls for the development of advanced and innovative electrical energy storage systems [1]. Among energy storage systems, rechargeable lead–acid batteries and lithium-ion batteries (LIBs) are currently the most widely used technologies in portable electronics, transportation, and stationary energy storage. Nevertheless, due to the slow progress of technologies that can improve the energy density and cycle life of these two types of batteries, as well as cost and safety concerns, there is an urgent need for energy storage technologies "beyond lead–acid and lithium-ion batteries," i.e. replacing advanced battery systems to meet the emerging market demand due to rapid growth in electric vehicles and grid energy storage. Thus, there is a high demand for batteries with the advantages of low cost, environmentally friendly, high energy/power density, and prolonged lifetime.

Compared with state-of-the-art nonaqueous LIBs, newly developed water rechargeable batteries, especially aqueous rechargeable multivalent metal ion batteries (ARMMBs) using multielectron transfer electrode reactions, have outstanding advantages, including cost-effectiveness, high safety, environmental friendliness, and high capacity and rate capability. ARMMBs typically consist of a polyvalent metal anode, an aqueous electrolyte, and a cathode to hold polyvalent metal ions. Due to better electrochemical properties and advances in the development of ARMMB nanostructured materials, research on ARMMBs has gradually increased in recent years. The working mechanism of ARMMBs is different from that of polyvalent metal alkaline batteries (Figure 7.1a,b, using zinc-ion batteries and

Towards Next Generation Energy Storage Technologies: From Fundamentals to Commercial Applications, First Edition. Edited by Minghua Chen.

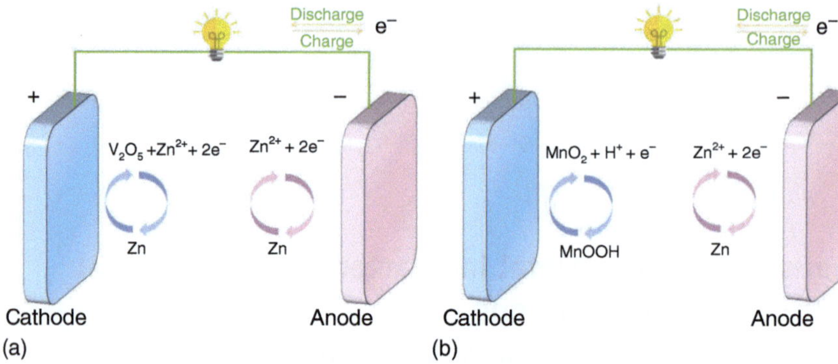

Figure 7.1 Schematic diagrams of the basic structure and working principle of (a) Zn-ion batteries (taking V_2O_5 cathode as an example) and (b) Zn–alkaline batteries (taking MnO_2 cathode as an example).

zinc–alkaline batteries as examples). In addition, the high ionic conductivity of the water electrolyte (for nonaqueous electrolytes ≈ 1 and $10^{-3} - 10^{-2}\,\text{S cm}^{-1}$) favors the high rate capability and high power density of ARMMBs. Therefore, due to their better electrochemical performance and ability to form nanostructured materials, ARMMBs have gained huge research interests. However, their insufficient energy density and limited life hinder their grid-scale application in this modern world. Therefore, identification of electrode materials with high reversible capacity and long cycle stability is crucial for the practical application of ARMMBs [2].

7.2 Classification Based on Energy Storage Mechanism

7.2.1 Insertion/Extraction Mechanism of Ions

Ion insertion, particularly zinc-ion insertion, is the key mechanism in aqueous zinc-ion batteries (AZIBs). As early as 2012, the insertion of Zn^{2+} into the tunnel of α-MnO_2 in AZIBs was reported. Although Zn^{2+} ions have a similar ionic radius (0.74 Å) to Li^+ ions (0.76 Å), their strong interaction with H_2O molecules through hydrogen bonding results in a large radius (4.3 Å) for hydrated zinc ions. Consequently, to accommodate the hydrated zinc ions, significant layered or tunneled spacing is required. Although the desolvation process can reduce the size of the hydrated ions, it requires additional energy, which negatively affects the reaction kinetics. Therefore, typically, crystals with expansive layered or tunneled distances can facilitate the insertion/extraction of hydrated zinc ions, whereas materials with constrained layered or tunneled spaces allow for the insertion/extraction of Zn^{2+} ions (Figure 7.2a) [3]. It is worth noting that the considerable size of the hydrated zinc ions and the intense electrostatic repulsion between the divalent Zn^{2+} ions and the host materials can facilitate the phase transformation of cathode materials upon Zn^{2+} insertion. The tunneled β-MnO_2 transforms into layered-type zinc buserite during the initial discharging, followed by the reversible insertion and extraction of Zn^{2+} in the resulting structure [10].

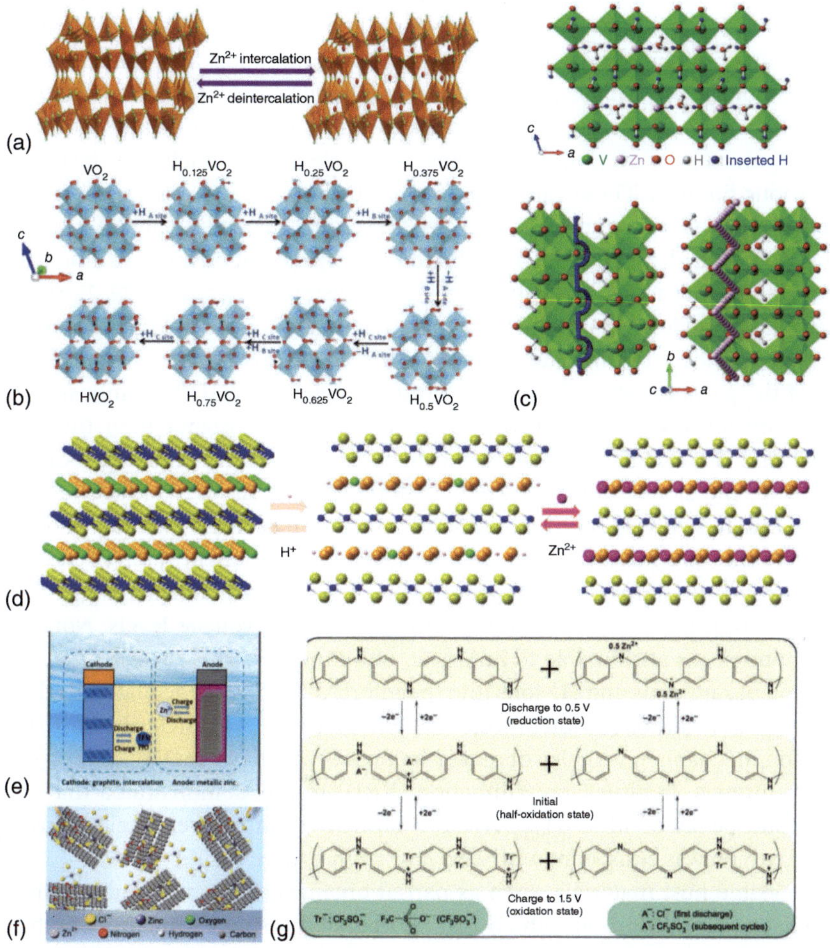

Figure 7.2 Schematic of zinc-ion insertion mechanism in (a) tunneled VO_2 cathode. Source: Reproduced with permission from Ding et al. [3]. © 2018/John Wiley & Sons. (b) H^+ insertion path in VO_2 cathode. Source: Reproduced with permission from Wang et al. [4]. © 2019/John Wiley & Sons. (c) H^+/Zn^{2+} co-insertion mechanism of VO_2 cathode. Source: Reproduced with permission from Zhu et al. [5]. © 2020/Elsevier. (d) H^+/Zn^{2+} co-insertion mechanism of δ-MnO_2 cathode. Source: Reproduced with permission from Wang et al. [6]. © 2019/American Chemical Society. (e) Schematic illustration of Zn/graphite battery with anion insertion. Source: Reproduced with permission from Zhang et al. [7]. © 2020/Elsevier. (f) Schematic illustration of anion adsorption and insertion mechanism in few-layered graphite. Source: Reproduced with permission from Guo et al. [8]. © 2020, John Wiley & Sons. (g) Energy storage mechanism of Zn/PANI battery during discharge and charge process. Source: Reproduced with permission from Wan et al. [9]. © 2018/John Wiley & Sons.

Besides Zn^{2+} incorporation, H^+ ions can also penetrate the interlayer spaces or tunnels within cathode materials. Wang et al. proposed that the energy storage process in Zn/VO_2 using $ZnSO_4$ as the electrolyte depends on the intercalation and movement of H^+ ions through these tunnels (Figure 7.2b) [4]. They highlighted the significance of the energy dynamics of H^+ incorporation and the solvation energy of anions in electrolytes, which strongly influence the competition between Zn^{2+} and H^+ incorporation. Similarly, instances of H^+ being inserted and extracted from MnO_2 during its cycling process have been observed, which is a deviation from previous findings. These variations can be attributed to the differences in the thermodynamics and kinetics of H^+ and Zn^{2+} insertion, depending on the cathode material's size and shape, as well as distinct electrolyte compositions. In this context, Huang and colleagues determined that the diffusion barrier for H^+ (1.21 eV) in hydrated VO_2 cathodes is marginally less than that for Zn^{2+} (1.41 eV) [5]. Therefore, the Zn^{2+}/H^+ co-insertion mechanism was observed in this system (Figure 7.2c). Results of the galvanostatic intermittent titration technique (GITT) demonstrated that the diffusion coefficients of ions at high voltages are higher than those at low voltages. The rapidly diffusing H^+ can potentially control the high-voltage region, while the slowly diffusing Zn^{2+} predominates in the low-voltage region. Similarly, Zhi and coworkers proposed a co-insertion mechanism of dual ions in δ-MnO_2, by Na^+ pre-insertion into the interlayer (Figure 7.2d) [6]. GITT results revealed distinct ion diffusion coefficients for the two voltage platforms, suggesting the presence of different types of ion insertion at varying voltage levels. Notably, there were no peaks attributable to $ZnMn_2O_4$ and MnOOH. This suggests that the co-insertion of Zn^{2+} ions and H^+ ions into the interlayer and cathode materials does not follow a chemical conversion process. In addition to layered MnO_2, the Zn^{2+}/H^+ co-insertion mechanism was also observed in a novel phase of $MnO_2H_{0.16}(H_2O)_{0.27}$ with 1×1 tunnels. In situ X-ray diffraction (XRD) results demonstrated that the characteristic peaks of $MnO_2H_{0.16}(H_2O)_{0.27}$ shifted toward lower angles during discharge, which was consistent with Zn^{2+}/H^+ co-insertion.

Certain viable host materials, such as partial organic compounds and graphite, can facilitate anion insertion, which can be operated at high voltages. Graphite can be intercalated by various anions, including $TFSI^-$, TFO^-, BF_4^-, $AlCl_4^-$, and PF_6^-, during the charging process due to its redox-amphoteric nature. Using a "water-in-salt" electrolyte, the stable voltage range of graphite can be extended to $0.5 \sim 2.5$ V, allowing anions ($TFSI^-$ and TFO^-) to be stably inserted into the graphite cathode. Simultaneously, Zn^{2+} can be reduced to Zn metal on the Zn anode (Figure 7.2e) [7]. In addition, it has also been identified that the adsorption of anions plays a role in enhancing capacity. During charging, $(ZnCl_4)_2^-$ moves to the surface of N-doped few-layered graphite and gets adsorbed within the pores at a low voltage. Subsequently, $(ZnCl_4)_2^-$ is intercalated into graphite's interlayer spaces, and Cl^- gets separated from $(ZnCl_4)_2^-$, proceeding to be intercalated into the cathode at an increased potential (Figure 7.2f) [8].

The anion insertion mechanism is also detected in polyaniline (PANI) cathodes. In contrast to graphite, anion insertion/extraction in a PANI cathode is concurrent with Zn^{2+} extraction/insertion during cycling [9]. The as-synthesized partially

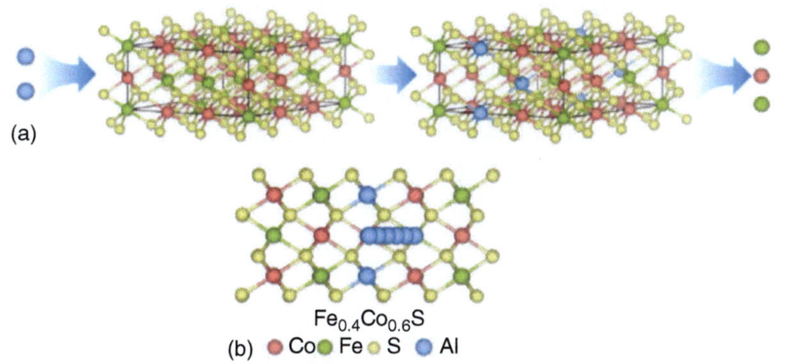

Figure 7.3 (a) Schematic of the conversion reaction between Al^{3+} and $Fe_{0.4}Co_{0.6}S$. (b) Schematic of the conversion reaction between Al^{3+} and $Fe_{0.4}Co_{0.6}S$. Source: Reproduced with permission from Zheng et al. [11]. © 2022/American Chemical Society.

oxidized PANI typically contains undoped (=N— and —NH—) and doped (=NH^+— and —NH^+—) nitrogen, and anions such as $TFSI^-$, Cl^-, and SO_4^{2-} are adsorbed near the —NH^+— and =NH^+— regions. Therefore, during the discharge process, anions near the doped nitrogen are removed from PANI, and —NH^+— is reduced to —NH— to facilitate the release of electrical energy (Figure 7.2g). Concurrently, =N— in the PANI chain is reduced to —N^-— and interacts with Zn^{2+}.

Meanwhile, rechargeable aqueous aluminum-ion batteries (AAIBs), using low-cost and inherently safe Al metal, are regarded as promising energy storage devices after LIBs. By vulcanizing Prussian blue analogues (PBAs), Zheng et al. realized the synergistic promotion of the Al^{3+} insertion/extraction process through the solid solution phase and cavity structure [11]. Figure 7.3 shows the schematic of the conversion reaction between Al^{3+} and $Fe_{0.4}Co_{0.6}S$ and the migration path of Al^{3+} in $Fe_{0.4}Co_{0.6}S$. The solid solution phase is conducive to the formation energy and diffusion kinetics of Al^{3+}, while the yolk–shell structure is conducive to the insertion/extraction space of Al^{3+}. These observations can provide new insights into enhancing the storage performance of aluminum in sulfide cathode materials.

Several studies have reported electrochemical insertion/extraction reactions of Mg ions in $FePO_4$ with a redox potential of approximately 2.4 V(versus Mg^{2+}/Mg) for $Mg_{0.5}FePO_4/FePO_4$, in which volume expansion is similar to that of the $LiFePO_4/FePO_4$ system [12]. However, there are no experimental reports on the phase change behavior of $FePO_4$ during Mg^{2+} insertion/extraction reactions. Watanabe et al. [12] studied the electrochemical insertion/extraction reactions of Mg^{2+} in $FePO_4$. Compared with Mg^{2+} insertion, Mg^{2+} extraction from $MgxFePO_4$ is more difficult. XRD results show that the insertion/extraction of Mg ions in $FePO_4$ proceeds not by a two-phase reaction but by a single-phase reaction.

7.2.2 Chemical Conversion Mechanism

In contrast to ion insertion, chemical conversion reactions produce new substances during the discharge/charge process. They primarily occur in the aqueous Zn/MnO_2

system. Based on the findings of Liu et al., the characteristic peaks of MnOOH and the distinct distribution of Zn and Mn elements after discharge suggest that the energy storage mechanism is a chemical conversion reaction between α-MnO_2 and H^+. Flake-like Zn distribution can be attributable to the $ZnSO_4[Zn(OH)_2]_3{\cdot}xH_2O$ phase, which results from the reaction between OH^-, $ZnSO_4$, and H_2O. These observations suggest the following energy storage mechanism:

Cathode:

$$H_2O \leftrightarrow H^+ + OH^- \tag{7.1}$$

$$MnO_2 + H^+ + e^- \leftrightarrow MnOOH \tag{7.2}$$

$$\frac{1}{2}Zn^{2+} + OH^- + \frac{1}{6}ZnSO_4 + \frac{x}{6}H_2O \leftrightarrow \frac{1}{6}ZnSO_4[Zn(OH)_2]_3 + xH_2O \tag{7.3}$$

In some instances, cathode materials undergo a co-insertion process of Zn^{2+} and H^+, followed by a chemical conversion reaction between host materials and the inserted ions. Studies have documented that the discharge process of α-MnO_2@CFP involves two distinct stages, as revealed by discharge GITT profiles (Figure 7.4a,b) [13]. As shown in the figure, the total overvoltage in region I is approximately 10 times lower than that in region II, yet the equilibrium voltage in region I is comparable to that in region II. Moreover, the sequential appearance of MnOOH and $ZnMn_2O_4$ in ex situ XRD strongly indicates that MnO_2 undergoes H^+ insertion followed by Zn^{2+} insertion during the discharge stage, which is in line with GITT results (the ion diffusion rate of Zn^{2+} is much slower than that of H^+ due to its larger ion size) (Figure 7.4c). A comparable co-insertion and conversion mechanism was also observed in layered δ-MnO_2. According to Liu et al., a combined charge storage mechanism is present in δ-MnO_2, involving Zn^{2+} insertion into the interlayer at a high voltage, followed by H^+ insertion and a chemical conversion reaction between H^+ and MnO_2 (Figure 7.4d) [14].

Recently, Chao et al. designed an electrolytic Zn-Mn system that follows the mechanism of dissolution/deposition of MnO_2/Mn^{2+} (Figure 7.4e) [15]. In the initial charge process, the MnO_2 active material was in-situ-synthesized via the following reaction mechanism:

$$Mn^{2+} + 2H_2O \rightarrow MnO_2 + 4H^+ + 2e^- \tag{7.4}$$

The discharge process followed a new three-step reaction mechanism:

$$MnO_2 + 4H^+ + 2e^- \rightarrow Mn^{2+} + 2H_2O \ \ (D1:\ 1.7 - 2.0\,V) \tag{7.5}$$

$$MnO_2 + H^+ + e^- \rightarrow MnOOH \ \ (D2:\ 1.4 - 1.7\,V) \tag{7.6}$$

$$MnO_2 + 0.5\,Zn^{2+} + e^- \rightarrow Zn_{0.5}\,MnO_2 \ \ (D3:\ 0.8 - 1.4\,V) \tag{7.7}$$

Previous studies have shown that cations within the host crystals always contribute to the alteration of the valence state in chemical transformation reactions. However, in some V-based compounds, anionic redox reactions have been documented recently. For instance, the N element in VN_xO_y can facilitate a pseudocapacitive surface redox reaction through anionic redox between N^{3-} and N^{2-}. Furthermore, Niu et al. suggested an aqueous Zn/$VOPO_4$ system with a high

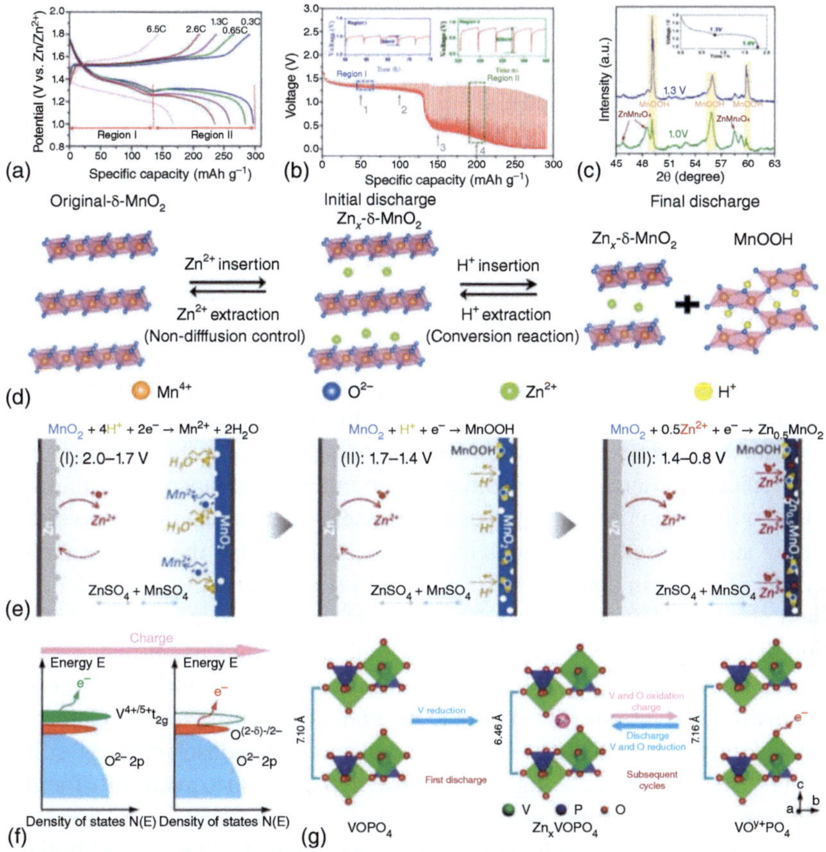

Figure 7.4 (a) Charge and discharge curves at different rates of α-MnO_2. (b) GITT profiles of the Zn/MnO_2@CFP battery. (c) Ex situ XRD of MnO_2@CFP at depth of discharge at 1.3 and 1.0 V. Source: Reproduced with permission Sun et al. [13]. © 2017/American Chemical Society. (d) Schematic of joint charge storage of Zn^{2+} insertion and chemical conversion. Source: Reproduced with permission from Jin et al. [14]. © 2019/John Wiley & Sons. (e) Schematic of galvanostatic discharge, including D1, D2, and D3 process. Source: Reproduced with permission Chao et al. [15]. © 2019/John Wiley & Sons. (f) DOS results of $VOPO_4$ during the charge process. (g) Schematic of oxygen redox reaction mechanism of Zn/$VOPO_4$ battery. Source: Reproduced with permission from Wan et al. [16]. © 2019/John Wiley & Sons.

average voltage that demonstrated a highly reversible anionic redox upon cycling [16]. Analysis of density of states (DOS) indicated that upon complete oxidation of vanadium with an increase in voltage, an electron gets transferred from oxygen, leading to the formation of a localized electron hole (Figure 7.4f). This implies that the oxygen redox reaction can occur at an increased output voltage. In line with these findings, electrochemical testing showed that $VOPO_4$ can enable a reversible oxygen redox reaction between O^{2-} and O^{1-} across the voltage spectrum of 1.7–2.1 V. Similarly, the cationic redox transition involving V^{3+} and V^{2+} can take place within the voltage range of 1.7 to 0.8 V (Figure 7.4g).

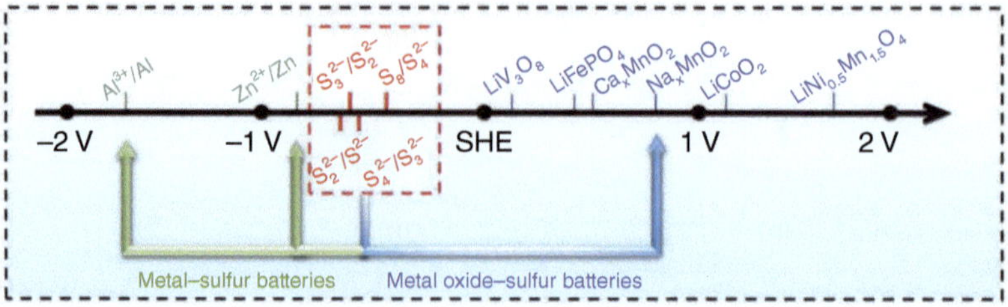

Figure 7.5 Development of aqueous Zn-S batteries of the redox potential of some redox couples in sulfur-based aqueous batteries. Source: Reproduced with permission from Hu et al. [17]. © 2020/Royal Society of Chemistry.

Sulfur, abundant and environmentally benign, is well suited for cutting-edge energy storage applications. Remarkably, it can achieve a substantial specific capacity of 1672 mAh g^{-1} via a two-electron redox process, significantly surpassing the capacities typical of intercalation electrodes. The rapid mass transport attributable to the high ionic conductivity of aqueous electrolytes enhances the rapid conversion dynamics of sulfur-based compounds in AMIBs. In particular, the transformation mechanisms of sulfur in aqueous conditions span solid–liquid (S-L), liquid–liquid (L-L), and solid–solid (S-S) redox processes, which proceed more quickly than solid–liquid–solid (S-L-S) redox reactions observed in nonaqueous settings.

Figure 7.5 shows the redox potential of sulfur, which is around −0.5 V versus Standard hydrogen electrode (SHE). Due to its moderate redox potential, elemental sulfur can adaptively coordinate with these metal anodes to construct metal–sulfur batteries [17]. Al metal has a high volumetric capacity (8040 mAh cm^{-3}) and a relatively low redox potential and therefore has been selected as an anode to design aqueous Al-S batteries with sulfur electrodes. Stuart Licht et al. [18] initially designed an aqueous Al-S battery containing a sulfur cathode (or polysulfide catholyte) and a Al metal anode. This Al-S battery showed a S-L reaction redox, as depicted by Eqs. (7.9) and (7.10):

$$\mathrm{Al + 3S + 3KOH + 3H_2O \rightarrow 2Al(OH)_3 + 3KHS} \tag{7.8}$$

$$\mathrm{2Al + S_4^{2-} + 2OH^- + 4H_2O \rightarrow 2Al(OH)_3 + 4HS^-} \tag{7.9}$$

The transformation reaction of halogen elements (chlorine, iodine, and bromine) has garnered significant attention due to their vast availability. Among them, the iodine-centered reaction, which is nontoxic and cost-effective and possesses rapid reaction kinetics, is considered a promising energy storage solution; therefore, research on iodine-based water system batteries has been increasing, comprising flow batteries and static batteries. Over the past few decades, iodine/polyiodides have been coupled with numerous counter-electrodes (such as Zn, Al, and Fe) to prepare AMIBs [19].

These batteries typically follow a reaction route from I^- to I_2, which delivers a specific capacity of 211 mAh g^{-1} (based on the mass of iodine). Recently, the I_2/I^+ redox reaction has been activated by introducing halogen ions (Cl^-) or molecular solvents to form interhalogens with I^+. The $I^-/I_2/I^+$ reaction is related to a four-electron transfer, which has a higher theoretical capacity (422 mAh g^{-1}) than the two-electron transfer reaction (211 mAh g^{-1}).

7.3 Highly Stable and Energetic Cathodes

7.3.1 Mn-Based Materials

Due to the advantages of manganese-based compounds such as abundant availability, environmental friendliness, and rich redox reactions, these compounds have emerged as promising options for cathode materials in batteries that use either monovalent or multivalent ions. In particular, MnO_2, with its diverse crystal forms and microstructures, presents a significant theoretical specific capacity (308 mAh g^{-1}, based on the single-electron redox reaction between Mn^{4+} and Mn^{3+}) and a moderate discharge potential. Octahedral MnO_6 units that share edges form ribbons of limited width that are linked together, creating tunnel structures. The dimensions of these tunnels depend on the number of MnO_6 octahedra (Figure 7.6a) [20]. The following types of structures are found: tunnel type: β-MnO_2 grows along the *c*-axis to form [1 × 1] tunnels, α-MnO_2 manifests a large [2 × 2]

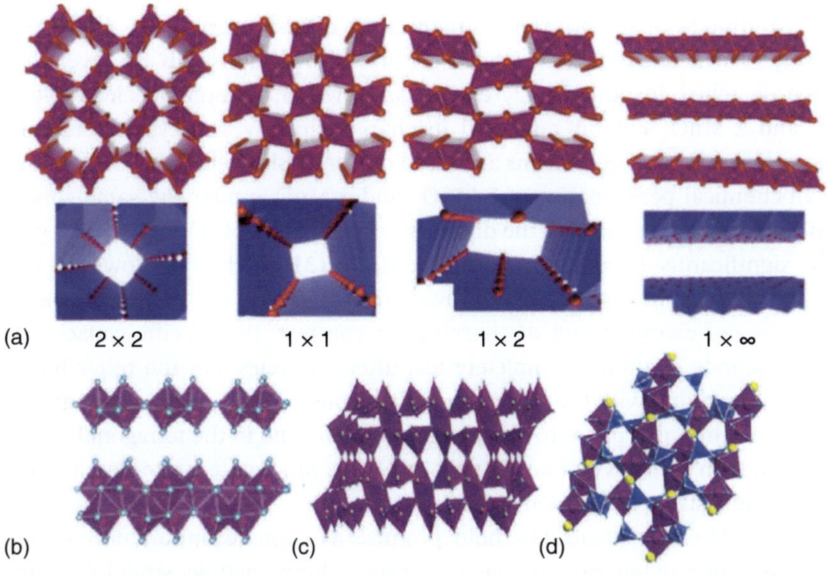

Figure 7.6 (a) Schematic diagram of various crystal structures of α-MnO_2, β-MnO_2, γ-MnO_2, and δ-MnO_2, respectively. Source: Reproduced with permission from Liu et al. [20]. © 2018/John Wiley & Sons. (b–d) Schematic diagram of three typical crystal structures of layered $Mg_xV_2O_5$. Source: Reproduced with permission from Ming et al. [21]. © 2018/American Chemical Society.; tunneled VO_2. Source: Reproduced with permission from Ding et al. [3]. © 2018/John Wiley & Sons.; and NASICON structured $Na_3V_2(PO_4)_3$. Source: Reproduced with permission from Li et al. [22]. © 2016/Elsevier.

tunnel structure, γ-MnO_2 features [1×1] and [1×2] hybrid tunnels along the *b*-axis, and todorokite MnO_2 has larger [3×3] tunnels with a similar structure to α-MnO_2; layered type: δ-MnO_2 presents large interlayer spacing (~7 Å); and spinel type: λ-MnO_2 or $ZnMn_2O_4$. These distinct crystal structures result in varying electrochemical properties, even with dissimilar reaction mechanisms [23]. Among them, α-MnO_2, δ-MnO_2, and λ-MnO_2 are the most studied conversion cathodes in aqueous Mg-ion batteries (AMIBs). α-MnO_2 has a 2×2 tunneling structure that provides a large ion transport channel (≈0.5 nm), much larger than the size of Mg^{2+} (0.076 nm), which facilitates the diffusion of magnesium ions. Besides α-MnO_2, layered MnO_2 consisting of stacked MnO_6 octahedra and structured water shows great potential for application in AMIBs. Its large 2D diffusion path with an interlayer distance of 7 Å facilitates Mg^{2+} diffusion. Although a Mg-ion battery with α-MnO_2 as the positive electrode has a high initial discharge capacity, its specific capacity decays extremely fast after the cycle starts. This is because Mg^{2+} ions do not enter the interior of the α-MnO_2 lattice during the discharge process, but the conversion of crystalline α-MnO_2 to amorphous Mg_xMnO_2 occurs, which eventually leads to the irreversible collapse of the cathode tunneling structure. Subsequently, Miyayama et al. improved the electrical conductivity of α-MnO_2 and achieved high initial capacity by compounding α-MnO_2 with acetylene black. The introduced K^+ stabilized the α-MnO_2 tunneling structure and improved the cyclic stability of the material. The layered δ-MnO_2 and spinel-type λ-MnO_2 can provide two-dimensional and three-dimensional diffusion channels for Mg^{2+}, respectively. When aqueous electrolytes are used, Mg^{2+} ions can be reversibly embedded/detached in both MnO_2 cathodes. However, in organic electrolytes, δ-MnO_2 and λ-MnO_2 exhibit restricted discharge capacity and cycling stability because their discharge mechanisms are based on conversion reactions. Comparing the electrochemical performance of δ-MnO_2 and α-MnO_2 under the same conditions, Miyayama et al. found that the discharge capacity of δ-MnO_2 is 109 mAh g^{-1}, which is significantly lower than that of α-MnO_2 (310 mAh g^{-1}). However, the cycling performance of δ-MnO_2 is improved, retaining 50% of its initial capacity after 25 cycles of charging and discharging. In contrast, the specific capacity of a α-MnO_2 cathode is almost completely lost after 20 cycles. On the other hand, Fenter et al. found that with the insertion of Mg^{2+} ions, λ-MnO_2 undergoes a phase transition from the cubic phase to the tetragonal phase, and in the tetragonal-phase λ-MnO_2, some Mn^{4+} enters octahedral sites and some Mn^{3+} enters tetrahedral sites, resulting in the restricted diffusion of Mn^{2+}.

In addition, Mn-based materials hold promise as cathode options for AZIBs, exhibiting a range of energy storage processes. These include straightforward Zn^{2+} insertion; simultaneous insertion of Zn^{2+} and H^+, followed by mixed insertion/conversion actions; and ultimately, mechanisms of dissolution and deposition. Variations in these mechanisms are influenced by several factors, such as electrolyte composition and additives, along with the cathode material's crystalline structure and fabrication method. To enhance the ion transport and electrical characteristics across different forms of MnO_2, techniques such as construction of nanostructures, creation of oxygen vacancies, doping with metal ions, and application of conductive carbon are utilized, either individually or in synergy. However, due to the commonality of MnO_2, the disproportionation reaction of Mn^{3+} can lead to severe capacity fading during the cycling.

Besides MnO_2, other manganese-based oxide materials are also considered promising candidates for AZIB cathodes. Some researchers proposed that the use of other manganese-based oxides than MnO_2 may provide a deeper understanding of the zinc-ion storage mechanism, such as Mn_2O_3 and Mn_3O_4. However, the lower valence of Mn in these materials leads to lower theoretical capacity. In addition, pure spinel $ZnMn_2O_4$ is considered inactive for Zn^{2+} energy storage due to its limited 3D tunnels. In order to address the intrinsic drawbacks of Mn-based materials, a wide range of modification strategies such as electrolyte additives, ion defects, metal ion pre-intercalation, and conductive composites have been developed in recent years, which will be discussed below.

7.3.2 V-Based Materials

Since the discovery of the reversible multivalent cation intercalation reaction, numerous vanadium-based cathode materials with intercalation capabilities have been developed, endowing them with significant discharge capacity along with high coulombic efficiency (CE). The crystal structure of V-based oxides is primarily classified into the following three types: layer-type, tunnel-type, and NASICON-type structures, which determines the number of active sites for electrochemical Zn^{2+} insertion (Figure 7.6b–d) [3, 21, 22]. Layer-type and tunnel-type polymorphs offer extensive spatial dimensions with rapid intercalation kinetics. However, constant insertion and extraction of multivalent cations, accompanied by numerous electron transfers, lead to significant structural modifications and lattice strains, which are detrimental to the host cathode. To ensure the structural stability of vanadium oxides during cycling, a limiting cutoff voltage is often employed to prevent deep charge and discharge, at the cost of reversible capacity. Recently, some emerging cathode materials for multivalent ion migration, such as pillared layered XV_2O_5 (X = Zn, Mg, Ca, Na, etc.), have been developed, especially in aqueous Zn-ion batteries. $Zn_{0.25}V_2O_5{\cdot}nH_2O$, with interlayer Zn^{2+} and water molecules as pillars, shows more than one Zn^{2+} insertion/extraction per unit (a capacity up to 300 mAh g^{-1}) and a capacity retention of over 80% after 1000 cycles at 8 C. The pre-inserted Zn^{2+} stabilizes the initial structure, while the nanobelt reduces the ion diffusion path and provides numerous reaction active sites. Furthermore, upon immersion of $Zn_{0.25}V_2O_5{\cdot}nH_2O$ in the electrolyte, water molecules can be inserted into the structure to form an additional monolayer between the layers, functioning like a "lubricant."

This effect, observed in both Mg-ion and Al-ion batteries, is recognized as the charge shielding effect. It describes the dampening of interactions by offsetting the charge of multivalent ions. Thanks to the cooperative action of Mg^{2+} ions and structural water, bilayer $Mg_{0.3}V_2O_5{\cdot}1.1H_2O$ shows excellent rate capability and unmatched cycle life, maintaining 80.0% of its capacity after 10,000 cycles.

Similarly, the presence of pre-inserted Mg ions enhances the electronic conductivity and serves as "pillars" to keep the layered structure intact, whereas structural water helps increase the lattice distance and prevent Mg ions from directly interacting with the oxygen atoms in the lattice. However, in nonaqueous electrolytes, structural water might induce adverse reactions with the Mg anode; thus, a more harmonized electrode–electrolyte framework is needed to advance the use of $xV_2O_5{\cdot}nH_2O$ in Mg-ion batteries. The intrinsic challenges of V_2O_5 cathodes, such as poor electronic conductivity and sluggish ion diffusion, have driven researchers to refine vanadium oxides through ion doping, shape control, or surface treatments. For example, integration of conductive materials (like graphene, amorphous carbon, and conductive polymers) to create a biphasic mixture helps alleviate stress from continuous ion insertion and provides numerous active sites. Despite the strong charge density and polarity of Zn ions, layered vanadium oxides and their variants have, to date, shown exceptional results in aqueous Zn-ion batteries. On the contrary, the performance of vanadium oxides in other multivalent metal-ion batteries, such as those using magnesium, aluminum, and calcium ions, remains largely unsatisfactory, possibly due to the lack of a suitable electrolyte that facilitates reversible metal deposition without corrosion or formation of a passive layer on the metal anode. It is critically important to acknowledge the toxicity and harmful effects of vanadium on human health, highlighting the necessity for V recycling.

VS_2 is a layered transition metal sulfide that shows metallic conductivity, in which layers are connected by relatively weak van der Waals forces. This unique layer structure is conducive to the diffusion of ions between the layers, and its layered structure properties are also suitable for the embedding of magnesium ions. An organic sulfide is obviously an n-type organic material because of the doping of sulfur element, with the chemical formula R–S–S–R (R = benzene ring and five-membered ring, etc.), and its magnesium storage mechanism is based on reversible breakage and generation of disulfide bonds (S–S).

7.3.3 Organic Compounds

Compared with traditional metal oxide materials, using organic compounds as electrodes has numerous advantages, such as molecular-level controllability and structural diversity. Among them, quinone electrodes have recently attracted the attention of researchers as they are stable in aqueous electrolytes. Their soft molecular structure, which is held together by weak van der Waals force, can adapt to the minimum volume change through facile molecular reorientation. However, the dissolution of active materials in bare quinone electrodes is detrimental to the lifespan of the battery configuration. Using a Nafion separator (fluorine-containing membrane), calix[4]quinone cathode can achieve an impressive capacity of 335 $mAh\,g^{-1}$ and an enhanced cell lifespan of 87% retention after 1000 cycles at 500 $mA\,g^{-1}$. Unfortunately, the Nafion separator greatly increases the battery cost, thus with less flexibility in practical applications.

It is well accepted that conducting polymers possess high conductivity due to long π-electron-conjugated systems. However, the PANI cathode needs a high proton concentration to accelerate the redox process in an aqueous electrolyte. Deprotonation of PANI during the cycle can lead to rapid capacity decay of the cell according to previous research. For instance, studies showed that the lifespan of a pure PANI electrode supported by Ni foam for AZIBs is unsatisfactory (88% capacity retention only after 50 cycles). Therefore, modification strategies need to be developed to maintain the electrochemical activity of the PANI cathode in mild aqueous systems (e.g. doping strategy).

7.3.4 Prussian Blue Analogues

Recently, PBAs have attracted increasing attention in various energy storage devices that use monovalent and multivalent metal ion batteries due to their open framework structures, excellent structural stability, and numerous active sites. PBAs, which can be expressed using the general formula $A_xM1[M2(CN)_6]_y{\cdot}nH_2O$ (A represents alkaline metal ions M1 and M2 represents transition metal ions), have a typical face-centered cubic structure with a Fm3m space group, and two metal ions (M1 and M2) are linked together by CN ligands. Recently, some researchers proposed AZIBs based on zinc hexacyanoferrate and copper hexacyanoferrate cathodes and reported high output voltage (more than 1.6 V in general). Nevertheless, the cycling capability and zinc retention capacity of these PBAs often fall short of expectations. This deficiency can be attributable to the chemical stability of PBAs, which is highly susceptible to the ions being intercalated. More importantly, the capacity of PBA cathodes is essentially constrained by the amount of Fe^{2+} present within PBA crystals, as this dictates the number of sites available for reactions. While the development of composites and modifications to the surface chemistry may improve the lifespan of the cell, the limited capacity remains a significant hurdle for meeting the demands of real-world applications. Thus, the pursuit of new PBA cathodes with a higher specific capacity is indeed warranted.

7.3.5 Other Cathodes

Apart from the above four types of cathode materials, the use of some other materials such as Mo-based materials and Co-based materials has been reported in aqueous batteries. Although MoO_3 nanomaterials are considered versatile electrodes for energy storage systems due to the +6 valence state of Mo, the dissolution of Mo during repeated Zn^{2+} insertion and extraction has a great impact on non-negligible capacity fading. Besides, pristine MoS_2 has a slow Zn^{2+} diffusion process and leads to an undesirable specific capacity. From an economic point of view, although Co-based materials deliver an acceptable electrochemical performance, the high cost of using Co in the cathode and electrolyte additives makes the application of Co-based materials unfeasible compared with Mn-based and V-based materials.

Up to now, various materials have been explored for aqueous cathodes, but some challenges to their applications have been reported, such as dissolution of electrode materials, sluggish ion migration, inadequate active sites, and unexpected byproducts, which require further investigation for industrial applications. Recently, there have been numerous reports using different design strategies, including doping, ion vacancy, and surface coating to optimize the electrochemical performance of cathodes. Some studies have used a combination of design strategies to achieve high-electrochemical-performance cathode materials.

7.4 Strategies for Dendrite-Free Metal Anodes

7.4.1 Optimization Strategy of Zn Anodes

Zinc is a highly desirable material for use in batteries due to its high capacity and standard reduction potential. It has a hydrogen evolution potential of around 1.2 V, which minimizes the generation of hydrogen gas. However, the practical use of zinc-ion batteries faces some challenges. One major issue with Zn is the uneven deposition and dissolution of zinc ions. This can lead to the formation of dendrites, which are needle-like structures that can penetrate the separator in the battery. These dendrites have the potential to cause short circuits, compromising the safety and performance of the battery. Another issue associated with Zn is the variable diffusion of Zn^{2+} ions. This uneven diffusion results in variable electrostatic interactions within the battery, leading to anode corrosion and hydrogen evolution. The corrosion of the anode can further contribute to dendrite formation and the deterioration of the battery's cycling stability. Furthermore, Zn^{2+} ions have a tendency to react with excess OH^- ions in the battery, resulting in the formation of by-products. These by-products hinder the reuse of Zn, limiting its potential for recycling and sustainability. Furthermore, the formation of insoluble passivation layers, such as ZnO or $Zn(OH)_2$, on the electrode surface is another challenge in using Zn in batteries. These passivation layers reduce the reaction area, leading to passivation. They further reduce the specific surface area of the electrode. After long-term cycling, the density of the electrode will increase, resulting in the polarization of the battery system and the generation of zinc dendrites, ultimately reducing the cycling stability of the Zn anode.

When evaluating the stability of a zinc anode in batteries, it is crucial to consider key parameters such as areal capacity, current density, state of charge, zinc utilization per cycle (depth of discharge, DOD), and CE. Optimizing these parameters and exploring thinner Zn anodes can improve energy density and cycling performance. It has been reported that optimal battery cycle life and safety can be achieved with a state of charge of 20% and DOD of 80%. However, operating at high DOD increases the risk of side reactions. In the case of AZIBs, the cathode typically has a mass loading of approximately 2 mg cm^{-2}, while the thickness of the zinc foil used as the anode is 80 μm. It is important to avoid excess anode material as it can reduce energy density and affect CE. It is worth mentioning that the actual DOD per cycle is often lower than 1%, which can give a misleading impression of a 100% CE. To improve energy density, research suggests exploring the use of thinner Zn anodes, with a thickness of ≤30 μm, to achieve higher DOD. This approach is comparable with traditional lithium metal anodes and commercial nickel–zinc anodes. Furthermore, AZIBs should meet the goal of operating for at least 2000 cycles with a cumulative capacity of 10 Ah cm^{-2}, assuming a 100% CE, and maintaining at least 80% Zn^{2+} transfer per cycle.

As mentioned above, the distribution of Zn^{2+} ions on planar or two-dimensional Zn foil is nonuniform, leading to the formation of Zn dendrites and self-corrosion. Nonetheless, employing a porous 3D substrate material can prevent the localized generation of electric current and facilitate the uniform deposition of Zn^{2+} ions on the anode. Notably, Archer et al. outlined an epitaxial mechanism that determines the nucleation and reversibility of Zn anodes. The graphene material, with an atomic arrangement similar to that of the Zn plane, is used as the substrate for Zn epitaxial electrodeposition. A uniform metal coating grows parallel to the metal anode in the charging process, and when the metal coating is removed, the substrate material remains undamaged in the discharging process. Although this mechanism is debatable, this type of epitaxial Zn anode material shows continuous reactivity and superior reversible stability compared with conventional Zn anode materials [24]. Epitaxial Zn anode materials, such as graphene, carbon cloth, carbon nanotube paper, and yarn, have shown better reversible stability than general Zn anode materials. Another promising approach to modifying Zn anode interfaces is using 3D metal-organic frameworks (MOFs), particularly ZIF-8 organic frameworks. MOFs have shown progress in modifying the Zn anode interface. By subjecting MOFs to heat treatment, an N-doped porous carbon layer can be formed, which enables the uniform distribution of Zn^{2+} ions on the anode surface. Furthermore, the coating of MOFs also creates a desolvation effect, which leads to the formation of a saturated electrolyte near the innermost layer of the Zn foil. This promotes the desolvation process and reduces the accumulation of by-products (Figure 7.7a–c) [25].

To promote the uniform deposition of Zn^{2+} ions in the reaction area of Zn anodes, a functional interfacial buffer layer can be created. One approach is to use a polyamide (PA) coating inspired by brighteners. This coating provides hydrogen bond (HB) groups and coordination sites, which facilitate controlled Zn^{2+} deposition. The presence of the PA layer enhances the nucleation overpotential and leads to the formation of compact, small-sized nuclei. In addition to promoting uniform Zn^{2+} deposition, the PA layer reduces the permeability of O_2 and H_2O. By decreasing the likelihood of side reactions in the battery system, the PA coating helps improve the overall performance and stability of the Zn anode. Similarly, other materials such as Ag, TiO_2, nanoporous $CaCO_3$, ZrO_2, and Al_2O_3 have been proven to serve similar functions. These materials form a passivation layer on the surface of the Zn anode, preventing charge exchange with the electrolyte and promoting uniform Zn^{2+} deposition. One promising material for Zn anode modification is kaolin, a type of clay. Kaolin possesses a sieve-like functionality, offering numerous adsorption sites such as O—H and Si—O bonds. These sites guide the even deposition of Zn^{2+} ions and significantly reduce zinc corrosion and hydrogen evolution. The use of high-functional clay layered structures like kaolin in Zn anode modification is considered a promising direction for enhancing the performance of Zn anodes. By

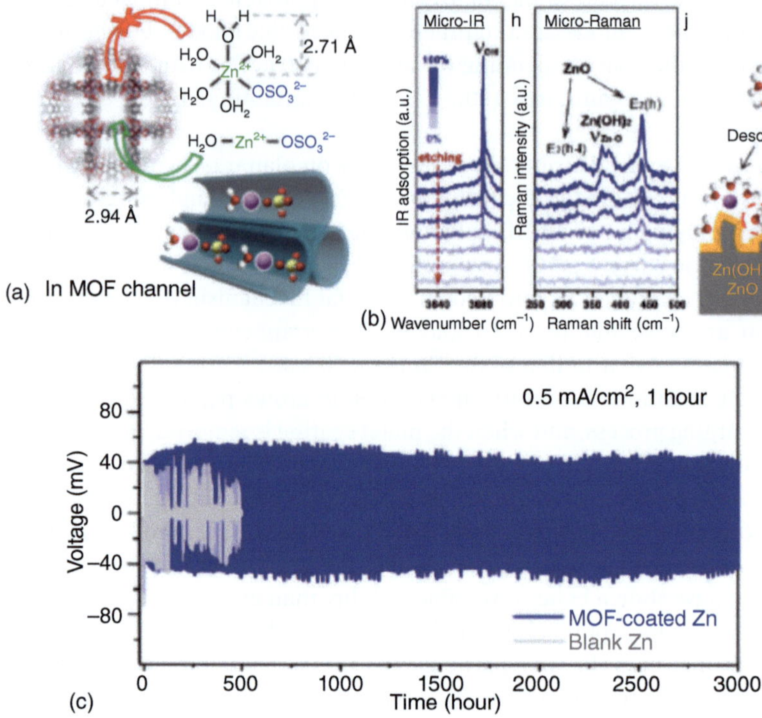

Figure 7.7 (a) Schematic illustration of highly coordinated ion complexes of H_2O-Zn^{2+} OSO_3^{2-} migrating through MOF channels. (b) Micro-IR and Raman spectra of bare Zn anodes and MOF-coated Zn anodes. (c) Long-term cyclic stability of symmetric cells in 2 M $ZnSO_4$ electrolyte at 0.5 mA cm^{-2} for one hour. Source: Reproduced with permission from Yang et al.[25]. © 2020/John Wiley & Sons.

providing a stable and uniform deposition surface for Zn^{2+} ions, these materials help improve the efficiency and durability of Zn-based batteries.

7.4.2 Anode Materials of Mg-Ion Batteries

Unlike Li and Zn metal anodes, Mg^{2+} deposits in hemispherical or cluster forms on Mg anodes, which is advantageous as this avoids magnesium dendrite formation. Nevertheless, the corrosion and passivation also reduce the CE, leading to a reduced energy density and specific capacity. At low current densities, the separation of undissolved metal parts can lead to reduced efficiency of Mg anodes. To improve Mg anode utilization, a porous $FeVO_4$ structure with a large specific surface area has been proposed, enhancing electrolyte permeability and electrochemical performance, especially when coated with carbon. This kind of anode material exhibits a large specific surface area that improves electrolyte permeability into electrode material. Polyimide, a promising organic anode material for aqueous Mg-ion batteries, facilitates rapid Mg^{2+} migration and maintains potential difference. Furthermore, an aqueous Mg-ion battery using a polypyromellitic dianhydride anode and $Mg_xLiV_2(PO_4)_3$ cathode can exhibit excellent cycle stability. Nonetheless, when it comes to battery capacity, inorganic materials generally exhibit superior capacity and a wider voltage window than polyimide. To fully exploit battery performance, the VO_2 nanowire anode material synthesized using the solvothermal method can be used to develop a more robust structure during the reaction process, effectively preventing the formation of MgO by-products. This material can facilitate a highly reversible insertion and extraction process of Mg^{2+} in $MgSO_4$, $Mg(NO_3)_2$, and $MgCl_2$ electrolyte.

7.5 Strategies for Designing Electrolytes

Compared with cathodes and anodes, electrolytes receive less attention while discussing the practical applications of ARMMBs. However, electrolyte plays an important role in achieving high-performance ARMMBs. Therefore, it is necessary to have an in-depth understanding of the electrolytes used in ARMMBs, including the composition and structure of the electrolyte, in situ state of the electrolyte, relationship between the electrolyte and electrochemical properties, and practical considerations of the electrolyte. In this context, electrolytes can be categorized into aqueous and nonaqueous types based on the solution environment. Furthermore, electrolytes are morphologically classified into liquid and solid electrolytes. A S-L mixed electrolyte, which is a combination of the liquid and solid electrolyte, is generally classified as a semisolid battery, a quasi-solid battery (gel electrolyte), or a solid battery according to its liquid content.

7.5.1 Liquid Electrolytes

7.5.1.1 Aqueous Liquid Electrolytes

Aqueous batteries primarily use water as an ideal solvent for solutions, which addresses safety concerns in the manufacturing and production process while also curtailing resource consumption and production expenses. Aqueous secondary batteries have recently garnered significant attention due to these characteristics. In comparison with nonaqueous electrolyte solutions, aqueous electrolytes possess the following advantages: (i) cost-effective and accessible: the price of solvent water is markedly lower than that of nonaqueous electrolytes; (ii) typically showing ionic conductivity that is 1–2 orders of magnitude greater than that of organic liquid electrolytes; (iii) free from the safety risks associated with the flammability of the electrolyte; (iv) resistant to overcharging, with oxidation–reduction products being hydrogen and oxygen, which do not contaminate the electrolyte solution; (v) requiring minimal operating environment specifications, thereby reducing production and technical costs; (vi) presenting with significantly reduced demands for the diaphragm; and (vii) being environmentally friendly [26]. Despite the increase in research on aqueous electrolytes due to these advantages, some drawbacks continue to impede their development. These primarily include: (i) they have a relatively narrow electrochemical stability window (1.23 V), which is primarily restricted by

H_2O electrolysis. Consequently, the selection of electrode materials and operating voltage play a vital role in the further advancement of aqueous secondary batteries in high-energy-density energy storage; (ii) there is a possibility of proton activity in the electrolyte prior to the intercalation and delamination reaction of metal ions, leading to side reactions; (iii) the electrode material's surface cannot form an effective passivation film in the aqueous solution, nor can the electrode structure be adequately protected, which curtails the battery's cycle life.

Aqueous zinc battery electrolytes chiefly include aqueous zinc salt solutions (such as zinc sulfate, zinc nitrate, and zinc acetate). However, aqueous electrolytes contain highly reactive water that can cause unnecessary electrochemical side reactions and uncontrollable S-L interface reactions, which will affect battery energy storage. In addition, intense interactions between water molecules and anions in concentrated electrolytes result in an electrolyte acquiring the antifreezing property. The red shift of the B—F bond with an increased $Zn(BF_4)^2$ concentration confirms the intense interaction between water molecules and BF_4^-, as shown in Figure 7.8a. A previous study showed that 4 M $Zn(BF_4)_2$ electrolyte with a freezing point of −122 °C and an ionic conductivity of 1.47 mS cm^{-1} at −70 °C enabled a Zn//tetrachlorobenzoquinone battery to operate at 95 °C [27]. A supersoluble $ZnCl_2/ZnBr_2/Zn(OAc)_2$ aqueous electrolyte with a record concentration of 75 M overcomes limited physical solubility. Nonpolarized protons in acetate anions prevent the overgrowth and precipitation of ionic oligomers, to result in an antifreezing

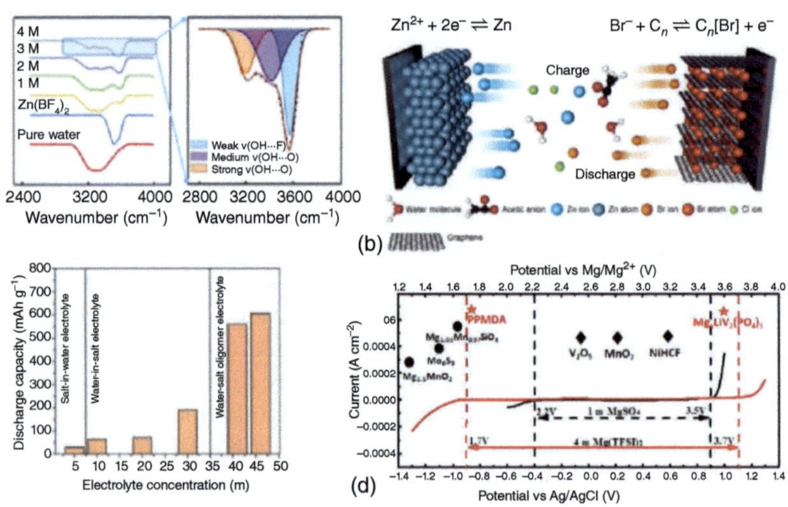

Figure 7.8 (a) FTIR spectra for O—H bond in $Zn(BF_4)_2$-based electrolyte. Source: Reproduced with permission from Sun et al. [27]. © 2021/The Royal Society of Chemistry. (b) Schematic of Zn/GFF|| $WSOE_{45}$-1||PGA battery. (c) Discharge capacity for dual-ion battery with 5–46 M electrolyte. Source: Reproduced with permission from Cai et al. [28]. © 2021/John Wiley & Sons. (d) Electrochemical stability window of 4 M $Mg(TFSI)_2$ aqueous electrolytes. Source: Reproduced with permission from Wang et al. [29]. © 2017/American Chemical Society.

45 m $ZnBr_{0.5}Cl_{1.5}$ + 1 m $Zn(OAc)_2$ water–salt oligomer electrolyte (WSOE45-1) of Tg~−70 to −60 °C. The high discharge capacity and CE of Zn/graphene fiber fabric (GFF)||$WSOE_{45}$-1|| graphene cathode (PGA) battery is attributable to a suppressed "shuttle effect" and improved intercalation of Br in the highly concentrated electrolyte (Figure 7.8b,c) [28].

A specific quantity of inorganic magnesium salts is used as electrolytes and dissolved into aqueous solvents to prepare an electrolyte that can have widespread use in AMIBs. Unique inorganic magnesium salts also possess varying attributes regarding the performance of aqueous ion batteries. Zhang et al. compared three diverse electrolytes, $MgSO_4$, $MgCl_2$, and $Mg(NO_3)_2$, and determined that $Mg(NO_3)_2$ electrolytes showed a higher Mg^{2+} diffusion coefficient and a lower charge transfer resistance [30]. Nonetheless, the diminished electrochemical window of inorganic salts curtails the functionality of batteries. The voltage range can be expanded by increasing the concentration of the electrolyte to suppress the activity of water. Wang et al. developed an electrolytic liquid system with a high concentration of 4 M $Mg(TFSI)_2$ and an electrochemical window of 2.0 V, which promptly facilitated the reversible deposition of Mg^{2+} (Figure 7.8d) [29]. Furthermore, given that the pH level of the measured electrolyte is nearly 7, this type of electrolyte has unparalleled advantages in terms of non-corrosiveness, nonvolatility, and strong conductivity, making it highly suitable for large-scale applications in AMIBs.

7.5.1.2 Ionic Liquid Electrolytes

Ionic liquids are known as "green solvents" due to their unique properties. They are also referred to as designer solvents since they can be selectively combined with specific cations and anions to synthesize ionic liquids tailored to specific requirements. It is presumed that their nonflammable nature addresses the safety concern of flammable electrolytes in organic systems. However, their properties, such as viscosity, hinder the diffusion behavior of metal ions in the electrolyte, leading to suboptimal compatibility with interfaces. To overcome these challenges, numerous researchers have integrated ionic liquids with other electrolytes. In particular, in polymer solid electrolytes, ionic liquids help mitigate ionic conductivity issues resulting from polymer solid electrolyte interface problems. For instance, they can effectively address the issue of electrode interface compatibility when paired with inorganic solid electrolytes [31]. Meanwhile, ionic liquids are also used in gel electrolytes.

Good temperature adaptability of aqueous electrolytes is a crucial requirement. Alterations of anions cause significant changes in the overall physicochemical properties of the aqueous electrolyte of Zn salts. For example, the weak interaction between OTf- and Zn^{2+} leads to a strong interaction between Zn^{2+} and H_2O in $Zn(OTf)_2$ electrolyte. in $Zn(OTf)_2$ solutions, the weak interaction of OTf^- with Zn^{2+} and H_2O, respectively, gives a strong interaction of Zn^{2+} with H_2O. Besides affecting temperature adaptation of the Zn metal electrode, the selection of the anion in Zn salt affects its stability as well. The strong interaction of the anion with Zn^{2+} decreases the coordination number for Zn^{2+} with H_2O. Based on this, Patil et al. [32] reported that $ZnSO_4$ solutions, in which H_2O in the solvation sheath of

Zn^{2+} is extruded by SO_4^{2-}, have more free water than $Zn(TFSI)_2$ solutions in which the solvation sheath of Zn^{2+} contains no anions. Therefore, the $Zn(TFSI)_2$ solution with a lower amount of free water improves the cycling stability of the constructed Zn symmetric cells compared with $ZnSO_4$ solution.

Adding inorganic additives to electrolytes has been shown to be one of the effective methods to achieve the controlled behavior of Zn plating. For example, graphitic carbon nitride (C_3N_4) quantum dots strongly interact with Zn^{2+} and regulate the solvation structure of Zn^{2+} as $[Zn(C_3N_4)(H_2O)_4]^{2+}$ has the best binding energy. Driven by coulombic force, C_3N_4 quantum dots are involved in dynamic construction and self-repairing of interphase on the Zn surface during repeated stripping/plating without consumption. The regenerative interphase featuring zincophilic pores enables single Zn^{2+}-ion conduction and blocks active water, leading to an even Zn deposition [33]. A stable $Zn_3(PO_4)_2{\cdot}4H_2O$ SEI in situ forms because of the reaction between local water-decomposition-derived OH and the additive $Zn(H_2PO_4)_2$.

Density functional theory (DFT) computation based on an optimum Zn^{2+} diffusion pathway tagged from Zn^0 to Zn^6 in SEI shows that the SEI enables rapid Zn^{2+} transportation (Figure 7.9a). Furthermore, Zn^{2+} diffusion energy barrier in a channel within a SEI of 0.66 eV is less than that for other multivalent cations, such as Mg^{2+} diffusion in $FePO_4$ or V_2O_5, confirming the superiority of the SEI designed for ZMBs (Figure 7.9b) [34]. By adding a trace amount of $Zn(NO_3)_2$,

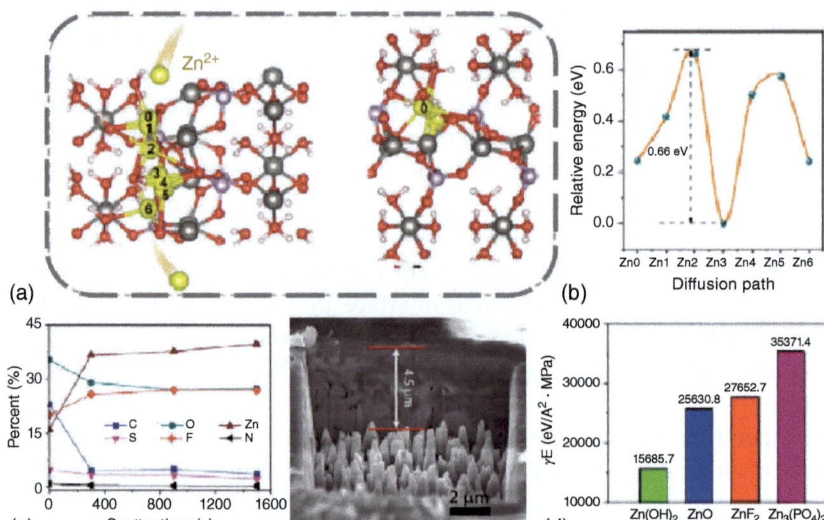

Figure 7.9 Electrolyte additive strategy for AZIBs. (a) DFT computations for optimum Zn-ion diffusion pathway in the SEI layer and (b) migration energy barrier. Source: Reproduced with permission from Zeng et al. [34]. © 2021/John Wiley & Sons. (c) Composition of SEI on Zn following 50 plating/stripping cycles in $Zn(OTf)_2$-$Zn(NO_3)_2$ electrolyte. Source: Reproduced with permission from Xiong et al. [31]. © 2021/John Wiley & Sons. (d) Capability of suppressing Zn dendrite for different SEI compounds. Source: Reproduced with permission from Li et al. [35]. © 2021/John Wiley & Sons.

robust inorganic ZnF_2-$Zn_5(CO_3(OH)_6$-$Zn(NO_3)_2)_2$-organic bilayer SEI is fabricated. As shown in Figure 7.9c, element distribution for SEI following 50 cycles shows that the dominant composition gradually changes from organic components to inorganic ZnF_2 while approaching the inner layer, confirming the bilayer structure of SEI. This ionically conductive and electrically insulating SEI suppresses water decomposition and significant volume change of the inner part. A Zn//MnO_2 battery with this electrolyte showed a high energy density of 168 Wh kg^{-1} and a capacity retention of 96.5% following 700 cycles at 10 C [35]. Furthermore, Zn dendrite suppression ability (CE) of SEI compounds are compared. $Zn_3(PO_4)_2$ and ZnF_2 have higher interfacial energy with Zn metal, in which Zn atoms preferentially move parallel to the interface layer instead of penetrating into the SEI (Figure 7.9d). Given the significant dendrite suppression ability of $Zn_3(PO_4)_2$ and accelerated Zn^{2+} transport kinetics of ZnF_2, a $Zn_3(PO_4)_2/ZnF_2$-rich SEI results from the unstable KPF_6 additive. Moisture-sensitive KPF_6 decomposes and takes part in the formation of the multifunctional interphase and optimizes the cyclability of AZIBs.

Davidson et al. [36] observed the development of highly directional fractal structures, primarily composed of magnesium, in Grignard reagent solutions and discovered that the formation of nanowires, as opposed to dendrites, can be promoted by the introduction of molecular Lewis bases. Among various electroplating techniques examined, none has led to the formation of evenly layered magnesium depositions. The process is significantly influenced by cation-desolvation-induced exchange current, which plays a pivotal role in the deposition dynamics of magnesium. Developing electrolytes with a weak coordination featuring monochloride cations or large anion-based Li-dimethoxyethane solutions, along with magnesium surface modifications using strongly negative halogen ions and a gallium–magnesium alloy coating, can facilitate even magnesium layers at increased current densities.

In the 1980s, Koura [37] used the molten salt electrolyte $AlCl_3$-NaCl to pair with a FeS_2 cathode material and a metallic aluminum anode and for the first time successfully developed reversible rechargeable AAIBs. This study showed that the decomposition of the cathode and the release of Cl_2 in the electrolyte hinder the advancement of AAIBs during cycling. Later, Reynolds and Dymek [38] for the first time used an ionic liquid instead of a molten salt electrolyte in AAIBs. They showed that the ionic liquid Al/$AICl_3$-MEICl (1-methyl-3-ethylimidazolium chloride) hinders the formation of an oxide layer on the surface of metallic aluminum and enhances the movement of Al^{3+} ions within the electrolyte. Subsequently, Archer et al. [39] used Al/$AICl_3$-[MEI]Cl as the electrolyte, V_2O_5 as the cathode, and metallic aluminum as the anode to assemble AAIBs with a stable electrochemical behavior, which caught the attention of later researchers. In 2015, Dai et al. [40] also used ionic liquid Al/$AICl_3$-[MEI]Cl as the electrolyte, a three-dimensional graphitic-foam cathode, and an aluminum metal anode. They designed a battery cell with excellent performance at high speeds, attaining a discharge voltage plateau of 2 V. The CE for the deposition and removal of aluminum in the anode was recorded at 98%. They also demonstrated that the battery could sustain exceptionally long charging and discharging stability (exceeding 7500 cycles) at a current of 4 A g^{-1}

and ambient temperature. Furthermore, throughout the charging and discharging process, $AlCl_4^-$ ions in the ionic liquid electrolyte were intercalated and expelled from graphite.

Li et al. [41] discovered that Al^{3+} (ionic liquid $AlCl_3$/[EMIm]Cl electrolyte) intercalates into MoS_2. Initially, they constructed a rechargeable aluminum-ion battery featuring a cathode made of MoS_2 microspheres, an aluminum anode, and an ionic liquid as the electrolyte, which demonstrated outstanding electrochemical properties. In this battery, Al^{3+} ions were integrated into the electrode material, alongside a phase change at the interface, as verified via ex situ X-ray photoelectron spectroscopy and XRD etching methods across different charging and discharging conditions. Evidently, ionic liquids are crucial components in electrolytes, and it is presumed that their use in electrolyte formulations will increase in the future to enhance the quality of electrolytes.

7.5.1.3 Organic Liquid Electrolytes

An organic liquid electrolyte is an electrolyte solution formed by dissolving an appropriate metal salt in an organic aprotic solvent, with nonaqueous electrolyte solutions being the majority. The liquid electrolyte solution serves as a carrier within the battery, providing a route for the transfer of ions between the cathode and anode electrodes. However, the challenges associated with traditional organic liquid electrolytes are as follows: (i) reactions with the metal electrode interface, leading to self-discharge of the battery; (ii) persistent metal dendritic growth during charge and discharge, damaging the battery structure; and (iii) consumption and leakage of flammable electrolytes, resulting in a series of safety incidents. In response to these issues, besides replacing liquid electrolytes with solid electrolytes, modification strategies for widely used organic electrolyte solutions are also essential. For nonaqueous liquid electrolytes, the primary strategies that currently aim at addressing the abovementioned issues include: (i) adjustment of electrolyte chemical composition, (ii) incorporation of flame retardants, (iii) artificial SEI film to stabilize the anode interface, and (iv) composite metal anode.

Organic additives are significant in regulating the electrolyte solvation structure or leading to protective SEI on a zinc anode. In particular, sorbitol and 2,3,4,5-tetrahydrothiophene-1,1-dioxide reduce the desolvation energy of the electrolyte and can dominantly adsorb on the zinc anode for rendering dendrite growth. In addition to the salt-decomposition-derived SEI design strategy, a stable poly-dopamine (PDA) SEI can be obtained by in situ electrochemical polymerization of organic dopamine (DA) additive as illustrated in Figure 7.10a. The Zn anode cycled in the electrolyte with the addition of DA exhibits "smooth" morphology, while a "rough" surface with obvious dendrite is generated in the electrolyte without DA. The hydrophilic, adhesive PDA SEI leads to uniform Zn deposition, boosts Zn^{2+} transport kinetics, and suppresses parasitic reactions [42]. This pioneering research has led to a better understanding of the crucial and practical applications of additives in stabilizing Zn electrodes.

After dissolving different magnesium salts in various aprotic organic solvents and analyzing the solubility and ionic conductivity of various magnesium salts,

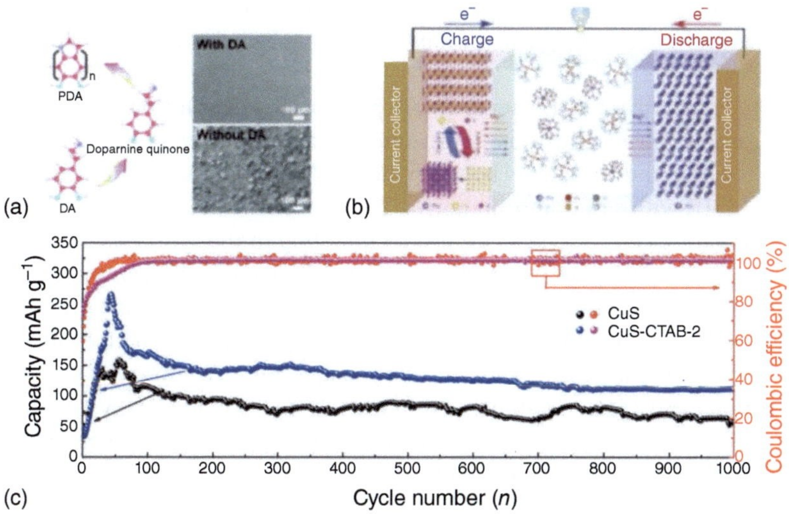

Figure 7.10 (a) Illustration for PDA polymerization and SEM images of Zn electrodes cycled in electrolyte with/without DA. Source: Zeng et al. [42]. © 2021/Royal Society of Chemistry. (b) Schematic diagram of the magnesium ion batteries with the CuS/Mg electrode system and the $Mg[B(hfip)_4]_2$/DME electrolyte. (c) Cycling performance and Coulombic efficiencies at current densities of fluorinated borate magnesium ion electrolyte. Source: Reproduced with permission from Shen et al. [43]. © 2019/John Wiley & Sons.

they concluded that the magnesium-ion deposition and dissolution behavior can be observed in $Mg(CF_3SO_3)$ with dimethylformamide and dimethylacetamide. Based on this study, Liebenow et al. [44] further delved into the migration behavior of magnesium ions and found that magnesium outperforms other organic solutions (such as magnesium halides) in terms of conductivity, oxidation stability, electrodeposition, and reoxidation in organic borate solution. In the same year, based on Gregory's report, Aurbach et al. [45] used Mo_6S_8 as the cathode material, magnesium as the anode material, and a $Mg(AlCl_2BuEt)_2$-THF solution as the electrolyte and developed a secondary magnesium battery using an optimal ratio of Grignard reagent (MgR_2, R = ethyl or butyl) to Lewis acid ($AlEtCl_2$ or $AlCl_3$). The electrochemical performance of the resulting battery was exceptional, with a magnesium deposition/dissolution cycle efficiency of nearly 100%, an electrochemical stability window of 2.5 V, and a capacity loss of only 15% after 2000 charge/discharge cycles, indicating good compatibility among battery components. Furthermore, studies have confirmed that aluminum-based Lewis acids surpass other Lewis acids in performance, thus accelerating the exploration and innovation of electrolytes for magnesium-ion batteries. Recently, Shen et al. [43] introduced a copper sulfide compound as an energy-dense cathode for magnesium battery systems using a non-corrosive magnesium-ion electrolyte. As shown in Figure 7.10b, a fluorinated borate magnesium-ion electrolyte was used to boost the stability and adaptability of the electrode–electrolyte boundary. As a result, the fluorinated borate magnesium ion electrolyte displayed an excellent electrochemical performance in Figure 7.10c.This research introduced groundbreaking approaches for developing high-efficiency rechargeable magnesium-ion batteries and other batteries based on multivalent ions with enhanced electrode/electrolyte integration.

Wang et al. [46] demonstrated the reversible deposition and dissolution of calcium ions in an organic electrolyte at room temperature. The main formation product, calcium, resulted from the interaction between the anode and the electrolyte, specifically $Ca(BH_4)_2$-THF, with a minor formation of CaH_2. This reaction mechanism was common across various electrolyte solutions, culminating in the synthesis of compounds like $CaCO_3$, $Ca(OH)_2$, and elemental calcium, which typically obstructed the electrochemical process. Li et al. developed a novel, high-efficiency electrolyte for Ca-ion batteries, (tetrafluoroisopropyl) calcium borate $Ca[B(hfip)_4]_2$, which enabled the reversible deposition of calcium ions at room temperature. The results of their study showed that the oxidation stability of the electrolyte is as high as 4.5 V and the ionic conductivity reaches 8 mS cm^{-1}. This discovery has opened a new avenue for room-temperature rechargeable Ca-ion batteries.

7.5.2 Solid-State Electrolytes

Typically, solid-state electrolytes are developed from metal salts and a supporting electrolyte framework, which may include a variety of polymer monomers for polymer-based electrolytes or inorganic oxides for ceramic electrolytes. Depending on the structure of the electrolyte, solid-state electrolytes are classified as follows: solid polymer electrolytes, inorganic solid electrolytes, and composite solid electrolytes made from inorganic and polymer materials. Solid electrolytes in batteries using multivalent metals need to have the following characteristics: (i) high ionic conductivity, (ii) excellent chemical and electrochemical robustness, (iii) efficient contact between the electrode and electrolyte, (iv) substantial mechanical strength, (v) sufficient flexibility, (vi) safety, and (vii) affordability.

7.5.3 Gel Electrolytes

Gel electrolytes, also referred to as quasi-solid electrolytes, show unique ion transport processes that distinguish them from solid electrolytes. They are comprised of a polymeric network infused with liquid components (typically organic electrolytes), integrating the advantages of both liquid and solid states. The liquid part contributes to achieving high ionic conductivity and ensuring stable interfaces, whereas the polymeric framework provides the necessary mechanical robustness and adaptability. Initiators and crosslinkers play a key role in this context, causing the polymer to entwine and form a cohesive three-dimensional network. During the gelation process, the polymer solution transitions into a gel polymer, which, unlike solid electrolytes, does not flow. Nonetheless, its ion transport mechanism closely resembles that of liquid electrolytes due to the inclusion of liquid elements within its architecture. Therefore, gel electrolytes embody the characteristics of both solid states – such as form retention, structural integrity, and flexibility – and liquids, characterized by efficient ion transport and conductivity. The synthesis of gel electrolytes is classified into two approaches: physical and chemical. Physical synthesis involves forming a network through polymers interlocking or partially crystallizing, resulting in microscopic phase separation where the separated phase does not support ion conduction. Chemical synthesis, on the other hand, involves connecting polymer chains via covalent bonds through thermal or photo-induced polymerization, resulting in a gel polymer electrolyte with a stable crosslink density that does not crystallize under varying conditions.

In applications such as flexible electronics, where devices function in varied environments, gel and quasi-solid electrolytes are pivotal due to their flexibility, mechanical strength, and environmental resilience. Through the establishment of either physically or chemically crosslinked, self-sustaining gel matrices, the mechanical integrity of quasi-solid electrolytes can be significantly improved. This enhances not only their mechanical properties but also their electrochemical efficacy. To make further advances in this field, innovative electrolyte materials with compatible separators and electrolytes are crucial for batteries to operate across extensive temperature ranges and maintain anode stability.

7.6 Design Strategies for Extreme Temperatures

ARMMBs are considered a highly viable option for energy storage and wearable technology, thanks to their safety features, affordability, substantial theoretical capacity, and eco-friendliness. Nonetheless, the versatile application scenarios of AZIBs pose significant challenges to their functionality at very high or low temperatures. The mechanisms behind the failure of AZIBs at high and low temperatures are not well understood, which hinders the development of effective adaptation strategies. This analysis offers a thorough overview from theoretical concepts to practical applications and examines the causes of performance declines at extreme temperatures through thermodynamic and kinetic analyses, including crucial characteristics such as ion movement, electrode surface oxidation–reduction reactions, and polarization. It evaluates the difficulties encountered by AZIBs under cold/warm conditions, with a focus on issues such as electrolyte freezing, increased polarization, reduced ionic conductivity, detrimental side reactions, dissolution of materials, and thermal runaway risks. It then outlines successful approaches to navigating these issues and enhancing electrochemical performance. Finally, it explores the broader challenges batteries face under extreme temperature conditions and suggests directions for future advancements.

7.6.1 Mechanisms and Challenges at Extreme Temperatures

Electrochemical energy storage devices convert and store electrochemical energy through a reversible electrochemical reaction, which takes place via the flow of electrons and ions. In the following, the impact of temperature on battery performance is thoroughly examined from the perspectives of thermodynamics and dynamics.

In thermodynamics, Gibbs free energy (ΔG) represents a type of thermal potential energy that characterizes the thermodynamic properties of electrochemical systems, which is determined by the following equation:

$$\Delta G = \Delta H - T\Delta S = -nFE \tag{7.10}$$

where T is temperature, n is the actual charge transferred by the metal ion (in electrons), F is the Faraday constant, E is the open-circuit voltage of the battery, and ΔH and ΔS are enthalpy change and entropy change, respectively. ΔS is defined by the following equation:

$$\Delta S = S_{\mathrm{B}} - S_{\mathrm{A}} = \int_{\mathrm{A}}^{\mathrm{B}} \frac{Q}{T}\mathrm{d}T \tag{7.11}$$

where S represents entropy and Q denotes the heat exchanged at an absolute temperature. Taking into account Eqs. (7.1) and (7.2), it becomes apparent that the battery's output voltage is influenced by its temperature, enthalpy shift, and entropy variation within the system.

From a kinetic standpoint, the discharge process for zinc ions generally involves the following steps: (i) migration of solvated zinc ions from the electrolyte to the electrode's surface (stage 1); (ii) occurrence of oxidation–reduction reactions on the surface of the electrode (stage 2); and (iii) diffusion of zinc ions within the electrode's material (stage 3). Migration in stage 1 is characterized by ionic conductivity; redox reactions on the cathode's surface during stage 2 are determined by adsorption/desolvation dynamics; and the behavior of stage 3 depends on the ionic diffusion coefficient inside the electrode. Each of these phases is sensitive to temperature changes, especially notable declines in temperature. In stage 1, quick movement of solvated ions occurs within the electrolyte, setting the stage for the forthcoming electrochemical reactions. The ability of the electrolyte to transport ions is reflected in its conductivity, which can be expressed by the following equation:

$$\sigma = \sum n_i \mu_i z_i e \sigma \tag{7.12}$$

where μ_i represents the ion migration number of different ions, and e, z_i, and n_i represent the unit charge, charge valence, and the number of ions, respectively. Ion mobility is expressed as follows:

$$\mu = \frac{1}{6\pi\eta\gamma} \tag{7.13}$$

where η represents the viscosity of the electrolyte and γ signifies the radius of the solvated ion. As the temperature drops, the viscosity of the electrolyte increases, resulting in reduced ionic conductivity in colder environments. According to Eqs. (7.12) and (7.13), it is evident that the transport dynamics of the liquid component are strongly associated with the composition, concentration, and solvation formation of the electrolyte. This insight points toward potential solutions for addressing the challenges faced by AZIBs under extreme temperature conditions (Figure 7.11) [47].

In the second stage, it is presumed that the cathode material has a suitable adsorption energy barrier. Overly high adsorption energy barriers may obstruct

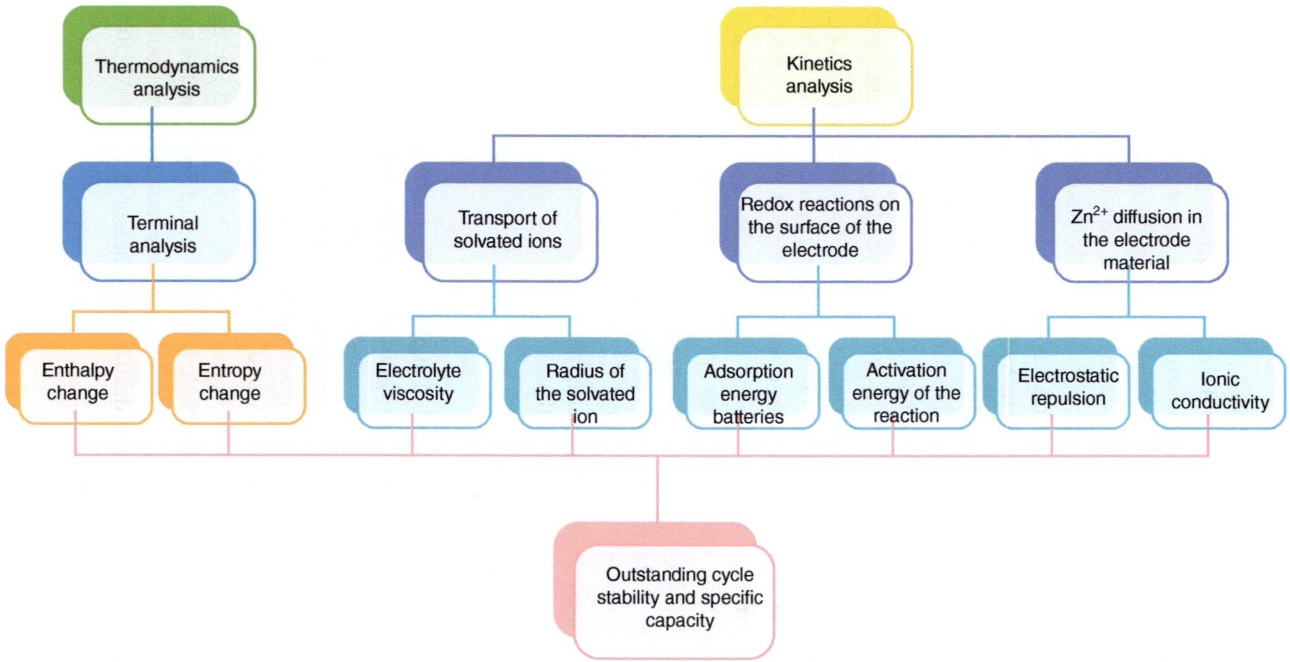

Figure 7.11 Thermodynamic and kinetic parameters related to the electrochemical performances of AZIBs. Source: Reproduced with permission from Chen et al. [47]. © 2022/Elsevier.

the intercalation of Zn^{2+}, causing a decrease in the number of active sites of the cathode material. Conversely, extremely low adsorption energy barriers may lead to an excess build-up of intercalated Zn^{2+} within the cathode material, sparking irreversible phase alterations. The desolvation process of hydrated Zn^{2+} during charging and discharging is recognized as a pivotal factor affecting electrode kinetics. Typically, Zn^{2+} and the polar water in the electrolyte construct systematic and structured hydrated zinc ions. The architecture of a hydrated Zn^{2+} can be classified into two spheres: the interior and the exterior. The interior sphere comprises Zn^{2+} and six H_2O molecules linked by hydration coordination bonds. In the exterior sphere, free water molecules distanced from the inner hydrated shell get connected through HBs. To achieve Zn deposition at the anode–electrolyte boundary of AZIBs, Zn^{2+} needs to be dissociated from hydrated Zn^{2+}. This desolvation step is commonly identified as a rate-limiting process for Zn deposition. The substantial coulombic interaction between Zn^{2+} and its enveloping H_2O solvent sheath triggers significant desolvation losses and decelerates the electrochemical reaction kinetics. Meanwhile, at the cathode–electrolyte interface of AZIBs, due to the restrictive tunneling and layered structures of certain cathodes, hydrated Zn^{2+} is required to undergo complete or partial desolvation before proceeding to subsequent electrochemical reactions. In addition, desolvation energy is primarily determined by cations and electrolytes. Smaller cations lead to higher radii of hydrated ions, thereby demanding high energy barriers for desolvation. More specifically, the smaller ionic radius of Zn^{2+} in contrast to typical metal carriers (Li^+, Na^+, and K^+) consequently results in the larger radius of hydrated Zn^{2+}, necessitating a higher desolvation energy barrier. This increases the rate of interfacial reaction kinetics and mass transfer kinetics. The relationship between the reaction rate and reaction activation energy can be expressed as follows:

$$v = kc \exp\left(\frac{-E_a}{RT}\right) \tag{7.14}$$

where v represents the reaction speed, c signifies the concentration of reactive particles, E_a stands for the activation energy of the reaction, and k is the pre-factor. This equation clearly shows a positive correlation between the rate of reaction and temperature, implying a decrease in the reaction rate when the temperature drops.

In the third stage, multivalent ion diffusion within the cathode material primarily depends on the size and electrostatic repulsion of guest ions. For most cathode materials, ion diffusion usually acts as the controlling step that sets the pace for electrode reaction kinetics. The considerable size of hydrated Zn^{2+} and the substantial electrostatic repulsion between Zn^{2+} and the cathode material contribute to the sluggish transport kinetics of Zn^{2+}. The diffusion coefficient $D_{Zn}{}^{2+}$ of AZIBs can be expressed by the following equation:

$$D_{Zn^{2+}} = \frac{R^2T^2}{2A^4F^4n^4C^2\sigma^2} \tag{7.15}$$

where R, T, A, F, n, C, and σ represent gas constant, absolute temperature, effective area of the hydrogel electrolyte, Faraday's constant, number of electrons transferred per formula unit, ion concentration, and Warburg coefficient, respectively. This equation clearly shows that temperature and ion concentration considerably affect the ion diffusion coefficient. Ion diffusion in the cathode occurs simultaneously with electron transfer, which also influences the kinetics of electrode reactions. Notably, electron transport primarily depends on the inherent conductivity of the cathode material. However, many cathode materials exhibit poor inherent conductivity due to a broad band gap and low-energy DOS, resulting in disappointingly low electrochemical performance. When batteries experience extreme temperatures, these issues may worsen.

Battery electrochemical dynamics are closely associated with electrode polarization, including ohmic polarization (η_o), electrochemical polarization (η_a), and concentration polarization (η_c). Such polarization adversely affects the battery voltage at harsh temperatures, thus destabilizing the electrochemical system. Consider, for example, polarization at low temperatures: a drop in temperature slows down the Zn^{2+} diffusion kinetics in the electrolyte, preventing the timely arrival of Zn^{2+} at the cathode surface to replace depleted Zn^{2+} ions, thus intensifying concentration polarization. Lower temperatures reduce the electrolyte's ionic conductivity, causing an upswing in ohmic polarization. Furthermore, colder temperatures stall electrochemical reactions and lead to a surplus of electrons on the cathode surface, increasing electrochemical polarization. Apart from polarization's effect on potential, polarization resistance, attributable to ohmic polarization, concentration polarization, and electrochemical polarization, is also prevalent in batteries. This internal resistance generates heat when current flows within the battery, increasing the thermal runaway risk at high temperatures. Moreover, the potential shift caused by polarization resistance, ohmic resistance, and so on further affects power density. As a result, minimizing polarization is crucial to mitigate the performance deterioration of batteries at extreme temperatures.

7.6.2 Challenges of AZIBs at Low Temperatures

Components of a battery can be primarily categorized into three types: (i) electrolyte, (ii) cathode, and (iii) anode. When AZIBs function at low temperatures, the primary challenge for the electrolyte is that its freezing process can compromise the battery's integrity. When Eqs. (7.7) and (7.8) are taken into account, the polarization process mentioned earlier at low temperatures leads to a negative shift in the discharge operating voltage. Furthermore, the viscosity of the electrolyte also increases at low temperatures, making the diffusion of mass-transferring ions challenging and augmenting the desolvation barrier. The reduced kinetics degrades electrochemical performance, resulting in a reduced cycle life at low temperatures.

In AZIBs, the transfer of Zn^{2+} ions between the cathode and anode through the electrolyte is a critical function. At lower temperatures, the solidified electrolyte hinders the ion flow and disrupts the bond between the electrode and electrolyte. Preparing electrolytes that can withstand freezing conditions is a key strategy for the operation of AZIBs at sub-zero temperatures. The freezing threshold of the electrolyte is largely determined by the number of HBs it contains. With decreasing temperatures, water molecules shift from a tetrahedral coordination to layered hexagonal ice structures, a transition driven by new HB formations. This inhibits the movement of both water molecules and ions within the electrolyte. The ability of solute molecules in an electrolyte to bond with water molecules through HBs can be determined by two metrics: (i) the ratio of bound water's mass to total water's mass (Wbw/Ww) and (ii) the ratio of bound water molecules to the number of solute molecules (Nbw/Ns). Bound water refers to the water molecules that establish one or more HBs with solute molecules. As solution concentration increases, so do the Wbw/Ww and Nbw/Nw ratios, leading to a higher proportion of bound water. Given that bound water is less prone to forming and expanding ice, targeting the breakdown of HBs among free water and boosting the volume of bound water emerges as a viable approach to reducing electrolyte freezing points. To modify the electrolyte for enhanced performance, several tactics can be applied: (i) altering the electrolyte's composition to introduce a plethora of ions (including both cations and anions) that bind to water, thereby interrupting the HBs among water molecules and averting freezing at low temperatures; (ii) minimizing the number of water molecule HBs by introducing hydrophilic groups (like –COOH, –OH) into the electrolyte; and (iii) using organic electrolytes, which lack HBs among free water molecules, thus leading to a reduced freezing point.

Besides electrolytes, the condition of electrodes at reduced temperatures plays a crucial role in maintaining the operation of batteries. Nevertheless, studies on how low temperatures affect AZIB electrodes are scarce. At colder temperatures, electrodes exhibit an increased charge transfer resistance, which slows down the reaction rates of the cathode materials. In addition, both ion mobility and electrical conductivity of cathode materials are decreased when temperatures drop. Furthermore, problems that cathodes encounter at ambient temperature might become more severe under colder conditions. For example, the recurrent insertion and removal of ions could damage the cathode's crystal structure, and the potent electrostatic forces repelling between the cathode and Zn^{2+} hinder ion movement. To ensure cathode materials perform well electrochemically at lower temperatures, strategies such as reducing the energy required for reactions and diffusion, diminishing charge transfer resistance, and improving electronic conductivity prove to be beneficial. Various design methods including carbon coating, introducing defects, pre-insertion of ions, and employing topological insulators have been explored to tackle the issues cathodes face in colder environments.

Until now, zinc has been widely utilized as an anode in energy storage devices though its broader application is hindered by certain undesirable reactions. The major obstacles facing zinc anodes include the generation of zinc dendrites and unwanted reactions at the interface between the electrode and electrolyte. In particular, the irregular surface of the zinc anode leads to a nonuniform distribution of zinc ions. Zinc ions tend to accumulate in areas with higher concentrations of Zn^{2+}, leading to the development of bumps. Furthermore, the increased local electric fields at these bumps on the surface of the zinc electrode promote the clustering of Zn^{2+} ions, which accelerates dendrite growth. Regarding side reactions, such as the corrosion of zinc anodes, the hydrogen evolution reaction (HER) is the primary issue. The gases generated by the HER increase the battery's internal pressure, creating a significant safety risk. Specifically, when zinc anodes are in contact with the electrolyte, several microprotocells are established, with zinc serving as the anode and impurities (primarily carbon) as the cathode. These microprotocells produce OH during HER, resulting in the accumulation of insoluble by-products on the surface of the zinc anode. Although zinc deposition is more uniform and side reactions are reduced at colder temperatures due to decreased water activity, issues with dendrite formation and corrosion pose significant challenges to achieving long cycle lives for AZIBs at these temperatures. Using uniform zinc nucleation as a strategy to limit Zn dendrite growth on the anode is proposed. However, the association between Zn nucleation overpotential and dendrite formation is contentious. Some experts argue that a high nucleation overpotential suggests a strong nucleation force, leading to more refined zinc deposits and a desirable crystal structure. An alternative perspective suggests that a low nucleation overpotential promotes uniform Zn deposition [48]. Except for the nucleation overpotential, uniform interfacial electric field, more nucleation sites, and speedy Zn^{2+} transfer all affect Zn deposition [49, 50].

7.6.3 Challenges of AZIBs at High Temperatures

The difficulties encountered by AZIBs under high temperature conditions are not the same as those experienced at colder temperatures. Unlike typical organic batteries, AZIBs show superior thermal stability when exposed to high temperatures. Nonetheless, the operation of AZIBs naturally generates heat. This thermal energy primarily originates from three sources: (i) losses due to polarization (including ohmic, electrochemical, and concentration types), (ii) battery electrochemical reactions alongside entropy shifts from side reactions, and (iii) the blending process occurring as the battery settles [51]. With an increase in the temperature, the molecular activity within the electrolyte, the kinetics of ion movement, and the rates of electrochemical reactions all increase. Consequently, issues such as hydrogen production from side reactions, corrosion, degradation of interfaces, and self-discharge hasten the battery's aging process, leading to its failure. In addition, charge and discharge cycles at high temperatures can induce local hotspots due to the insufficient cooling of the sealed battery, further diminishing the battery's longevity. In more extreme cases, prolonged exposure to high heat may induce thermal runaway, a series of escalating chain reactions triggered by an interior short circuit.

At high temperatures, the electrolyte's water content quickly evaporates, resulting in salt precipitation and an increase in the battery's internal pressure. Furthermore, electrolyte components may undergo thermal decomposition when exposed to high heat. Thus, improving the electrolyte's water retention and thermal resistance is beneficial for the battery's performance at high temperatures. The cathode experiences irreversible alterations, such as structural collapse and dissolution, under these conditions. High temperatures can further induce significant electrochemical side reactions and polarization at the interface between the electrode and electrolyte. Hence, it becomes especially important to investigate the ways to enhance the cathode material's stability. Addressing the challenge of maintaining the cycle stability of zinc anodes at high temperatures is critically important. The increased activity of water molecules within the electrolyte at high temperatures accelerates side reactions and the formation of dendrites on zinc anodes. Consequently, adjusting zinc anodes is essential to achieve satisfactory performance under high thermal conditions.

7.6.4 Mechanism and Strategies of AZIBs at Low Temperatures

Altering the electrolyte composition to reduce its freezing point has been demonstrated to be an efficient strategy. Cations and anions in the electrolyte serve as oxygen acceptors and hydrogen acceptors, respectively, both of which can form HBs with water and thereby reduce the freezing point. The objective of this section is to introduce oxygen and hydrogen acceptors into the electrolyte to reduce the freezing point by disrupting free water HBs.

A high concentration of Zn^{2+} acts as an oxygen receiver and actively engages with bipolar water molecules, breaking up the network of HBs among free water molecules, thus reducing the freezing point to −114 °C. This disruption leads to a deviation in the high freezing point of water from the expected values by the law of chalcogen hydrides [52]. The electrolyte's zinc ions form a structured hydration shell through strong electrostatic forces with water molecules. These interactions are analyzed using DFT, which reveals that the binding energy of water to itself is 5.1 kcal mol^{-1}, significantly less than the 98.5 kcal mol^{-1} binding energy between Zn^{2+} and water. The lower binding energy for various hydration structures of Zn^{2+} suggests they are in a stable, low-energy configuration, which complicates their dissociation into ions and water. In a 0 M $ZnCl_2$ solution, water molecules primarily exist as 4-HB structures, forming a tetrahedral coordination essential for ice crystallization [53]. Increasing the $ZnCl_2$ concentration leads to a progressive reduction in these 4-HB water molecules. Using 7.5 M $ZnCl_2$ as a cryogenic electrolyte allows for designing a battery that operates reliably through 2000 cycles at −70 °C with a current density of 0.2 A g^{-1}.

The presence of an anion in the electrolyte, serving as a recipient for hydrogen, plays a crucial role in affecting the HBs with free water. DFT studies show, for instance, that the SO_4^{2-} anion has an electrostatic potential of 10.29 eV, making hydrogen atoms in water more prone to attaching to the oxygen in SO_4^{2-} than to the oxygen in water molecules. This interaction with 5 M SO_4^{2-} leads to the formation of HBs with water molecules, which disrupts the HBs among free water molecules and significantly reduces the freezing point. Furthermore, due to its high electronegativity, the fluorine atom enables the BF_4 anion to form weaker HBs with water, acting as an acceptor of hydrogen ions. This process challenges the HBs between water molecules within the electrolyte, considerably reducing the strong water–water HBs present. The $Zn(ClO_4)_2$ electrolyte, featuring a chaotropic anion, also exhibits superior resistance to low temperatures. The freezing point of a 3 M $Zn(ClO_4)_2$ solution can drop to −74 °C, greatly outperforming similar concentrations of $ZnSO_4$ and $Zn(CF_3SO_3)_2$. Furthermore, at −50 °C, 3 M $Zn(ClO_4)_2$ maintains an impressive ionic conductivity of 4.23 mS cm^{-1}.

In aqueous electrolytes, ice formation at sub-zero temperatures poses a significant barrier to the electrochemical efficiency of water-based batteries. As a result, a practical approach could be to substitute these aqueous electrolytes with their non-aqueous counterparts with a lower freezing point. Notably, organic solvents such as ethyl acetate, ethylene carbonate, dimethyl carbonate, and diethyl carbonate are routinely used as low-temperature electrolytes for LIBs, thanks to their low freezing points and high ionic conductivities.

7.7 Practical Progress in Grid-Scale Energy Storage and Wearable Devices

Over the years, ARMMBs have undergone significant advancements in order to meet the requirements of next-generation energy storage systems. To ensure the successful commercialization of ARMMBs, three key criteria have been recommended: safety, low cost, and high performance. Safety is of utmost importance in ARMMBs, particularly when compared with the safety concerns associated with the existing LIBs. In addition, ARMMBs also offer a cost advantage over LIBs. They use abundant resources, such as Zn, Ca, and Mg, which are more readily available and less expensive than the materials used in LIBs. In addition, the manufacturing process of ARMMBs is simpler and requires less complex infrastructure, leading to lower production costs. In terms of performance, ARMMBs are expected to surpass LIBs in lifespan and power density. Considering these characteristics, it is presumed that ARMMBs find significant applications in large-scale energy storage systems and wearable devices in the future.

Electrical grids can utilize energy storage systems for functions such as maintaining frequency and responding swiftly to shifts in demand, which contributes to the grid's dependability, steadiness, and economic efficiency. Activities such as leveling peak loads and redistributing energy demand are essential for optimizing grid performance. In massive storage solutions, safety is a critical concern, and characteristics such as specific energy ($Wh\,kg^{-1}$) and peak power capability may not be as crucial as in portable devices. Instead, the emphasis for grid-scale applications is on affordability of the setup, longevity, energy conversion efficiency, and scalability, ranging from a few kWh to several hundred MWh. The battery technologies used in grid services should stabilize peak demands, manage load variations, and regulate voltage and frequency. They should also be ready for backup during power deficits. The concept of load leveling involves energy accumulation during off-peak periods and its redistribution during peak usage. Maintaining the grid's balance in terms of voltage and frequency relies on the ability of the battery system to absorb or supply power rapidly. Emergency power storage is ideally instantaneous, capable of immediate full-power output in times of significant power deficits. Compared with conventional storage technologies that operate on longer timescales (ranging from several seconds up to minutes), battery storage offers a rapid response (less than 20 ms) for grid stabilization. Given the instantaneous and sometimes unpredictable nature of power demand and supply on the grid, battery storage stands as a critical component for securing grid integrity and safety.

The importance of grid energy storage systems cannot be overemphasized when it comes to ensuring a stable equilibrium between electricity supply and demand within power networks. Such storage solutions require characteristics such as swift responsiveness, durability in terms of cycle life, affordability, and efficiency in both power and energy capacities. There is a pressing need for innovative battery technologies that can satisfy these overall high performance standards. While lead–acid batteries are the most economical option available, their use is limited by their restricted depth of discharge, reduced cycle longevity, and suboptimal battery effectiveness. Furthermore, the operational capacity of Na-S and flow battery systems is often compromised by their slow reaction times and lower energy efficiencies. This is why ARMMBs are emerging as a favorable choice for integration into large-scale grid systems due to their impressive safety credentials, budget-friendliness, and potent power density attributes. Nevertheless, the challenge with ARMMBs is their relatively modest energy density, a consequence of their confined output voltage range. For large-scale applications, the mass or volumetric energy density is considered a crucial metric, which is directly correlated to the voltage and capacity parameters of the device in question. In brief, there are three strategic approaches to amplifying the energy density of ARMMBs: (i) broadening the electrochemical stability window to boost output voltage, (ii) investigating new substrate materials or chemical reactions that facilitate greater electron exchange with minimal weight or space demands, and (iii) enhancing electrolyte utilization and minimizing its degradation or loss.

Besides these issues, it is crucial to thoroughly assess self-discharge rates to successfully implement the emerging ARMMB technologies. Recent strides in developing electrode materials have yielded a diverse array of options for ARMMBs, showcasing their promising electrochemical capabilities. It is presumed that advances in the development of innovative electrode materials will significantly enhance the performance of ARMMBs in the foreseeable future. Continued efforts in improving safety, reducing costs, and optimizing performance will propel the path to commercializing high-efficiency ARMMBs. Although LIBs may continue to lead as the primary energy source for consumer electronics, vehicles, drones, and robots over the next ten years, the pursuit of more affordable (achieving over a 50% reduction in costs) and safer alternatives to post-lithium-ion technologies is set to escalate shortly.

Recently, there has been a surge of interest in both flexible electronics and energy storage sectors toward wearable energy storage devices (WESDs) designed to provide power amidst physical movements. Considerable progress has been made in WESDs due to innovations in the domains of battery/capacitor chemistry, electrode materials, and electrolytes – enhancing energy density, extending cycling lifetimes, and improving charge/discharge rates. Concurrently, the flexibility and wearability aspects have also seen significant advances with the introduction of novel device architectures such as island and fiber configurations, development of elastic gel-based electrolytes, and implementation of suitable sealing layers.

Despite these technological leaps, the actual integration of WESDs into daily human–computer interaction, particularly for varied body locations, lags behind optimal requirements. Challenges persist in terms of flexibility, energy density, safety, user comfort, and biomedical compatibility. Energy density and flexibility remain critical metrics for the endurance of wearable batteries in flexible devices and their ability to withstand mechanical bending. In terms of safety, it is essential to consider the risk these batteries pose to the human body with repetitive flexing. Comfortability is a cutting-edge concept that encompasses the need for wearables to be breathable and hydrophilic. The placement of WESDs on the body – whether on the head, chest, or wrist – influences the necessary safety and flexibility levels required by the electronics. Using these evaluation criteria collectively helps in fine-tuning the specific needs of a variety of flexible electronic applications.

Based on these requirements, ARMMBs are considered promising energy storage systems for powering wearable electronics. In particular, aqueous Zn-MnO_2 batteries are considered a solution for omnidirectional deformable batteries due to their high safety, low cost, and high energy density. To boost the integration of both uniaxially and multidirectionally flexible batteries in powering wearable devices, greater emphasis must be placed on addressing the potential structural breakdown of electrode materials due to continuous flexing. The constant bending may cause electrode active substances to detach from the current collector, leading to a rapid decline in power delivery. Two primary forces contribute to this detachment: the in-plane tension within the cathode framework and the lateral force from separator movement. A promising strategy involves the enhancement or alteration of binder molecules, such as polyvinylidene fluoride, by adjusting the polymer chain length and varying the side chain types to incorporate elasticity and improve resilience. Using a metallic anode can lead to mechanical bending, which can promote dendrite growth on the anode, potentially leading to battery failure if a dendrite punctures the separators. This issue arises due to a significant increase in the curvature radius at areas of bending, distorting the electric field and causing an uneven ion flow. Moreover, the thickness of the metallic anode should be minimized while retaining high energy capacity, necessitating an extensive depth of discharge for the metal anode.

References

1 Pan, Z., Liu, X., Yang, J. et al. (2021). Aqueous rechargeable multivalent metal-ion batteries: advances and challenges. *Advanced Energy Materials* 11 (24): 2100608.

2 Fu, N., Xu, Y.-T., Zhang, S. et al. (2022). Electrode materials for aqueous multivalent metal-ion batteries: current status and future prospect. *Journal of Energy Chemistry* 67: 563–584.

3 Ding, J., Du, Z., Gu, L. et al. (2018). Ultrafast Zn^{2+} intercalation and deintercalation in vanadium dioxide. *Advanced Materials* 30 (26): 1800762.

4 Wang, C., Wei, S., Chen, S. et al. (2019). Delaminating vanadium carbides for zinc-ion storage: hydrate precipitation and H^+/Zn^{2+} Co-action mechanism. *Small Methods* 3 (12): 1900495.

5 Zhu, K., Wu, T., Sun, S. et al. (2020). Synergistic H^+/Zn^{2+} dual ion insertion mechanism in high-capacity and ultra-stable hydrated VO_2 cathode for aqueous Zn-ion batteries. *Energy Storage Materials* 29: 60–70.

6 Wang, D., Wang, L., Liang, G. et al. (2019). A superior δ-MnO_2 cathode and a self-healing Zn-δ-MnO_2 battery. *ACS Nano* 13 (9): 10643–10652.

7 Zhang, H., Liu, X., Qin, B. et al. (2020). Electrochemical intercalation of anions in graphite for high-voltage aqueous zinc battery. *Journal of Power Sources* 449: 227594.

8 Guo, Q., Kim, K.i., Jiang, H. et al. (2020). A high-potential anion-insertion carbon cathode for aqueous zinc dual-ion battery. *Advanced Functional Materials* 30: (38).

9 Wan, F., Zhang, L. et al. (2018). An aqueous rechargeable zinc-organic battery with hybrid mechanism. *Advanced Functional Materials* 28: (45).

10 Xingyu, Z., Xinqi, L., Li, Y. et al. (2021). Challenges and design strategies for high performance aqueous zinc ion batteries. *Energy Storage Materials* 42: 533–569.

11 Zheng, J., Ju, S., Xia, G. et al. (2022). Co-construction of solid solution phase and void space in yolk–shell $Fe_{0.4}Co_{0.6}S$@N-doped carbon to enhance cycling capacity and rate capability for aluminum-ion batteries. *ACS Applied Materials & Interfaces* 14 (6): 8076–8085.

12 Aruto, W., Kentaro, Y., Orikasa, Y. et al. (2020). Reaction mechanism of electrochemical insertion/extraction of magnesium ions in olivine-type $FePO_4$. *Solid State Ionics* 349: 115311.

13 Sun, W., Wang, F., Hou, S. et al. (2017). Zn/MnO_2 battery chemistry with H^+ and Zn^{2+} coinsertion. *Journal of the American Chemical Society* 139 (29): 9775–9778.

14 Jin, Y., Zou, L., Liu, L. et al. (2019). Joint charge storage for high-rate aqueous zinc–manganese dioxide batteries. *Advanced Materials* 31 (29): e1900567.

15 Chao, D., Zhou, W., Ye, C. et al. (2019). An electrolytic Zn–MnO_2 battery for high-voltage and scalable energy storage. *Angewandte Chemie International Edition* 58 (23): 7823–7828.

16 Wan, F., Zhang, Y., Zhang, L. et al. (2019). Reversible oxygen redox chemistry in aqueous zinc-ion batteries. *Angewandte Chemie International Edition* 58 (21): 7062–7067.

17 Hu, Z., Guo, Y., Jin, H. et al. (2020). A rechargeable aqueous aluminum–sulfur battery through acid activation in water-in-salt electrolyte. *Chemical Communications* 56 (13): 2023–2026.

18 Peramunage, D. and Licht, S. (1993). A solid sulfur cathode for aqueous batteries. *Science* 261.

19 Wei, C., Song, J., Wang, Y. et al. (2023). Recent development of aqueous multivalent-ion batteries based on conversion chemistry. *Advanced Functional Materials* 33 (45): 2304223.

20 Liu, B., Sun, Y., Liu, L. et al. (2018). Advances in manganese-based oxides cathodic electrocatalysts for Li–air batteries. *Advanced Functional Materials* 28 (15): 1704973.

21 Ming, F., Liang, H., Lei, Y. et al. (2018). Layered $Mg_xV_2O_5{\cdot}nH_2O$ as cathode material for high-performance aqueous zinc ion batteries. *ACS Energy Letters* 3 (10): 2602–2609.

22 Li, G., Yang, Z., Jiang, Y. et al. (2016). Towards polyvalent ion batteries: a zinc-ion battery based on NASICON structured $Na_3V2(PO_4)_3$. *Nano Energy* 25: 211–217.

23 Chen, S., Zhao, D., Chen, L. et al. (2021). Emerging intercalation cathode materials for multivalent metal-ion batteries: status and challenges. *Small Structures* 2 (11): 2100082.

24 Zheng, J., Zhao, Q., Tang, T. et al. (2019). Reversible epitaxial electrodeposition of metals in battery anodes. *Science* 366 (6465): 645–648.

25 Yang, H., Chang, Z., Qiao, Y. et al. (2020). Constructing a super-saturated electrolyte front surface for stable rechargeable aqueous zinc batteries. *Angewandte Chemie International Edition* 132 (24): 9463–9467.

26 Zhang, S., Long, T., Zhang, H.Z. et al. (2022). Electrolytes for multivalent metal-ion batteries: current status and future prospect. *ChemSusChem* 15 (21): e202200999.

27 Sun, T., Yuan, X., Wang, K. et al. (2021). An ultralow-temperature aqueous zinc-ion battery. *Journal of Materials A* 9 (11): 7042–7047.

28 Cai, S., Chu, X., Liu, C. et al. (2021). Water-salt oligomers enable supersoluble electrolytes for high-performance aqueous batteries. *Advanced Materials* 33 (13): 2007470.

29 Wang, F., Fan, X., and Gao, T. (2017). High-voltage aqueous magnesium ion batteries. *ACS Central Science* 3 (10): 1121–1128.

30 Zhang, H., Cao, D., and Bai, X. (2019). High rate performance of aqueous magnesium-ion batteries based on the δ-MnO_2@ carbon molecular sieves composite as the cathode and nanowire VO_2 as the anode. *Journal of Power Sources* 444: 227299.

31 Li, D., Cao, L., Deng, T. et al. (2020). Design of a solid electrolyte interphase for aqueous Zn batteries. *Angewandte Chemie International Edition* 60 (23): 13035–13041.

32 Patil, N., de la Cruz, C., Ciurduc, D. et al. (2021). An ultrahigh performance zinc-organic battery using poly (catechol) cathode in $Zn(TFSI)_2$-based concentrated aqueous electrolytes. *Advanced Energy Materials* 11 (26): 2100939.

33 Zhang, W., Dong, M., Jiang, K. et al. (2022). Self-repairing interphase reconstructed in each cycle for highly reversible aqueous zinc batteries. *Nature Communications* 13 (1): 5348.

34 Zeng, X., Mao, J., Hao, J. et al. (2021). Electrolyte design for in situ construction of highly Zn^{2+}-conductive solid electrolyte interphase to enable high-performance aqueous Zn-ion batteries under practical conditions. *Advanced Materials* 33 (11): 2007416.

35 Li, D., Cao, L., Deng, T. et al. (2021). Design of a solid electrolyte interphase for aqueous Zn batteries. *Angewandte Chemie International Edition* 60 (23): 13035–13041.

36 Davidson, R., Verma, A., Santos, D. et al. (2018). Formation of magnesium dendrites during electrodeposition. *ACS Energy Letters* 4 (2): 375–376.

37 Koura, N. (1980). A preliminary investigation for an Al/$AlCl_3$-NaCl/FeS_2 secondary cell. *Journal of the Electrochemical Society* 127 (7): 1529.

38 Reynolds, G. and Dymek, C. Jr., (1985). Primary and secondary room temperature molten salt electrochemical cells. *Journal of Power Sources* 15 (2-3): 109–118.

39 Jayaprakash, N., Das, S., Archer, L.A. et al. (2011). The rechargeable aluminum-ion battery. *Chemical Communications* 47 (47): 12610–12612.

40 Lin, M.-C., Gong, M., Lu, B. et al. (2015). An ultrafast rechargeable aluminium-ion battery. *Nature* 520 (7547): 324–328.

41 Li, Z., Niu, B., Li, J., and Kang, F. (2018). Rechargeable aluminum-ion battery based on MoS_2 microsphere cathode. *ACS applied materials & interfaces* 10 (11): 9451–9459.

42 Zeng, X., Xie, K., Liu, S. et al. (2021). Bio-inspired design of an in situ multi-functional polymeric solid–electrolyte interphase for Zn metal anode cycling at 30 mA cm^{-2} and 30 mA h cm^{-2}. *Energy & Environmental Science* 14 (11): 5947–5957.

43 Shen, Y., Wang, Y., Miao, Y. et al. (2020). High-energy interlayer-expanded copper sulfide cathode material in non-corrosive electrolyte for rechargeable magnesium batteries. *Advance Materials* 32 (4): 1905524.

44 Liebenow, C., Yang, Z., and Lobitz, P. (2000). The electrodeposition of magnesium using solutions of organomagnesium halides, amidomagnesium halides and magnesium organoborates. *Electrochemistry communications* 2 (9): 641–645.

45 Aurbach, D., Lu, Z. et al. (2000). Prototype systems for rechargeable magnesium batteries. *Nature* 407 (6805): 724–727.

46 Wang, D., Gao, X., Chen, Y. et al. (2018). Plating and stripping calcium in an organic electrolyte. *Nature materials* 17 (1): 16–20.

47 Chen, M., Xie, S. et al. (2022). Aqueous zinc-ion batteries at extreme temperature: mechanisms, challenges, and strategies. *Energy Storage Materials* 51: 683–718.

48 Wang, D., Lv, D., Peng, H. et al. (2022). Site-selective adsorption on ZnF_2/Ag coated Zn for advanced aqueous zinc–metal batteries at low temperature. *Nano Letters* 22 (4): 1750–1758.

49 Jia, H., Wang, Z., Tawiah, B. et al. (2020). Recent advances in zinc anodes for high-performance aqueous Zn-ion batteries. *Nano Energy* 70: 104523.

50 Gao, X., Wu, H., Li, W. et al. (2020). H+-Insertion boosted α-MnO_2 for an aqueous Zn-ion battery. *Small* 16 (5): 1905842.

51 Srinivasan, V. and Wang, C. (2002). Analysis of electrochemical and thermal behavior of Li-ion cells. *Journal of The Electrochemical Society* 150 (1): A98.

52 Zhang, Q., Ma, Y., Lu, Y. et al. (2020). Modulating electrolyte structure for ultralow temperature aqueous zinc batteries. *Nature communications* 11 (1): 4463.

53 Zhu, K., Li, Z., Sun, Z. et al. (2022). Inorganic electrolyte for low-temperature aqueous sodium ion batteries. *Small* 18 (14): 2107662.

8

Li-O_2 and Li-CO_2 Batteries

Fan Wang[1] *and Minghua Chen*[2]

[1]*Harbin University of Science and Technology, School of Electrical and Electronic Engineering, No. 52 Xuefu Road, Harbin, 150080, China*
[2]*Harbin University of Science and Technology, School of Electrical and Electronic Engineering, Key Laboratory of Engineering Dielectric and Applications (Ministry of Education), No. 52 Xuefu Road, Harbin, 150080, China*

8.1 Introduction

In the early twenty-first century, the global economy experienced steep growth, resulting in a gradual enhancement in human living standards. However, this progress came at the cost of substantial energy consumption, leading to severe environmental challenges. Currently, more than 80% of the energy essential for global economic advancements is derived from the combustion of traditional fossil fuels, including coal, oil, and natural gas [1]. Notably, these fossil fuel resources are nonrenewable, and their combustion produces a large amount of carbon dioxide, dust, and acid gas, which causes problems such as the "greenhouse effect" and air pollution on a global scale [2]. Therefore, research and development of green, pollution-free, and renewable new energy has become an immediate need that has to be addressed with the goals of optimizing the allocation of energy resources, addressing environmental issues, and developing a green economy.

Recently, advances in science and technology have propelled significant progress in the development of various new forms of renewable energy, including wind energy, solar energy, tidal energy, and nuclear energy, which have gained traction due to their clean, safe, and environmentally friendly characteristics. However, large-scale production of renewable energy requires huge infrastructure and high cost and is easily affected by climate and geographical location, making it highly limited in practical applications. In particular, as portable electronic devices become ubiquitous and vehicles utilizing new forms of energy experience rapid development, their practical applications demand adherence to criteria such as compact size, increased portability, superior energy density, and ability to reuse. The development of new energy conversion devices has become a key breakthrough to solving these problems. Common electrical energy storage and conversion devices

Towards Next Generation Energy Storage Technologies: From Fundamentals to Commercial Applications, First Edition. Edited by Minghua Chen.

include supercapacitors [3], fuel cells [4], solar cells [5], and secondary batteries, which have been used in different fields. Among them, the lithium-ion secondary battery stands out for its high efficiency, simplicity, and robust reliability, making it the most extensively researched energy storage device in the field of electrical energy storage. It can seamlessly convert energy between electrical and chemical forms through reversible electrochemical redox reactions, fulfilling the essential goal of energy storage and conversion. Since their commercial launch by Sony in 1991, lithium-ion batteries have found an extremely wide range of applications in various fields, including portable electronic devices, pure electric vehicles, and hybrid electric vehicles. However, the energy density of lithium-ion batteries using intercalated materials is not sufficient for long-distance transportation (i.e. 300 km).

Recently, interest in Li-O_2 and Li-CO_2 batteries has increased due to their remarkable theoretical capacities and energy densities, making them particularly well suited for applications in remote transportation. Abraham and Jiang designed the first rechargeable Li-O_2 battery in 1996, although it drew little interest until Bruce and coworkers showed the recharge ability of Li_2O_2. Li-O_2 cells possess significant strength because oxygen gas can be acquired from ambient air. Li ions react with oxygen as follows: $O_2 + 2Li^+ + 2e^- \rightarrow Li_2O_2$, and Li_2O_2 is the discharge product with an equilibrium potential of 2.96 V and a high specific energy of 3600 Wh kg^{-1}. Li-CO_2 batteries are the optimal choice due to their relatively high discharge potential (2.8 V) and impressive theoretical specific energy density (1876 Wh kg^{-1}), as derived from the reaction of $4Li + 3CO_2 \rightarrow 2Li_2CO_3 + C$. Typically, both Li-$O_2$ and Li-CO_2 batteries use Li metal as anode because Li possesses the largest capacity (3860 mAh g^{-1}) and the most negative electrochemical potential (−3.04 V) versus standard hydrogen electrodes. These observations highlight the exceptional potential of Li-O_2 and Li-CO_2 batteries in the field of next-generation energy storage systems for pure electric vehicles and hybrid electric vehicles.

Li-O_2 and Li-CO_2 batteries have achieved significant breakthroughs in the past decade. Nevertheless, they face some serious challenges that need to be addressed to achieve practical viability: (i) the reaction mechanisms of these batteries are unclear; (ii) the types and catalytic activity of cathode material/catalyst are relatively limited; (iii) the stability of electrolytes is insufficient; and (iv) metal anodes are slowly degraded during reactions. These undesired chemical characteristics are internalized in the primary reaction of Li-O_2 and Li-CO_2 batteries, resulting in limited cycling stability and suboptimal energy conversion efficiency. To address these issues, a comprehensive understanding of the reaction mechanism of Li-O_2 and Li-CO_2 batteries is required, and the root causes of these issues need to be revealed. Recently, a few review articles have focused on the progress of Li-O_2 and Li-CO_2 batteries. Xie et al. reviewed catalytic materials that can be applied in different battery reactions. Zhou et al. summarized the development course, challenges, and research directions of rechargeable Li-CO_2 batteries. Nevertheless, a summary of the electrochemical reaction mechanism, the protection of the Li anode, and the effect of electrolyte stability is limited in these articles. Hence, the most recent advances in all aspects of Li-O_2 and Li-CO_2 batteries are discussed in this chapter, including the reaction mechanism, the design strategy of the cathode material/catalyst, the construction of the metal anode–electrolyte interface, the influence of the electrolyte on the reaction mechanism, and the external factors that affect its reaction during the discharge–charge process (working atmosphere, temperature, current density). Finally, the opportunities, challenges, and future directions of Li-O_2 and Li-CO_2 batteries are proposed.

8.2 The Mechanism for Li-O_2 and Li-CO_2 Batteries

8.2.1 Classification of Li-O_2 Batteries

From a structural point of view, Li-O_2 batteries are primarily composed of four components: cathode, electrolyte, separator, and lithium anode. Li-O_2 batteries of different systems differ in their structure. Cathode materials are generally porous, with good electrical conductivity and excellent catalytic performance. At present, cathodes primarily consist of carbon-based materials such as carbon nanotubes (CNTs), graphene, and porous carbon, and metal compounds such as metal oxides, carbides, and nitrides, along with their various composites. Electrolytes are categorized based on their physical properties into liquid, gel, and solid electrolytes. The separator can be a polyethylene film, polypropylene film, or a glass fiber separator with good adsorption to the electrolyte. In addition, solid electrolytes can not only serve as lithium-ion conductors but also act as separators due to their poor electronic conductivity. As the anode part of the Li-O_2 battery, the lithium source can be metallic lithium or a lithium alloy.

Generally, based on the electrolytes used in the battery, Li-O_2 batteries can be classified into four types, as shown in Figure 8.1 [6]: nonaqueous electrolyte systems (Figure 8.1a), aqueous electrolyte systems (Figure 8.1b), solid-state electrolyte (SSE) systems (Figure 8.1c), and organic–aqueous hybrid electrolyte systems (Figure 8.1d).

8.2.1.1 Nonaqueous Electrolyte Li-O_2 Batteries

Organic Li-O_2 batteries are also referred to as nonaqueous Li-O_2 batteries. The cathode is a porous material, the anode is metal Li, and the electrolyte is an organic electrolyte, such as carbonate electrolytes, ether electrolytes, nitrile electrolytes, amide electrolytes, sulfone electrolytes, and ionic liquids. The organic Li-O_2 battery was first proposed by Abraham in 1996, in which an organic polymer electrolyte was used, which addressed the corrosion problem of anode lithium to a certain extent [7]. Although the charging/discharging progress of organic Li-O_2 batteries is relatively complex, it is currently the major research direction because of their simple structure, convenient fabrication, and relatively stable performance. The overall reaction process is shown below [8]:

$$2Li + O_2 \leftrightarrow Li_2O_2, (E^0 = 2.96\,\text{V versus Li}^+/\text{Li}) \quad (8.1)$$

The open-circuit voltage of an organic Li-O_2 battery is 2.96 V. The discharging phase corresponds to the oxygen reduction reaction (ORR), where metal Li and O_2 react to generate solid Li_2O_2 and deposit on the porous cathode material; the charging phase corresponds to the oxygen evolution reaction (OER), where deposition of

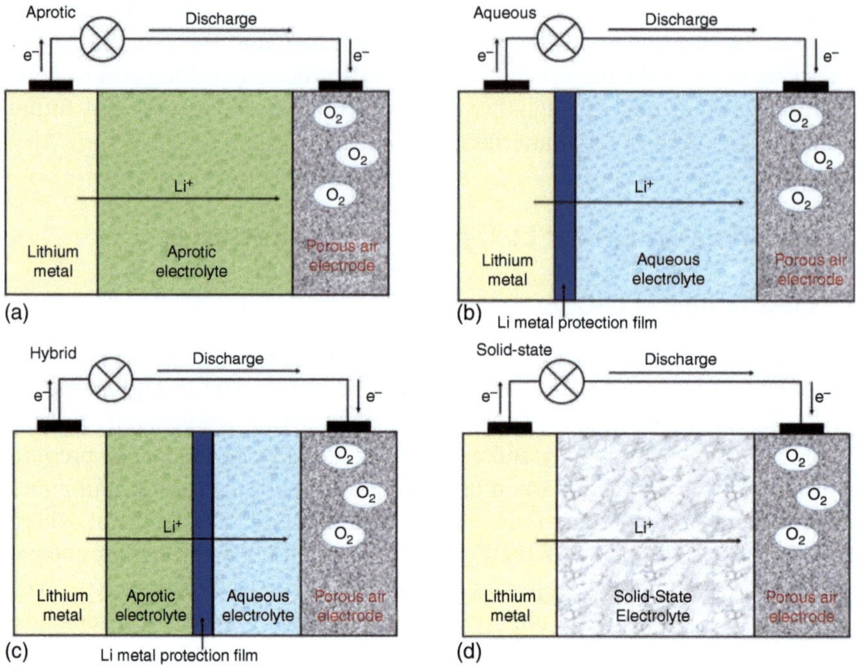

Figure 8.1 Schematic illustration of the four different architectures of Li-O_2 and Li-CO_2 batteries. Source: Reproduced with permission from Lu et al. [6]. © 2014/American Chemical Society.

Li_2O_2 on the cathode material gradually decomposes into Li^+ and O_2, to realize the reversible cycle of the battery.

Compared with the other three Li-O_2 battery systems, organic electrolytes have relatively ideal lithium ion conductivity, high lithium salt solubility, and good protection and compatibility for lithium anodes [9]. Usually, organic electrolytes react with the Li surface to form a solid electrolyte interface (SEI) film, which protects the anode to a certain extent. However, they also have some disadvantages that cannot be ignored, such as poor stability during the charging and discharging process, being prone to side reactions, having limited ability to dissolve and transport discharge products Li_2O_2 and oxygen, and some electrolytes being highly toxic and volatile, all of which limit the performance of organic Li-O_2 batteries to a certain extent. In addition, the discharge product of organic Li-O_2 batteries is solid Li_2O_2, which is deposited on the cathode during the ORR process, causing the blockage of the pores of the cathode material; moreover, solid Li_2O_2 is difficult to decompose during the OER process, which usually causes relatively high charging overpotential. Therefore, while designing cathode materials, porosity and a substantial specific surface area should be taken into account to facilitate the deposition and storage of Li_2O_2 during the discharge phase [10]. The cathode material should also possess proficient OER catalytic capabilities to effectively decompose Li_2O_2, thereby reducing battery overvoltage, which is crucial for enhancing energy efficiency and cycle capacity in the battery.

8.2.1.2 Aqueous Electrolyte Li-O_2 Batteries

The aqueous Li-O_2 battery was first proposed by E. L. Littauer in 1976 [11]. Due to technological limitations at that time, metal Li was easily corroded by the aqueous solution to generate H_2, resulting in extremely poor electrochemical performance of the battery. Therefore, the biggest problem facing aqueous Li-O_2 batteries is constructing a lithium anode that is stable in water. In 2004, Visco et al. proposed a three-layer composite Li-O_2 battery for the first time. They used water-stable NASICON lithium-ion solid glass ceramics (LATP) as a protective layer, which prevented direct contact between lithium and water-based electrolyte, thus solving the problem of the lithium anode being corroded by the water-based electrolyte [12].

As shown in Figure 8.1b, the structure of the aqueous Li-O_2 battery has four parts: lithium anode, lithium-ion solid conductor separator, aqueous electrolyte, and porous cathode [13]. According to the acidity and alkalinity of the aqueous electrolyte, its electrochemical reaction process is as follows [14]:

$$\text{Alkaline solution: } 2\text{Li} + 1/2\text{O}_2 + 2\text{H}_2\text{O} \rightarrow 2\text{LiOH} \tag{8.2}$$

$$\text{Acidic solution: } 2\text{Li} + 1/2\text{O}_2 + 2\text{H}^+ \rightarrow 2\text{Li}^+ + \text{H}_2\text{O} \tag{8.3}$$

The discharge by-product in aqueous Li-O_2 batteries is soluble LiOH, characterized by its restricted solubility within the aqueous electrolyte. When the concentration of LiOH in the solution reaches saturation, crystallization occurs at the bottom of the electrolyte, forming a precipitate. Nevertheless, the solubility of LiOH in the electrolyte can well solve the problem of discharge products blocking the pores of the cathode material, so that the discharge process can continue. Although the addition of a lithium-ion solid conductor separator can prevent the lithium anode from being corroded by moisture in the electrolyte and gas molecules such as N_2 and CO_2 in the air, the lithium-ion solid conductor separator itself has a relatively low lithium-ion conductivity. The unstable contact of the anode and the high impedance of the contact interface limit the research and development of aqueous Li-O_2 batteries to a certain extent.

8.2.1.3 All-Solid-State Li-O_2 Battery

The structure of an all-solid-state Li-O_2 battery is relatively simple, as shown in Figure 8.1c, which is mainly composed of three parts: lithium anode, SSE, and porous cathode [15]. The SSE is the core of the all-solid-state Li-O_2 battery, acting not only as a lithium-ion conductor but also as a separator. Therefore, its physicochemical properties directly influence the overall electrochemical performance of the battery. Based on their composition, solid electrolytes can be categorized into two major types: solid polymer electrolytes and solid inorganic electrolytes. Solid polymer electrolytes are formed by dispersing the polymer material in the solvent, simultaneously adding lithium salt, and finally curing it. Solid inorganic electrolytes are prepared by mixing inorganic salt precursors in proportion or mixing organic salt precursors for gelation treatment, and then subjecting them to high-temperature calcination, high-energy ball milling, compaction, crystallization, and other processes. Some examples of solid inorganic electrolytes include sodium super ion conductor (NASICON) type, lithium super ion conductor (LISICON) type, perovskite type, and garnet type [16]. The reaction process of an all-solid-state Li-O_2 battery is as follows:

$$2Li + O_2 \leftrightarrow Li_2O_2, (E^0 = 3.10\,V\text{ versus } Li^+/Li) \tag{8.4}$$

The research on all-solid-state Li-O_2 batteries was first reported by Kumar et al. in 2010. These batteries used lithium metal as the anode, a solid electrolyte made of glass ceramics and polymer ceramic materials as the lithium-ion conductor, and a cathode prepared by carbon material and glass ceramic powder that conducts ions [17]. In the temperature range of 30–105 °C, these batteries showed excellent thermal stability and rechargeability. At current densities ranging from 0.05 to 0.25 mA cm^{-2}, they showed the ability to undergo 40 cycles with minimal polarization evident in the charge and discharge curves. However, as the temperature decreased, the ionic conductivity of the solid electrolyte decreased significantly, leading to a synchronous decay of the discharge capacity.

In contrast to various Li-O_2 battery systems, all-solid-state Li-O_2 batteries offer notable advantages. The solid electrolyte has relatively high stability, thereby markedly enhancing battery safety and reducing the likelihood of side reactions compared with conventional liquid electrolytes. It prevents the influence of water, CO_2, N_2, and other gases in the air on the battery, thereby effectively protecting the lithium anode and improving the rechargeable capacity and circulation capacity of the battery. It also prevents the formation of Li dendrites and reduce the risk of battery short circuits [18]. However, all-solid-state Li-O_2 batteries face some challenges. Further enhancements are required, such as improving the lithium-ion conductivity of the SSE, addressing the relatively high interface impedance between the electrolyte and the cathode/anode, and optimizing the rate performance and energy density, which are not currently ideal [19].

8.2.1.4 Organic–Aqueous Hybrid Electrolyte Li-O_2 Batteries

In aqueous Li-O_2 batteries, the incorporation of solid electrolytes effectively addresses the corrosion issue of Li in the anode, yet it presents two drawbacks. Firstly, direct contact between the anode lithium and the solid electrolyte triggers a reaction, compromising the stability of the solid electrolyte. Secondly, contact between the anode lithium and the solid electrolyte lacks complete bonding, resulting in a relatively high interface resistance between the two. In organic Li-O_2 batteries, the solubility of the discharge product Li_2O_2 in the organic electrolyte is low, and its deposition gradually blocks the pores of the cathode, resulting in the termination of the battery discharge process. In addition, Li_2O_2 and LiO_2 produced in organic Li-O_2 batteries have strong oxidizing properties, leading to the decomposition of the electrolyte and carbon-containing cathode materials [20]. To address these difficulties, Zhou Haoshen et al. [21] proposed an organic–aqueous hybrid electrolyte Li-O_2 battery in 2010. The structure of the battery included a

Li cathode, organic electrolyte, LISICON glass separator, aqueous electrolyte, and porous cathode. The organic electrolyte was 1 M $LiClO_4$ dissolved in ethylene carbonate/dimethyl carbonate, and the aqueous electrolyte was 1 M KOH solution in a volume of 10 ml. The battery was continuously discharged in the air for 500 hours, and the test data showed that the organic–aqueous mixed electrolyte Li-O_2 battery has characteristics similar to fuel cells, with an ability to continuously reduce O_2. During deep discharge, the battery showed an impressive specific capacity of up to 50,000 mAh g^{-1}, accompanied by excellent ORR performance. The structure of an organic–aqueous mixed electrolyte Li-O_2 battery is shown in Figure 8.1d, which is primarily composed of a lithium anode, a solid lithium ion conductor, and a cathode. One side of the lithium anode is filled with an organic electrolyte, and the cathode side is filled with an aqueous electrolyte. The reaction process of the organic–aqueous mixed electrolyte system Li-O_2 battery is as follows [22]:

$$\text{Alkaline solution: } 4Li + O_2 + 2H_2O \rightarrow 4LiOH \tag{8.5}$$

$$\text{Acidic solution: } 4Li + O_2 + 4H^+ \rightarrow 4Li^+ + 2H_2O \tag{8.6}$$

An organic–aqueous hybrid electrolyte Li-O_2 battery has the characteristics of both an organic Li-O_2 battery and an aqueous Li-O_2 battery. Similar to organic electrolytes, it can solve the problem of instability of the solid lithium-ion conductor due to the contact between Li and solid lithium-ion conductor. It can also fully infiltrate the surface of the lithium anode and the solid lithium-ion conductor, reducing the interface impedance. Moreover, it can also react with the lithium surface to form a stable SEI film, which can protect the lithium anode to a certain extent. Similar to aqueous electrolytes, the product formed during the discharge process is soluble LiOH, which can effectively solve the problem of cathode pore blockage. Due to the facile decomposition of LiOH, hybrid Li-O_2 batteries have a relatively low charging platform and minimal overpotential. However, organic–aqueous mixed electrolyte Li-O_2 batteries encounter challenges such as a complex battery structure and relatively high material costs. The lithium-ion conductivity of solid lithium-ion conductors needs to be improved, and their stability in a mixed electrolyte environment is lacking, resulting in poor cycle performance. Furthermore, solid lithium-ion conductors require high mechanical strength and compactness to effectively prevent electrolyte penetration and damage.

8.2.2 Reaction Mechanism of Charge and Discharge Process of Organic Li-O_2 Batteries

The electrochemical reaction of organic Li-O_2 batteries usually involves two processes: the reduction reaction of O_2 to Li_2O_2 during discharge and the oxidation reaction of Li_2O_2 to precipitate O_2 during charging. In the ORR and OER stages, the formation of intermediate products (O^{2-} or LiO_2) means that the mechanism of charge–discharge reactions is more complicated [23]. Therefore, during the reaction stages of Li-O_2 batteries, the formation and decomposition of end products and intermediate products usually involve multistep reactions [24].

8.2.2.1 Discharge Reaction (ORR) Mechanism of Organic Li-O_2 Batteries

The discharge mechanism of Li-O_2 batteries is closely associated with factors such as current density, overpotential, and whether the intermediate product LiO_2 is dissolved in the electrolyte or adhered to the electrode surface [25]. Under high current density and overpotential, O_2 is reduced to Li_2O_2, which is deposited on the electrode surface in the form of a thin film [26]. In contrast, at lower overpotential, whether Li_2O_2 is deposited on the electrode in the form of a thin film or nucleates and grows into large particles depends on the solubility of LiO_2 in the electrolyte [27]. As the first step in the discharge process, O_2 dissolved in the electrolyte acquires an electron on the electrode surface and then combines with Li^+ to form LiO_2. Depending on the difference in the solvation degree of the cation produced by the solvent to LiO_2, LiO_2 may be dissolved in the solvent to form $Li^+{}_{(sol)}$ and $O^{2-}{}_{(sol)}$, or it may be adsorbed on the electrode surface to form LiO_2*. These different processes have an obvious impact on the subsequent discharge process. Therefore, based on the strength of Li^+ solvation in the electrolyte, the discharge process of organic Li-O_2 batteries, i.e. the growth process of Li_2O_2, can be classified into two categories: solution growth mechanism and electrode surface growth mechanism [24, 25b, 27].

Solution Growth Mechanism In the first step, $O_{2(sol)}$ dissolved in the electrolyte acquires an electron on the electrode surface and then combines with Li^+ to form the intermediate product LiO_2. In a strong Li^+ solvation solution, LiO_2 breaks away from the electrode surface and dissolves into the electrolyte, and $Li^+{}_{(sol)}$ and $O^{2-}{}_{(sol)}$ are formed (Eq. (8.7)). In the second step, $Li^+{}_{(sol)}$ and $O^{2-}{}_{(sol)}$ in the electrolyte undergoes a disproportionation reaction to form Li_2O_2 and O_2 (Eq. (8.8)). They can also continue to acquire an electron on the electrode surface and then be reduced to Li_2O_2 (Eq. (8.9)). The electrochemical reaction steps of the solution growth mechanism are as follows:

$$Li^+_{(sol)} + e^- + O_{2(sol)} \rightarrow Li^+_{(sol)} + O^{2-}_{(sol)} \tag{8.7}$$

$$2\left(Li^+_{(sol)} + O^{2-}{}_{(sol)}\right) \rightarrow Li_2O_2 + O_2 \tag{8.8}$$

$$2Li^+{}_{(sol)} + e^- + O^{2-}{}_{(sol)} \rightarrow Li_2O_2 \tag{8.9}$$

Mechanism of Electrode Surface Growth In the first step, similar to the first mechanism, $O_{2(sol)}$ dissolved in the electrolyte acquires an electron on the electrode surface and then combines with Li^+ to form an intermediate product LiO_2. But in a weak Li^+ solvation solution, LiO_2 is adsorbed on the electrode surface, forming LiO_2* (Eq. (8.10)). In the second step, LiO_2* adsorbed on the electrode surface undergoes a disproportionation reaction to generate Li_2O_2* and O_2 (Eq. (8.11)); it also acquires an electron and Li^+ from the electrode and electrolyte, respectively, and is reduced to Li_2O_2* (Eq. (8.12)). The reaction steps of the electrode surface growth mechanism are as follows [24]:

$$Li^+ + e^- + O_{2(sol)} \rightarrow LiO_2{}^* \tag{8.10}$$

$$2LiO_2{}^* \rightarrow Li_2O_2{}^* + O_2 \tag{8.11}$$

$$LiO_2{}^* + Li^+ + e^- \rightarrow Li_2O_2{}^* \tag{8.12}$$

Other ORR Mechanisms In addition, studies reported that during the discharge process of Li-O_2 batteries, O_2 can be reduced to Li_2O through a four-electron transfer process, which can be expressed as follows [28]:

$$Li_2O_2 + 2e^- + 2Li^+ \rightarrow 2Li_2O \tag{8.13}$$

$$4Li^+ + 4e^- + O_2 \rightarrow 2Li_2O \tag{8.14}$$

When the discharge product is Li_2O, the number of transferred electrons is the highest ($4e^-$), exceeding LiO_2 ($1e^-$) and Li_2O_2 ($2e^-$), which means that the reactions presented in Eqs. (8.13) and (8.14) can provide more charges, and the highest energy density is achieved. However, the production of Li_2O is generally irreversible, and it is difficult to decompose, which can easily result in a high charging overpotential, which in turn leads to the side reaction of the electrode during the charging process, and the released product is not O_2 [29]. Therefore, taking all these factors into account, the most suitable discharge product for Li-O_2 batteries is Li_2O_2.

8.2.2.2 The Charging Reaction (OER) Mechanism of Organic Li-O_2 Batteries

The OER stage corresponds to the oxidative decomposition of Li_2O_2 and the process of releasing O_2. Since Li_2O_2 decomposition is easily affected by the type of electrolyte, the formation mechanism of Li_2O_2, its deposition morphology, and the side reactions caused by the decomposition of the electrolyte and carbon materials, the charging process of Li-O_2 batteries is relatively complicated. Based on the decomposition steps of Li_2O_2, it can be classified into the following three types:

Charging Process, OER Mechanism 1 [30] During charging, Li_2O_2 decomposes at its two-phase interface with the electrolyte, releasing O_2 and Li^+ from the Li_2O_2 surface (Eq. (8.15)). The whole decomposition reaction is completed in one step, during which no intermediate product (LiO_2) is formed, as follows:

$$Li_2O_2 \rightarrow O_2 + 2Li^+ + 2e^- \tag{8.15}$$

Charging Process, OER Mechanism 2 [31] During charging, Li_2O_2 first decomposes into Li^+ and intermediate LiO_2 groups (Eq. (8.16)). LiO_2 then continues to decompose, producing Li^+ and releasing O_2 (Eq. (8.17)). OER steps are as follows:

$$Li_2O_2 \rightarrow LiO_2 + Li^+ + e^- \tag{8.16}$$

$$LiO_2 \rightarrow O_2 + Li^+ + e^- \tag{8.17}$$

Charging Process, OER Mechanism 3 [32] Driven by the solid solution reaction of $Li_{2-x}O_2$ in the lithium-deficient phase, the oxidative decomposition of Li_2O_2 and the release of O_2 include two stages. In the first stage, amorphous Li_2O_2 is first oxidized in the voltage range of 2.8–3.4 V, and some Li^+ vacancies are generated on the Li_2O_2 interface, forming $Li_{2-x}O_2$ in the Li-deficient phase (Eq. (8.18)). In the second stage, crystalline Li_2O_2 undergoes oxidative decomposition in the voltage range of 3.4–3.9 V to precipitate O_2, which is driven by the solid solution reaction of $Li_{2-x}O_2$ (Eq. (8.19)).

$$Li_2O_2 \rightarrow Li_{2-x}O_2 + xLi^+ + xe^- \quad (8.18)$$

$$Li_{2-x}O_2 \rightarrow O_2 + (2-x)Li^+ + (2-x)e^- \quad (8.19)$$

8.2.3 The Mechanism of Li-CO_2 Batteries

Electrochemical reactions in Li-CO_2 batteries are complex, encompassing various interfacial reactions between gaseous active materials, electrolytes, catalysts, and reaction products. The working principle of a Li-CO_2 battery is similar to that of a Li-O_2 battery. During the discharge process, lithium at the anode releases electrons, forming Li^+ ions that migrate toward the cathode due to the electric field. At the cathode–electrolyte interface, CO_2 dissolved in the electrolyte obtains electrons from the cathode and further combines with Li^+ to form a discharge product, i.e. the reduction process of CO_2. During charging, the discharge products are decomposed and CO_2 and Li^+ are released, which is the evolution of CO_2. The reaction process varies depending on the discharge product of the battery. Based on the different discharge products, the following three reaction mechanisms can be summarized:

8.2.3.1 $CO_2 \leftrightarrow Li_2CO_3 + C$ Reaction Mechanism

Based on the inaugural operation of the first Li-CO_2 battery, researchers found that the primary components of the discharge product are Li_2CO_3 and carbon (C) through various characterization methods. Building upon the discharge products Li_2CO_3 and C of Li-CO_2 batteries and the established disproportionation mechanism of lithium superoxide in Li-O_2 batteries, Chen et al. [33] proposed a reasonable disproportionation reaction mechanism during the discharge process of a Li-CO_2 battery:

$$2CO_2 + 2e^- \rightarrow C_2O_4^{2-} \quad (8.20)$$

$$C_2O_4^{2-} \rightarrow CO_2^{2-} + CO_2 \quad (8.21)$$

$$C_2O_4^{2-} + CO_2^{2-} \rightarrow 2CO_3^{2-} + C \quad (8.22)$$

$$CO_3^{2-} + 2Li^+ \rightarrow Li_2CO_3 \quad (8.23)$$

CO_2 is first reduced to $C_2O_4^{2-}$ on the cathode surface through a one-electron process (Eq. (8.20)). The unstable $C_2O_4^{2-}$ undergoes a two-step reaction to decompose into CO_3^{2-} and C (Eqs. (8.21) and (8.22)), and CO_3^{2-} combines with Li^+ to form Li_2CO_3 crystals (Eq. (8.23)). Zhou et al. [34] used the in situ Raman characterization technique to observe the discharge process of the Li-CO_2 battery, and their experimental findings were in line with the hypothesis of Chen et al. The reaction mechanism of this discharge process has been widely accepted. In the charging process, Zhou et al. [35] used a Li_2CO_3-C electrode to simulate the cathode of the Li-CO_2 battery after discharge and proved that the charging process was realized by the typical reaction of Li_2CO_3 and C through the surface-enhanced Raman characterization technique and gas chromatography–mass spectrometry. Therefore, the charging reaction of CO_2 can be expressed as follows [36]:

$$2Li_2CO_3 + C \rightarrow 3CO_2 + 4Li^+ + 4e^- \quad (8.24)$$

Combined with the reaction mechanism of the charge–discharge process, Li-CO_2 batteries are generally considered to follow four-electron reactions. Nevertheless, the precise reaction mechanism remains contentious. This reversible process necessitates a charging voltage exceeding 4 V, a condition prone to inducing numerous side reactions.

8.2.3.2 $CO_2 \leftrightarrow Li_2C_2O_4$ Reaction Mechanism

Lithium oxalate ($Li_2C_2O_4$) has been analyzed as an intermediate discharge product in previous investigations of Li-CO_2 batteries [33, 34]. Notably, some studies have demonstrated that Mo_2C catalysts can stabilize $Li_2C_2O_4$, act as an intermediary product [37], and prevent its subsequent transformation into Li_2CO_3 and C, thus reducing the charging voltage. This, in turn, facilitates the occurrence of the reversible reaction, thereby achieving a reversible Li-CO_2 battery. Chen et al. [33] showed that after discharge, amorphous $Li_2C_2O_4$-Mo_2C is formed, which is easier to decompose than Li_2CO_3. The formation of amorphous $Li_2C_2O_4$-Mo_2C can be attributable to accelerated reaction kinetics, lower voltage plateaus, and a higher likelihood of the occurrence of charging reactions. The discharge process of the battery is shown in Eqs. (8.25) and (8.26):

$$2CO_2 + 2e^- \rightarrow C_2O_4^{2-} \quad (8.25)$$

$$C_2O_4^{2-} + 2Li^+ + Mo_2C \rightarrow Li_2C_2O_4 - Mo_2C \quad (8.26)$$

Subsequently, Zhou et al. used Mo_2C nanoparticle cathodes to prepare a quasi-solid-state, flexible, fiber-shaped Li-CO_2 battery characterized by a low overpotential and high energy efficiency (Figure 8.2). The synergy between the CNT substrate and Mo_2C contributes to achieving a low charging voltage of less than 3.4 V. Finally, density functional theory (DFT) calculation shows that the electrons generated by low-valence Mo atoms, facilitated by the Mo-O coupling bridge in Mo_2C, play a stabilizing role for the formation of amorphous intermediate product $Li_2C_2O_4$. Compared with Li_2CO_3, which has high stability, thermodynamically unstable $Li_2C_2O_4$ is more susceptible to decomposition, as shown in Eq. (8.27):

$$Li_2C_2O_4 \rightarrow 2Li^+ + 2CO_2 + 2e^- \quad (8.27)$$

In summary, the reversible reaction process of Li-CO_2 batteries can be divided into two categories: using Li_2CO_3 and amorphous C as the final discharge products, which accounts for the majority; and termination by the fixation of intermediate products, which applies only to batteries with Mo_2C and Mo_2N cathode catalysts. No other cathode catalysts than Mo_2C and Mo_2N have been found to follow this phenomenon.

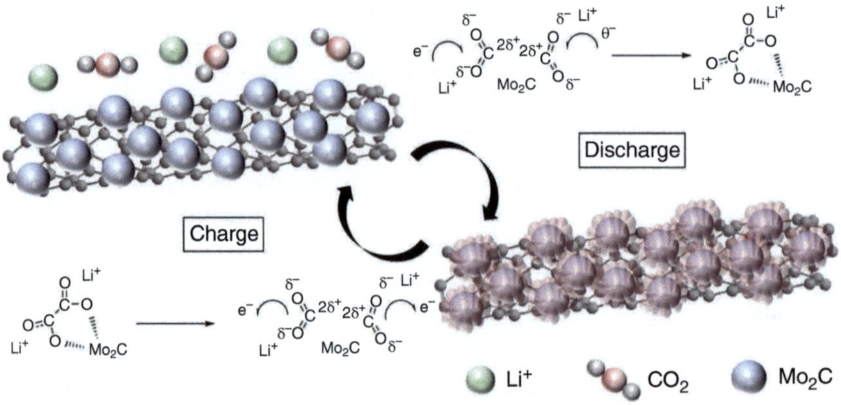

Figure 8.2 Diagram of Mo_2C/CNT cathode during the discharge–charge process in a Li-CO_2 battery. Source: Reproduced with permission from Hou et al. [33]. © 2017/John Wiley & Sons.

8.2.3.3 CO_2↔CO Reaction Mechanism

In the field of electrochemical CO_2 reduction research, various commercial chemicals can serve as indirect precursors for producing carbon-based materials. This involves the conversion of CO_2 to carbon monoxide (CO), followed by subsequent reduction processes. It is widely acknowledged that the direct synthesis of CO typically requires inexpensive or low-selectivity catalysts, thus making CO a crucial industrial feedstock [38]. Consequently, Li-CO_2 battery systems, which produce CO gas, are considered pivotal platforms for future endeavors in harnessing valuable chemicals and fuels. Wang et al. [39] realized CO gas as the discharge product of Li-CO_2 batteries using a 3D porous fractal zinc cathode. These observations not only showcase the feasibility of deriving high-value carbon-based commodities from Li-CO_2 battery systems but also underscore the need for further investigation into the specific reaction mechanism.

8.3 Design Strategy of Cathode Materials

Based on the reaction mechanism of Li-O_2 and Li-CO_2 batteries, the air cathode has four main functions: (i) to catalyze the formation and decomposition of discharge products at the active site; (ii) to transfer lithium ions and gases to the active site through porous channels; (iii) to provide storage space for discharge products; and (iv) to support the growth and morphological evolution of discharge products induced on the electrode surface. The air cathode plays a crucial role in determining both the energy storage and power capacity of Li-O_2 and Li-CO_2 batteries. Numerous studies highlight that optimizing the cathode – focusing on factors such as catalytic activity, surface properties, and the adsorption capacity of lithium ions and gases – can significantly enhance the performance of Li-O_2 and Li-CO_2 batteries.

Guided by the principles of air cathode selection, various cathode structures and catalyst materials have been extensively investigated. The following section delves into the research advancements in the cathodes of Li-O_2 and Li-CO_2 batteries, covering aspects such as the selection of efficient catalysts, structural design, and the development of functional cathodes.

8.3.1 Cathode Structural Design

An ideal cathode should have the following characteristics: high electronic and ionic conductivity, reasonable structural design, fast gas diffusion, high catalytic activity (including CO_2/O_2 capture, limiting the accumulation of discharge products, and promoting the decomposition of discharge products), and high electrochemical stability. The cathode usually consists of porous carbon, catalyst, and binder, whose compositions can vary, which in turn significantly affects the conductivity, porosity, catalytic activity, and electrolyte wettability of the cathode, resulting in variable electrochemical performance. Specific surface area and porosity of an electrode material are crucial factors, as the battery's total capacity and energy density depend on the amount of the discharge product that can be stored in the porous electrode structure [40]. When an aqueous solution is used as the electrolyte in Li-O_2 and Li-CO_2 batteries, the discharge product gets dissolved in the electrolyte. When a nonaqueous aprotic electrolyte is used, the solubility of the discharge product is low, and it gets deposited on the porous structure. Therefore, the cathode structure in a nonaqueous aprotic electrolyte system is more strict, i.e. it provides a larger pore volume to deposit solid discharge products and prevents CO_2/O_2 as well as electrolyte diffusion pathways from being clogged. In this regard, a structure characterized by high porosity and an extensive specific surface area can provide abundant catalytically active sites and effective mass transfer channels, facilitating the dissolution and diffusion of CO_2/O_2 in the electrolyte (Figure 8.3). Thus, in all Li-O_2 and Li-CO_2 batteries studied to date, porous carbon has been used as the cathode material. The porous structure, which consists of macropores, mesopores, and micropores, is beneficial to the wetting of electrolytes, dissolution of CO_2/O_2, and mass transfer between CO_2/O_2 and electrolytes [40]. Among them, mesopores and macropores contribute the most to the capacity of Li-O_2 and Li-CO_2 batteries, providing sufficient space for the deposition of solid products. Micropores transport gases effectively, but they are easily blocked by concrete discharge products and thus have a small contribution to capacity. Hence, microstructure significantly influences the electrochemical performance of Li-O_2 and Li-CO_2 batteries. However, theoretical calculations have proved that an overlarge pore structure reduces the volumetric energy density of the cathode [41]. Hence, the optimized pore size should be verified in future studies. Furthermore, the cathode catalyst constructed using the traditional coating method has various drawbacks such as increased resistance, buried active site, low adhesion between the catalyst and collector, and easy separation. Therefore, designing self-supported electrode materials can avoid the abovementioned issues, promote charge and ion transfer, and enhance the contact between the electrocatalyst and electrolyte, improving the durability

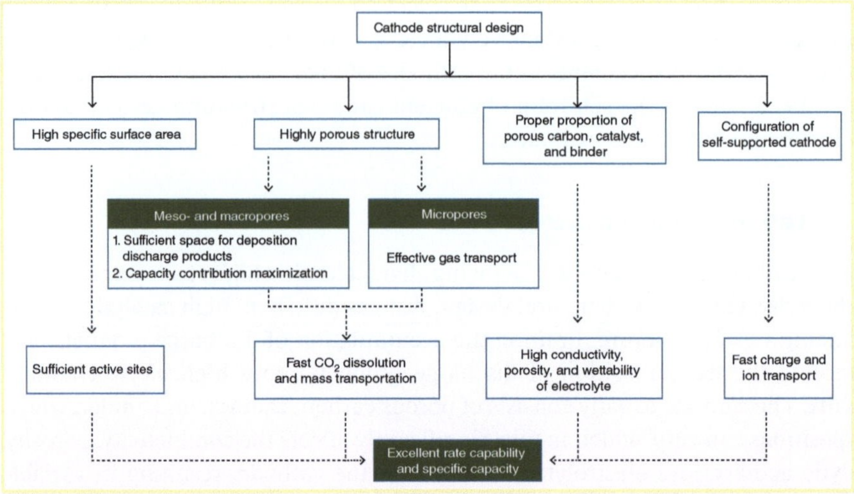

Figure 8.3 Schematic diagram of the influence of different factors on battery performance in the design of cathode structure. Source: Reproduced with permission Hou et al. [33]. © 2017/John Wiley & Sons.

and catalytic activity. In addition, the thickness of the cathode material is critical to the electrochemical performance; a thicker electrode provides a lower specific capacity due to the slow diffusion of CO_2/O_2 in the electrode immersed in the electrolyte.

8.3.2 Selection of Efficient Catalysts

Due to the slow formation and decomposition of solid discharge products (e.g. Li_2O_2 and Li_2CO_3), many types of electrocatalysts have been developed to improve the performance of nonaqueous Li-O_2 and Li-CO_2 batteries, including some cathode materials that are successfully used in electrocatalysis and other metal–air batteries. At present, solid catalysts and liquid catalysts are primarily used as electrocatalysts.

8.3.2.1 Carbon Materials and Their Derivatives

Carbonaceous materials are widely used in electrochemical storage devices due to their high conductivity, large surface area, controllable pore structure, and low cost, especially in emerging Li-O_2 and Li-CO_2 batteries. They are primarily used as metal-free catalysts or catalyst carriers for research. Different types of carbon materials have different physical, electrical, and chemical properties. Carbon atoms may be sp^2 or sp^3 hybrids. Carbon materials having sp^3-hybridized carbon with tetrahedral bonding have low electrical conductivity and poor density, which hinders their applications as catalysts. Graphite, graphene, CNTs, and other sp^2-hybridized carbon materials have high electrical conductivity and are suitable for electrode applications. To date, various types of carbon materials such as CNTs and graphene have been explored as cathode materials for Li-O_2 and

Li-CO_2 batteries [42]. Takechi et al. used KB as the cathode in the first primary Li-O_2/CO_2 batteries, which showed a high discharge capacity of 5860 mAh g^{-1}. The higher capacity was attributable to the porous structure and high surface area of carbon materials. Then, Liu and coworkers reported rechargeable Li-O_2/CO_2 (2 : 1, volume ratio) and Li-CO_2 batteries with discharging specific capacities of 1808 and 1032 mAh g^{-1}, respectively, using KB as the cathode. In addition, a primary Li-CO_2 battery as a platform for capturing CO_2 and producing electrical energy with Super-P-based cathodes showed a high discharge capacity of around 2500 mAh g^{-1} at moderate temperatures, as reported by Archer et al. Furthermore, CNTs and graphene are considered the most magnetic materials due to their high Young's modulus and tensile strength, large surface area, porous 3D networks, and abundant channels [43]. What is more, these features ensure electrode reactions with rapid charge and mass transfer processes and thus promote reaction kinetics, which can be reflected in the overpotential. Zhang and coworkers reported for the first time the introduction of graphene into Li-CO_2 batteries, resulting in a high discharge capacity of up to 14,774 mAh g^{-1}. This cathode material showed excellent electrical conductivity, porous structure, and catalytic activity, which can result in efficient diffusion channels, sufficient space, and active sites for O_2 and CO_2 utilization and capture [43b].

Although CNTs improve the electrochemical performance of Li-O_2 and Li-CO_2 batteries, due to their finite catalytic activity, the batteries have high overpotential and cycle reversibility. It is widely reported that heteroatom doping can adjust the electronic structure and alter the Fermi level of the material, which greatly influences the adsorption mode and energy between the gas molecules and the surface of the carbon material, thereby significantly improving the gas reduction kinetics [44]. In addition, heteroatom doping promotes the decomposition of solid discharge products in Li-O_2 and Li-CO_2 batteries. It is a widely used strategy to enhance the catalytic activity of Li-O_2 and Li-CO_2 batteries (such as N, P, S, and B). In particular, nitrogen has been shown to be an effective functional component or dopant to enhance activity. Nitrogen exists in the graphite network in the form of graphite, pyridine, and pyrrole. In a recent development, Li and his colleagues designed, for the first time, flexible fiber-shaped Li-CO_2 batteries with ultralong cycle life, high rate capability, and large specific capacity, using periodic bamboo-like N-doped CNT fibers as flexible, durable metal-free catalysts for both CO_2 reduction and evolution reactions. Due to high N doping, especially in the exposure of high-percentage N-6 doping configuration on the surface, of which N-6 sites can be the most efficient chemical state for preliminary conversion of CO_2 and act as an efficient bridge to drive rapid electron transfer between reactants and catalysts, the Li-CO_2 battery showed outstanding electrochemical performance with a high discharge capacity of 23,328 mAh g^{-1}, a low potential gap up to 1.96 V, and a stability over 360 cycles with good flexibility. These observations demonstrate that heteroatom doping can significantly enhance the catalytic activity of Li-O_2 and Li-CO_2 batteries.

Although heteroatom doping can significantly improve the catalytic activity of carbon materials and the cycling stability of Li-O_2 and Li-CO_2 batteries, these batteries show a high overpotential while using carbon materials as the cathodic catalyst, resulting in lower energy efficiency and poor cycling stability, especially in high-density rates. Surprisingly, Jin et al. used carbon quantum dots (CQDs) supported by holey graphene (CQD/hG) as the cathode catalyst to prepare a high-performance composite catalyst in Li-CO_2 batteries. This efficient, dual-function, metal-free catalyst was supported by CQDs with high defect content, which could support the rapid formation and decomposition of Li_2CO_3 with an overpotential of 1.02 V at 100 mA g^{-1}. These batteries showed good performance with a reversible capacity of 12,300 mAh g^{-1} under 500 mA g^{-1}, achieving excellent long-term stability as high as 235 cycles with a cutoff capacity of 500 mAh g^{-1} at 1 A g^{-1}. The excellent battery performance was attributable to the synergistic effects arising from the catalytic activity of CQDs with a high defect content and the improved electron–electrolyte transport facilitated by holey graphene sheets [45].

In conclusion, carbon-based materials play a vital role as cathode components in lithium–oxygen batteries, due to their significant specific surface area, outstanding electrical conductivity, and easy-to-build porous structure. However, due to their lack of catalytic activity, pure carbon electrodes cannot be used as sole cathode materials in high-performance lithium–oxygen batteries. Therefore, to make up for this lack of catalytic activity, finding suitable catalysts to combine with carbon-based materials is inevitable. Some of the common cathode catalysts include noble metals and their oxides, various transition metal compounds, etc., which will be described in detail below.

8.3.2.2 Precious Metals and Precious Metal Oxides

It is well known that precious metals and their composite materials are considered effective catalysts for chemical reactions due to their unique electronic configuration and good chemical/electrochemical stability. Their inherently semi-filled anti-bonding state can provide the catalyst and reactants with appropriate adsorption strength, thereby improving the reaction kinetics [46]. Ruthenium (Ru) is a widely used precious metal catalyst that shows superior catalytic activity in electrocatalysis. Zhou et al. reported the application of Ru nanoparticles deposited on Super P (Ru@Super P) as a cathode catalyst in Li-CO_2 batteries (Figure 8.4a) [35]. These batteries, featuring Ru@Super P as the cathode, showed markedly decreased charge potential and sustained operation over 80 cycles at a consistent capacity of 1000 mAh g^{-1} at 100 mA g^{-1}, which is the most robust cycling stability reported among Li-CO_2 batteries. To understand the contribution of Ru nanoparticles in the discharge–charge process, we conducted in situ surface-enhanced Raman spectroscopy and gas chromatography–mass spectrometry analyses to explore the transformation of discharge products on both Au and Au-free electrodes. The results of the analyses showed that in the absence of Au, carbon did not participate in the charging reaction, as shown in Figure 8.4b (reaction (a)), due to an irreversible reaction route – self-decomposition of Li_2CO_3 – that was often involved in the charging process of Li-CO_2 batteries. Meanwhile, at high potentials, superoxide radicals would trigger electrolyte decomposition, which would require a large overpotential. In addition, the by-products generated by electrolyte decomposition were likely to be attached to the electrode, so this process could not benefit from the

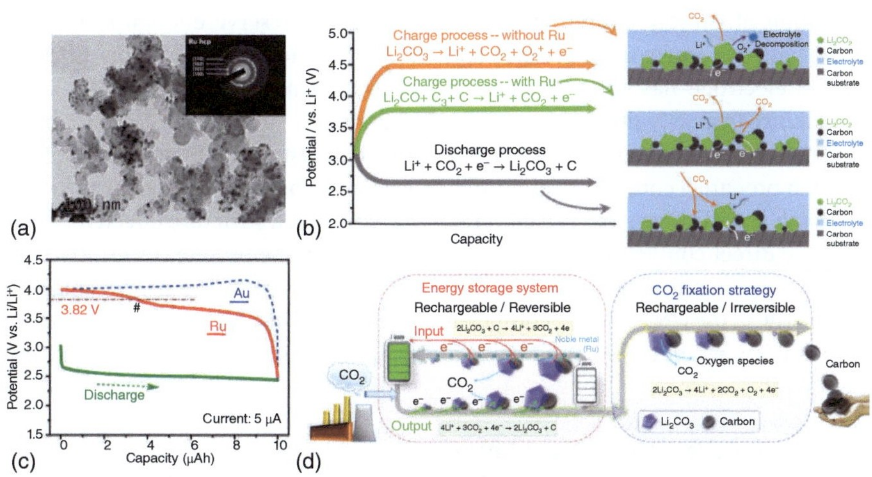

Figure 8.4 (a) Transmission electron microscope (TEM) image and selected area electron diffraction (SAED) pattern of Ru nanoparticles (inset). (b) Schematic diagram of the reaction mechanism of the Li-CO_2 battery. Source: Yang et al. [35] © 2017/Royal Society of Chemistry. (c) The charge profile was conducted on the Au- and Ru-based cathode. (d) Schematic of the achievement of an energy storage system and a CO_2 fixation strategy via the Li-CO_2 battery technology. Source: Reproduced with permission from Qiao et al. [34]. © 2017/Elsevier.

reversibility of Li-CO_2 batteries. Nevertheless, with the participation of Ru nanoparticle catalysts, the charging potential of Li-CO_2 batteries was as low as 3.8–4.2 V. This indicated that Ru nanoparticles could promote the reversible reaction between Li_2CO_3 and carbon (reaction (b)), rather than the self-decomposition of Li_2CO_3, thereby reducing the charge potential and avoiding electrolyte decomposition. This finding helped us understand why a Ru catalyst could lead to better reversibility. Later, these authors introduced an innovative approach for either CO_2 fixation or energy storage. They found that the Ru cathode had a low charging voltage (3.6 V) in Li-CO_2 batteries (Figure 8.4c), and in situ Raman spectroscopy results suggested that during the charging process, the discharge product Li_2CO_3 reversibly reacted with carbon to produce the initial reactants CO_2 and Li^+ on the Ru cathode. In contrast, only Li_2CO_3 was electrochemically decomposed into Li^+, $O_2{\cdot}^-$, and CO_2 molecules on the Au electrode, while the carbon obtained remained fixed. Based on completely different decomposition mechanisms of different electrodes, an energy storage system and a CO_2 fixation strategy were proposed through Li-CO_2 electrochemical reactions (Figure 8.4d) [34].

Tu et al. [47] designed a unique array-type cathode material on a nickel foam frame, which is primarily composed of an open carbon submicron tube array and Au nanoparticles loaded in the tube (Au@CST), as shown in Figure 8.5a. Au nanoparticles serve a dual purpose by enhancing electrode conductivity and offering numerous catalytic active sites, which can induce the growth of Li_2O_2 in the form of thin layers in the open CST inner tube (as shown in Figure 8.5b). This growth mode can not only make Li_2O_2 easier to decompose, but also effectively prevent the deactivation of carbon submicron tube electrodes and reserve enough space and channels for the transport of reactive species. Electrochemical tests show that the Au@CST electrode has good rate performance (1000 mA g^{-1}, 1208 mAh g^{-1}) and cycle stability (capacity limit 500 mAh g^{-1}, 112 cycles).

Jian et al. [48] uniformly coated a layer of RuO_2 catalyst on CNTs using the sol-gel method to prepare a core–shell structure CNT@RuO_2 composite material and used it as a catalyst for lithium–oxygen batteries. This core–shell structure effectively prevented the direct contact between CNTs and the discharge product Li_2O_2, thereby reducing the generation of the by-product Li_2CO_3. Due to the efficient OER catalytic ability of RuO_2, the CNT@RuO_2 composite electrode showed high charge–discharge energy efficiency (about 79%) and low overvoltage, and good rate performance and cycle stability.

8.3.2.3 Non-precious Metal-Based Catalysts

Non-noble transition metals (e.g. Ni, Co, Mn, Cu, Fe, and Mo) and their composites are probably the most interesting type of catalysts due to their following advantages: abundance, low cost, environmental friendliness, and especially, unique tunable structure as well as multivalence characteristics, which enable them to form different oxides with various crystal structures and excellent electrochemical catalytic activity. To date, various transition-metal-based catalysts with superior catalytic activities and cycle stabilities have been designed. Recently, Zhou et al. reported a composite material with nickel nanoparticles highly dispersed on nitrogen-doped graphene via the hydrothermal method and subsequent annealing treatments, with a superior discharge capacity of 17,625 mAh g^{-1} in Li-CO_2 batteries. The battery had a stable discharge platform of ≈2.82 V at a cutoff capacity of 1000 mAh g^{-1}. These values were attributable to the abundant large pores and channels of the nitrogen-doped graphene cathode, which provided sufficient space for the deposition of discharge products, thus enhancing the stability of the discharge platform

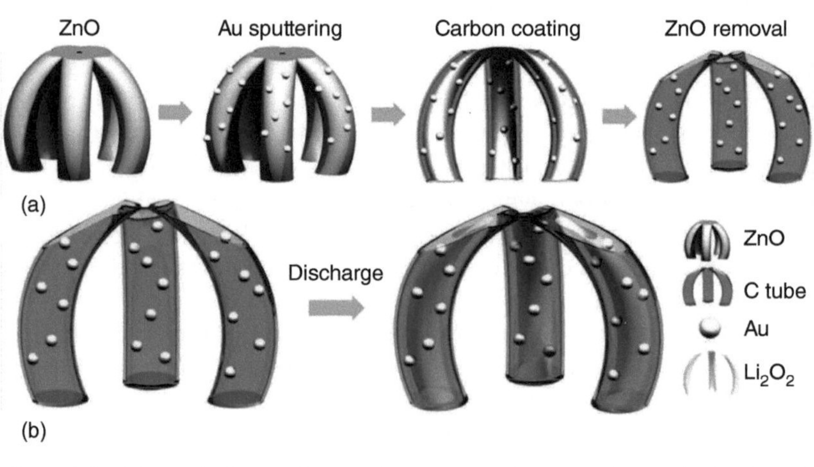

Figure 8.5 Schematic illustration of (a) the preparation procedure of Au@CST on Ni substrate and (b) Li_2O_2 growth inside Au@CST. Source: Reproduced with permission from Tu et al. [47]. © 2016/John Wiley & Sons.

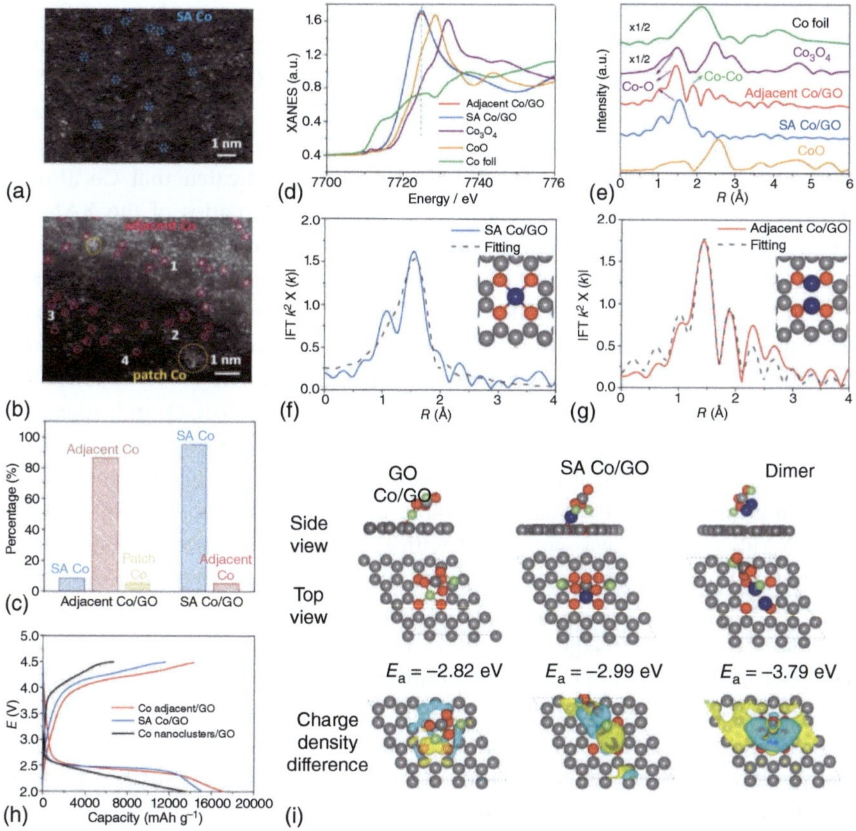

Figure 8.6 High angle angular dark field-scanning transmission electron microscopy (HAADF-STEM) image of (a) SA Co/GO and (b) adjacent Co. (c) Statistical distribution map of different types of Co atoms. (d) XANES spectra and (e) Fourier transforms at the Co K-edge of SA Co/GO, adjacent Co/GO, Co_3O_4, CoO, and Co foil. (f, g) Corresponding fittings of the EXAFS spectra of SA Co/GO and adjacent Co/GO in R space, respectively. The insets of (f) and (g) are, respectively, schematics of SA Co/GO and adjacent Co/GO. (h) Galvanostatic discharge–charge profiles of SA Co/GO, adjacent Co/GO, and Co nanoclusters/GO at a current density of 0.1 A g^{-1}. (i) Side view, top view, and charge density difference of Li_2CO_3 adsorbed on graphene oxide, single cobalt atom, and dimer cobalt, respectively. The purple, green, gray, and red balls represent Co, Li, C, and O, respectively. Source: Zhang et al. [49]. © 2019/John Wiley & Sons.

and improving the reversible discharge capacity. The unique porous structure ensured the uniform transportation of CO_2 and the expansion of the electrolyte, which promoted the decomposition of the discharge products and the improvement in reversibility. Zhang et al. reported that the highly loaded adjacent Co single atoms (≈5.3%) were anchored on graphene oxide (adjacent Co/GO) (Figure 8.6a) and single Co atom/GO (SA Co/GO) (Figure 8.6b) as electrocatalyst for Li-CO_2 batteries [49]. Due to the obvious difference in z-contrast between Co and C, it can be easily determined that the small bright spots in the figure were Co atoms, which were evenly distributed. The statistical analysis of 100 adjacent Co samples showed that the distance between adjacent bright Co dots ranges from ≈1.20 to 2.54 Å. The average Co–Co distance was about 1.79 Å, indicating that some atoms form Co-dimers. The X-ray absorption near-edge structure (XANES) spectra of SA Co/GO and adjacent Co/GO (Figure 8.6c) showed that the absorption edge peak positions of both SA Co/GO and adjacent Co/GO are located near those of Co foil, and the positive shift toward those of CoO and Co_3O_4 indicated that Co atoms are dispersed in SA Co/GO and adjacent Co/GO. The intensities of the XANES spectra of adjacent Co/GO and SA Co/GO were close to that of CoO and Co_3O_4. This indicated that the atomic form of Co interacts with O to form a Co—O bond. The Fourier-transformed k3-weighted extended X-ray absorption fine structure (EXAFS) spectra of SA Co/GO and the adjacent Co/GO, CoO, Co_3O_4, and Co foil (Figure 8.6d) showed that there is no obvious peak from 2.0 to 3.0 Å. It could be speculated that Co atoms are atomically dispersed in SA Co/GO and adjacent Co/GO. SA Co/GO had a significant peak at 1.53 Å, which was attributable to the six-coordinate Co—O bond of the nearest neighbor in CoO_6. The peak observed at 1.90 Å in adjacent Co/GO was attributable to Co—Co bonds. The SA-Co/GO R-space EXAFS spectrum was consistent with the structure shown in Figure 8.6e (optimized by DFT). Therefore, the Co-O structure was again proved in SA Co/GO. The fitting structures of adjacent Co/GO (Figure 8.6f) also showed that the dimer Co structure exists in adjacent Co/GO. This further proved that the Co atoms in SA Co/GO and the adjacent Co/GO are atomically dispersed. The adjacent Co/GO showed a higher discharge capacity of 17,358 mAh g^{-1} at 100 mA g^{-1} (14,878 mAh g^{-1} for SA Co/GO and 13,240 mAh g^{-1} for Co nanoclusters/GO) (Figure 8.6g), with the overpotential hardly increasing after 100 cycles. DFT results further proved that the dimer Co/GO adsorption energy is −3.79 eV (−2.99 eV for SA Co/GO and −2.82 eV for GO), which had the strongest ability to capture the discharge product Li_2CO_3, as shown in Figure 8.6h. Therefore, during discharge, Li_2CO_3 might be deposited around the dimer Co, and Li_2CO_3 was catalytically decomposed on the adjacent Co/GO during charging. The difference in charge density also showed that there is a strong interaction between Li_2CO_3 and dimer Co. These results suggested that in Li-CO_2 batteries, adjacent Co atoms possess a synergistic effect, which provides a larger active site for the adsorption of Li_2CO_3 and exhibits better cycle performance. Furthermore, experimental and theoretical calculations implied that Co-Co and Co-O structures in adjacent Co/GO show a synergistic effect, forming continuous active sites for the strong adsorption and reversible decomposition of Li_2CO_3. Meanwhile, the introduction of adjacent Co into the catalyst provided a new research direction for the design of electrode materials.

8.3.2.4 Metal-Organic Frames and Their Derivatives

Metal-organic frames (MOFs) are generally centered on metal ions and form periodic crystalline structures through self-assembly. MOFs were initially used in gas separation and drug release, but more recently, they have been used in energy storage and conversion. The application of MOFs and MOF derivatives in Li-O_2 and Li-CO_2 batteries can be divided into three categories, and one among them is the direct application of MOF precursors. In 2014, Li [50] first applied a MOF precursor in a Li-O_2 battery due to the high specific surface area, open metal site, and regular microstructure of the MOF precursor. The pore structure of the MOF precursor can enrich, adsorb, and transform the oxygen molecules, but due to its poor conductivity, it should be used in combination with conductive carbon materials. In this study, they studied the performance of five MOF precursors in Li-O_2 batteries, including MoF-5, HKUST-1, and M-MoF-74 (M = Mg, Mn, Co). Among them, Co-MoF-74 showed the highest oxygen enrichment capacity, reaching a value of 18.4 mg g^{-1}. The oxygen enrichment ability of Mn-MOF-74 was slightly lower than that of Co-MOF-74 (15.1 mg g^{-1}), but its specific discharge capacity was up to 9420 mAh g^{-1}, indicating that the pore structure and open metal site of the MOF precursor material influence the performance of Li-O_2 batteries. The diversity of MOF precursors makes it possible to use them as catalysts. Lee et al. [51] prepared a MnCo-MOF-74 bimetallic organic framework as an effective cathode catalyst, presuming that the bimetallic organic framework could achieve better catalytic performance than the single-metal framework. The experimental results showed that MnCo-MOF-74 has better catalytic performance than single-metal framework. At a current density of 200 mA g^{-1}, the specific discharge capacity reached 11,150 mAh g^{-1}, with a limited capacity of 1000 mAh g^{-1}, and the stable cycle could be sustained for 44 cycles. Recently, Wang and colleagues successfully applied MOFs in a Li-CO_2 battery system, confirming their catalytic activity in Li-CO_2 electrochemistry. Their findings highlighted the substantial potential of MOFs in enhancing the performance of Li-CO_2 batteries by increasing discharge capacity and reducing polarization potential. To determine the activity of the transition metal center, they conducted a comparative analysis of six MOFs with different transition metal centers. The results revealed that among these MOF materials, manganese-based MOFs showed the lowest charge potential (about 3.96 V for Mn_2(dobdc) and 4.19 V for Mn(bdc)). Furthermore, Wang et al. synthesized nano-MnO particles dispersed in graphene-wrapped N-doped 3D carbon frameworks (MnO@NC-G), as shown in Figure 8.7a–c, via pyrolyzed GO-wrapped MOF containing Mn(II) species, as the cathode material for Li-CO_2 batteries [52]. These batteries achieved low voltage hysteresis, high rate capability, long cycle life, and high discharge capacity, far superior to those of other Mn(II)-based catalysts (Figure 8.7d,e). More importantly, after 206 cycles, the cathode material continued to operate stably after replacing the Li anode and electrolyte with the new ones (Figure 8.7f). The excellent performance was attributable to highly dispersed MnO nanoparticles that accelerated the decomposition of Li_2CO_3, conductive N-doped carbon framework that allowed the deposition of discharge products, and interconnected 2D graphene nanosheets that maintained the structural integrity of the electrode and rapid electron transmission. These findings provided valuable design principles for manufacturing high-performance Li-O_2 and Li-CO_2 electrodes. In addition, DFT calculations indicated that the C—F covalent bond formed by doping a carbon material with an F atom can change the hybrid state of carbon from sp^2 to sp^3, so it has high electrical conductivity and thermodynamic stability. Boron-doped graphene has a smaller diffusion barrier and smaller structural deformation. Therefore, F and B are considered potential carbon-doped elements.

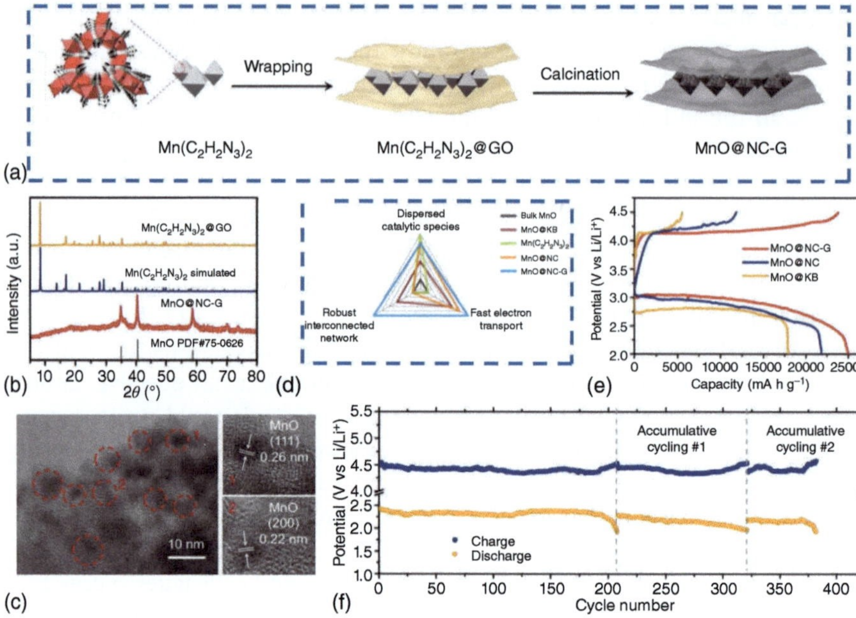

Figure 8.7 (a) The fabrication process of MnO@NC-G. Reproduced with permission. (b) XRD of the obtained samples. (c) High resolution transmission electron microscope (HRTEM) image of MnO@NC-G. (d) Comparison of their material properties. (e) Discharge–charge profiles of various cathodes. (f) Cycling performance of MnO@NC-G. Source: Li et al. [52]. © 2019/Royal Society of Chemistry.

8.3.2.5 Soluble Catalysts

Although solid catalysts show excellent catalytic performance in Li-O_2 and Li-CO_2 batteries, their inherent disadvantages limit the electrochemical reaction between Li metals and O_2/CO_2, particularly in terms of the electrochemical degradation of solid discharge products. Most importantly, the catalytic effect is reduced by the incomplete contact between the active surface of the catalyst and the solid-state discharge product. In addition, the active sites on the surface of solid catalysts are easily passivated by other by-products. Therefore, researchers have proposed to use soluble catalysts instead of solid catalysts to further improve the electrochemical performance of Li-O_2 and Li-CO_2 batteries, such as redox mediators (RM) and chemical catalysts (Figure 8.8). In general, a satisfying RM capable of decomposing solid discharge products should meet the following criteria: (i) it should have high solubility and should not react with electrolyte components; (ii) it should resist the attack of active intermediates such as superoxide radicals and should have a higher decomposition voltage in the working voltage range; and (iii) its redox potential should be higher than the potential when CO_2 is reduced in CO_2RR and lower than the decomposition potential of discharge products in CO_2ER, and the oxidation/reduction reaction of RMs^+/RMs redox couple should be complete. Similar to the catalytic decomposition mechanism observed in Li-O_2 batteries, the process of decomposition

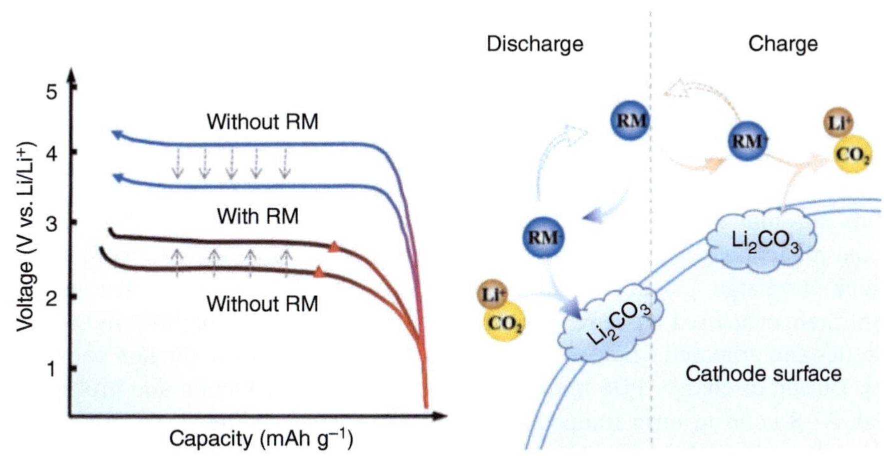

Figure 8.8 Mechanism of a soluble catalyst in the decomposition/formation of Li_2CO_3.

of Li_2CO_3 in Li-CO_2 batteries can be classified into the following steps:

$$RM \rightarrow RM^{o+} + oe^- \tag{8.28}$$

$$RM^{o+} + nLi_2CO_3 \rightarrow RM + mLi + +pCO_2 \tag{8.29}$$

During the charging process, when the voltage reaches the oxidation potential of RM, electrons are transferred to the cathode, achieving the desired oxidation state RM^{o+} (Eq. (8.29)). Then, RM^{o+} further oxidizes the insulated Li_2CO_3 into Li+ and CO_2, and the oxidized RM^{o+} undergoes reduction to form RM (Eq. (8.30)). Therefore, the apparent voltage during charging reflects the oxidation potential of the added RM, not the decomposition potential of Li_2CO_3, and the overpotential of the Li-CO_2 battery can be significantly reduced. So far, there are a few research reports on the application of RMs in Li-CO_2 batteries. For example, Zhou et al. first proposed the introduction of a RM (LiBr) into rechargeable Li-CO_2 batteries. During the charging progress, the Br^- ion preferentially lost electrons to form Br^{3-}, which in turn formed Br_2. Br_2 interacted with Li_2CO_3 and carbon to release CO_2 and reduce it to Br^- [53]. Thanks to the moderate redox potential (4.0 V versus Li/Li^+) of Br^{3-}/Br_2 couplers and fast diffusion kinetics, LiBr-involved Li-CO_2 batteries achieved high energy efficiency, stable rate capability, and cycle durability. Moreover, Li et al. developed a gel polymer electrolyte (GPE)-based Li-CO_2 battery using a novel pencil-trace cathode and 0.0025 mol l^{-1} binuclear cobalt phthalocyanine (Bi-CoPc)-containing GPE (Bi-CoPc-GPE) [54]. Bi-CoPc-GPE acted as a RM, which consisted of a polymer matrix with a liquid electrolyte and showed good ionic conductivity, leak-proof performance, and effective protection for lithium anode. It could promote the decomposition of Li_2CO_3 at low overpotential. In addition, the discharge products Li_2CO_3 were in the form of nanosheets, which were conducive to ion/electron transfer and CO_2 diffusion. Therefore, this polymer Li-CO_2

battery had the advantages of high discharge capacity, low overpotential, and cycle stability.

In contrast to the common reaction mechanisms of RMs, Gallant et al. introduced 2-ethoxyethylamine (EEA) into the Li-CO_2 cell, which could capture CO_2 prior to the reduction of the EEA-CO_2 (C—N bond) adduct [53]. During discharge, EEA-CO_2 underwent reduction at the carbon cathode surface through the selective cleavage of C—N bonds within these adducts to generate Li_2CO_3 and C and recoverable EEA molecules (Eq. (8.31)). Therefore, the discharge voltage was determined by the reduction potential of the amine adduct. After EEA, the battery showed a high discharge potential and discharge capacity. Furthermore, Arumugam Manthiram et al. used diphenyl disulfide (PDS) as an electrolyte additive in Li-CO_2 batteries and proposed a new solution-mediated carbon dioxide reduction pathway [55]. During discharge, PDS underwent electrochemical reduction and broke the weak S—S bond to form thiophenol anion (TP^-). The CO_2 molecules dissolved in the electrolyte adhered to TP^- in the solution to form the soluble intermediate S-phenylthiocarbonate (SPC^-). The C=O and C—O bonds in SPC^- were sequentially reduced through a series of one-electron reduction reactions to produce O^{2-} in the electrolyte, amorphous C in the positive electrode, and then TP^-. Then, O^{2-} reacted with CO_2 molecules in the electrolyte to form ${CO_3}^{2-}$. Finally, Li^+ in the electrolyte combined with ${CO_3}^{2-}$ to form Li_2CO_3. During the charging process, TP^- decomposed the discharge products through the reverse process to the above. The solution-mediated pathway promoted the reversible formation and decomposition of Li_2CO_3 and amorphous carbon during the cycle. Li-CO_2 batteries that used PDS additives showed significantly improved capacity, energy efficiency, and cycle life. But to date, soluble catalysts have only been utilized in Li-CO_2 batteries, and their application in other metal–gas batteries requires further exploration by researchers.

$$RNHCOO^- + RNH^{3+} + 2CO_2 + 4Li^+ + 4e^- \rightarrow 2RNH_3 + C + 2Li_2CO_3 \tag{8.30}$$

8.4 Electrolyte and Electrolyte Stability

The instability of the current electrolyte system is one of the most severe challenges faced by Li-O_2 and Li-CO_2 batteries, significantly affecting the working mechanism, morphology, and crystallinity of the discharge products. The electrolyte used in Li-O_2 and Li-CO_2 batteries should meet the following requirements: high gas solubility and diffusivity; high stability; and low volatility and viscosity, which provide the electrolyte with rapid ion transport ability and nontoxic nature. Furthermore, in flexible Li-O_2 and Li-CO_2 batteries, the electrolyte also needs to adapt to various deformations to maintain regular operation. However, to date, no stable electrolyte has been found in this system. Finding a new organic electrolyte with good electrochemical stability is one of the key challenges that should be addressed for the commercial application of lithium–oxygen batteries.

8.4.1 Nonaqueous Aprotic Electrolyte System

8.4.1.1 Variety of Solvents

Previous research on nonaqueous Li-O_2 and Li-CO_2 battery electrolytes has focused on electrolytes that commonly use carbonic acid in Li-ion batteries, e.g. propylene carbonate, vinyl carbonate, dimethyl carbonate, and methyl ethyl carbonate. However, it has been reported that these carbonate electrolytes are unstable in Li-O_2 and Li-CO_2 batteries and decompose irreversibly and produce by-products. Bruce et al. [8] proved that in the discharge process of Li-O_2 batteries using propylene carbonate electrolyte, the solvent was attacked by oxygen reduction products, and Li_2CO_3, $C_3H_6(OCO_2Li)_2$, CH_3CO_2Li, HCO_2Li, CO_2, and H_2O were generated. The charging process is only the oxidation reaction of Li_2CO_3, CH_3CO_2Li, HCO_2Li, and $C_3H_6(OCO_2Li)_2$. Li_2O_2 is not reversible in the process of battery charging and discharging. Furthermore, researchers have shown that carbonate-based electrolytes are unstable in the presence of highly reactive oxygen-containing substances (i.e. $O_2{\cdot}^-$, $O_2{}^{2-}$). The $O_2{\cdot}^-$ species, which could be generated in the reaction of Li-O_2/CO_2, was considered a strong Lewis base and an efficient electron-transfer agent, which can decompose polar electrolyte solvents through nucleation attack (proton substitution or abstraction) [56]. Superoxide radicals could function as nucleophiles to attack the *o*-alkyl carbon (methylene) of organic carbonate molecules, causing the electrolyte solvent to undergo a ring-opening reaction and produce various irreversible by-products.

The ether electrolyte has good compatibility with lithium metal, low vapor content, high oxygen solubility, high oxygen diffusivity, safety, and low cost. In addition, it has the advantages of having a large electrochemical window (greater than 4.0 V), stability to $O_2{\cdot}^-$, low volatility, and good wettability and is widely used as an organic solvent for Li-O_2 and Li-CO_2 batteries. Read et al. [57] first confirmed in 2006 that ether solvents used as electrolytes for Li-O_2 batteries are beneficial to the stability and rate performance of the battery system. A large number of studies have shown that the primary discharge product of Li-O_2 batteries based on ether electrolytes is Li_2O_2, as shown in Figure 8.9. Later, Sun et al. [58] successfully discovered that the Li-O_2 battery electrolyte composed of tetraethylene glycol dimethyl ether (TEGDME) as the solvent and lithium trifluoromethyl sulfonate ($LiCF_3SO_3$) as the lithium salt can lead to the stable cycle of the battery system and achieve a discharge capacity as high as 5000 mAh g^{-1}. This achievement has promoted the research of Li-O_2 and Li-CO_2 battery electrolyte to the next stage. Although ethers are relatively stable against nucleophilic attack, the long-term stability of ether electrolytes is their Achilles' heel, since they tend to auto-oxidize to unstable peroxides under the action of oxygen-containing radicals. Professor McCloskey [59] used the in situ differential mass spectrometry technology to study the performance of lithium–oxygen batteries with dimethyl ether (DME) as a solvent. This study showed that the oxygen released by charging is only 60% of the oxygen consumed by discharging, which may be due to the side reaction of Li_2O_2 and DME. Subsequently, Bruce et al. [60] comprehensively used XRD, infrared spectroscopy, Raman spectroscopy, and nuclear magnetic resonance spectroscopy to confirm that the decomposition products of ethers are primarily Li_2CO_3, HCO_2Li, CO_2, and H_2O. Moreover, they also found

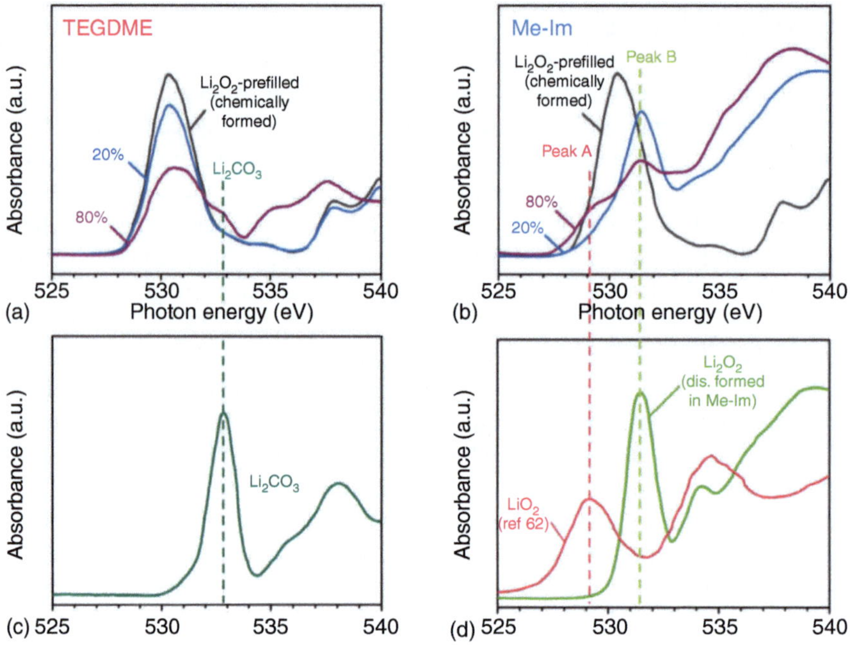

Figure 8.9 O K-Edge XANES total-electron-yield spectra of various charging Stages for Li_2O_2-prefilled electrodes. Source: Reproduced with permission from Wang et al. [56]/Elsevier.

that even changing the types of ether solvents and lithium salts could not prevent the formation of these by-products that affect the electrochemical efficiency. In addition, ethers are also prone to slow auto-oxidation, which is not conducive to the cycle stability of Li-O_2 and Li-CO_2 batteries. Theoretical calculations have shown that fluorine substitution can be performed at the β—C position of ethers to inhibit C—H bond breakage, thus improving the auto-oxidation stability of ethers. Nazar et al. replaced the hydrogen in the β-C position with methyl (–CH_3) group, thus eliminating the possibility of methylene hydrogen substitution, and obtained 2,3-dimethyl-2,3-dimethoxy-butane. The electrolyte solvent used inhibited the formation of lithium formate, dimethyl oxalate, and lithium carbonate during the discharge process. In situ electrochemical mass spectrometry also showed that no CO_2 was generated during the battery charging process. The cycle stability of batteries can be significantly improved using the electrolyte and non-carbon TiC to assemble lithium–air batteries.

In order to improve the actual cycle performance of organic electrolyte system Li-O_2 and Li-CO_2 batteries, researchers also tried other solvents such as DMF, dimethyl sulfoxide (DMSO), ethylmethanesulfonate (EMS) , acetonitrile, and ionic liquids. Bruce and Peng et al., who have been committed to the in-depth study of DMSO as an organic solvent for Li-O_2 batteries, in 2012 used 0.1 M $LiClO_4$/DMSO as the electrolyte and a nanoporous Au electrode as the oxygen cathode. The results of their study showed that the assembled Li-O_2 battery can cycle for 100 cycles without significant attenuation [61]. Furthermore, they also used infrared spectroscopy, Raman spectroscopy, and other characterization methods to confirm the reversible formation and decomposition of the main discharge product Li_2O_2, and no obvious decomposition of the electrolyte was detected. However, when they replaced the porous gold electrode plate with a carbon material as the oxygen electrode, under the same electrolyte system, a very obvious electrolyte decomposition was detected, even during the first charging and discharging process. It can be inferred that DMSO is not a suitable electrolyte solvent for carbon materials as the oxygen cathode. In addition, there is another problem that cannot be ignored. The anode metal Li can directly react with DMSO. Therefore, when DMSO is used as the solvent for Li-O_2 and Li-CO_2 batteries, it is necessary to protect the anode metal Li by spreading.

Recently, hydrophobic ionic liquids have attracted attention due to their high stability, high safety, wide electrochemical window, and extremely low volatility, which can be used in Li-O_2 and Li-CO_2 batteries. Mizuno et al. [62] studied the performance of Li-O_2 batteries based on ionic liquid *N*-methyl, propylpiperidine bis-trifluoromethane sulfonamide salt (PP13TFSA) as a solvent and found that although a small amount of by-product Li_2CO_3 is generated, Li_2O_2 is the primary discharge product. Therefore, improving parameters such as the viscosity of ionic liquids and the diffusion coefficient of oxygen is the breakthrough point to truly realize the application of ionic liquids in lithium–oxygen batteries.

8.4.1.2 Compatibility of Lithium Salts

Lithium salts that have been extensively studied include lithium hexafluorophosphate ($LiPF_6$), trifluoromethane sulfonamide (LiTFSI), lithium trifluoromethane sulfonate ($LiCF_3SO_3$), lithium perchlorate ($LiClO_4$), etc. As an indispensable part of the electrolyte, the stability of lithium salts also affects the reversibility and capacity of Li-O_2 and Li-CO_2 batteries. Oswald et al. [63] found that $LiPF_6$ reacts with Li_2O_2 to form LiF when it is in contact. Recently, McCloskey et al. [64] also found that anion properties in lithium salts affect the discharge capacity of Li-O_2 batteries. According to Person's hard and soft acid–base theory, anions with a high number of donors can improve the solubility of LiO_2, thereby increasing the discharge capacity of the battery. Furthermore, many experiments and theoretical calculations have confirmed that the matching of lithium salts and solvents can help reduce the occurrence of side reactions. For example, Du et al. studied the decomposition of $LiPF_6$ in different solvents (i.e. trimethylsilane substituted with triethylene glycol and TEGDME) and found that $LiPF_6$ decomposes into HF in the former solvent, but not in the latter. $LiPF_6$ has a good match with TEGDME. Therefore, the correct choice of lithium salt cannot be ignored. Qiao et al. conducted a comparative experiment, in which when the electrolyte did not contain lithium ions, it showed a charge potential of 3.49 V. However, once Li ions were introduced, the charge potential increased to 4.2 V. Raman spectroscopy further proved that the discharge product was peroxodicarbonate ($C_2O_6^{2-}$) without lithium ions. After adding lithium ions, the discharge product was converted into hard-to-decompose

Li_2CO_3, and the charging overpotential was relatively high. Subsequently, they developed a Li-O_2/CO_2 battery with an aggregated contact ion pair (CIP) structure into the electrolyte. In dilute electrolytes, the interaction between $[Li(DMSO)_4]^+$ and ($TFSI^-$) was relatively weak. Therefore, unstable $C_2O_6^{2-}$ anions would combine with Li^+ and finally form Li_2CO_3 (Figure 8.10a) [65]. To prevent the

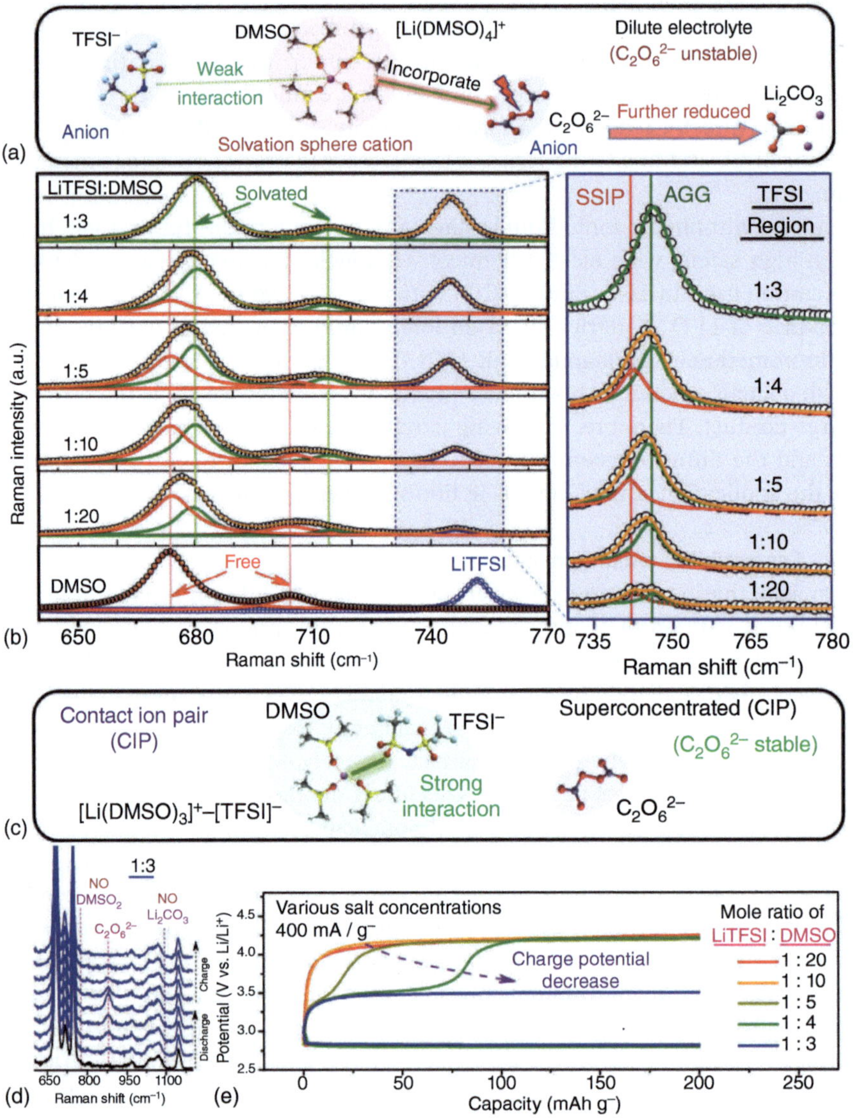

Figure 8.10 (a) Schematic illustrations of dilute LiTFSI/DMSO-based electrolyte structure. (b) Raman spectra of LiTFSI/DMSO solutions. (c) Schematic illustration of electrolyte with $[Li(DMSO)_3]^+$-$[TFSI^-]$. (d) In situ Raman spectra (1 : 3 mole ratio). (e) Voltage profiles of Li-O_2/CO_2 (1 : 1, volume ratio) cells operating with different mole ratios. Source: Reproduced with permission from Qiao et al. [65]. © 2018/Royal Society of Chemistry.

combination of Li^+ and $C_2O_6{}^{2-}$, they proposed a superconcentrated electrolyte with a DMSO-solvated CIP ($[Li(DMSO)_3]^+$-$[TFSI^-]$) by increasing the salt concentration (Figure 8.10b). In the superconcentrated electrolyte with a 1 : 3 mole ratio (LiTFSI: DMSO), the ion association mode entirely evolved into $[Li(DMSO)_3]^+$-$[TFSI^-]$ CIP (Figure 8.10c), which effectively stabilized $C_2O_6{}^{2-}$ and prevented the formation of Li_2CO_3 (Figure 8.10d). The appropriate introduction of electrolytes into aggregated CIP gave the Li-O_2/CO_2 battery a low charge potential (3.5 V) (Figure 8.10e). This strategy can make up for the inherent shortcomings and provide greater feasibility for Li-CO_2 batteries.

8.4.2 Aqueous Aprotic Electrolyte System

For aprotic electrolytes, many types of liquid or solid ion conductors can be used, while aqueous electrolytes are limited to acidic or alkaline solutions. Since aqueous electrolytes are nonflammable, they avoid some safety hazards, which can cause serious trouble in the organic electrolyte system. They provide the ideal medium for flexible electrochemistry, namely water, which can be tuned to generate fuel and chemicals via proton-coupled electron transfer. Furthermore, aqueous electrolytes also have high ionic conductivity, high boiling point, low volatility, and low viscosity. However, their operating potential is limited to a narrow range to maintain water stability. The major advantage of operating Li-O_2 and Li-CO_2 batteries in an acidic solution is that the formation of insoluble products can be avoided. In comparison, violent reactions and severe corrosion of Li anodes in the acidic electrolyte can cause serious safety hazards and shorten the lifespan. Therefore, alkaline electrolytes such as KOH, NaOH, and LiOH have been used in Li-O_2 and Li-CO_2 batteries due to their relatively stable environment and superior ionic conductivity. Among alkaline solutions, KOH-based electrolytes are more commonly used in flexible Li-O_2 and Li-CO_2 batteries due to their higher stability compared with NaOH- and LiOH-based electrolytes. When using metals with higher activity such as Li as anodes, a protective metal ion conductive membrane that allows the migration of metal ions only between the electrode and aqueous electrolyte is needed, which prevents the metal anode from violently reacting with water. For example, a LISICON glass electrolyte film $Li_{1+x+y}Al_xTi_{2-x}Si_yP_{3-y}O_{12}$ (LATP), which has good Li^+ conductivity ($\approx 3\times10^{-4}$ S cm^{-1} at 25 °C) and water impermeability, is used in hybrid Li–air batteries. Notably, in aqueous higher-activity Li-O_2 and Li-CO_2 batteries, H_2O as a solvent does not impose limitations on battery performance, providing a significant advantage over the aprotic system. Nevertheless, due to distinct electrochemical reactions involving Li metal and O_2/CO_2, the energy density of aqueous Li-O_2 and Li-CO_2 systems is much lower than that of nonaqueous systems.

8.4.3 Solid-State Electrolyte System

As mentioned earlier , although liquid electrolytes are most widely used in Li-O_2 and Li-CO_2 batteries, they have some disadvantages in practical applications.

For example, the parasitic reaction between the electrode material and the liquid electrolyte produces uneven SEI, which hinders the migration of cations, consumes the active material on the boundary, and reduces the battery life. Because they have an unstable SEI and low mechanical strength, they involve in uncontrollable side reactions with the electrode. In addition, liquid-electrolyte-based batteries also present issues such as electrolyte leakage, metal dendrite, volatilization, flammability, and other safety problems, which greatly limit their application. To overcome these drawbacks of liquid electrolytes, quasi-solid-state electrolytes and SSEs with reaction mechanisms similar to those of aprotic electrolyte systems are proposed to replace traditional liquid electrolytes, which can effectively address the problems of organic electrolyte solution instability, flammability, toxicity, and volatility. A SSE should meet the following criteria: (i) high chemical stability and compatibility with metal anodes and cathodes; (ii) wide voltage window; (iii) high thermal and mechanical stability; and (iv) high ionic conductivity and low interfacial resistance. Currently, the semi-SSE lithium–oxygen battery that has been under research has the same structure as the all-solid-state lithium-ion battery, the main difference being that the open positive electrode is in contact with the air. The basic electrochemical reaction mechanism of the semi-SSE lithium–oxygen battery is consistent with that of the organic electrolyte system. SSEs can be divided into three main categories: glass ceramics, polymers, and their mixtures. In research on lithium-ion batteries, inorganic electrolytes are primarily sulfides and oxides, including Li_2S-P_2S_5 system, LISICON-type structure system, NASICON-type structure system, perovskite structure system, and garnet structure system. The synthesis method of this type of ceramic-phase solid electrolyte diaphragm aims primarily to obtain the ceramic phase through high-temperature solid-phase sintering of the precursor and then crush, press, and sinter to achieve the ceramic diaphragm. In the field of lithium–oxygen batteries, the most widely used are LAGP (Li_2O-Al_2O_3-GeO_2-P_2O_5) and LATP. Polymer electrolytes are usually made of high-molecular-weight polymers such as poly(methyl methacrylate), poly(vinylidene fluoride), poly(ethyleneoxide), and polyacrylonitrile, after adding a solvent to disperse, and then mixed with lithium salt or fast ion conductor powder. Glass ceramic SSEs have the advantages of high ionic conductivity ($10^{-5} - 10^{-2}$ S cm^{-1}), strong mechanical rigidity, and good flame retardancy.

In 1976, Goodenough [66] et al. reported that NASICON $Na_{1-x}Zr_2P_{3-x}Si_xO_{12}$, in the crystal, can allow sodium ions to transport along the c-axis and has very high ion conductivity. Since lithium and sodium have similar properties, sodium ions can be directly replaced by lithium ions to form LISICON. In general, the elemental composition of LISICON is $LiM_2(PO_4)_3$, where the M sites are composed of Ge, Zr, and Ti and can be doped with Al, Ga, Fe, Sc, etc. Among many solid electrolytes, LATP has been widely reported to have an ionic conductivity as high as 1.09×10^{-3} S cm^{-1} at room temperature [67]. However, when LATP is in direct contact with lithium metal, metallic lithium is highly reducing, which causes the reduction reaction of Ti^{4+}, resulting in the stability of the solid-state ceramic separator and a decrease in ionic conductivity. Recently, LAGP has been considered one of the most potential alternatives to LATP. The LAGP solid ceramic separator shows excellent stability in contact with lithium metal and has a working potential window of up to 6 V (versus Li/Li^+) and excellent lithium ion conductivity. However, its brittleness and the ability to grow metal dendrites along the grain boundaries of SSEs limit its practical application. Polymer SSEs overcome the shortcoming of poor flexibility of glass ceramic SSEs but have low ionic conductivity ($10^{-6} - 10^{-4}$ S cm^{-1}). Thus, a hybrid electrolyte system can avoid the intrinsic weaknesses of inorganic electrolytes and organic electrolytes.

In 2010, Kumar et al. first reported an all-solid-state lithium-oxygen battery [17]. The primary components of the battery are a metal lithium anode, glass ceramic and polymer ceramic material composite solid electrolyte, high-specific-surface-area carbon, and glass ceramic composite oxygen electrode. In the temperature range of 30–105 °C, the battery showed excellent thermodynamic stability and good reversibility. Compared with the organic system, this system used a solid electrolyte instead of an organic electrolyte. Although its working principle was the same as that of organic lithium–oxygen batteries, the introduction of the solid electrolyte alleviated the problems caused by electrolyte leakage and improved the safety of the battery. Furthermore, the electrochemical window of solid electrolytes is wider than that of organic electrolytes, and they have high temperature resistance. Therefore, the stable operating range of the battery can be increased. However, the low ionic conductivity of SSEs slows down the transport rate of lithium ions, resulting in poor rate performance of batteries. Furthermore, the contact between the solid electrolyte and the positive and negative electrodes of the lithium–oxygen battery is poor, resulting in extremely high interfacial impedance of the battery, which deteriorates the electrochemical performance of the battery. Therefore, effectively improving the ionic conductivity of SSEs and reducing the solid–solid interface resistance inside the battery is the core issue in the development of all-solid-state lithium–oxygen batteries.

In 2017, Chen et al. presented a solid-state Li-CO_2 battery without a liquid component, using a SiO_2-incorporated composite polymer electrolyte (CPE) comprised of poly(methacrylate)/poly(ethylene glycol)–$LiClO_4$–3%$SiO_2$$_{2.18}$. This CPE combined favorable mechanical properties with acceptable ionic conductivity (7.14 *10^2 mS cm^{-1} at 55 °C). It not only shielded the Li anode from CO_2-induced corrosion but also offered opportunities for integrated cathode preparation. Its incorporation into the Li-CO_2 battery resulted in notable flexibility and improved electrochemical performance, including an extended cycle life and excellent rate capability. Seeking polymer electrolytes with high ionic conductivity at room temperature, Xia et al., inspired by Chen's work, developed a GPE consisting of a polymer host, lithium bis(trifluoromethane sulfonyl) imide salt (LiTFSI), and tetraglyme (TEGDME). The polymer matrix was filled with the conventional LiTFSI/TEGDME liquid electrolyte [68]. Morphological observations, electrochemical impedance spectroscopy (EIS), linear sweep voltammetry (LSV), and thermogravimetric (TG) studies (Figure 8.11a–c) revealed that the GPE had compact structural characteristics, high ionic conductivity (0.5 mS cm^{-1}), excellent electrochemical stability (4.5 V), and thermal stability (below 100 °C), as shown in

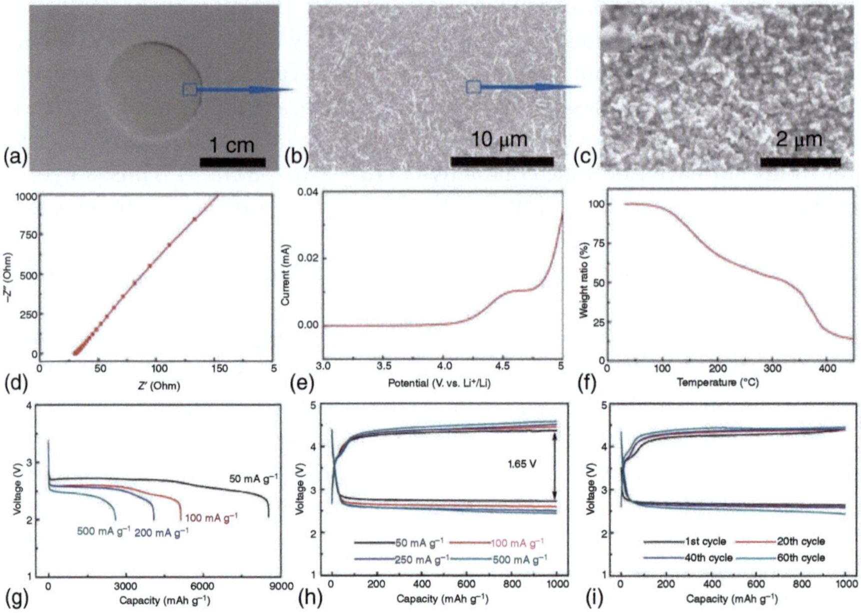

Figure 8.11 (a) Optical photograph, (b) and (c) SEM images, (d) EIS curve, (e) LSV, and (f) TG analysis for GPE. (g) Rate performance, (h) discharge–charge curves, and (i) cycle performance of the Li-CO_2 battery based on GPE. Source: Li et al. [68]. © 2017/John Wiley & Sons.

Figure 8.11d–f. Consequently, the Li-CO_2 battery using GPE and a CNT cathode showed significantly enhanced rate performance, prolonged recyclability, and reduced potential polarization (Figure 8.11g–i). This improvement in battery performance is attributable to the formation of particle-shaped Li_2CO_3 with poor crystallinity at the GPE/CNT interface.

The use of solid electrolytes can effectively avoid the leakage, volatilization, and combustion of the electrolyte in an open system. In addition, the dense ceramic structure encapsulates the lithium chip to avoid the corrosion of H_2O, CO_2, N_2, etc., brought in during the test. Compared with traditional polymer separators, lithium dendrites can easily penetrate the short circuit caused by the separator during the charging process, while solid electrolyte separators have high mechanical strength and can effectively prevent the separator from being pierced. In addition, solid electrolytes have a wide window of electrochemical activity, are stable to oxygen and superoxide radicals, and are also beneficial to reduce side reactions during battery cycling.

8.5 Stable Anode/Electrolyte Interface Construction

Using metallic Li as an anode is considered the most optimal choice for all Li-metal-based batteries, given its exceptionally high theoretical specific capacity

($3860\,mAh\,g^{-1}$) and lowest electrochemical potential ($-3.040\,V$ versus SHE). However, structural and interfacial instability in metal anode/aprotic electrolyte systems significantly affects the safety and overall performance of the battery. In electrochemical systems using gas and a lithium anode, as seen in Li-O_2 batteries, the interaction between the Li metal anode and dissolved O_2 is inevitable. Furthermore, due to the difference in the conduction rate of lithium ions on the surface or the inhomogeneity of the lithium surface during the cycle, lithium ions are deposited unevenly on the surface of lithium metal, forming lithium dendrites. Once the dendrites are broken, irreversible “dead lithium” is formed, resulting in a decrease in the rate of utilization of lithium metal on the negative plate. More seriously, lithium dendrites may penetrate the separator and cause a short circuit in the battery, which poses a safety hazard. Recently, considerable efforts have been dedicated to establishing a stable protective layer on lithium metal characterized by excellent Li^+ conductivity, uniform morphology, and precisely defined components. These initiatives encompass treatments applied to lithium metal for forming artificial coatings on its surface or adjustments to the concentration and composition of the electrolyte through additive incorporation, resulting in significant accomplishments. This section predominantly explores the three-phase boundary reaction involving metallic lithium, the electrolyte, and gas, with a focus on elucidating how the properties of the formed discharge product passivation layer affect battery performance.

Recently, Asadi et al. presented a rechargeable Li-O_2 battery with an exceptionally long cycle life in a simulated air atmosphere. The battery used a Li_2CO_3/C coating layer to protect the Li metal anode [69], formed through cycling of Li-CO_2 batteries. After 10 consecutive discharge and charge cycles of the Li-CO_2 battery, a rod-shaped network structure layer evenly covered the surface of Li. The absence of catalysts, such as MoS_2, rendered the reaction irreversible, leading to the continuous accumulation of Li_2CO_3 and C on the anode. Remarkably, the protected Li anode showed a high lithium retention of 99.97%, as confirmed by exhaustive stripping tests after 51 cycles in an air environment. Mao et al. found that the Li anode in a nonfunctional Li-CO_2 battery had completely transformed into LiOH, resulting in the failure of the battery [70]. Given these observations, a strategic pretreatment of Li metal to establish a specific protective layer with both efficient Li^+ ion conductivity and high lithium retention emerges as an optimal approach. This approach not only mitigates potential safety hazards but also enables the achievement of a Li-CO2 battery with outstanding cyclability.

In addition, since Li-O_2 and Li-CO_2 batteries adopt a semi-open system, the reaction between the negative plate metal lithium and the electrolyte and O_2/CO_2 is more complicated. Recently, Shui et al. [71] used X-ray synchrotron radiation and three-dimensional microtomography to confirm that the LiOH layer continuously accumulates on the surface of the negative metal lithium during the charging and discharging process of the lithium–oxygen battery and that there are many microscopic scales on this layer. The tunnel facilitates the contact between the organic electrolyte and lithium metal. This study provides good guidelines for understanding the protection of lithium in lithium–oxygen electrochemical anodes. Many research groups have used $LiNO_3$ as an additive to form a stable SEI film in lithium–oxygen batteries. Walker et al. [72] studied the cycle stability of lithium–oxygen batteries using $LiNO_3$/*N*,*N*-dimethylacetamide (DMA) as the electrolyte and showed that the battery system can be cycled stably for more than 2000 hours, which is beneficial to the stable SEI film formed on the surface of metal lithium in the presence of $LiNO_3$.

On the other hand, developing new high-capacity negative plate materials as an alternative to metal lithium is also an important topic in the research of Li-O_2 and Li-CO_2 batteries. B. Scrosati et al. [73] used a lithium/silicon/carbon compound instead of metal lithium as the negative plate of lithium–oxygen batteries, combined with $LiCF_3SO_3$/TEGDME electrolyte and Super P positive electrode, to assemble a lithium–oxygen battery with small polarization and good cycle performance. The discharge voltage of the battery is around 2.4 V, and the estimated energy density is as high as 980 Wh kg^{-1}, which is far higher than the 384 Wh kg^{-1} output of commercial $LiCoO_2$/graphite lithium-ion batteries, confirming the feasibility of replacing lithium metal negative electrodes with lithium-free metal negative electrodes.

The use of separators with specific functions can also effectively protect the lithium anode. Choi et al. reported an inexpensive, nonporous polyurethane separator [74]. Although the separator has no pore structure, it has good electrolyte wettability and excellent Li-ion conductivity. As shown in Figure 8.12, the use of this separator can successfully prevent pollutants such as O_2 and H_2O from directly contacting the lithium anode, avoiding the corrosion of the lithium anode. In addition, the use of a solid electrolyte can also inhibit the growth of lithium dendrites, and it acts as a protective layer to prevent the passage of pollutants. However, in the SSE system, the solid–solid interfacial impedance of the battery is relatively high, and only a few materials have been tested in lithium–oxygen batteries so far.

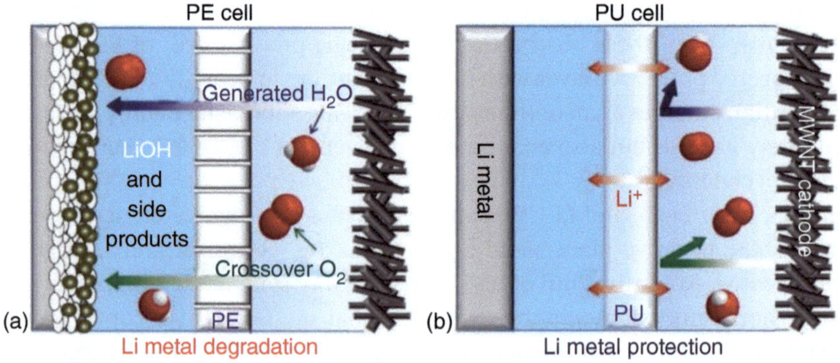

Figure 8.12 Schematic diagram of a lithium–oxygen battery with (a) porous polyethylene separator and (b) a nonporous polyurethane separator (PU) as the separator.
Source: Reproduced with permission from Kim et al. [74]. © 2016/John Wiley & Sons.

Although these innovative works have helped achieve great progress in understanding the interfacial chemistry involving metallic Li-O_2/CO_2 and organic electrolytes, further knowledge is needed. Finding other effective methods to inhibit the growth of lithium dendrites and developing new high-capacity negative plate materials are the prerequisites for realizing the practical application of Li-O_2 and Li-CO_2 batteries. Therefore, we believe that exploring the properties of the Li metal/electrolyte interface will be an important topic for future research.

8.6 Application Potential Analysis

Craft an innovative blueprint for a sustainable and decarbonized future by engineering efficient energy storage and conversion systems. These systems will serve as a green alternative to traditional fossil fuels, propelling the twenty-first-century global community towards a cleaner, more energy-efficient era. Among them, Li-O_2 and Li-CO_2 batteries with high energy density show higher theoretical energy density than other traditional energy storage systems (such as lithium-ion, lead–acid, and Ni-MH batteries) and are promising candidates for large-scale energy storage. However, the development of Li-O_2 and Li-CO_2 batteries is still in its infancy. Although tremendous progress has been made in recent years and some advanced battery models have been fabricated in the laboratory, significant challenges remain. The following fundamental challenges need to be overcome before practical application:

(1) The electrochemical performances (capacity, rate capability, cycle life, efficiency, etc.) of the battery need to be improved further, which involves developing and designing rational cathodes, exploring accommodative electrolytes, and protecting the lithium anode.
(2) As Li-O_2 and Li-CO_2 batteries operate in an open system, parasitic reactions caused by contamination (H_2O, N_2, etc.) that occur in particular on the anode put sand in the wheels of the development of these batteries. Some approaches addressing this issue, such as using a special electrolyte and additives, coating a protective layer, and soaking with organic solvent, have been found. In addition, the application of a flexible SSE may alleviate these problems to a large extent; however, low ionic conductivity is another serious problem. Furthermore, there have been very few investigations toward the detailed characterization of the reaction dynamics of solid-based systems. Therefore, the development of flexible SSEs with high ionic conductivity and accommodation for all battery components is critical.
(3) Intrinsic descriptors of Li-O_2 and Li-CO_2 battery systems need to be determined. In the distant future, materials scientists may rely on supercomputers to calculate all physical parameters of all fundamental reactions under all conditions, thereby greatly reducing the need for large-scale experiments. However, this approach is considered too demanding, or even impractical, for current computers. Therefore, artificial intelligence can be used in this approach, which may

contribute to the evaluation of the physical and chemical properties of materials and promote the development of Li-O_2 and Li-CO_2 batteries.

(4) The application of lithium batteries in other fields should be explored, e.g. Mars exploration and subsea operations. Interstellar missions involving unmanned rovers, such as NASA's Curiosity and Perseverance, have been considered references for this insight. Lithium-ion batteries have been used to provide energy for the rover's operation. However, the low specific energy of lithium-ion batteries leads to increased mission cost and payload, requiring the development of alternative battery chemistries with higher energy storage capacity and efficient operation. In addition, the Martian atmosphere contains about 95.6% carbon dioxide, which provides natural working conditions for Li-CO_2 batteries, so lithium–carbon dioxide batteries have become the most attractive option for Mars interstellar missions.

In summary, although Li-O_2 and Li-CO_2 batteries have made significant progress in energy storage and conversion, these novel electrochemical systems are in their infancy and need a lot of research. However, we firmly believe that the development of practical Li-O_2 and Li-CO_2 batteries will revolutionize the field of electrochemical power sources, which can surpass the performance of most of the current battery technologies with ultrahigh energy density, but with a much lower cost and lower environmental impact. Furthermore, it will play an important role in building an environment-friendly society.

References

1 Martin, E., and Marc, T. (2006). The 21st Century Electric Car, Tesla Motors, 17.

2 Nazri, G.-A. and Gianfranco, P. (2008). *Lithium Batteries: Science and Technology*. Springer Science & Business Media.

3 Scrosati, B. (2002). *Advances in Lithium-Ion Batteries*. Springer.

4 Linden, D. (1984). *Handbook of Batteries and Fuel Cells*. New York.

5 Vinodkumar, E., Rotem, M., Ran, E. et al. (2011). Challenges in the development of advanced Li-ion batteries: a review. *Energy & Environmental Science* 4 (9): 3243–3262.

6 Lu, J., Li, L., Jin-Bum, P. et al. (2014). Aprotic and aqueous Li-O_2 batteries. *Chemical reviews* 114 (11): 5611–5640.

7 Abraham, K. and Jiang, Z. (1996). A polymer electrolyte-based rechargeable lithium/oxygen battery. *Journal of the Electrochemical Society* 143: 1.

8 Ogasawara, T., Aurélie, D., Michael, H. et al. (2006). Rechargeable Li_2O_2 electrode for lithium batteries. *Journal of the American Chemical Society* 128 (4): 1390–1393.

9 Balaish, M., Alexander, K., and Yair, E.-E. (2014). A critical review on lithium-air battery electrolytes. *Physical Chemistry Chemical Physics* 16 (7): 2801–2822.

10 Wang, L., Xu, D., Xu, J. et al. (2012). Graphene oxide gel-derived, free-standing, hierarchically porous carbon for high-capacity and high-rate rechargeable Li-O_2 batteries. *Advanced Functional Materials* 22 (17): 3699–3705.

11 Littauer, E. and Keh, C. (1976). Anodic behavior of lithium in aqueous electrolytes: ii. Mechanical passivation. *Journal of the Electrochemical Society* 123 (7): 964.

12 Visco, S., Nimon, E., Katz, B., et al. (2004). The 12th International Meeting on Lithium Batteries. Extended Abstracts, 53.

13 Puech, L., Cantau, C., Vinatier, P. et al. (2012). Elaboration and characterization of a free standing LiSICON membrane for aqueous lithium-air battery. *Journal of Power Sources* 214: 330–336.

14 Bhatt, M., Geaney, H., Nolan, M. et al. (2014). Key scientific challenges in current rechargeable non-aqueous Li-O_2 batteries: experiment and theory. *Physical Chemistry Chemical Physics* 16 (24): 12093–12130.

15 Li, L., Chen, C., and Yu, A. (2017). New electrochemical energy storage systems based on metallic lithium anode-the research status, problems and challenges of lithium-sulfur, lithium-oxygen and all solid state batteries. *Science China Chemistry* 60: 1402–1412.

16 Manthiram, A., Yu, X., and Wang, S. (2017). Lithium battery chemistries enabled by solid-state electrolytes. *Nature Reviews Materials* 2 (4): 1–16.

17 Kumar, B., Kumar, J., Leese, R. et al. (2009). A solid-state, rechargeable, long cycle life lithium-air battery. *Journal of The Electrochemical Society* 157 (1): A50.

18 Takada, K., Inada, T., Kajiyama, A. et al. (2004). Solid state batteries with sulfide-based solid electrolytes. *Solid State Ionics* 172 (1–4): 25–30.

19 Wang, J., Li, Y., and Sun, X. (2013). Challenges and opportunities of nanostructured materials for aprotic rechargeable lithium-air batteries. *Nano Energy* b2 (4): 443–467.

20 Black, R., Lee, J., Adams, B. et al. (2013). The role of catalysts and peroxide oxidation in lithium-oxygen batteries. *Angewandte Chemie International Edition* 52 (1): 392–396.

21 Wang, Y. and Zhou, H. (2010). A lithium-air battery with a potential to continuously reduce O_2 from air for delivering energy. *Journal of Power Sources* 195 (1): 358–361.

22 Yoo, E. and Zhou, H. (2011). Li-air rechargeable battery based on metal-free graphene nanosheet catalysts. *ACS Nano* 5 (4): 3020–3026.

23 Liu, Y., Wang, L., Cao, L. et al. (2017). Understanding and suppressing side reactions in Li-air batteries. *Materials Chemistry Frontiers* 1 (12): 2495–2510.

24 Aurbach, D., McCloskey, B., Nazar, L. et al. (2016). Advances in understanding mechanisms underpinning lithium-air batteries. *Nature Energy* 1 (9): 1–11.

25 (a) Horstmann, B., Gallant, B., Mitchell, R. et al. (2013). Rate-dependent morphology of Li_2O_2 growth in Li-O_2 batteries. *The Journal of Physical Chemistry Letters* 4 (24): 4217–4222. (b) Wang, F., Li, Y., Xia, X.X. et al. (2021). Metal-CO_2 electrochemistry: From CO_2 recycling to energy storage. *Advanced Energy Materials* 11 (25): 2100667. (c) Wang, F., Niu, S.W., Liang, X.Q. et al. (2022).

Phosphorus incorporation activates the basal plane of tungsten disulfide for efficient hydrogen evolution catalysis. *Nano Research* 15: 2855–2861.

26 Wang, G., Huang, L., Huang, W. et al. (2015). Nanostructured porous RuO_2/MnO_2 as a highly efficient catalyst for high-rate Li-O_2 batteries. *Nanoscale* 7 (48): 20614–20624.

27 Aetukuri, N., McCloskey, B., García, J. et al. (2015). Solvating additives drive solution-mediated electrochemistry and enhance toroid growth in non-aqueous Li-O_2 batteries. *Nature chemistry* 7 (1): 50–56.

28 Kraytsberg, A. and Ein-Eli, Y. (2011). Review on Li-air batteries-opportunities, limitations and perspective. *Journal of Power Sources* 196 (3): 886–893.

29 Lu, Z., Zhou, Y., Dai, W. et al. (2017). Recent advances in understanding of the mechanism and control of Li_2O_2 formation in aprotic Li-O_2 batteries. *Chemical Society Reviews* 46 (19): 6046–6072.

30 Luo, L., Liu, B., Song, S. et al. (2017). Revealing the reaction mechanisms of Li-O_2 batteries using environmental transmission electron microscopy. *Nature Nanotechnology* 12 (6): 535–539.

31 Peng, Z., Freunberger, S., Hardwick, L. et al. (2011). Oxygen reactions in a non-aqueous Li^+ electrolyte. *Angewandte Chemie International Edition* 50 (28): 6351–6355.

32 Ganapathy, S., Adams, B., Stenou, G. et al. (2014). Nature of Li_2O_2 oxidation in a Li-O_2 battery revealed by operando X-ray diffraction. *Journal of the American Chemical Society* 136 (46): 16335–16344.

33 Hou, Y., Wang, J., Liu, L. et al. (2017). Mo_2C/CNT: an efficient catalyst for rechargeable Li-CO_2 batteries. *Advanced Functional Materials* 27 (27): 1700564.

34 Qiao, Y., Yi, J., Wu, S. et al. (2017). Li-CO_2 electrochemistry: a new strategy for CO_2 fixation and energy storage. *Joule* 1 (2): 359–370.

35 Yang, S., Qiao, Y., He, P. et al. (2017). A reversible lithium-CO_2 battery with Ru nanoparticles as a cathode catalyst. *Energy & Environmental Science* 10 (4): 972–978.

36 Jiao, Y., Qin, J., Sari, H. et al. (2021). Recent progress and prospects of Li-CO_2 batteries: mechanisms, catalysts and electrolytes. *Energy Storage Materials* 34: 148–170.

37 (a) Xing, Y., Yang, Y., Li, D. et al. (2018). Crumpled Ir nanosheets fully covered on porous carbon nanofibers for long-life rechargeable lithium-CO_2 batteries. *Advanced Materials* 30 (51): 1803124. (b) Chen, M.H., Meng, H.Y., Wang, F. et al. (2023). Dispersed Mn_2Co_2C Nanoparticles in interconnected nitrogen-doped carbon framework as cathode catalysts for efficient and long-life Li-CO_2 batteries. *Chemical Engineering Journal* 455: 140564.

38 Montoya, J., Seitz, L., Chakthranont, P. et al. (2017). Materials for solar fuels and chemicals. *Nature Materials* 16 (1): 70–81.

39 Xie, J., Liu, Q., Huang, Y. et al. (2018). A porous Zn cathode for Li-CO_2 batteries generating fuel-gas CO. *Journal of Materials Chemistry A* 6 (28): 13952–13958.

40 Lee, J., Tai, K., Cao, R. et al. (2011). Metal-air batteries with high energy density: Li-air versus Zn-air. *Advanced Energy Materials* 1 (1): 34–50.

41 Zheng, J., Liang, R., Hendrickson, M. et al. (2008). Theoretical energy density of Li-air batteries. *Journal of the Electrochemical Society* 155 (6): A432.

42 (a) Liu, C., Younesi, R., Tai, C. et al. (2016). 3-D binder-free graphene foam as a cathode for high capacity Li-O_2 batteries. *Journal of Materials Chemistry A* 4 (25): 9767–9773. (b) Li, Q., Xu, P., Gao, W. et al. (2014). Graphene/graphene-tube nanocomposites templated from cage-containing metal-organic frameworks for oxygen reduction in Li-O_2 batteries. *Advanced Materials* 26 (9): 1378–1386.

43 (a) Zhang, X., Zhang, Q., Zhang, Z. et al. (2015). Rechargeable Li-CO_2 batteries with carbon nanotubes as air cathodes. *Chemical Communications* 51 (78): 14636–14639. (b) Zhang, Z., Zhang, Q., Chen, Y. et al. (2015). The first introduction of graphene to rechargeable Li-CO_2 batteries. *Angewandte Chemie* 127 (22): 6650–6653.

44 Ci, L., Song, L., Jin, C. et al. (2010). Atomic layers of hybridized boron nitride and graphene domains. *Nature materials* 9 (5): 430–435.

45 Hu, C., Yu, C., Li, M. et al. (2015). Nitrogen-doped carbon dots decorated on graphene: a novel all-carbon hybrid electrocatalyst for enhanced oxygen reduction reaction. *Chemical communications* 51 (16): 3419–3422.

46 Lu, J., Jung, L., Luo, X. et al. (2016). A lithium-oxygen battery based on lithium superoxide. *Nature* 529 (7586): 377–382.

47 Tu, F., Hu, J., Xie, J. et al. (2016). Au-decorated cracked carbon tube arrays as binder-free catalytic cathode enabling guided Li_2O_2 inner growth for high-performance Li-O_2 batteries. *Advanced Functional Materials* 26 (42): 7725–7732.

48 Jian, Z., Liu, P., Li, F. et al. (2014). Core-shell-structured CNT@ RuO_2 composite as a high-performance cathode catalyst for rechargeable Li-O_2 batteries. *Angewandte Chemie International Edition* 53 (2): 442–446.

49 Zhang, B., Jiao, Y., Chao, D. et al. (2019). Targeted synergy between adjacent Co atoms on graphene oxide as an efficient new electrocatalyst for Li-CO_2 batteries. *Advanced Functional Materials* 29 (49): 1904206.

50 Li, J., Kuppler, R., and Zhou, H. (2009). Selective gas adsorption and separation in metal-organic frameworks. *Chemical Society Reviews* 38 (5): 1477–1504.

51 Lei, Z., Xue, Y., Chen, W. et al. (2018). New opportunities for energy-related CO_2 conversion. *Advanced Energy Materials* 8: 1801587–1801617.

52 Li, S., Liu, Y., Zhou, J. et al. (2019). Monodispersed MnO nanoparticles in graphene-an interconnected N-doped 3D carbon framework as a highly efficient gas cathode in Li-CO_2 batteries. *Energy & Environmental Science* 12 (3): 1046–1054.

53 Liu, B., Sun, Y., Liu, L. et al. (2019). Recent advances in understanding Li-CO_2 electrochemistry. *Energy & Environmental Science* 12 (3): 887–922.

54 Li, J., Zhao, H., Qi, H. et al. (2019). Drawing a pencil-trace cathode for a high-performance polymer-based Li-CO_2 battery with redox mediator. *Advanced Functional Materials* 29 (11): 1806863.

55 Pipes, R., Bhargav, A., and Manthiram, A. (2019). Phenyl disulfide additive for solution-mediated carbon dioxide utilization in Li-CO_2 batteries. *Advanced Energy Materials* 9 (21): 1900453.

56 Wang, Y., Lai, N., Lu, Y. et al. (2018). A solvent-controlled oxidation mechanism of Li_2O_2 in lithium-oxygen batteries. *Joule* 2 (11): 2364–2380.

57 Read, J. (2005). Ether-based electrolytes for the lithium/oxygen organic electrolyte battery. *Journal of the Electrochemical Society* 153 (1): A96.

58 Jung, H., Hassoun, J., Park, J. et al. (2012). An improved high-performance lithium-air battery. *Nature chemistry* 4 (7): 579–585.

59 Bryantsev, V., Giordani, V., Walker, W. et al. (2011). Predicting solvent stability in aprotic electrolyte Li-air batteries: nucleophilic substitution by the superoxide anion radical ($O_2^{\bullet -}$). *The Journal of Physical Chemistry A* 115 (44): 12399–12409.

60 Freunberger, S., Chen, Y., Drewett, N.E. et al. (2011). The lithium-oxygen battery with ether-based electrolytes. *Angewandte Chemie International Edition* 50 (37): 8609–8613.

61 Peng, Z., Freunberger, S., Chen, Y. et al. (2012). A reversible and higher-rate Li-O_2 battery. *Science* 337 (6094): 563–566.

62 Mizuno, F., Takechi, K., Higashi, S. et al. (2013). Cathode reaction mechanism of non-aqueous Li-O_2 batteries with highly oxygen radical stable electrolyte solvent. *Journal of Power Sources* 228: 47–56.

63 Oswald, S., Mikhailova, D., Scheiba, F. et al. (2011). XPS investigations of electrolyte/electrode interactions for various Li-ion battery materials. *Analytical and Bioanalytical Chemistry* 400: 691–696.

64 Burke, C., Pande, V., Khetan, A. et al. (2015). Enhancing electrochemical intermediate solvation through electrolyte anion selection to increase nonaqueous Li-O_2 battery capacity. *Proceedings of the National Academy of Sciences* 112 (30): 9293–9298.

65 Qiao, Y., Yi, J., Guo, S. et al. (2018). Li_2CO_3-free Li-O_2/CO_2 battery with peroxide discharge product. *Energy & Environmental Science* 11 (5): 1211–1217.

66 Goodenough, J., Hong, H., and Kafalas, J. (1976). Fast Na^+-ion transport in skeleton structures. *Materials Research Bulletin* 11 (2): 203–220.

67 Xu, X., Wen, Z., Yang, X. et al. (2008). Dense nanostructured solid electrolyte with high Li-ion conductivity by spark plasma sintering technique. *Materials Research Bulletin* 43 (8–9): 2334–2341.

68 Li, C., Guo, Z., Yang, B. et al. (2017). A rechargeable Li-CO_2 battery with a gel polymer electrolyte. *Angewandte Chemie International Edition* 56 (31): 9126–9130.

69 Asadi, M., Sayahpour, B., Abbasi, P. et al. (2018). A lithium-oxygen battery with a long cycle life in an air-like atmosphere. *Nature* 555 (7697): 502–506.

70 Mao, Y., Tang, C., Tang, Z. et al. (2019). Long-life Li-CO_2 cells with ultrafine IrO_2-decorated few-layered δ-MnO_2 enabling amorphous Li_2CO_3 growth. *Energy Storage Materials* b18: 405–413.

71 Shui, J., Okasinski, J., Kenesei, P. et al. (2013). Reversibility of anodic lithium in rechargeable lithium-oxygen batteries. *Nature Communications* 4 (1): 2255.

72 Walker, W., Giordani, V., Uddin, J. et al. (2013). A rechargeable Li-O_2 battery using a lithium nitrate/N, N-dimethylacetamide electrolyte. *Journal of the American Chemical Society* 135 (6): 2076–2079.

73 Hassoun, J., Jung, H., Lee, D. et al. (2012). A metal-free, lithium-ion oxygen battery: a step forward to safety in lithium-air batteries. *Nano Letters* 12 (11): 5775–5779.

74 Kim, B., Kim, J., Min, J. et al. (2016). A moisture-and oxygen-impermeable separator for aprotic Li-O_2 batteries. *Advanced Functional Materials* 26 (11): 1747–1756.

9

Supercapacitors

Jiawei Zhang[1] and Minghua Chen[2]

[1] Harbin University of Science and Technology, School of Electrical and Electronic Engineering, 52 Xuefu Road, Harbin, 150080, China
[2] Harbin University of Science and Technology, School of Electrical and Electronic Engineering, Key Laboratory of Engineering Dielectric and Applications (Ministry of Education), 52 Xuefu Road, Harbin, 150080, China

9.1 Brief Introduction

The concept of supercapacitors (SCs) was proposed in the middle of the nineteenth century, which has a long history and evolution. The first SC was proposed and developed by Becker without using electrical double-layer capacitors (EDLCs) [1]. Later, SOHIO expanded another version of SC in 1966 and patented the same [2, 3]. In 1982, Pinnacle Research Institute (PRI) developed the first military SC: PRI ultracapacitors. Moreover, Maxwell Laboratories focused on the superior power density of SCs and designed "Boost Caps". Furthermore, Evans proposed a tantalum electrolytic capacitor that combined electrolytic and electrochemical properties with high voltage; however, the application of this capacitor was limited to military applications [4, 5]. In recent years, a new type of capacitor, the battery–SC hybrid capacitor (HC), has been developed, which has advantages of both battery and SC with respect to performance, which will be discussed in Chapter 10. Nowadays, SCs play an important role in the field of energy storage, and researchers aim to enhance their characteristics, optimize their electrochemical performance, and decrease their production costs. In general, SCs show a higher energy density and specific capacity than conventional capacitors, and a higher power density and superior cycling stability than lithium-ion batteries. This way, SCs play a crucial role in the field of energy storage. Therefore, in this chapter, an in-depth discussion about SCs is presented, including their energy storage mechanism, applications in commercial systems, and electrolyte and device configuration.

9.2 Energy Storage Mechanism

The energy storage performance of SCs can be significantly improved by increasing the flight speed. At present, SCs are generally composed of electrodes, current

Towards Next Generation Energy Storage Technologies: From Fundamentals to Commercial Applications, First Edition. Edited by Minghua Chen.

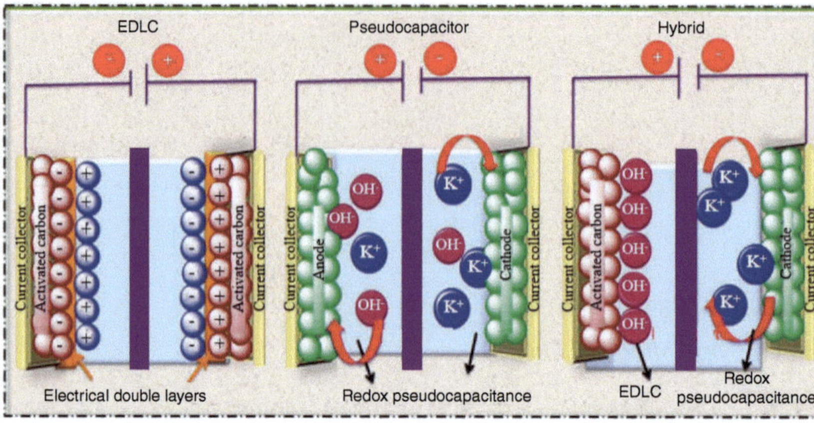

Figure 9.1 Energy storage mechanism of the supercapacitor. Source: Ray and Saruhan [6]/MDPI/CC BY 4.0.

collectors, and electrolytes, as shown in Figure 9.1. According to their distinct structural configurations, SCs can be classified into symmetric SCs, which have two exactly same electrodes, and asymmetric SCs, which have two dissimilar electrodes (different materials or different energy storage mechanisms). Based on the charge storage mechanism, SCs can be classified into three types [6]: (i) EDLCs, (ii) pseudocapacitors (PCs), and (iii) HCs. The detailed electrochemical processes are explained below:

9.2.1 EDLCs

EDLCs can be considered holding their place between the battery and the capacitor. They do not show the change of electrode materials and redox reaction. Therefore, EDLCs exhibit faster charging–discharging processes, longer cycling life, and wider working temperature range. The concept of EDLCs was first proposed by Helmholtz, the German physicist, and EDLCs can be considered as a new kind of capacitor. As we all know, there will be excess charges of opposite signs at the interface between a solid-state electrode and a liquid electrode, which will cause a potential difference between the phases. Then, if the two electrodes are immersed in the electrolyte and a voltage lower than the electrochemical stable windows (ESW) of the electrolyte is applied, the positive and negative ions in the electrolyte will move rapidly to the cathode and anode, respectively, under the action of the applied electric field and will be separated at the two poles. A tight charge layer is formed at the surface of each electrode, i.e. an electric double layer (EDL). This EDL is similar to the polarized charge generated by the dielectric in traditional capacitors under the action of an electric field, thereby producing a capacitive effect. Furthermore, the compact EDL is similar to a flat plate capacitor, but due to the small distance (<0.5 nm) between the charge layers, it is smaller than that of a traditional capacitor; hence, it has a higher capacity than traditional capacitors. In principle, EDLCs deliver a high power density (>10 kW kg^{-1}) and superior cycling stability, which is attributable to

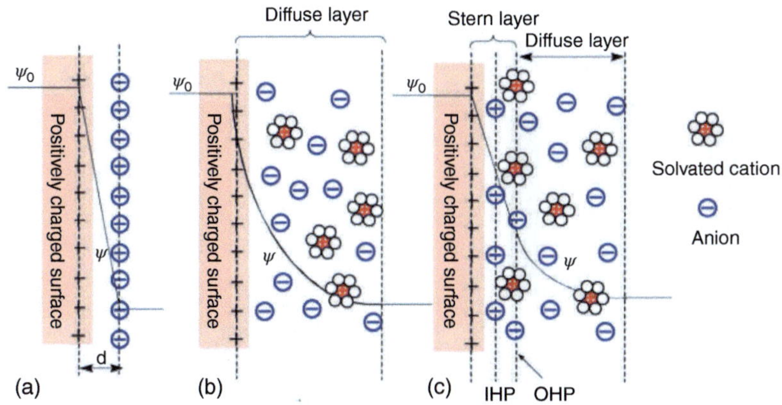

Figure 9.2 Schematic of EDLCs. (a) the Helmholtz model, (b) the Gouy–Chapman model, and (c) the Stern model, showing the inner Helmholtz plane (IHP) and outer Helmholtz plane (OHP) Source: Reproduced with permission from Zhang and Zhao [7]/© 2009/ The Royal Society of Chemistry.

the charge storage mechanism of electrostatic interactions at the interface between the electrode and electrolyte ions.

As shown in Figure 9.2a, according to the model proposed by Helmholtz, opposite charges are layered at the electrode–electrolyte interface and separated by an atomic distance, similar to two-plate, conventional capacitors [7]. The Helmholtz model was modified by Gouy and Chapman into a diffusive layer model, as shown in Figure 9.2b. The capacitance of two separated charge arrays increases with their distance. It is generally believed that EDLCs have the same capacitive behavior as parallel plate capacitors. Therefore, when the point charge is close to the electrode surface, the magnitude of capacitance increases with an increase in the point charge. Furthermore, Stern proposed a model combining the Helmholtz and Gouy–Chapman models to suggest two regions of the particle distribution – the inner region called the Stern layer (dense layer) and the diffuse layer (Figure 9.2c). Ions (usually hydrated) are an effective adsorbate for the conductor. Besides, anions can be regarded as specific ions adsorbed at the compact layer (inner Helmholtz planes) and counterions as non-specific adsorbate ions (outer Helmholtz planes). Based on the Gouy–Chapman model, it is easy to understand the diffusive layer: the kinetic energy of counterions results in a diffuse double layer affected by the thickness.

The capacitance of EDLCs is contributed by two regions: the Stern layer, denoted as C_H, and the diffusive region, denoted as C_{diff}. The value of capacitance can be calculated using the following mathematical equation:

$$\frac{1}{C_{dl}} = \frac{1}{H} + \frac{1}{C_{diff}} \tag{9.1}$$

The electrical conductance of an EDLC at a planar electrode surface is controlled by the electrical field across the electrode and electrolyte and the chemical affinities between them. It is strongly dependent on the porous structure and surface area of electrode materials as the energy storage mechanism of EDLCs is based on

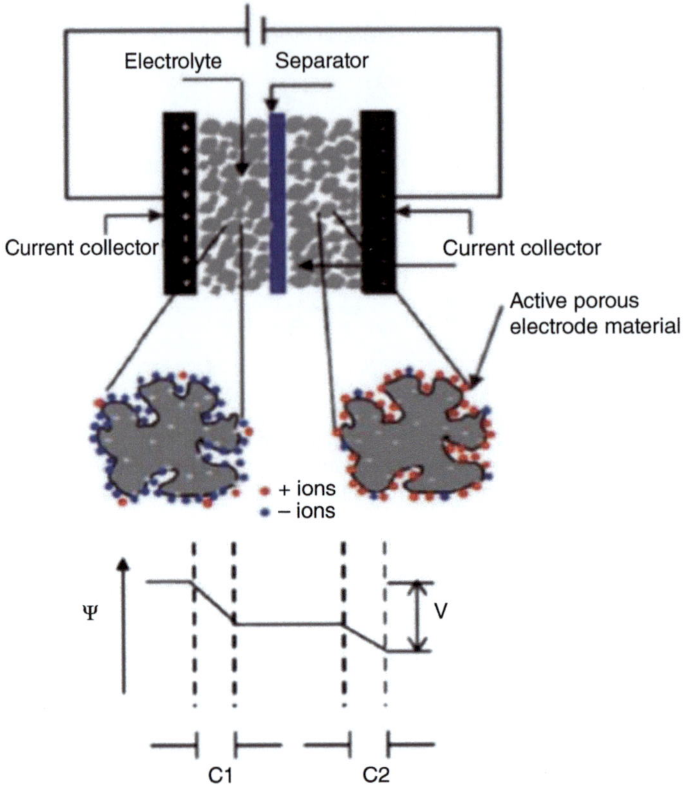

Figure 9.3 Schematic of an EDLC made from porous electrode materials.
Source: Reproduced with permission from Zhang and Zhao [7]/© 2009/The Royal Society of Chemistry.

the adsorption–desorption of electrolyte ions. Thus, particle transport in a closed system is controlled by the affinity of the curved mass transfer path, the limitations of the area in the pore, the electrical phenomena associated with the solution, and the wettability of the solution to the pore surface. Figure 9.3 shows the preparation mechanism of an EDLC using porous carbon conductive materials. The specific capacitance of an EDLC can be calculated as follows:

$$C = \frac{\varepsilon_r \varepsilon_0 \times A}{d} \tag{9.2}$$

where ε_r, ε_0, A, and d are vacuum dielectric constant of the electrolyte, dielectric constant of the electrolyte, available specific surface area (SSA), and the effective thickness of the EDL (Debye length). Specific capacitance (C) and available area (A) are closely related: the larger A, the higher C. Interestingly, many studies have shown that the relationship between A and C is not linear in practice. Some carbon-based electrodes with a larger surface area deliver a lower specific capacitance than that of bulk carbon materials because internal electrochemical processes are complicated in reality. Ions in electrolytes usually react with solvent molecules to form solvated ions and exist in the capacitor system, resulting in tiny pores being useless, thus making ion solvation an energy-consuming process.

Initially, it is suggested that effective capacitance can be generated by the accumulation of pores according to the solvent energy and traditional EDL theory. Nevertheless, Doron Aurbach [8, 9] investigated the electrochemical processes between activated carbon (AC) with different pore sizes and different electrolytes, and suggested that AC shows capacitance when the size of the pores and electrolyte ions is similar. It is worth mentioning that the size of solvation ions varies as the solvation shell and is not the same. Since then, numerous studies have demonstrated that the capacitance of ions in sub-nanopores increases abnormally. Moreover, capacitance reaches the maximum value when ion size is close to pore size; thus, the optimal pore size of various electrodes that can generate maximum capacitance in different electrolytes is different [10]. Chmiola et al. [11, 12] proposed a theory explaining the different capacitance phenomena under different pore sizes as they found an abnormal increase in capacitance in sub-nanopores. The difference between the size of pores and that of solvation ions can be classified into three types: (i) when the size of pores is two times larger than that of solvated ions, specific capacitance can be explained by the Stern model; (ii) when the size of pores is between one and two times that of solvated ions, the electrical layer cannot be formed due to the impact of the dense layer adjacent to the pore wall; and (iii) reducing the size of pores to less than that of solvated ions results in an abnormal increase in normalized capacitance. The solvent shell will be partially removed as ions squeeze through the pores. They concluded that the partial removal of solvent molecules and increased ion confinement lead to an unusual increase in capacitance. To further verify the effect of ion confinement on capacitance, the electrochemical performance of carbide-derived carbon (CDC) was investigated using ionic liquids (ILs) with precise ion size and no solvent shell formation as electrolytes [13]. Results showed that the maximum capacitance is reached when the pore size is close to the ion size. Increasing or decreasing porosity resulted in a significant decrease in capacitance.

Meanwhile, Huang et al. proposed a theory to enhance the understanding of unusual improvement in capacitance in micropores [14–16]. In carbon-based electrode materials, pore shape is highly contingent on synthesis methods, including cylindrical pores, slit pores, and so on. Among them, cylindrical pores are often used as theoretical models for gas physical sorption and impedance spectroscopy. Therefore, Huang et al. reported advanced electric double-cylinder capacitors (EDCCs), with the model hypothesis that mesopores in the electrode are cylindrical pores. In addition, the effect of curvature was introduced into the model. In EDCCs, the ions that enter the pores are solvated and adsorbed on pore walls, and capacitance can be evaluated based on the following equation:

$$C = \frac{2\pi\varepsilon_r\varepsilon_0 L}{\ln(b/a)} \text{ or } \frac{C}{A} = \frac{\varepsilon_r\varepsilon_0}{b\ln[b/(b-d)]} \tag{9.3}$$

where L is the pore length, and b and a are the radius of the outer and inner cylinders, respectively. In contrast to mesopores, the size of micropores is too small to generate a double cylinder and contribute to capacitance. Hence, Huang et al. proposed electric wire-in-cylinder capacitors (EWCCs) and showed that the center of ions in cylindrical micropores lines up along the pore axis to form EWCCs. The corresponding capacitance can be calculated as follows:

$$\frac{C}{A} = \frac{\varepsilon_r \varepsilon_0}{b \ln(b/a_0)} \tag{9.4}$$

where a_0 is the radius of the inner cylinder.

The electrochemical behavior of macropores is different from that of mesopores and micropores, which can be explained by the theory of classic EDLCs, as pores with inconspicuous curvature can be approximated as a plane. Combined with Huang's model, capacitance of porous carbon-based materials with negative curvature can be evaluated using the models of EDCCs, EWCCs, and EDLCs. However, not all carbon materials have negative curvature. For example, onion-like carbon and carbon nanotubes (CNTs) with special structures show a positive curvature in which ions can only adsorb on outer surfaces, and new theoretical models are required to calculate their capacitance. For onion-like carbon, an exohedral electric double-sphere capacitor (xEDSC) is formed due to the physical adsorption of solvated ions on the outer surface. Ion adsorption in CNTs is similar to that in onion-like carbon, but an exohedral electric double-cylinder capacitor (xEDCC) is formed due to the different architecture from that of onion-like carbon. The capacitance of xEDSCs and xEDCCs can be calculated as follows:

$$C = \frac{4\pi\varepsilon_r \varepsilon_0}{b - a} \text{ or } \frac{C}{A} = \frac{\varepsilon_r \varepsilon_0 (a + d)}{ad} \tag{9.5}$$

$$\frac{C}{A} = \frac{\varepsilon_r \varepsilon_0}{a \ln\left[\frac{a+d}{a}\right]} \tag{9.6}$$

where d is the thickness of the EDL, and a and b are the radius of the inner sphere/cylinder and outer sphere/cylinder, respectively.

In addition to macropores, mesopores, and micropores, slit-shaped pores appear in most carbonaceous materials. Feng et al. [17] investigated the distribution of K^+ in charged slit-shaped micropores using molecular dynamics simulation, proposed the sandwich capacitor model, and described the formation mechanism of 1.0~1.47 nm slit micropores. The sandwich capacitor consists of a layer of counterions that is adsorbed between two electrodes with the same polarity (pore size being $2b$) [18]. Meanwhile, the slit capacitor can be considered two parallel capacitors thanks to the two electrodes sharing the net charge of counterions. Thus, the capacitance of slit pores can be calculated as follows:

$$C_{tot} = 2C_s = 2\varepsilon_r \varepsilon_0 A/d_{eff} \text{ or } C_{tot}/2A = C_s/A = \varepsilon_r \varepsilon_0/(b - a_0) \tag{9.7}$$

where C_{tot} is the capacitance of the system, C_s is the capacitance of a single capacitor that consists of one electrode and one side of counterions, and d_{eff} is the effective distance between the electrode surface and counterions, which is equal to $(b - a_0)$, as described for EWCCs.

For EDCC and EWCC models, researchers successfully obtained a good fit with the mathematical results from Eqs. (9.6) and (9.7) for carbon materials and electrolytes [7]. Inconsistent increments in capacitance for pore sizes lower than

1 nm and slightly increased capacitance with pore sizes higher than 2 nm can be explained using EWCC and EDCC models. Moreover, the value of dielectric constant calculated by fitting the data into Eq. (9.7) is close to that of the vacuum value, which indicates that the desolation of hydrous ions takes place prior to their penetration of micropores. Despite these theoretical and experimental studies, EDL charge storage mechanisms in nanostructured surfaces have not been fully elucidated. Thus, there is a high demand for in-depth theoretical and experimental studies that accurately assess the charge storage mechanism inside micropores for the development of sustainable EDLCs.

9.2.2 Pseudocapacitors

The energy storage mechanism of PCs) is very different from that of transitional capacitors and EDLCs and is based on reversible or irreversible redox reactions taking place near the surface of the electrode. In general, the capacitance of PCs is higher than that of EDLCs due to the existence of charge transfer. The electrochemical performance of the electrode in PCs is strongly dependent on its surface area and the type of electrolyte. The majority of metal compounds (oxides, nitrides, carbides, etc.), conductive polymers (CPs) (such as polyaniline [PANI], polypyrrole [PPy], and poly(3,4-ethylenedioxythiophene) [PEDOT]), and metal-organic frameworks (MOFs) can be used as electrode materials in PCs. Carbonaceous materials are typically EDLCs; however, heteroatom-doped carbon can be used as the PC electrode since heteroatoms can provide activated sites for reduction/oxidation reaction, delivering a higher specific capacitance than un-doped carbon.

In the redox reaction, PCs store charges via reduction/oxidation processes. Moreover, it is conducive to activating redox pairs on the electrode surface during charging processes by applying an external potential. During discharging, a reversible chemical reaction occurs, resulting in redox-active surface groups being transferred back to their initial state. As a necessary part of every electrochemical energy storage device, electrolytes can provide an internal ionic circuit for redox reactions to proceed. According to the charge storage process, PCs are primarily classified into the following categories [19, 20]:

(1) *Underpotential deposition*: This phenomenon occurs when ions deposit on the interface between the electrode and electrolyte at a potential that is lower than the reversible reduction potential of ions (e.g. H^+ on Pt and Pd^{2+} on Au). Although capacitance is high via this phenomenon (~2200 $\mu F\,cm^{-2}$), the working potential window is narrow (0.3~0.6 V), resulting in a limited energy density.

(2) *Redox pseudocapacitance*: This phenomenon is the most common charge storage process in PCs. Faradaic reactions occur on the surface of the active electrode across which potential is being applied, e.g. many metal oxides (such as RuO_2) and all CPs. This process can be explained as reversible electrochemical adsorption. Based on redox pairs on the surface, redox PCs can deliver a capacitance of ~5000 $F\,cm^{-3}$.

(3) *Intercalation pseudocapacitance*: This is similar to the battery-like intercalation chemistry generated by the redox intercalation chemistry of activated electrodes; however, intercalation PCs show a much faster time scale than conventional battery electrodes, such as Nb_2O_5 and V_2O_5. In this process, besides capacitance contribution from intercalation pseudocapacitance, surface redox pseudocapacitance and EDLCs also exist, leading to an extended electrochemical potential window, thus achieving higher energy density than that of EDLCs.

9.2.2.1 Conductive-Polymer-Based Pseudocapacitors

Various CPs can be used as PC electrodes. Three types of Faradaic processes take place on the electrode surface for charge storage: (i) reversible physical adsorption/desorption, (ii) reversible ion doping/dedoping on the surface of the electrode, and (iii) redox reaction near the surface. Generally, the cycling life of various CPs with low resistance and significant pseudocapacitance is inferior to that of transition metal oxide (TMO) electrodes because of sluggish charge transfer, such as polypyrrole, polyacetylene, polythiophene polyaniline, and poly(phenylenevinylene). The energy storage mechanism of CP-based capacitors is attributable to storing and releasing charges during redox reactions. In the process of oxidation (or ion doping), activated ions of the electrolyte get inserted into the electrode or CP backbone; in the process of reduction (or ion dedoping), ions extract from the electrode material into the electrolyte. Moreover, the charging reaction is distributed throughout the electrode or the internal of electrode materials rather than near the surface of the electrode because EDLCs support the enhancement of specific capacitance. CPs primarily contain π–orbital conjugations that are responsible for good electronic conductivity in polymers. Therefore, CPs can be electrochemically reduced or oxidized by injecting or withdrawing electrons, respectively. The energy density of a CP-based electrode can reach $39\,Wh\,kg^{-1}$ at an excellent power density of $35\,kW\,kg^{-1}$. According to the types of carriers, CPs can be classified into two types: p-doped and n-doped [21].

PANI is an attractive and extensively studied electrode. Diaz et al. first reported the electrochemical behavior of PANI [22]. The systematic study of the electrochemical deposition of polypyrrole has been demonstrated as an effective method for the preparation of CPs via electro-polymerization. Since then, PANI has been regarded as the most promising CP in the field of electrochemical energy storage, thanks to its good polymer-forming ability via electro-polymerization and high electrochemical activity. As early as 1984, Feldberg et al. speculated that there is a proportional capacitive current contribution in the electrochemical processes of CPs [23]. The capacitive properties of CPs are further highlighted by cyclic voltammetry (CV) and electrochemical impedance. Based on Conway's groundbreaking works in establishing the concept of SCs [24], there is no doubt that PANI can be used as a potential electrode material for electrochemical capacitors with pseudocapacitance behavior.

CV curve of a PANI electrode shows three redox peaks in the voltage range of $-0.2{\sim}0.8\,V$ during the charging/discharging process, indicating that there are

Faradic reactions between leucoemeraldine (insulating) and protonated emeraldine (conducting), transition between *p*-benzoquinone and hydroquinone, and redox transition between emeraldine and pernigraniline. These redox peaks embedded in a high background current indicate the pseudocapacitive performance of PANI [25, 26].

A typical feature of SCs is that there is no charge–discharge plateau in galvanostatic charge–discharge (GCD) curves, as represented by the diagonal line in charge–discharge profiles. When using pure PANI materials as the cathode of lithium-ion batteries, GCD curves are similar to those of SCs, i.e. no plateau appears during charge–discharge processes. However, there are reports on PANI SCs in which 80% of the charge is delivered at 20% of the potential window. The electrochemical performance is comparable to the best performance of lithium-ion batteries; therefore, it is inappropriate to regard it as a SC. This does not mean that the latter systems cannot store energy, but wrong labeling (SC instead of battery) limits their potential for practical development.

9.2.2.2 Metal Compound Pseudocapacitors

Besides CPs, metal compounds, especially metal oxides, can also significantly enhance specific capacity and energy density, such as MnO_2, WO_3, VO_x, Fe_3O_4, RuO_2, $Co(OH)_2$, Co_3O_4, $Ni(OH)_2$, and NiO due to their improved thermodynamic and mechanical stability. Therefore, metal compounds have been extensively used as high-performance PC electrode materials [27, 28].

***RuO*$_2$** To identify the critical roles of RuO_2 in the field of electrochemical energy storage, its redox activity needs to be understood. RuO_2 electrodes have been investigated since the early 1970s as they have both electronic and ionic conductivity [29]. In 1971, a new electrochemical capacitor was proposed based on RuO_2 electrodes, named a PC, which involved a Faraday-type charge storage process. The theoretically estimated specific capacitance of RuO_2 was about 1400~2000 F g^{-1}. The mechanism of charge storage and discharge associated with the proton-coupled electron exchange process from mixed conducting hydrous RuO_2 is precisely shown in Eq. (9.8):

$$Ru^{IV} - O^{2-} + H^+ + e^- = Ru^{III} - OH^- \tag{9.8}$$

Thus, it is easy to infer that the oxide surface can be charged using two procedures: using a suitable electrode potential and tuning the pH of the medium [30].

***MnO*$_2$** MnO_2 has garnered significant attention as a SC electrode material as a result of its eminent characteristics such as affluent crystal structures (α, β, γ, δ, λ, etc.), excellent theoretical capacitance (1370 F g^{-1}), multiple valence states of manganese, low cost, high natural profusion, and good ecological compatibility. In 1999, Lee and Goodenough carried out a revolutionary work on manganese dioxide for SCs. The capacitive response of manganese dioxide is primarily pseudocapacitive, resulting from redox activities at the surface and in the bulk of the electrode. The adsorption of C^+ cations onto the surface of the electrode and embedment/de-embedment of C^+ cations in the bulk of MnO_2 are represented by the following reactions:

$$(MnO_2)_{surface} + e^- + C^+ \leftrightarrow (MnOOC)_{surface} \tag{9.9}$$

$$MnO_2 + e^- + C^+ \leftrightarrow MnOOC \tag{9.10}$$

Here, C^+ is the cation (K^+, Na^+, etc.) from the electrolyte [31].

Ni-/Co-Based Oxides and Hydroxides Very recently, much effort has been focused on Ni-/Co-based oxides and hydroxides, which can be used as high-performance pseudocapacitive electrode materials, including monometallic compounds (NiO, Co_3O_4, $Ni(OH)_2$, $Co(OH)_2$, etc.) and binary metal compounds ($NiCo_2O_4$, $NiCo_2S_4$, and NiCo LDH (layered double hydroxide)). Previous studies have shown that binary metal compounds exhibit higher electronic conductivity. The charge storage processes of Ni-/Co-based materials are generally observed within a very narrow voltage window, which is similar to that of the $Ni(OH)_2$ electrode (it is categorized as a battery-like electrode). However, a potential independent feature has been observed: the difference between Ni-/Co-based compounds and RuO_2 or MnO_2. Herein, the Faradaic electrochemical processes of the battery-like electrode cannot be described using the concept of "capacitance" (F). Ni-/Co-based oxides or hydroxide materials can only be coupled with carbonaceous materials to prepare composites for assembling hybrid SCs. Thus, when converted to the unit of capacity ($mAh\,g^{-1}$), the super-high capacitance value achieved ($>1000\,F\,g^{-1}$) cannot facilitate the high utilization of electrode materials. However, reports exploring the kinetics properties of these materials have rarely been published. At present, it is not reasonable to assume that these materials share identical kinetics as $Ni(OH)_2$. In $Ni(OH)_2$, the charge is completely controlled by ion diffusion within the crystalline framework.

Others The electrochemical profile of some metal oxides, such as TiO_2(B), Na_2O_5, V_2O_5, and MoO_3 is different from that of RuO_2, MnO_2, and Ni/Co oxides whose energy storage is based on a redox reaction between the electrode and electrolyte. Some metal oxides store energy through the intercalation of electrolyte ions into the lattice of the electrode, called intercalation pseudocapacitance behavior. Typically, in 2005, Kavan and coworkers [32] proposed the pseudocapacitive Li^+ intercalation/de-intercalation in TiO_2(B). They synthesized phase-pure TiO_2(B) with fibrous architecture derived from amorphous TiO_2 precursors. Then, they proposed lithium-ion insertion/extraction, which is dominated by pseudocapacitive Faradaic behavior. In CV curves at various scan rates, the oxidation and reduction peaks of the as-synthesized TiO_2(B) showed linear variations in scan rates, i.e. $b = 1$, suggesting typical pseudocapacitive Li^+ intercalation/de-intercalation. Based on this unusual behavior in terms of the crystal structure of TiO_2(B), it is presumed that novel pseudocapacitive intercalation can be ascribed to various freely accessible parallel channels for Li^+ transport, perpendicular to the (010) face.

9.3 Electrode Materials

9.3.1 Activated-Carbon-Based Electrode

Carbonaceous materials have gained attention in the field of energy storage due to their high conductivity, excellent structural stability, and morphological diversity, which can meet the demands of different applications. As a typical carbon-based material, AC is a promising material due to its higher SSA, i.e. higher than $1000\,m^2\,g^{-1}$, highly porous structure, low cost, and high availability of raw materials. Therefore, AC has been widely used as an adsorbent that can remove impurities from gases and liquids. The origins of AC can be dated back to the twentieth century when powder-like AC with specific properties was synthesized by industrial production processes. Thereafter, the development of protective gas masks has fostered the research and design of AC. Nowadays, AC with different structures and morphology, including fibers, pellets, cloths, felts, and so on, has been developed using advanced process technologies to widen its usefulness in more demanding applications, especially in SCs. Thus, in this section, we will introduce the fundamentals of AC in SCs, such as synthesis methods, advantages, and recent research progress. Although other porous materials exhibit a more controlled porosity, such as templated carbon, CDC, and MOFs, the simplicity and well-established synthesis methods of AC mean that it continues to be the major candidate for SCs.

9.3.1.1 Activation of AC

The activation mechanisms of AC can be classified into two categories: physical process and chemical process. In physical processes, there are usually two steps: first, the organic precursor is subjected to pyrolysis in an inert atmosphere at 400–900 °C to form carbon, and then, an oxidation gas is passed through it to create pores and enlarge the surface area. Chemical processes are relatively simple and can be realized in one step, in which a mixture of the precursor and the activating agent (acid, base, or some salt) is annealed at 450–900 °C in a specific atmosphere. In some cases, a combination of both physical and chemical processes can also be used.

Various organic compounds, polymers, and plants can be used as raw materials to synthesize AC. Coal tar, wood, straw, rice, PANI, and PPy are promising candidates as uncarbonized feedstock. The chemical and physical properties of AC are highly dependent on the intrinsic structure and contents of the precursor. Advanced carbon-based materials, such as CNTs, carbon fibers, and carbon based on templated and CDC, are not considered as belonging to AC. However, they can be converted to AC by post-synthesis activation processes, which have been confirmed as effective methods to improve their porous characteristic. In the following, we will introduce various activation methods.

Physical Activation As mentioned previously, an organic precursor is first calcined in an Ar_2 or N_2 atmosphere at relatively low temperatures of ~400–900 °C to remove the volatile content, followed by partial gasification using an oxidizing gas at 350–1000 °C. Initially, the highly active oxygen in the activating agent reacts with the volatile content within the closed space, resulting in the opening of some closed pores. Subsequently, a more porous structure is generated by the escape of carbon, during which CO and CO_2 gas are formed by the chemical reaction between carbon and the oxidizing agent. Moreover, the number and size of pores depend on the intrinsic properties of the precursor, the type of oxidizing gas (CO_2, air, and steam are commonly used as activating agents), and the temperature of activation. Based on these factors, AC with different porosities and surface structures (i.e. the type and number of surface functional groups) can be designed for various applications. Generally, the size and number of pores increase with an increase in activation time and treatment temperature.

Some inevitable problems arise when using air or oxygen as the primary activating agent because of the oxidization reaction of carbon with oxygen. For example, the extremely high oxidation reaction rate makes it difficult to control the degree of activation, resulting in excessive burn-off and reducing the yield of AC. However, the lower cost and lower level of activation than other activating agents make air and oxygen promising agents in many cases. As described earlier, steam and CO_2 are more preferred as activating agents than oxygen and air. Moreover, the reactivity of steam is superior to that of CO_2, so that the gasification process can be induced at lower temperatures. The structural properties of the as-prepared AC, including surface area, pore volume, and pore size, generally increase with the extent of burn-off in both gasifying agents. However, some studies raised concerns regarding the features of pores generated by different activating agents. Some of them have indicated that AC activated by steam shows narrower and more extensive microscopic structure than that underwent CO_2 activation, whereas others have reported lower micropore volumes but larger external surface areas in AC subjected to steam activation. Nevertheless, AC with a surface area lower than $1800\,m^2\,g^{-1}$ and a pore volume of up to $\sim 0.9\,cm^3\,g^{-1}$ is normally obtained.

Chemical Activation In chemical activation processes, the precursor is first mixed with chemical reagents, such as $ZnCl_2$, KOH, NaOH, H_3PO_4, K_2CO_3, and $FeCl_3$, in a certain proportion to achieve AC with specific properties. In this method, carbonization and activation occur simultaneously, which have a great impact on the decomposition and structure of the organic precursor, leading to the formation of abundant porous structures and functionalities on the as-prepared carbon surface. Among the chemical reagents that have been used as activating reagents of AC, $ZnCl_2$ and KOH are the most promising due to their low cost and suitable pore size.

The mechanism of activation for KOH takes place according to the following chemical reactions:

$$6KOH + 2C \rightarrow 2K + 3H_2 + 2K_2CO_3 \tag{9.11}$$

$$K_2CO_3 \rightarrow K_2O + CO_2 \tag{9.12}$$

$$C + K_2O \rightarrow 2K + CO \tag{9.13}$$

$$CO_2 + C \rightarrow 2CO \tag{9.14}$$

$$K_2CO_3 + 2C \rightarrow 2K + 3CO \tag{9.15}$$

During these chemical processes, metal K is first produced due to the reduction of carbon. Then, it enters the carbon framework, resulting in volume expansion. Pores of various sizes are formed due to the escape of metal K and carbon atoms under high-temperature annealing. For chemical activation, the temperature and mass ratio of carbon to the activating agent are important to regulate the morphology and structure of AC. In summary, chemical activation can produce AC with highly developed porosity and high SSA.

Microwave-Assisted Activation Activation of carbon materials can be achieved as long as the energy (heating) is enough. In common heating techniques (tube furnace), the heat source is placed outside the carbon source. It generates an inward positive temperature gradient, leading to an inhomogeneous and distorted pore structure of AC (Figure 9.4a) [33]. On the contrary, in microwave-assisted activation, the temperature gradient occurs in the reverse direction, which leads to a faster reaction (Figure 9.4b). Microwave heating is directly related to internal heating of the material. Here, energy is transferred from the interaction between molecules or atoms, representing the transformation of electromagnetic energy into thermal energy. The advantages of microwave-induced activation include volumetric and internal heating, fast energy transfer, instant start-up and shutdown, and improved efficiency. Unfortunately, this method is limited to lab-scale applications, and therefore, it is under continuous exploration.

9.3.1.2 Development of AC for Supercapacitors

Nowadays, AC is a key electrode material in commercial SCs. It provides a highly porous structure at the nano- to micro-scale, resulting in a higher specific capacitance than traditional electrolytic capacitors. It is possible to construct the cell with thousands of Farads by increasing the size of the cell because the contact surface between the porous carbon electrode and the electrolyte is large enough. Various precursors, including coal, wood, straw, food, and shell, have been investigated to produce AC with different pore sizes and micro-morphologies, aiming to achieve enhanced energy storage performance.

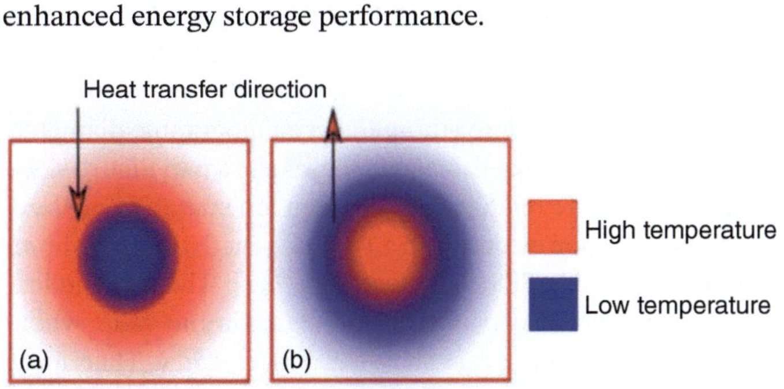

Figure 9.4 Schematic of temperature distribution, in (a) common heating technique and (b) microwave heating. Source: Dos et al. [33]/MDPI/CC BY 4.0.

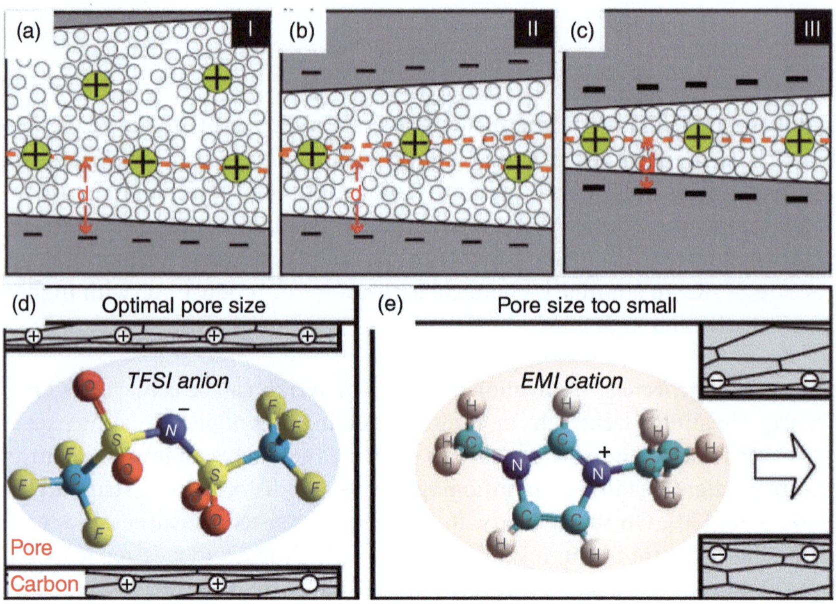

Figure 9.5 Schematic of solvated ions residing in pores with the distance between adjacent pore walls (a) higher than 2 nm, (b) between 1 and 2 nm, and (c) less than 1 nm. Source: Adapted from Chmiola et al. [11] (d,e) Relationship between ion size and pore size. Source: Reproduced with permission from Largeot et al. [13]/© 2013/American Chemical Society.

Increasing the SSA and porosity is an effective strategy for improving the energy storage ability due to the energy storage mechanism of EDLCs. Moreover, pore size, pore shape, and pore connectivity can also influence the diffusion and transfer of ions/electrons in an AC electrode. For example, Chmiola et al. proposed a model to explain the storage behavior of electrolyte ions in pores of different sizes [11], as shown in Figure 9.5a–c. Capacitance is dominated by the compact layer, in which solvated ions are located on adjacent pore walls, as shown in Figure 9.5a, in which pore size is two times the size of solvated ions. When pore size is equivalent to that of the size of the solvated ions or twice, specific capacitance is largely decreasing due to the impingement of compact layers. Moreover, the surface area is low and not enough to provide high double-layer capacitance, as shown in Figure 9.5b. The opposite trend was observed when the pore size became smaller than the size of solvated ions. Decreasing the pore size to a value approaching the crystallographic diameter of the ion led to a 100% increase in normalized capacitance, as shown in Figure 9.5c. Subsequently, Largeot et al. showed the relationship between ion size and pore size in EDLCs based on a TiC-derived carbon electrode in a 1-ethyl-3-methylimidazolium bis(trifluoromethylsulfonyl)imide [EMI-TFSI] IL electrolyte [13]. They concluded that volumetric capacity and gravimetric capacity reach the maximum value when the pore size is almost the same as the ion size, as shown in Figure 9.5d,e. Therefore, using AC with an appropriate pore size is an effective strategy to enhance charge storage ability.

The type of precursor is the crucial factor in the determination of pore size and micro-morphology, resulting in differential SC performance. For example, wood from trees or plants is the richest source of precursor and consists of various other elements (oxygen, nitrogen, sodium, calcium, etc.). Thus, it has been regarded as a promising feedstock in the preparation of AC. Prabakar et al. synthesized AC from birchwood via KOH activation and high-temperature annealing processes [34]. Pore size distribution and surface area of AC can be regulated by changing the activation temperature. AC prepared at 800 °C showed the maximum specific capacitance due to its highest surface area and pore volume among all as-prepared samples. Tian et al. reported *Eucommia* wood tar-based AC, and the number and structure of pores were regulated by the mass ratio of wood to KOH activator, namely 1 : 1, 1 : 3, 1 : 5, and 1 : 7, denoted as AR1-2, AR3-2, AR5-2, and AR7-2, respectively [35]. They found that AR3-2 with a surface area of $2912.1\,m^2\,g^{-1}$ delivers a higher specific capacitance of $510.54\,F\,g^{-1}$ than AR-5-2 with a larger surface area ($3354.8\,m^2\,g^{-1}$). Inconsistent increases in SSA and capacitance can be attributable to a more excellent pore structure of AR3-2, which shows a high percentage of mesopores.

Besides wood, dead leaves and flowers are promising candidates to synthesize AC. Recently, Anant et al. reported the synthesis of a porous AC (PELAC) electrode of SCs using *Phyllanthus emblica* leaves [36]. PELAC achieved a high SSA of $1244\,m^2\,g^{-1}$ with superior capacitance ($336\,F\,g^{-1}$ at $1\,A\,g^{-1}$) and excellent cycling retention. Similarly, pineapple leaf fibers were used to fabricate AC (PAC) via carbonization combined with Co^{2+}/KOH and CO_2 activation, which showed excellent rate performance ($186\,F\,g^{-1}$ even at $60\,A\,g^{-1}$) [37]. In addition, Pant et al. synthesized three-dimensional AC using koiralo flowers, which delivers a significant specific capacitance of $165\,F\,g^{-1}$ at $0.5\,A\,g^{-1}$ in 6 M KOH [38].

Carbon–carbon symmetric SCs have significant power density, excellent cycling lifespan, and superior reliability; however, they have limited energy density and serious self-discharge properties. Although the energy density of currently available materials is acceptable for applications such as emergency doors, memory backup, and energy recovery, it is popularly perceived as the main impediment to the market growth of SCs. Therefore, huge research efforts have been devoted to the development and design of carbon-based electrodes with improved energy density and reasonable cost. Nowadays, almost all the SC manufacturers fabricate AC using the coconut shell precursor. Kuraray supplies most of the SC carbon with low ash content (<1%) and impurities (<100 ppm). Initially, the cost of this product was up to $150–200 per kg due to bottlenecks in the production processes. Over time, it has dropped to $15 per kg. Such a low cost is a huge barrier to the development of other types of porous carbon that can be used in electrode materials of SCs.

In summary, there are two potential opportunities for the further development of carbon-based materials in the SC market: the energy density of new carbon is more than twice that of conventional AC, and it may find a market niche; and the lower cost of new carbon can be achieved; however, the energy storage performance is comparable to that of coconut shell carbon. There is a substantial gap between the $4 per kg cost of commodity AC and the $15 per kg cost of SC-grade carbon. While process control and purification requirements may increase the cost of producing SC carbon to compete with coconut shell, the opportunity still exists; furthermore, both sugar-derived carbon and high-purity coal-derived carbon may also be viable. However, future attempts to develop premium SC carbon in the performance range already demonstrated will fail like they had in the past.

9.3.2 Conductive-Polymer-Based Electrodes

It has been demonstrated that the energy storage mechanism of CPs follows that of PCs. Unlike metals that conduct electricity through the directional movement of electrons, CPs conduct electricity via a conjugated bond system along the polymer backbone. Among CPs, PANI and PPy are the most popular electrode materials due to their lower price, environmentally friendly nature, and easier preparation strategies. P-doping of PANI and PPy can be achieved, but not n-doping, as the n-doping potential is lower than the reduction potential of the common electrolyte. Thus, both of them are used as the cathode material for SCs.

9.3.2.1 Preparation of CPs

Chemical Synthesis Synthesis of CPs at liquid–liquid interfaces is a common chemical strategy, e.g. interfacial polymerization (heterogeneous polymerization). The interface between two immiscible liquids provides a unique environment where physical and chemical properties are distinct from those of the corresponding single liquid. This field has shown slow progress due to the unknown interfacial structure and potential distribution across the interface; hence, there is tremendous scope for the synthesis of diverse classes of materials and applications. Liquid–liquid interfaces are involved in interfacial nucleation and growth of nanomaterials. In general, the polymer monomer exists in the organic part, and oxidizing agents are present in the aqueous part. Polymers are generated at the interface when the organic and aqueous parts mix. A huge quantity of polymers can be synthesized by selecting different monomers and immiscible solvents. This strategy has been developed for designing various CPs with different morphologies, such as PPy nanoneedles, PANI nanofibers, and PANI nanobelts. In addition, homogeneous polymerization is another approach to synthesizing CPs. In this method, both monomers and the resulting polymer are dissolved in the media or swollen in the case of crosslinked gels. In the case of homogeneous polymerization, the synthesis is performed either in aprotic solvents, such as DMSO, or in water (at room temperature).

Electrochemical Synthesis Compared with chemical methods, electrochemical synthesis is more conducive to environmental protection. Meanwhile, the rate of polymerization is faster and the stability of products is more superior than those of chemical methods. Electrochemical polymerization methods can be classified into the following categories: (i) potentiodynamic method: a typical process to achieve potentiodynamic polymerization is CV, in which current response is recorded using the variations in time-dependent external potential; (ii) galvanostatic method: a technology wherein a constant current density is applied on the substrate for a certain amount of time or voltage; the micro-morphology and thickness of CPs can be regulated by changing the time and current density of galvanostatic deposition; and (iii) potentiostatic method; there are only a few examples to prepare PANI or PPy via this method because it is difficult to quantify overpotential losses.

9.3.2.2 Development of CPs for Supercapacitors

Nowadays, many researchers have shown interest in the field of CPs in SCs. For example, a pristine PANI electrode was prepared by self-stabilized dispersion polymerization [39]. It delivered a gravimetric capacitance of 437.8 F g^{-1} at a current density of 0.5 A g^{-1} in H_2SO_4–hydroquinone electrolyte, which can be used as a redox-active electrolyte. He et al. in situ grew PPy onto the interior of wood aerogel to fabricate an electrode with a three-dimensional network, which showed a superior energy density of 20 Wh kg^{-1} at a power density of 161.6 W kg^{-1} in a solid-state symmetrical SC device [40].

Unfortunately, CPs may swell and shrink in the process of intercalation and de-intercalation, which often results in low cycling stability when they are used in SCs. The chain structure, electrical conductivity, mechanical stability, processability, and mitigation of mechanical stress of CPs can be improved by compositing with other species, which improves their cycling stability. High-stability and high-conductivity carbon nanomaterials can be used to prepare carbon–polymer nanocomposites with enhanced electrochemical performance. For example, Elanthamilan et al. reported a polydopamine-/PANI-decorated CNT composite (PDA/PANI@CNT) via an in situ polymerization route for hybrid SC application [41]. This composite delivered a significant specific capacitance of 970 F g^{-1} at a scan rate of 5 mV s^{-1}, which is higher than that of PDA and PDA/PANI electrodes. Wang et al. developed a hierarchical porous PPy-coated carbon cloth electrode via the electrochemical deposition method using nano-silica as the self-sacrificing template [42]. This porous structure electrode showed an impressive specific capacitance of 670.2 F g^{-1} at 1 A g^{-1} and excellent cycling stability with a capacitance retention of 90.3% after 10,000 charge–discharge cycles.

Metal hydroxides and PANI composites have recently emerged as composite electrode materials for SCs. For instance, Li et al. assembled the asymmetric micro-SCs with PPy@NiCo-LDH@ITO NWs as the cathode, PPy@FeOOH@ITO NWs as the anode, and a water-in-salt (WIS) electrolyte as the electrolyte, giving an extended voltage window of up to 2.5 V and a superior capacitance of 63.04 mF cm^{-2} at 0.4 mA cm^{-2}, as shown in Figure 9.6a,b [44]. Yang et al. designed a NiCo-LDH@PANI hollow structure electrode, where NiCo-LDH nanosheets were derived from Co-MOF to increase the contact area with the electrolyte and PANI was used as the conductive network to provide charge transfer pathways [43]. The unique hollow electrode showed an excellent capacitance of 1315 C g^{-1} at 1 A g^{-1} and 951 C g^{-1} at 20 A g^{-1}, as shown in Figure 9.6. CP–metal oxide composites have been extensively fabricated in the past decade, and enhanced conductivities have been reported. Manganese oxide (MnO_2) has primarily been coupled with PANI for SC applications. It is a low-cost, highly abundant, and environmentally

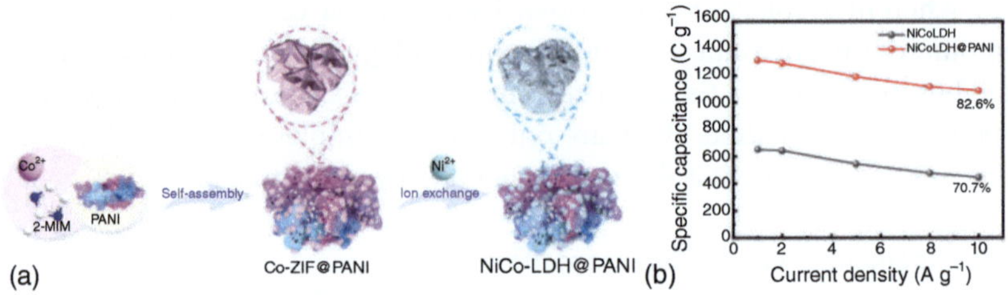

Figure 9.6 (a) Schematic of the synthesis route of NiCo-LDH@PANI. (b) Specific capacities of NiCo-LDH and NiCo-LDH@PANI at different current densities. Source: Reproduced with permission from Yang et al. [43]/© 2023/The Royal Society of Chemistry.

benign material with excellent electrical and capacitive properties. Other metal oxides such as NiO, Fe_3O_4, ZnO, and TiO_2 are also favorable for hybridization with CPs. Recently, Zhou et al. have investigated the electrochemical performance of $ZnCoO_x$/C-PANI composites [45]. They reported that the synergistic effects between PANI and $ZnCoO_x$/C are conducive to reducing the ion transport distance, thus improving the energy storage stability. The as-prepared electrode delivers a remarkable specific capacitance of 1055 F g^{-1} at 1 A g^{-1} with an outstanding cycling stability of 85.7% capacitance retention after 10,000 cycles.

CPs have been extensively investigated as SCs, including PPy, PANI, PTh, and their derivatives and composites, as they are easier to prepare, show morphological diversity, are of low cost, demonstrate superior electrical conductivity, and are structurally flexible. Although CPs exhibit numerous advantages, their insignificant structural stability and sluggish chemical kinetics lead to a lower energy density and power density when pristine CPs are used as electrodes. Thus, designing hybrid CPs with carbon-based materials and metal compounds is an effective strategy, which can show enhanced charge storage ability. However, achieving uniform solubility of metal compounds and carbon-based materials with a CP matrix is difficult. Therefore, a suitable hybridization ratio needs to be determined, and innovative approaches need to be developed. In addition, researchers are also investigating other polymers from an energy storage perspective. Although the gap between research and commercial applications of SCs appears to be huge, it is presumed that with continuous development, next-generation SCs with good flexibility, higher specific capacity, good cyclic stability, and high energy and power densities can be realized. Therefore, significant efforts should be made to improve the applications of SC electrode materials so they can be profitably employed in flexible and compressible devices.

9.3.3 Metal Compounds

Compared with carbon-based and CP materials, metal compounds can store higher charges due to the energy storage mechanism of metal compounds in redox reactions. In the past decades, the primary electrode materials in SCs based on metal compounds are as follows: metal oxides/hydroxides, sulfides, selenides, phosphides, and MXenes. However, many challenges such as low energy density, poor electrical conductivity, and high internal resistance on the surface of the material should be effectively addressed to achieve practical applications of metal-compound-based electrodes. In the following, we briefly introduce the synthesis methods and focus on the development of metal-compound-based electrodes in SCs.

9.3.3.1 Preparation of Metal Compounds

Thermal Treatment Thermal treatment methods are promising techniques for preparing metal compounds as they are time-saving and simple processes. Thus, they are used to synthesize metal compounds with the assistance of reaction additives and specific gases. For example, ammonia gas is used to prepare nitrides, and sulfur/selenium powder is suitable for generating sulfides/selenides. Zhang et al. prepared Ni-CoN@NC via nitridation of $NiCo_2O_4$ nanoparticles at 600 °C using NH_3, resulting in an improved capacitance [46]. Fu et al. designed core–shell Co-doped NiS_2 by the sulfidation of Co-doped Ni-MOF using sulfur powder as the sulfur source and heating at a high temperature under the protection of nitrogen [47]. Benefiting from the unique core–shell structure, bimetallic sulfides showed significant energy storage properties with a specific capacitance of 332.8 mAh g^{-1} at 1 A g^{-1}.

Electrochemical Deposition Electrochemical decomposition, also called electrodeposition or electroplating, is a simple and powerful method to prepare metal compounds on a conductive substrate. Moreover, the desired morphology and size of the as-prepared metal compound can be achieved by adjusting one or more electrodeposition factors, such as time, potential, and concentration of the solution. Zhang et al. prepared various MnO_2@rGO composites by adjusting deposition time [48]. They observed that with an increase in deposition time, the manganese dioxide nanostructure starts to grow and become denser, which is not conducive to the full contact reaction between the electrode and the electrolyte.

Atomic Layer Deposition Atomic layer deposition (ALD) is an attractive technique to prepare metal compounds, especially metal oxides such as TiO_2, MoO_3, and Al_2O_3. Compared with other methods, the thickness and composition of amorphous materials can be adjusted by tuning the number of cycles and organometallic sources in an ALD process. Recently, Pawar et al. designed $NiCo_2O_4$ nanosheets coated with an ultrathin MoS_2 layer in a plasma-enhanced ALD reactor [49]. An optimal thickness of MoS_2 resulted in a conducting sulfide layer to maximize the performance of SCs.

Solution-Processed Methods Solution-processed methods, such as hydrothermal method, solvothermal method, sol-gel method, and coprecipitation, are the most common methods to prepare metal compounds. Compared with ALD, chemical vapor deposition (CVD), and other methods, solution-processed methods are efficient in preparing amorphous materials with different morphologies. Teli et al. deposited manganese vanadium molybdenum oxide (MnVMo oxide) on nickel foam using a one-step hydrothermal process [50]. The nanostructure of MnVMo oxide can be changed from nanoparticle-assembled sheets to rectangular nanosheets by varying the deposition time. As a result, the MnVMo oxide electrode shows a promising areal capacitance of 814.5 mF cm^{-2} at 3 mA cm^{-2} in 1 M KOH electrolyte.

9.3.3.2 Development of Metal Compounds in Supercapacitors

Because of their redox nature, TMOs are a viable material as electrodes for SCs. Their unique characteristics are attributable to their variety of oxidation states, which is advantageous for achieving a rapid Faradaic reaction. This characteristic enables them to shuttle between different oxidation states during the charging and discharging process and also results in the rapid entry and exit of electrolyte ions into the oxide lattice. Many metal oxides, including Co_3O_4, RuO_2, MnO, NiO, and MoO_3, have been explored for their electrochemical behavior in SCs. For example, Chen synthesized CoO nanoparticles using 3D carbon foam as the substrate, resulting in increased an specific capacitance of 221 F g^{-1} at 1 A g^{-1} [51]. Wang et al. found that MoO_3 nanobelts show excellent SC performance based on the insertion and extraction of H^+ ions [52]. An asymmetric SC that consisted of a MoO_3 nanobelt cathode and a carbon nanosheet anode showed a good energy density of 25.69 Wh kg^{-1} at a power density of 1482.25 W kg^{-1}. Basu et al. prepared phase-pure monoclinic VO_2 nanoporous fibers with an excellent coulombic efficiency of up to 98.2% in micro-SCs [53]. Similarly, Shamim and coworkers reported VO_2 microparticles with a monoclinic phase via a one-step hydrothermal method [54]. The symmetric interdigitated VO_2 micro-SC can achieve a wide voltage window of 1.4 V, which is the highest operating voltage reported for symmetric SCs based on pure vanadium oxide electrodes in aqueous inorganic electrolytes.

In general, metal sulfides show higher redox reactivity and conductivity than metal oxides, resulting in excellent energy storage performance in SCs. Furthermore, some studies have demonstrated that metal sulfides can provide superior energy density due to their abundant active sites and rapid reaction kinetics. Moreover, the electronegativity of sulfides is lower than that of their oxide counterparts, which is conducive to the flexibility and structural integrity of electrode materials. Nickel sulfides, such as NiS_2, Ni_3S_4, and Ni_3S_2, are promising electrode materials due to their high theoretical capacity. For example, Muhammad et al. studied the energy storage ability of nickel sulfide nanoplates@GO/NF [55]. The binder-free electrode delivers a maximum capacitance of 1745.50 F g^{-1} at 5 mA cm^{-2}, which is much higher than that of most metal oxides. Similarly, NiS/NF materials with different morphologies can be prepared using an optimized solvothermal method [56]. The morphologies influence electron/ion transport pathways, and the contact area between the electrode and the electrolyte, leading to a higher between capacitance and rate performance. The NiS/NF electrode with a sphere-shaped structure shows the highest rate performance of 694.0 F g^{-1} at 1 A g^{-1} and 288 F g^{-1} at 6 A g^{-1}. Other metal sulfides, such as WS_2, VS_2, CuS, CoS, MoS_2, and FeS_2, have shown great potential for applications in SCs.

The attractive properties of metal nitrides such as excellent electrical conductivity, catalytic activity, and chemical properties play a unique role in the field of energy storage. The properties of nitrides are dominated by covalent, ionic, and metallic bonds, primarily bonds between transition metals and nitrogen atoms that expand the lattice of the parent metal and shrink the metallic d-band. A low d-band and a high density-of-state near the Fermi level furnish noble metals with a behavior similar to that of metal nitrides, which makes them promising candidates for SCs. Li et al. prepared CoN-Ni_3N/N-C nanosheets with a honeycomb-like structure using a solvothermal method combined with high-temperature nitriding [57]. CoN-Ni_3N/N-C can achieve an excellent capacitance of $1.2\,F\,cm^{-2}$, which is much higher than that of Ni_3N/CC ($0.35\,F\,cm^{-2}$), CoN/CC ($0.07\,F\,cm^{-2}$), and Ni-Co LDH/CC ($0.96\,F\,cm^{-2}$) electrodes. The remarkable electrochemical performance originated from enhanced intrinsic conductivity and increased concentration of active sites. Mohan et al. proposed a new vanadium nitride@porous carbon composite [58]. This new strategy was adopted to control the interfacial characteristics of the nanocomposite for superior charge transport properties, which led to improved electrochemical performance. Highly porous carbon nanobelts with large active surface sites provide considerably high charge separation, whereas uniformly distributed VN nanoparticles produce efficient redox reactions and sufficiently enhance the overall electrochemical properties of the composite electrode. Additionally, the in situ deposition of VN nanostructures on carbon nanobelts results in excellent interfacial properties with very low charge transfer resistance for efficient supercapacitive performance.

To further improve the energy storage ability of metal compounds in SCs, some strategies have been developed, including heteroatom doping, heterostructure, morphology design, polymetallic metal compounds, and composite formation. For heteroatom doping, the doped heteroatoms can be regarded as impurities, which can be introduced into lattice sites, resulting in lattice distortion and charge redistribution. In some cases, heteroatom doping generates vacancies and enlarges the electrochemical active area, leading to enhanced electrochemical properties. Lin et al. prepared pristine and S-doped NiCoP nanocrystals and discussed the impact of doping on the electrochemical properties of nanocrystals [59]. XRD and HR-TEM results indicate that doping does not change the crystal phase of NiCoP, but does affect the lattice structure, which demonstrates lattice contraction and distortions. The S-doped NiCoP electrode shows an excellent specific capacitance of $2229.9\,F\,g^{-1}$ at $1\,A\,g^{-1}$. Vikraman et al. investigated the intrinsic MoS_2 and transition-metal-doped (Fe, Cu, Ni) MoS_2 and showed the effect of heteroatoms on electrochemical performance [60]. Field emission scanning electron microscopy (FESEM) analyses indicated that the micro-morphologies of various doped samples are different. Cauliflower-like structured nanograins were obtained by pure MoS_2; however, porous curled-leaf-like structure for Cu-doped MoS_2, coarse-like morphological properties for Fe-doped MoS_2, and nanoarray structure with a smaller size for Ni-doped MoS_2 were reported. Combined with Brunauer–Emmett–Teller (BET) results, a high SSA of $17.32\,m^2\,g^{-1}$ was achieved by Cu-doped MoS_2 (that of pure MoS_2 is $2.54\,m^2\,g^{-1}$), which is consistent with FESEM outcomes. Due to the surface modification and interlayer spacing enhancement via metal doping, the porous nature and surface area characteristics of metal-doped MoS_2 are plausibly altered. In addition, due to the uniaxial strain variation between the lattices via metal doping, the surface porous behavior is characteristically changed. As expected,

Cu-doped MoS_2 electrode achieved a maximum specific capacitance of 353 $F\,g^{-1}$ at 1 $A\,g^{-1}$ applied current.

Constructing heterostructures is an effective method to optimize the electrical conductivity, structural flexibility, band gap, and charge carrier mobility of electrode materials. Moreover, the formation of active sites and defects in the grain boundary of heterostructure materials results in multiple redox activity, superior ionic conductivity, and short diffusion path. Therefore, studies on nano-heterostructure electrode materials need to be performed in order to explore their high power density and applications in the field of energy storage. Recently, Gao et al. reported the preparation of dual redox-active sites of $Ni_2P/NiSe_2$ heterostructures inlaid on an N-C substrate (N-C) via in situ phosphoselenization [61]. CV curves at different scan rates and GCD curves at different current densities indicated that N-C@$Ni_2P/NiSe_2$ has a higher current response and longer discharging time than individual electrodes, implying a faster redox reaction kinetics and greatly improved specific capacitance (Figure 9.7a,b), which is mainly attributable to its abundant electroactive sites and the optimized electronic structure of the heterointerface (Figure 9.7c). Evidently, N-C@$Ni_2P/NiSe_2$ shows a specific areal capacitance of up to 4660 $mF\,cm^{-2}$ (2118 $F\,g^{-1}$) at 1 $mA\,cm^{-2}$ and retains a specific areal capacitance of 3600 $mF\,cm^{-2}$ (1636 $F\,g^{-1}$) at 20 $mA\,cm^{-2}$ with a rate capability as high as

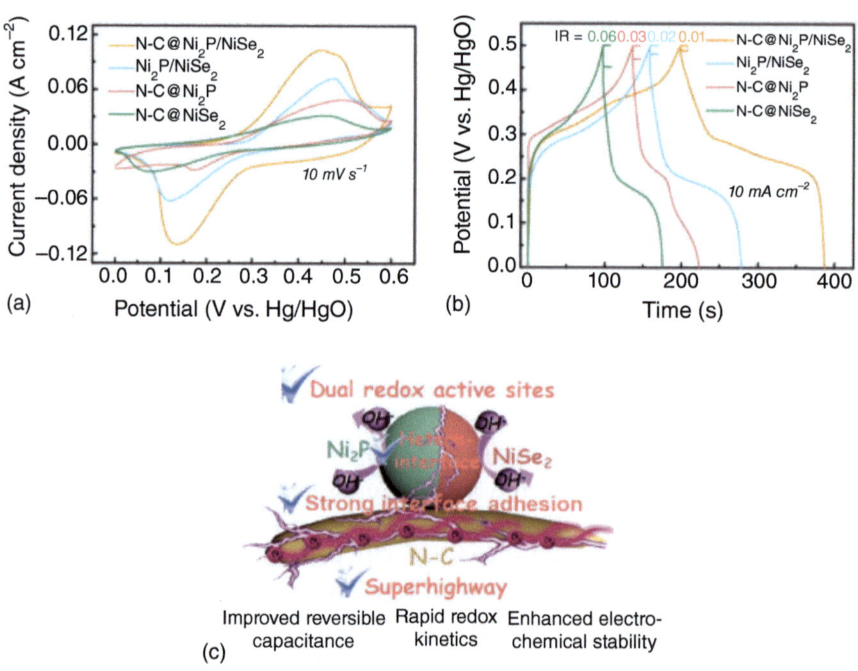

Figure 9.7 (a) CV curves and (b) GCD curves of the $Ni_2P/NiSe_2$ electrode. (c) Optimized mechanism of the $Ni_2P/NiSe_2$ electrode. Source: Reproduced with permission from Gao et al. [61]/© 2022/John Wiley & Sons.

77.3%, which is higher than those of $Ni_2P/NiSe_2$ (45.0%), N-C@Ni_2P (69.4%), and N-C@$NiSe_2$ (59.9%).

Electrode materials with high SSA and suitable pore size can achieve improved electrochemical performance. Therefore, it is important to regulate the morphology and phase structure of materials in SCs. Najib et al. designed ZnO materials with the structure of nanoparticles, nanoflowers, and nanourchins [62]. The concentration and location of defects in various structural samples are investigated using electron paramagnetic resonance (EPR) techniques. The results indicated that all three nano-morphologies (particles, urchins, and flowers) have core signals. However, the concentration of surface defects is strongly dependent on morphology; the highest concentration of defects is achieved by the nanoparticle structure, and the lowest is achieved by nanourchins. Real SC devices were assembled to understand the capacitive behavior of ZnO nano-morphologies. The values of specific capacitance are promising and deviate from ~330 to 65 $F\,g^{-1}$ at 0.1 C depending on the morphology of ZnO. Gao et al. designed $CuCo_2O_4$ nanoneedles, nanosheets, and nanorod arrays, by introducing different additives (urea, NH_4F, and examethylene-tetramine, respectively) [63]. According to their results, nanoneedle arrays with a length of 2 mm and a diameter of 20 nm were fabricated using urea as the alkali hydrolyzing agent. When adding NH_4F, rod-like $CuCo_2O_4$ arrays were observed. Moreover, when urea was replaced by hexamethylenetetramine and other reaction conditions remained unchanged, well-defined $CuCo_2O_4$ nanosheet arrays were obtained. When evaluated in a three-electrode system for SCs, $CuCo_2O_4$ nanoneedle arrays delivered the largest specific capacitance (1227.8 $F\,g^{-1}$ at 5 $mA\,cm^{-2}$) and optimal capacitance retention (95.4% after 1000 cycles) among the three samples. The remarkable electrochemical performance of $CuCo_2O_4$ nanoneedle arrays is attributable to the favorable one-dimensional mesoporous architecture, which can shorten the transport paths of electrons and facilitate the direct growing of electroactive materials on Ni foam with lower resistance.

In addition, various metallic phosphides, phosphates, and phosphites are found to be suitable as active materials for chemical SCs. Liu et al. constructed ultrafine porous nickel cobalt phosphide (NiCoP-P) nanosheets using aluminum (Al)-doped NiCo-layered double hydroxides as precursors, followed by the removal of Al species and subsequent phosphorization [64]. As a result of Al-induced rich active sites, ultrafine porous nanosheet structure, and rapid ion transport, the optimized NiCoP-P electrode showed a high specific capacity of 1059C g^{-1} at 1 $A\,g^{-1}$ and a capacitance retention of even at 50 $A\,g^{-1}$. An asymmetric SC showed a high energy density of 51.7 $Wh\,kg^{-1}$ and long-term cyclic stability (99% capacity retention after 10,000 cycles) in KOH/PVA sol electrolyte. Wu et al. prepared a zinc–cobalt–nickel phosphide composite derived from a Zn/Co MOF [65]. Due to the multimetal synergistic effect of Zn, Co, and Ni can promote the efficient utilization of active substances, and the unique nanosheet-like structure of the obtained ZnCoNiP-D metal phosphide electrode shows a high capacity of 770.4 $C\,g^{-1}$ at 1 $A\,g^{-1}$, an outstanding performance of 466.7 $C\,g^{-1}$ at 20 $A\,g^{-1}$, and a good capacity retention rate of 81.8 % after 5000 cycles.

Nevertheless, pristine metallic compounds limit the electrochemical performance of devices due to their inherent drawbacks, which can be overcome by using different optimization strategies such as unique heteroatom doping, modifications to heterostructure and morphology, designing polymetallic metal compounds, and composite formation. The cyclic performance of TMOs, TMSs, and TMNs is probably lower than that of pure EDLC-type materials, resulting in a limited cycle lifespan due to phase changes that induce internal lattice structure alterations throughout the charging–discharging process. However, compared with battery-type electrodes, these pseudocapacitive materials show low capacitance degradation over a large number of cycles in the electrochemical test. Another advantage of SCs is their higher charge–discharge rate and wide working temperature range, which makes them the most sustainable and resilient energy source.

9.4 Electrolytes

According to the configuration of SCs, electrolyte is important for the electrochemical performance of SCs, which has a great influence on cycling stability and working voltage window. In general, an electrolyte consists of a solute (mostly salt), which can provide active ions for energy storage, and a solvent (water or organic liquid), which is used to provide a path for the diffusion of active ions in the electrolyte. Therefore, the physical and chemical performances of an electrolyte are determined by the type and size of ions, as well as the interaction between ions and solvents. An ideal electrolyte for SCs should meet the following characteristics:

- ✓ Excellent electrochemical stability
- ✓ High concentration of free/dissociated ions along with high mobility and least ion pairing
- ✓ Availability at high purity
- ✓ Low volatility
- ✓ Wide working voltage window
- ✓ Low solvated ionic radius
- ✓ High conductivity giving high power delivery
- ✓ Low viscosity, low cost, and nontoxicity.

As shown in Figure 9.8, electrolytes can be primarily classified into liquid electrolytes and solid electrolytes. Based on the type of solvent used, liquid electrolytes can be further classified into aqueous electrolytes and nonaqueous electrolytes. Solid electrolytes can be classified into ceramic electrolytes, solid polymer electrolytes, and gel polymer electrolytes.

9.4.1 Liquid Electrolytes

Due to their low cost, nontoxicity, high dissociation constant, high dielectric constant, and superior ion conductivity, aqueous electrolytes have been subjected to extensive research as strong electrolytes. Moreover, as a result of the least ion

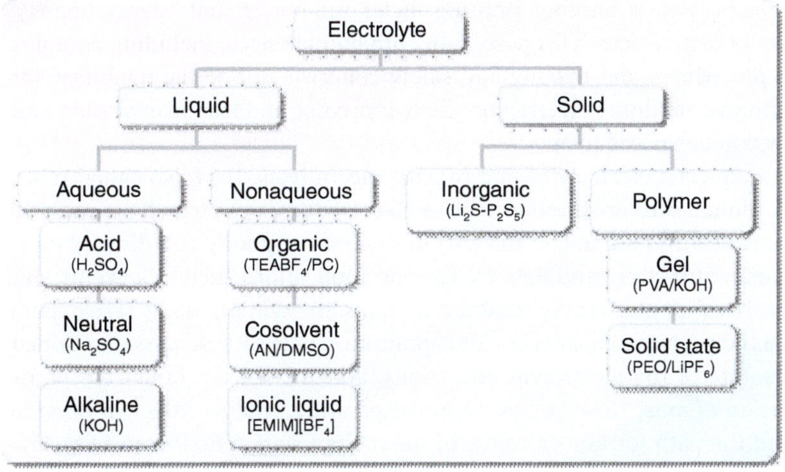

Figure 9.8 Classification of electrolytes for SCs.

pairing and smaller size of solvated ions, aqueous electrolytes show higher specific capacitance and power density. Aqueous electrolytes include acid electrolytes (e.g. H_2SO_4 and H_3PO_4), alkaline electrolytes (e.g. KOH, LiOH, and NaOH), and neutral electrolytes (e.g. LiCl, Li_3PO_4, and Na_2SO_4). Unfortunately, aqueous electrolytes generally show a narrow working voltage window, which is limited by the thermodynamic decomposition voltage of water (around 1.23 V). The equation of $E = CV^2/2$ (E is energy density, C is specific capacitance, and V is voltage window) shows that energy density is linear with the square of the voltage window. Thus, the narrow voltage window of an aqueous electrolyte results in a limited energy density.

To widen the voltage window, organic-based electrolytes are used. Most of the studies have focused on two typical organic solvents: acetonitrile (ACN) and propylene carbonate (PC), both of which have advantages and disadvantages in some cases. In general, ACN has low viscosity, low melting point, and high dielectric constant. Although ACN-based electrolytes have higher ion conductivity than PC-based electrolytes, due to the low toxicity and high flash point of PC-based electrolytes, researchers prefer PC-based electrolytes. The conductivity, solubility, stability, and safety of electrolyte salts have a great impact on organic electrolytes. At present, the most commonly used electrolyte salt is tetraethylammonium tetrafluoroborate (TEA-BF_4), which is widely used in commercial SCs. Other electrolyte salts such as TEMA-BF_4, $EMIBF_4$, TMPY-BF_4, and so on have been studied. The organic electrolyte 1 M Et_4NBF_4 is dissolved in PC and ACN, which has been regarded as the best candidate for SCs, especially for carbon-based electrode materials. However, lower specific capacitance is achieved by using organic electrolytes due to the larger size of solvated ions (it is hard to insert into pores in electrode materials) than in aqueous electrolytes. Thus, regulating porous structure is an effective strategy to improve energy storage capacitance. Besides, the insignificant ion conductivity of organic electrolytes is another limiting factor for power and energy density. Furthermore, organic electrolytes pose many other challenges, including complex purification procedures and toxicity and safety concerns due to flammability; for example, lithium-containing electrolytes are hygroscopic and lead to explosion and flammability when exposed to air.

Apart from aqueous electrolytes and organic electrolytes, ILs (also called room temperature molten salt) are another type of electrolyte system, which are entirely composed of cation and anion (no solvent). ILs consist of poorly coordinated ions, making them an attractive candidate for various applications such as solvents and "green electrolytes" or electrically conducting fluids. In addition, many researchers have devoted themselves to the design and application of ILs for SCs. As mentioned earlier, the solute of the electrolyte offers ions, and the solvent offers the paths for the diffusion of ions. However, at room temperature, molten salts can provide both ions and the path for ion transport in the molten state. The low toxicity, low volatility, and wide potential window of IL electrolytes have attracted extensive attention. Inevitably, IL electrolytes also have some disadvantages, such as high viscosity and low conductivity, which limit their application. Many cations and anions have been used in the synthesis of ILs, such as imidazolium and asymmetric aliphatic quaternary ammonium pyrrolidinium ions, tetrafluoroborate (BF_4^-), trifluoromethanesulfonate ($CF_3O_3S^-$), bis(trifluoromethanesulfonyl) imide ($TFSI^-$), bis(fluorosulfonyl) imide (FSI^-), and hexafluorophosphate (PF_6^-). In general, the physical and chemical properties of ILs are determined by the type of cations and anions; for example, pyrrolidinium and aliphatic quaternary ammonium and salts offer a wider potential window (>5 V) and lower conductivity than 1-ethyl-3-methylimidazolium-based ILs, which offer high conductivity ($10^{-2}\ S\ cm^{-1}$) at room temperature. The design and synthesis of an ideal IL with a wider voltage window and superior conductivity are still a challenge.

In other words, liquid electrolytes show significant energy storage ability. However, applications of liquid electrolytes are strongly limited by leakage, corrosion, and other problems. Hence, high-ion-conducting and mechanically and electrochemically stable polymer-based electrolytes are being developed to replace liquid electrolytes.

9.4.2 (Quasi)-Solid-State Electrolytes

Although rigid structure and thermal stability can be achieved by inorganic solid electrolytes, with respect to SCs, inorganic solid electrolytes have received less attention than polymer electrolytes due to their low elasticity and flexibility in applications. Polymer-based electrolytes dominate the field of solid-state electrolytes in SCs. A typical polymer-based electrolyte consists of a polymer matrix and suitable salts. The polymer matrix acts as an ion conductor that provides diffusion paths for ions and suitable salts, thus providing ions for energy storage. Polymer-based electrolytes have been widely investigated in SCs due to their superior ionic conductivity ($10^{-3}\ S\ cm^{-1}$ even at room temperature), easy synthesis, and high flexibility (expanding the application of SC). In the following, we provide a detailed description of the historical background and classification of polymer-based electrolytes.

In 1970s, Peter V. Wright showed that polyethylene oxide (PEO, an ion-conducting polymer) with sodium and potassium salts can be used as a solid electrical conductor [66, 67]. Subsequently, various polymer-based electrolytes with different polymer matrices and salts have been prepared and evaluated in many applications. Nowadays, polymer-based electrolytes are primarily classified into the following categories:

(1) *Polymer–salt complexes*: They belong to a type of solid polymer electrolyte without any solvent. They are fabricated using hybrid ionic salts with different polymer matrices (polypropylene oxide and polyethylene oxide).
(2) *Plasticized polymer electrolytes*: These types of electrolytes are prepared by mixing liquid plasticizers such as ethylene carbonate (EC), propylene carbonate (PC), and polyethylene glycol (PEG) with dry solid polymer electrolytes.
(3) *Rubbery electrolytes*: They are also referred to as "polymer-in-salt" materials and are prepared by mixing salts with a small number of polymers, viz. PEO and PPO.
(4) *Solvent-swollen polymers*: In these electrolytes, aqueous and nonaqueous solvents are used to swell the polymers, e.g. polyvinyl pyrrolidone (PVP) and polyvinyl alcohol (PVA), and dopant ionic solutes such as H_3PO_4 and H_2SO_4 are accommodated in the swollen lattice.
(5) *Polyelectrolytes*: Such electrolytes are also termed "polyions" or ionomers. In polyelectrolytes, the repeating unit is an electrolyte (salt) group or a self-ion-generating group attached to the main chain of the polymers. The self-generating group in polyelectrolytes is responsible for ionic conductivity.
(6) *Gel polymer electrolytes (GPEs)*: These electrolytes show significant ionic conductivity and are synthesized by mixing a large quantity of salt solution in a liquid plasticizer to a polymer matrix, resulting in achieving GPEs with unique features. In 1975, Fenullade and Perche [68] proposed that polymers can be plasticized using an aprotic solution (comprising an alkali metal salt). In this case, alkali metal salts can stick within the polymer matrix to form a gel system with high ionic conductivity; hence, it is suitable to be used as a solid-state electrolyte (even can replace the separator). Recently, many polymers have been applied to GPEs, including PEO, PVDF, PAN, PMMA, and PVDF-HFP. Usually, low-evaporation solvents, viz. dimethyl carbonate (DMC), EC, PC, diethyl carbonate (DEC), and dimethyl formamide (DMF), can be used as "plasticizers."
(7) *Composite polymer electrolytes*: They belong to two-phase polymer/gel electrolytes. The preparation process is similar for various composite polymer electrolytes, i.e. inorganic particles of micro-/nano-size disperse into the polymer matrix with continuous stirring. The introduction of inorganic particles improves the mechanical strength, interfacial activity, and ionic conductivity of GPEs. Moreover, the size and dispersion of these inorganic particles are crucial for improved physical properties.
(8) *Plastic crystal gel polymer electrolytes*: Some studies have reported that plastic crystalline materials can be used as solid-state solvents and have been applied

to SCs. Typically, succinonitrile (SN) is a promising plastic crystal with high ionic conductivity due to its high dielectric constant around 55 at room temperature. Moreover, plastic crystal gel polymer electrolytes are more resistant to mechanical pressure.

9.4.3 Optimizing the Electrolyte Properties

As mentioned earlier, the narrow voltage window is a major bottleneck for the application of SCs. To overcome this, three popular ways have been reported and proved to be effective, namely adjusting pH level, introducing redox mediators, and constructing a WIS electrolyte. On account of this, the enhanced mechanism and recent advances are summarized below.

9.4.3.1 Adjusting pH Level

According to the pH level, aqueous electrolytes can be classified into acidic, alkaline, and neutral. The major difference between electrolytes of different pH values is the amount of H^+ and OH^-, which greatly influences chemical reactions, such as redox reactions and electrolytic water reactions (oxidation evolution reaction [OER] and hydrogen evolution reaction [HER]). Nernst equations are described as follows:

$$+ \qquad 2H_2O \leftrightarrow O_2 + 4e^- + 4H^+ \qquad E_{ox} = 1.23 - 0.059\,\text{pH} \tag{9.16}$$

$$- \qquad H_2O + 2e^- \leftrightarrow H_2 + 2OH^- \qquad E_{red} = -0.059\,\text{pH} \tag{9.17}$$

We can conclude that the OER and HER are strongly related to the pH level of the solution. Therefore, adjusting the pH level is beneficial for expanding the ESW to achieve a high cell voltage of aqueous SCs. For example, Zhao et al. investigated the variations in voltage windows of the $NiCo_2O_4$/CC electrode in a neutral electrolyte (mixture of $LiSO_4$ and $CoSO_4$) and alkaline electrolyte (KOH) [69]. They found that the working voltage can be expanded to 1.2 V in a neutral electrolyte, which is 2.4 times higher than in an alkaline electrolyte (0.5 V). This neutral electrolyte can greatly boost the electrochemical characteristics of the $NiCo_2O_4$ electrode and play a role in achieving better characteristics in the field of energy storage when applied in other battery-type materials. Badhulika et al. discussed in detail working voltage windows in various pH levels ranging from 2.5 to 13.5 [70]. Their results indicated that the widest voltage window of 0∼−1.2 V is obtained by pH = 5.6 electrolyte; however, the highest specific capacitance is delivered by an electrolyte (pH = 2.5, Figure 9.9), demonstrating that the electrochemical performances of electrode materials are closely related to the pH level.

Many publications have confirmed that the performance of SCs is correlated with the pH level of the electrolyte, especially for voltage. However, effective methods to precisely adjust the pH value are less reported. Considering this, Slesinski et al. carried out an innovative approach to adjust pH levels [71]. In detail, via a modification with ammonia, the self-controlled pH gradient was introduced into this SC device, and a protective layer was formed on the electrode surface. In aqueous media, ammonia molecules not only increase local pH but also serve as protectors against

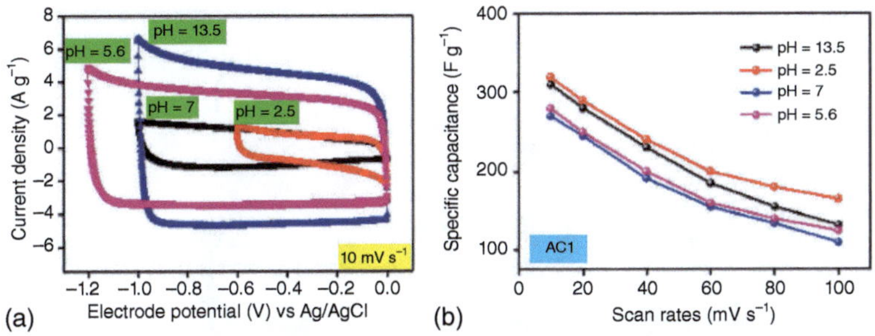

Figure 9.9 (a) Cyclic voltammograms of the electrode in different aqueous electrolytes 10 mV s^{-1}. (b) Rate performance of the electrode in different aqueous electrolytes. Source: Reproduced with permission from Slesinski et al. [70]/© 2021/Elsevier.

electrode oxidation or electrolyte decomposition, thereby enlarging the ESW of the aqueous electrolyte. Benefiting from this novel design, the as-synthesized aqueous SC system shows a maximum cell voltage of 1.8 V. Moreover, there are some works that use ion-exchange membranes to separate the cathode and anode, resulting in the cathode working under acidic conditions and the anode working under alkaline conditions.

In a word, the ESW and specific capacitance are strongly related to the pH level of the electrolyte. Usually, a wider voltage window is obtained by a neutral system due to the low activity of HER and OER. However, the above-mentioned modification measures cannot fulfill the demands of industrial manufacturing due to the complicated procedures and high costs involved. As a result, more convenient and practical methods to adjust the pH value of electrolytes are urgently needed.

9.4.3.2 Introducing Redox Mediators

Increasing the specific capacitance of SCs is another strategy to improve the energy density. Redox electrolytes have received significant attention in this field. Based on the electrolyte-enhanced mechanism, redox-type electrolytes can be divided into redox-additive electrolytes and redox-active electrolytes. In the former, one or two redox mediators are introduced into the electrolyte to optimize the electrochemical reaction kinetics on the interface between the electrode and electrolyte. In the latter, redox mediators are directly used as electrolytes to facilitate electron transfer. As a crucial component of redox electrolytes, redox mediators can be classified into organic (methylene blue, indigo carmine, and p-phenylenediamine) and inorganic (common redox additives, including transition metal ions, halide ions, and cyanide ions). Furthermore, redox mediators can also be grouped into single and dual redox mediators based on the diversity of species. When redox mediators are dissolved in an aqueous solvent, dissociative cations or anions are adsorbed on the electrode surface through either electrostatic or physical behavior and induce extra redox reactions to increase the charge storage capacity, as shown in Figure 9.10 [72]. Consequently, the assembled aqueous SC operating in the redox electrolyte substantially increases the energy density without sacrificing the power density.

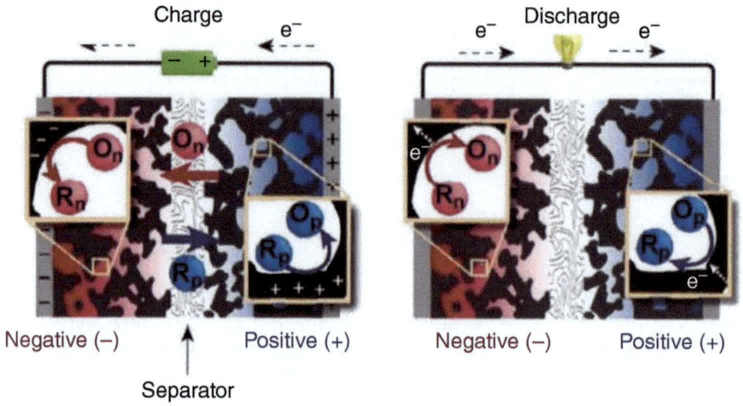

Figure 9.10 Schematic showing capacitive and faradic charge storage processes of redox couples in the electrolyte. Source: Reproduced with permission from Wang et al. [72]/ © 2020/American Chemical Society.

Besides increasing specific capacity, redox mediators can also expand the working voltage window of SCs. During the discharging process, oxidation (cathode side) and reduction (anode side) take place on the electrode surface, and vice versa is observed during the charging process. Each pair of redox reactions has a corresponding redox potential. The voltage window of SCs is expanded if a suitable redox mediator with a redox potential close to the H_2O decomposition potential is introduced. This expansion can be attributable to the fact that the reaction kinetics are faster than that of water electrolysis, leading to reduced water decomposition activity. Recently, Wang et al. have introduced tetrachloridehydroquinone (TCHQ), a redox mediator with a high redox potential, into organic electrolytes to achieve a broad electrochemical window [73]. Experimental and density functional theory (DFT) calculations proved that Lewis-based pyrrole N atoms in N-doped AC can act as electrocatalysis centers, further accelerating the redox reaction of TCHQ. As a result, the cell voltage can increase up to 2.7 V with an energy density of 35.7 Wh kg^{-1}, resulting in better electrochemical performance than commercial SCs. Moreover, the capacitance of a TCHQ-based electrolyte (65.5 F g^{-1}) is around 1.66 times higher than that of a blank electrolyte, as the redox reaction provides additional capacity. Lee et al. investigated the redox processes of $VOSO_4H_3PO_4$ gel electrolyte [74]. They found that the peaks at CV curves change from flat rounded to sharp redox peaks due to the redox reaction of VO^{2+}/VO_2^+. Furthermore, the Faradaic redox reaction of VO^{2+}/VO_2^+ with hydroxyl and carbonyl groups on the surface occurs via oxygen transfer. In particular, during the charging process of hydroxyl groups, VO^{2+} adsorbed on the electrode surface interacts with H^+ in the hydroxyl groups and forms a bond. Then, the oxygen atom is relocated and forms VO^{2+} on the electrode surface by liberating H^+ and e^-. Eventually, VO_2^+ diffuses into the bulk solution by replacing H^+ in the bulk solution. However, in the case of carbonyl groups, the adsorbed VO^{2+} forms a transition state with carbonyl groups. It then interacts with H_2O, while VO_2^+ diffuses back into the bulk electrolyte. Moreover, since the discharging process is the reverse of the charging process, functional hydroxyl and carbonyl groups on the electrode surface with the $VOSO_4$ electrolyte participate in the reversible Faradaic redox reaction of VO^{2+}/VO_2^{+}, leading to the enhanced energy storage performance of fibrous SCs.

In light of the enhanced mechanism, a dual redox mediator design can increase the voltage windows to the maximum, in which one mediator with low redox potential can be used in the anode to inhibit HER, and the other one with high redox potential can be used in the cathode to inhibit OER. For example, Aparna et al. reported a biomass-derived carbon-based SC by introducing a dual-redox mediator (KI and hydroquinone [HQ]) into H_2SO_4 electrolyte. The two-electrode test revealed that the working voltage can be up to 1.2 V and a specific capacitance of 228 F g^{-1}, both of which are higher than those of the H_2SO_4 electrolyte, with 0.8 V and 120 F g^{-1}, respectively. Subsequently, Abbasi et al. designed an asymmetric SC with a high energy density of 86.3 Wh kg^{-1} by adding KI and sodium anthraquinone-2-sulfonic acid (AQSA) into the electrolyte [75]. As a result of the beneficial effects of the enhancement of fast and reversible faradic reactions, this device displayed a maximum cell voltage of 1.7 V and an improved capacitance of 220 F g^{-1}.

In summary, introducing redox mediators into the electrolyte is an effective method to increase the SC voltage, as it is facile and less expensive, with high safety. Compared with other methods, introducing a redox mediator can maximize the energy density, because this strategy can simultaneously improve the capacity and expand the voltage window. Unfortunately, this strategy is limited by self-discharge caused by serious irreversible side reactions. Hence, it is a long way to find more proper redox mediators.

9.4.3.3 Constructing "Water-in-Salt" Electrolytes

WIS electrolyte is a relatively new modification strategy followed to increase the operating voltage, which can improve the stability of the solvent and inhibit side reactions (especially gas generation). WIS electrolytes have extremely high concentrations (near-saturated), resulting in the number of dissociative water molecules being much less than salt ions. Wang et al. first proposed the concept of WIS electrolytes in 2015 and applied a WIS electrolyte to aqueous lithium-ion batteries to realize the expansion of the working voltage window [76]. The most popular and most studied electrolyte salt is bis(trifluoromethane sulfonyl)imide (LiTFSI) due to its ultrahigh solubility in aqueous electrolytes and superior stability against water electrolysis. LiTFSI particles are ionized into Li^+ and $TFSI^-$ in a WIS electrolyte consisting of 21 molalities of LiTFSI and 1 kg water. With an increasing concentration of LiTFSI, the $TFSI^-$ conduction band minimum and the water valence band maximum shift to lower potentials. Then, the decomposition potential of $TFSI^-$ becomes lower than the HER potential, leading to the inhibition of water electrolysis. Moreover, water electrolysis activity near the electrode is also weakened due to the low availability of free water molecules, which is caused by the solvation of lithium ions, as shown in Figure 9.11. As a result, the ESW is expanded to 3.0 V.

Many studies have reported the electrochemical performance and energy storage mechanism of WIS. For example, 21 m LiTFSI WIS electrolyte was applied to a carbon-based SC, and the SC revealed a voltage of 2.2 V, which is considerably superior to the corresponding dilute aqueous electrolytes [77]. Bu proposed that 17 m $NaClO_4$ super-concentrated electrolyte can achieve a 2.3 V working voltage window in SCs [78]. Unfortunately, WIS electrolytes suffer from low conductivity and high viscosity, resulting in inferior rate performance and more "dead" volume of active electrode materials. To overcome these challenges, a cosolvent system was reported by mixing acetonitrile with the traditional WIS electrolyte (21 m LiTFSI/H_2O) to construct a novel "acetonitrile/water in salt" (AWIS) electrolyte. According to DFT-based molecular dynamics (DFTMD) simulations, acetonitrile molecules can easily coordinate to lithium ions for the remaining space of the Li^+ solvation sheath due to their smaller size and thus less steric interaction than $TFSI^-$ anions. Using the AWIS electrolyte in a model SC remarkably enhanced the high-rate performance, accompanied by a 2.4-times capacitance increase at 10 A g^{-1} with respect to the original WIS electrolyte. This AWIS electrolyte also enabled a stable long-term cycling capability of the model SC for over 14,000 cycles at a high operation voltage of 2.2 V.

To sum up, designing high-voltage aqueous SCs via constructing a WIS electrolyte is straightforward and convenient. Because of the simple operation procedure and admirable performance, this method has recently drawn considerable interest. However, inferior physicochemical features (e.g. conductivity and viscosity) and exorbitant price impede the further development of WIS electrolytes. Hence, enormous efforts should be made to optimize WIS electrolytes.

9.5 Conclusion

In summary, we introduced and summarized the latest progress in the fundamentals, electrode materials, and electrolytes in SCs. Carbon-based materials are the most

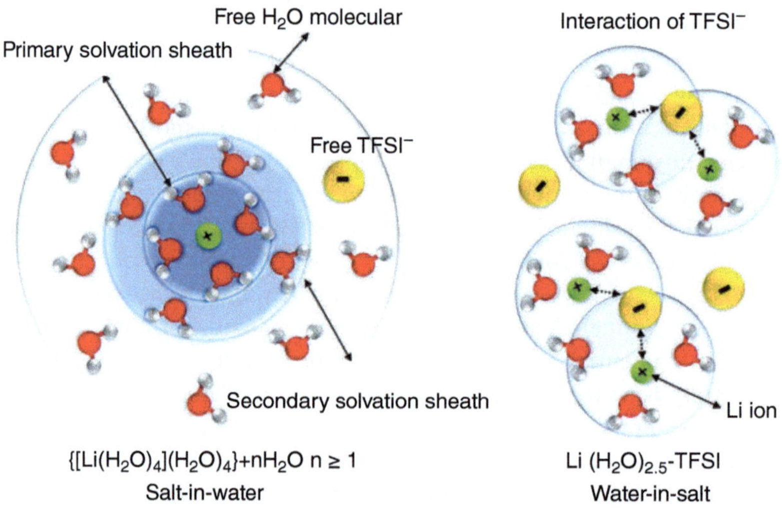

Figure 9.11 Evolution of the Li^+ primary solvation sheath in diluted and WIS solutions. Source: Reproduced with permission from Dou et al. [76]/© 2015/The Royal Society of Chemistry.

popular electrode materials for SCs. Many studies have focused on the optimization of pore structure (pore size, number, and shape) to enhance the utilization of electrode materials. Heteroatom doping including N, B, and S and introducing functional groups (–OH) have been proven to be effective strategies to improve specific pseudocapacitance, resulting in higher energy density. CPs are electroactive materials that are promising electrode material candidates for SCs. Besides, composite materials consisting of CPs and carbonaceous materials or metal compounds have shown excellent advantages due to the availability of plenty of empty spaces for ion storage. However, sometimes the space is too small to accommodate the ions, resulting in severe volume expansion. The electrochemical performance of metal compounds in SCs has also been investigated. Electrical conductivity, morphology, structure, and components are vital in energy storage processes.

There are two challenges to the further development of SCs, namely high cost and low energy density on the premise of no loss of ultralong cycling stability and excellent rate performance, which must be overcome to expand the applications of SCs. According to the equation of energy density $E = CV^2/2$, a wider working voltage window can improve the energy storage ability, leading to an enhanced energy density. Recently, two major strategies have been used to expand the voltage window: replacing aqueous electrolytes with ILs or organic electrolytes with higher decomposition voltage; and assembling asymmetric SCs with a capacitive electrode and a Faradaic electrode (introduced in Chapter 10) to widen the voltage window. In some cases, the working voltage windows can be increased to 2.2 V, which is far beyond the thermodynamic limit of 1.2 V in aqueous electrolytes. Moreover, the concept of aqueous and nonaqueous (one or more) cosolvent electrolytes has been proposed. This combined electrolyte system has not been applied to batteries or capacitors; thus, it should be further investigated for applications in SCs. In addition, solid-state SCs are promising candidates for energy storage devices because of their flexibility, safety, high power density, and easy integration with other energy storage systems. In this strategy, the design of quasi-solid-state or all-solid-state electrolytes is crucial.

In short, SCs play an important role in energy storage and conversion systems. They either complement batteries, increasing their efficiency and lifespan, or serve as energy solutions, where an extremely long cycling life and rapid power delivery are required, especially for fast-charging e-buses and electric tramcars.

References

1 Ho, J., Jow, T.R., and Boggs, S. (2010). Historical introduction to capacitor technology. *IEEE Electrical Insulation Magazine* 26: 20–25.

2 Schindall, J. (2007). The charge of the ultracapacitors. *IEEE Spectrum* 44: 42–46.

3 Boos, D.L. (1970). Electrolytic capacitor having carbon paste electrodes. U.S. Patent 3, 536,963, filed 29 May 1968 and issued 27 October 1970.

4 Evans Capacitor Company (2007). *The littlest big capacitor—an evans hybrid technical paper*. RI, USA: East Providence.

5 Evans, D.A. (1994). Containers with Anodes and Cathodes with Electrolytes. U.S. Patent 5369547, 29 November 1994. Evans D. A., Journal.

6 Ray, A. and Saruhan, B. (2021). Application of ionic liquids for batteries and supercapacitors. *Materials* 14: 2942.

7 Zhang, L.L. and Zhao, X.S. (2009). Carbon-based materials as supercapacitor electrodes. *Chemical Society Reviews* 38 (9): 2520–2531.

8 Salitra, G., Soffer, A., Eliad, L. et al. (2000). Carbon electrodes for double-layer capacitors i. relations between ion and pore dimensions. *Journal of Electrochemical Society* 147 (7): 2486.

9 Eliad, L., Salitra, G., Soffer, A. et al. (2001). Ion sieving effects in the electrical double layer of porous carbon electrodes: estimating effective ion size in electrolytic solutions. *The Journal of Physical Chemistry B* 105 (29): 6880–6887.

10 Vix, G.C., Frackowiak, E., Jurewicz, K. et al. (2005). Electrochemical energy storage in ordered porous carbon materials. *Carbon* 43 (6): 1293–1302.

11 Chmiola, J., Largeot, C., Taberna, P.L. et al. (2008). Desolvation of ions in subnanometer pores and its effect on capacitance and double-layer theory. *Angewandte Chemie International Edition* 47 (18): 3392–3395.

12 Chmiola, J., Yushin, G., Gogotsi, Y. et al. (2006). Anomalous increase in carbon capacitance at pore sizes less than 1 nanometer. *Science* 313 (5794): 1760–1763.

13 Largeot, C., Portet, C., Chmiola, J. et al. (2008). Relation between the ion size and pore size for an electric double-layer capacitor. *Journal of the American Chemical Society* 130 (9): 2730–2731.

14 Huang, J., Sumpter, B.G., and Meunier, V. (2008). A universal model for nanoporous carbon supercapacitors applicable to diverse pore regimes, carbon materials, and electrolytes. *Chemistry–A European Journal* 14 (22): 6614–6626.

15 Huang, J., Sumpter, B.G., and Meunier, V. (2008). Theoretical model for nanoporous carbon supercapacitors. *Angewandte Chemie International Edition* 120 (3): 530–534.

16 Huang, J., Sumpter, B.G., and Meunier, V. (2010). Curvature effects in carbon nanomaterials: exohedral versus endohedral supercapacitors. *Journal of Materials Research* 25 (8): 1525–1531.

17 Feng, G., Qiao, R., Huang, J. et al. (2010). Ion distribution in electrified micropores and its role in the anomalous enhancement of capacitance. *ACS Nano* 4 (4): 2382–2390.

18 Zhai, Z., Zhang, L., Du, T. et al. (2022). A review of carbon materials for supercapacitors. *Materials & Design* 221: 111017.

19 Park, H.W. and Roh, K.C. (2023). Recent advances in and perspectives on pseudocapacitive materials for supercapacitors–a review. *Journal of Power Sources* 557: 232558.

20 Shao, Y., El-Kady, M.F., and Sun, J. (2018). Design and mechanisms of asymmetric supercapacitors. *Chemical Review* 118 (18): 9233–9280.

21 Rudge, A., Davey, J., Raistrick, I. et al. (1994). Conducting polymers as active materials in electrochemical capacitors. *Journal of Power Sources* 47 (1–2): 89–107.

22 Diaz, A. and Logan, J. (1980). Electroactive polyaniline films. *Journal of Electroanalytical Chemistry and Interfacial Electrochemistry* 111 (1): 111–114.

23 Feldberg, S.W. (1984). Reinterpretation of polypyrrole electrochemistry. Consideration of capacitive currents in redox switching of conducting polymers. *Journal of the American Chemical Society* 106 (17): 4671–4674.

24 Conway, B.E. (1991). Transition from "supercapacitor" to "battery" behavior in electrochemical energy storage. *Journal of the Electrochemical Society* 138 (6): 1539.

25 Eftekhari, A., Li, L., and Yang, Y. (2017). Polyaniline supercapacitors. *Journal of Power Sources* 347: 86–107.

26 Karthikeyan, K., Amaresh, S., and Lee, S.N. (2014). High-power lithium-ion capacitor using $LiMnBO_3$-nanobead anode and polyaniline-nanofiber cathode with excellent cycle life. *ChemSusChem* 7 (8): 2310–2316.

27 Tundwal, A. (2024). Developments in conducting polymer-, metal oxide-, and carbon nanotube-based composite electrode materials for supercapacitors: a review. *RSC Advances* 14: 9406–9439.

28 Wang, Y., Guo, J., and Wang, T. (2015). Mesoporous transition metal oxides for supercapacitors. *Nanomaterials* 5 (4): 1667–1689.

29 Trasatti, S. and Buzzanca, G. (1971). Ruthenium dioxide: a new interesting electrode material. Solid state structure and electrochemical behaviour. *Journal of electroanalytical chemistry and interfacial electrochemistry* 29 (2): A1–A5.

30 Wang, G., Zhang, L., and Zhang, J. (2012). A review of electrode materials for electrochemical supercapacitors. *Chemical Society Review* 41 (2): 797–828.

31 Shi, F., Li, L., and Wang, X.L. (2014). Metal oxide/hydroxide-based materials for supercapacitors. *RSC Advances* 4 (79): 41910–41921.

32 Zukalová, M., Kalbác, M., and Kavan, L. (2005). Pseudocapacitive lithium storage in TiO_2 (B). *Chemical Materials* 17 (5): 1248–1255.

33 Dos, G.S., Larsson, S.H., Oliveira, H.P. et al. (2020). Sustainable biomass activated carbons as electrodes for battery and supercapacitors – a mini-review. *Nanomaterials* 10 (7): 1398.

34 Selvaraj, A.R., Chinnadurai, D., and Cho, I. (2022). Bio-waste wood-derived porous activated carbon with tuned microporosity for high performance supercapacitors. *Journal of Energy Storage* 52: 104928.

35 Tian, X.L., Yu, J.H., Qiu, L. et al. (2023). Structural changes and electrochemical properties of mesoporous activated carbon derived from *Eucommia ulmoides* wood tar by KOH activation for supercapacitor applications. *Industrial Crops and Products* 197: 116628.

36 Agrawal, A., Gaur, A., and Kumar, A. (2023). Fabrication of *Phyllanthus emblica* leaves derived high-performance activated carbon-based symmetric supercapacitor with excellent cyclic stability. *Journal of Energy Storage* 66: 107395.

37 Meekati, T., Pakawatpanurut, P., and Amornsakchai, T. (2023). High-performance activated carbon fiber derived from pineapple leaf fiber via Co(II)-assisted preparation revealed surprising capacitive-pseudocapacitive synergistic charge storage. *Journal of Energy Storage* 71: 108135.

38 Pant, B., Ojha, G.P., Acharya, J. et al. (2023). Preparation, characterization, and electrochemical performances of activated carbon derived from the flower of *Bauhinia variegata* L for supercapacitor applications. *Diamond and Related Materials* 136: 110040.

39 Anandhu, T.P., Mohan, R., Cherusseri, J. et al. (2022). High areal capacitance and enhanced cycling stability of binder-free, pristine polyaniline supercapacitor using hydroquinone as a redox additive. *Electrochimica Acta* 425: 140740.

40 He, W., Qiang, H., Liang, S. et al. (2022). Hierarchically porous wood aerogel/polypyrrole(PPy) composite thick electrode for supercapacitor. *Chemical Engineering Journal* 446: 137331.

41 Elanthamilan, E., Ganeshkumar, A., Wang, S.F. et al. (2023). Fabrication of polydopamine/polyaniline decorated multiwalled carbon nanotube composite as multifunctional electrode material for supercapacitor applications. *Synthetic Metals* 298: 117423.

42 Wang, S., Yu, X., Li, N. et al. (2023). Enhancing energy storage capacity of supercapacitors via constructing a porous PPy/carbon cloth electrode by a template-assisted method. *Journal of Energy Storage* 72: 108312.

43 Yang, W., Guo, H., Tian, J. et al. (2023). Integrated battery-capacitor storage system: polyaniline interwoven Co-ZIF derived hollow NiCo-LDH with high electrochemical properties for hybrid supercapacitors. *New Journal of Chemistry* 47: 16050–16058.

44 Li, X., Wang, X., Liu, G. et al. (2023). 2.5 V "water in salt" aqueous micro-supercapacitors based on polypyrrole-coated NiCo layered double hydroxides. *Chemical Engineering Journal* 452: 139140.

45 Zhou, B., Li, Z., Qin, D. et al. (2023). Construction of tunable ZnCoOx/C-PANI electrode materials for high-performance supercapacitors. *Journal of Alloys and Compounds* 956: 170327.

46 Zhang, J., Chen, J., Luo, Y. et al. (2020). MOFs-assisted synthesis of hierarchical porous nickel–cobalt nitride heterostructure for oxygen reduction reaction and supercapacitor. *ACS Sustainable Chemistry & Engineering* 8 (1): 382–392.

47 Fu, S., Ma, L., Gan, M. et al. (2022). Co-doped nickel sulfide (NiS_2) derived from bimetallic MOF for high-performance asymmetric supercapacitors. *Synthetic Metals* 283: 116972.

48 Zhang, M., Yang, D., Li, J. et al. (2020). Supercapacitor performances of MnO_2 and MnO_2/reduced graphene oxide prepared with various electrodeposition time. *Vacuum* 178: 109455.

49 Pawar, S.A., Patil, D.S., Nandi, D.K. et al. (2022). Cobalt-based metal oxide coated with ultrathin ALD-MoS_2 as an electrode material for supercapacitors. *Chemical Engineering Journal* 435: 135066.

50 Teli, A.M., Beknalkar, S.A., and Mane, S.M. (2023). Synergetic effect of ternary MnVMo-oxide electrode by hydrothermal method for high-performance asymmetric supercapacitor. *Journal of Energy Storage* 65: 107289.

51 Chen, H., Zhu, Y., Yan, X. et al. (2023). Porous CoO/carbon foam composites synthesized by solvothermal method for supercapacitor and enhanced microwave absorption applications. *Diamond and Related Materials* 138: 110270.

52 Wang, L., Gao, L., Wang, J. et al. (2019). MoO_3 nanobelts for high-performance asymmetric supercapacitor. *Journal of Materials Science* 54 (21): 13685–13693.

53 Basu, R., Ghosh, S., Bera, S. et al. (2019). Phase-pure VO_2 nanoporous structure for binder-free supercapacitor performances. *Scientific Reports* 9 (1): 4621.

54 Alhebshi, N.A., Vaseem, M., Minyawi, B.A. et al. (2022). Single and double layer of monoclinic VO_2 ink-based printed and interdigitated supercapacitors. *Energy Technology* 10 (11): 2200432.

55 Iqbal, M.F., Yousef, A.K.M., Hassan, A. et al. (2021). Significantly improved electrochemical characteristics of nickel sulfide nanoplates using graphene oxide thin film for supercapacitor applications. *Journal of Energy Storage* 33: 102091.

56 Al-Abawi, B.T., Parveen, N., Ansari, S.A. et al. (2022). Controllable synthesis of sphere-shaped interconnected interlinked binder-free nickel sulfide@nickel foam for high-performance supercapacitor applications. *Scientific Reports* 12 (1): 14413.

57 Li, K., Zhao, B., Zhang, H. et al. (2021). 3D Porous honeycomb-like CoN-Ni_3N/N-C nanosheets integrated electrode for high-energy-density flexible supercapacitor. *Advanced Functional Materials* 31 (28): 2103073.

58 Pallavolu, M.R., Anil, Y., Nallapureddy, R.R. et al. (2022). In-situ design of porous vanadium nitride@carbon nanobelts: a promising material for high-performance asymmetric supercapacitors. *Applied Surface Science* 575: 151734.

59 Lin, Y., Chen, X., Tuo, Y. et al. (2022). In-situ doping-induced lattice strain of NiCoP/S nanocrystals for robust wide pH hydrogen evolution electrocatalysis and supercapacitor. *Journal of Energy Chemistry* 70: 27–35.

60 Vikraman, D., Hussain, S., Karuppasamy, K. et al. (2022). Engineering the active sites tuned MoS_2 nanoarray structures by transition metal doping for hydrogen evolution and supercapacitor applications. *Journal of Alloys and Compounds* 893: 162271.

61 Gao, X., Zhang, Y., Yin, S. et al. (2022). Dual redox active sites N-C@Ni_2P/$NiSe_2$ heterostructure supercapacitor integrated with triboelectric nanogenerator toward efficient energy harvesting and storage. *Advanced Functional Materials* 32 (38): 2204833.

62 Najib, S., Bakan, F., Abdullayeva, N. et al. (2020). Tailoring morphology to control defect structures in ZnO electrodes for high-performance supercapacitor devices. *Nanoscale* 12 (30): 16162–16172.

63 Gao, Q., Wang, J., and Wang, J. Morphology-controllable synthesis of $CuCo_2O_4$ arrays on Ni foam as advanced electrodes for supercapacitors. *Journal of Alloys and Compounds* 789: 193–200.

64 Liu, R., Chen, L., Mo, F. et al. (2023). Porous cobalt-nickel phosphides prepared from Al-doped NiCo-LDH precursors for supercapacitor and electrocatalysis applications. *Chemical Engineering Journal* 455: 140545.

65 Yue, P., Zhang, Y., Wu, X. et al. (2023). Defective ZnCoNiP nanosheets derived from metal-organic-frameworks as electrodes for high-performance supercapacitors. *Journal of Energy Storage* 58: 106320.

66 Fenton, D.E. (1993). Complex of alkali metal ions with poly (ethylene oxide). *Polymer* 14: 589. Pistoia, G. Lithium batteries: new materials, developments and perspectives. OSTI ID: 6746344, Reference Number:ECN-95-0E0094; EDB-95-013937.

67 Wright, P.V. and Br. Polym. J. (1975). Electrical conductivity in ionic complexes of poly (ethylene oxide). *British Polymer Journal* 7: 319–327.

68 Feuillade, G., Perche, P. et al. (1975). Ion-conductive macromolecular gels and membranes for solid lithium cells. *Journal of Applied Electrochemistry* 5: 63–69.

69 Zhao, N., Fan, H., Zhang, M. et al. (2019). Investigating the large potential window of $NiCo_2O_4$ supercapacitors in neutral aqueous electrolyte. *Electrochimica Acta* 321: 134681.

70 Gunasekaran, S.S., Badhulika, S. et al. (2022). Effect of pH and activation on macroporous carbon derived from cocoa-pods for high performance aqueous supercapacitor application. *Materials Chemistry and Physics* 276: 125399.

71 Slesinski, A., Matei, C., Fic, K. et al. (2018). Self-buffered pH at carbon surfaces in aqueous supercapacitors. *Carbon* 129: 758–765.

72 Zhao, J. and Burke, A.F. (2021). Electrochemical capacitors: materials, technologies and performance. *Energy Storage Materials* 36: 31–55.

73 Wang, Z.F., Yi, Z., Yu, S.C. et al. (2022). High-voltage redox mediator of an organic electrolyte for supercapacitors by lewis base electrocatalysis. *ACS Applied Materials & Interfaces* 14 (21): 24497–24508.

74 Lee, S. and An, G.H. (2022). Reversible faradaic reactions involving redox mediators and oxygen-containing groups on carbon fiber electrode for high-performance flexible fibrous supercapacitors. *Journal of Energy Chemistry* 68: 1–11.

75 Abbasi, S., Hekmat, F., and Shahrokhian, S. (2023). Dual redox electrolytes for improving the performance of asymmetric supercapacitors constructed from heteroatom-doped green carbon spheres. *Journal of Alloys and Compounds* 957: 170452.

76 Suo, L., Borodin, O., and Gao, T. (2015). "Water-in-salt" electrolyte enables high-voltage aqueous lithium-ion chemistries. *Science* 350 (6263): 938–943.

77 Dou, Q., Lei, S., and Wang, D.W. (2018). Safe and high-rate supercapacitors based on an "acetonitrile/water in salt" hybrid electrolyte. *Energy & Environmental Science* 11 (11): 3212–3219.

78 Bu, X., Su, L., and Dou, Q. (2019). A low-cost "water-in-salt" electrolyte for a 2.3 V high-rate carbon-based supercapacitor. *Journal of Materials Chemistry A* 7 (13): 7541–7547.

10

Battery–Supercapacitor Hybrid Devices

Jiawei Zhang[1], Mingxing Zhang[1], and Minghua Chen[2]

[1]*Harbin University of Science and Technology, School of Electrical and Electronic Engineering, No. 52 Xuefu Road, Harbin, 150080, China*
[2]*Harbin University of Science and Technology, School of Electrical and Electronic Engineering, Key Laboratory of Engineering Dielectric and Applications (Ministry of Education), No. 52 Xuefu Road, Harbin, 150080, China*

10.1 Introduction

As discussed in previous chapters, electrochemical energy storage devices are important in the fields of energy conversion and electric vehicles. Among them, secondary batteries (dominated by lithium-ion batteries, LIBs) are attractive due to their high energy density; however, their power density is insignificant. On the contrary, supercapacitors (SCs) provide high power density and long cycling stability, but inferior energy density. An ideal energy storage system should offer both superior energy density and power density. A hybrid ion capacitor (HIC) is a new system that exhibits higher energy density than SCs and superior power density than batteries. Thus, HICs have a promising potential to be used in regenerative braking energy harvesting from trains, metros, and some heavy automotives. The energy per weight or per volume of a HIC is lower than those of sodium-ion batteries and LIBs due to the limitation of the energy storage mechanism; hence, sodium-ion batteries or LIBs cannot be replaced by HICs. A HIC (sodium or lithium) can deliver a specific energy density that is four to five times higher than that of a SC. This significant energy originates from two factors: the carbonate-based battery electrolyte that provides a wider working voltage window (around 4.2 V) than the acetonitrile-based organic electrolyte of a SC (around 2.7 V); and the electrode materials being both battery like and capacitive like, i.e. at least one electrode stores energy by insertion of ions. Therefore, HICs can be suitable to compete with SCs rather than batteries. Lithium-ion capacitors (LICs) were first designed in 2001 by Amatucci et al. [1] They are considered one of the most effective devices for storing energy and are often considered offspring of LIBs for several reasons. In addition, sodium-ion and potassium-ion capacitors (SIC and KIC, respectively) have also gained commercial interest as they are hybrid devices combining an ion battery

Towards Next Generation Energy Storage Technologies: From Fundamentals to Commercial Applications, First Edition. Edited by Minghua Chen.

with a traditional capacitor. This chapter will focus on the developments in LICs as the primary example of a hybrid capacitor, but it must be noted that there are many similarities between LICs, SICs, and KICs.

To date, due to their strong gravimetric functional potential and excellent electrochemical operation, a number of active materials, such as metal compounds, polyanions, and metalloid/metal compounds, have been produced in LICs for battery-type electrodes (Figure 10.1). However, their further production in LICs is constrained by the poor conductivity, wide volume variability, and strong polarization of these active materials. Carbonaceous materials have been extensively researched and used in energy storage devices such as LICs, due to their advantages such as low cost, sufficient availability, plenty of allotropes and transformations, and superior physical/chemical stability. Because of their large specific surface region, strong conductivity, and excellent usability in electrolytes, carbon materials were also inserted into electrodes to address such issues. Furthermore, carbon compounds can also specifically function as the active compounds of battery-type electrodes in LICs since they have active Li^+ intercalation/de-intercalation sites. Unlike battery-type electrodes, counter-electrodes of LICs are capacitor type. Consequently, numerous porous carbon materials with a wide specific surface area, such as activated carbon (AC) and graphene, are potential candidates for LICs' capacitor-type electrodes. Their capacitances rely primarily on the surface of carbon-based electrodes for the adsorption/desorption of ions. Therefore, porous structures with sufficient pore size distribution play an important role in the electrochemical production of carbon materials in LICs. Specific materials are typically inserted into carbonaceous materials to have pseudocapacity to further improve the capacitance of the capacitor-type electrode.

10.2 Classification Based on Energy Storage Mechanism

A LIC contains an organic electrolyte containing lithium ions and integrates the energy storage mechanism of LIB and electric double-layer capacitor (EDLC). A LIB-type electrode stores energy through electrochemical insertion. Since there are a lot of ions that can be embedded in the battery reaction, a LIB-type electrode material provides high capacity for a LIC, as shown in Figure 10.2. A EDLC electrode, on the other hand, stores energy through physical adsorption, which provides higher power density and better cycling stability for a LIC. Therefore, LICs are hybrid energy storage devices capable of providing relatively high energy density and power density with a long cycle life, which can better meet the overall requirements of the energy density and power density of power supply systems in practical applications.

10.2.1 The Basic Classifications of LICs

LICs can be classified into the following types according to the structure of the cathode and anode.

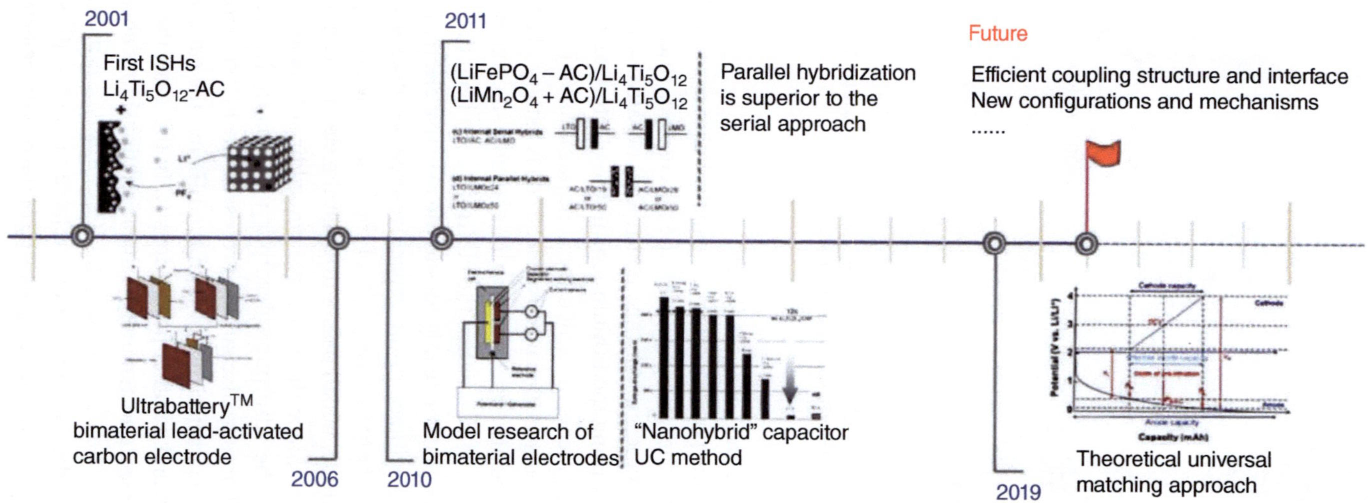

Figure 10.1 Development of lithium-ion hybrid capacitor. Source: Reproduced with permission from Xing et al. [2]. © 2022/John Wiley & Sons.

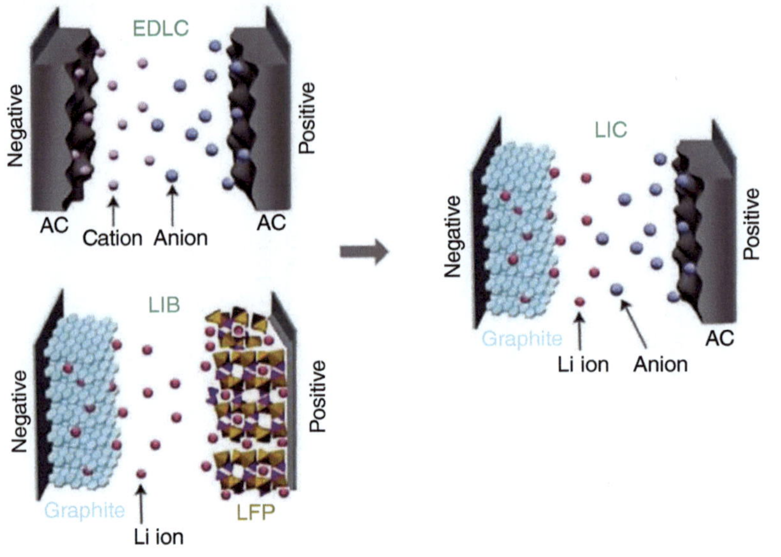

Figure 10.2 Conceptual presentation of fabrication with LICs. Source: Reproduced with permission from Soltani et al. [3]. © 2022/Elsevier.

10.2.1.1 Battery Cathode and Capacitor Anode

When a LIC is charged, the positive electrode usually behaves as a battery. Lithium ions are extracted from the electrode into the electrolyte, thus completing the chemical intercalation process. At this point, the negative electrode exhibits capacitive behavior, absorbing the lithium ions in the electrolyte, thus completing the physical adsorption process, in which the concentration of lithium ions in the electrolyte remains unchanged. The positive electrode is battery-like type, while the negative electrode is capacitive-like type. During charging, the battery-type positive electrode deliquesces Li^+, which enters the electrolyte. To keep the ion concentration constant, Li^+ in the electrolyte migrates to the negative electrode, generating a double electric layer capacitance in the capacitive-type negative electrode. On the contrary, during the discharge process, Li^+ of the double electric layer is released from the negative electrode, which passes through the electrolyte and is embedded in the positive electrode. Since the reversible embedding/de-embedding of Li^+ in LIB electrode side and the adsorption/desorption of ions in double electric layer electrode side during charging–discharging processes, the LIC constructed by these two electrodes has the characteristics of both the LIB and the capacitor to achieve a higher energy density and a higher power density. In this case, the capacitance of LIC varies with the voltage, and generally, a linear relationship is observed between voltage and capacitance. Thus, the capacity of LICs of this type is independent of the concentration of ions in the electrolyte and is related to the specific surface area of porous carbon. However, if the working potential of the negative electrode is too low, the solid electrolyte interface (SEI) membrane will grow rapidly, the output voltage of the LIC will become low, and the cycling stability will become worse. For example, the working potential of AC negative electrode in $LiFePO_4$ (+) | | AC (−), Li_2MnO_4 (+) | | AC (−) system should be higher than 2.0 V [4].

10.2.1.2 Capacitive Cathode and Battery Anode

The energy-storing type of LICs is the exact opposite of the previously discussed type, where the positive electrode shows capacitor behavior and the negative electrode shows battery behavior. When a LIC is charged, anions in the electrolyte are adsorbed to the positive electrode, while lithium ions are embedded in the negative electrode, resulting in a low concentration of lithium ions in the electrolyte. Therefore, the electrolyte not only acts as the carrier of lithium ions, but also serves as an important active substance. Therefore, the capacity of this type of LICs is related not only to the properties of the electrode (potential and capacity) but also to the concentration of ions in the electrolyte. Such as AC (+) | | $Li_4Ti_5O_{12}$ (−) or AC (+) | | Nb_2O_5 (−) is the representative of this type of energy storage type in these types of $Li_4Ti_5O_{12}$ and some constraint capacitor electrode material (such as Nb_2O_5, VN, V_2O_5, and TiC) of the discharge platform is higher. However, it can provide a longer cycle life and a higher power density, so it is considered an excellent electrode material. However, the energy density of LIC devices of this type is not too high. Generally, for an electrolyte with 1 m ion concentration, the energy density cannot exceed 30 Wh kg^{-1}.

10.2.1.3 Capacitive Cathode and Pre-lithium Battery Anode

The biggest difference of this type compared with the previous type is that the anode is pre-lithiated; apart from this, there is not much difference. The ion transport mechanism is more complex because two different types of transport phenomena occur during the charging and discharging process of this type of LIC. This type of LIC has AC (+) | | pre lithium hard carbon (−), biomass-derived AC (+) | | lithium change Li_2TiSiO_5 beforehand (−), etc. When the LIC is charged, the voltage increases from the open-circuit voltage to the maximum voltage, the flow process of lithium ions is the same as the previous type, and the concentration of lithium ions in the electrolyte also decreases constantly. However, when the LIC is discharged, the voltage increases from the lowest voltage to the open-circuit voltage, and lithium ions are desorbated from the positive electrode into the electrolyte, while lithium ions in the original electrolyte continue to flow into the negative electrode. During this process, the concentration of lithium ions in the electrolyte remains unchanged. This type of LIC has a higher voltage output and a higher energy density and power density as the anode is pre-lithiated, and an additional charge and discharge process is carried out.

10.2.1.4 Hybrid Cathode and Pre-lithium Battery Type Anode

In this LIC type, cathode is composite consisted of battery-like materials and capacitor-like materials with a optimized mass ratio; anode is LIB-like materials after pre-lithium, such as AC + $LiFePO_4$ (+) | | pre lithium hard carbon (−), AC + NMC (+) | | lithium pre-inserted hard carbon (−), etc. However, the power density and cycling life of this type of device are not as good as those of the previous types. In addition, its voltage distribution is nonlinear, independent of the capacity of the device, so it is not a capacitor type (Figure 10.3).

10.2.2 The Energy Storage Mechanism of LICs

Jin and Shen derived the basic energy storage mechanism of LICs by testing typical charge–discharge voltage curves and in situ nuclear magnetic resonance (NMR) of the AC(+)//pre-graphitized hard carbon (HC) (−) system. As shown in Figure 10.4, a LIC undergoes a two-step ion transport mechanism, with the first step below (open circuit voltage) OCV and the second step above OCV. In these two steps, anions and cations synergize and play a very important role. During the first charging process from OCV to the maximum voltage (I–II), PF_6^- starts to diffuse from the electrolyte to the cathode and Li^+ is simultaneously embedded in the anode. This results in an increasing concentration of Li^+ ions in the anode, while the concentration of ions in the electrolyte decreases. When discharging from the highest voltage to OCV (II–III), PF_6^- desorbs from the cathode and Li^+ ions are de-embedded from the anode into the electrolyte, during which time the concentration of Li^+ ions in the anode decreases while the concentration of ions in the electrolyte returns to its original state. As the voltage drops from OCV to the minimum voltage (III–IV), the Li^+ concentration in the anode continues to decrease again, while the Li^+ concentration in the cathode increases rapidly. The final charging process from the minimum voltage to OCV (IV–V) is the opposite of the discharging process from OCV to the minimum voltage. In general, the charge–discharge process of a LIC can be categorized as follows: the electrolyte consumption part (above OCV) and the Li^+ diffusion part (below OCV), which is supplied by the electrolyte and the Li^+ after pre-lithiation.

10.3 Key Scientific Problems

10.3.1 Electrode Design for LICs

The most basic components in a hybrid capacitor include capacitive electrodes, battery electrodes, and electrolytes, whose physical and chemical properties play a decisive role in the electrochemical performance of the device.

Capacitive electrodes: Capacitive electrodes are very important components of hybrid capacitors, whose capacity primarily results from the absorption and desorption of charge-storing ions at the surface interface of the material; thus, their specific capacity is usually lower than that of the battery-type electrode material, which is a limiting factor for the energy density of hybrid capacitors. In order to ensure high energy density of hybrid capacitors, capacitive electrodes need to be improved in terms of specific capacity, operating potential, high multiplicity performance, long cycling life, and low cost. Methods to increase the specific capacity of capacitive materials include optimization of the pore structure design and introduction of nitrogen doping into the oxidation potential to introduce redox active sites. Capacitive materials are usually carbon materials with a large specific surface area, including AC, carbon nanotubes (CNTs), graphene, and conductive polymers.

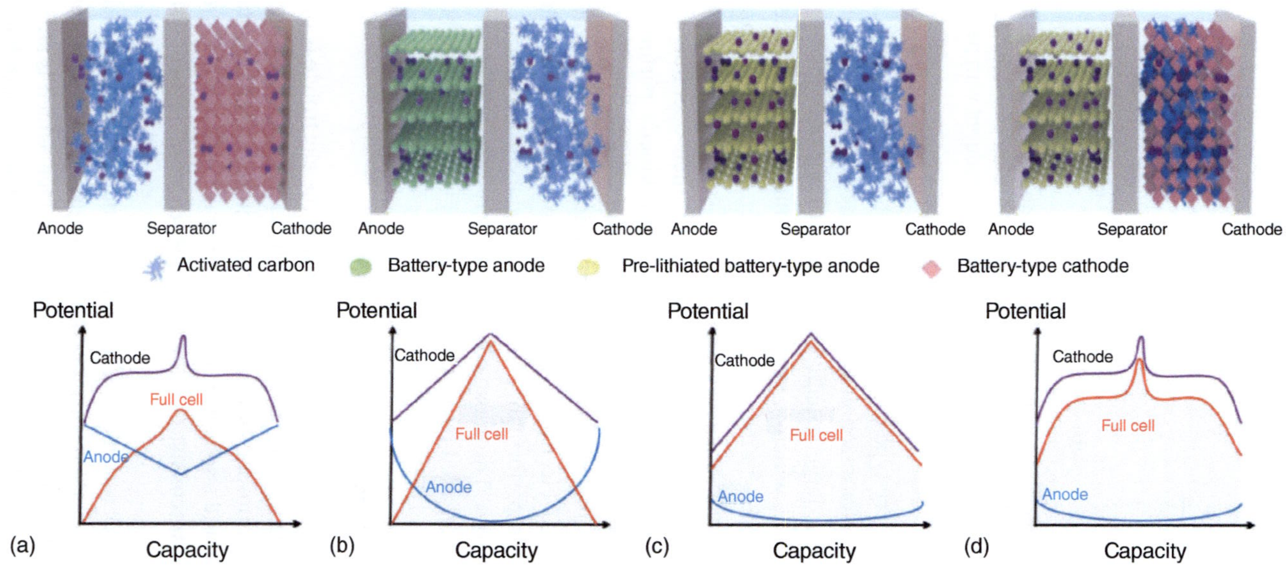

Figure 10.3 Structures and corresponding potential profiles of four typical energy storage systems commonly termed "lithium-ion capacitors": (a) battery-type cathode and capacitive anode; (b) capacitive cathode and battery-type anode; (c) capacitive cathode and pre-lithiated battery-type anode; and (d) hybrid cathode (capacitive and battery-type cathode) and pre-lithiated anode. Source: Reproduced with permission from Jin et al. [5]. © 2020/ Royal Society of Chemistry.

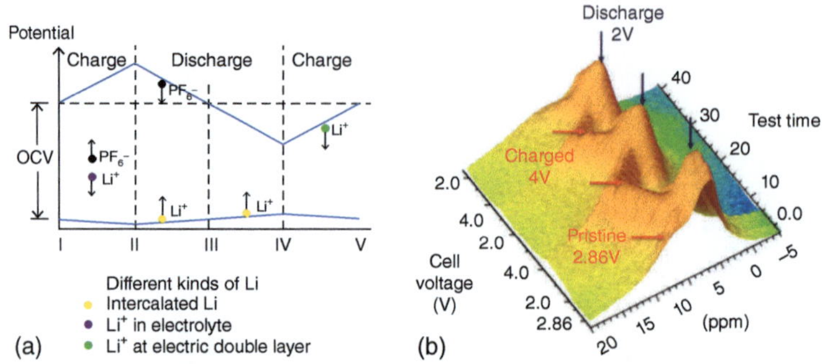

Figure 10.4 Energy storage mechanisms in LICs. (a) The charge and discharge profiles of the cathode and anode of the LIC and the corresponding ion diffusion processes; (b) in situ NMR spectra of the AC//HC system for several cycles. Source: Reproduced with permission from Jin et al. [6]. © 2020/John Wiley & Sons.

Battery-type electrodes: Battery-type materials, whose charge storage mechanism involves a bulk-phase redox reaction of charge-storing ions with the electrode material, typically have a large specific capacity, but their bulk-phase Faraday reaction kinetics are slow, which is generally the limiting factor in the power density of hybrid capacitors. To increase the power density of hybrid capacitors, the multiplicative performance and cycling stability of cell-based electrodes need to be improved without sacrificing high capacity, and their operating potential needs to be low. Common methods to improve the kinetic performance of battery-type materials are as follows: construction of highly efficient ion/electron hybrid conductors, including nanosizing, compounding with highly conductive materials, and integrated electrode structure design. Theoretically, electrode materials with high rate performance used in secondary batteries can also be used in hybrid capacitors. The most common battery-type materials used in hybrid capacitors are embedded reactive with large ion transport channels.

10.3.2 Electrolyte Design for LICs

LICs are similar to other energy storage systems in that the electrolyte is primarily responsible for charge transfer within the battery. The electrolyte consists of three primary components: a lithium salt as a charge carrier, additives to change the properties of the components, and a solvent to dissolve other components (Figure 10.5). In LIBs, the electrolyte can consist of either an aqueous or nonaqueous solution (organic solution).

Aqueous electrolytes offer a number of advantages, such as lower viscosity and higher ion mobility. However, there are more prominent disadvantages as well, such as lower specific energy, due to their lower operating voltage, which is attributable to the narrower electrochemical stable window of water-based electrolytes. Thus, applications such as LIBs for organic systems and hybrid SCs are widely recognized in many fields. Electrolyte plays a key role in charge transfer and balance between the two electrodes of LIBs and SCs and is indispensable. In all electrochemical

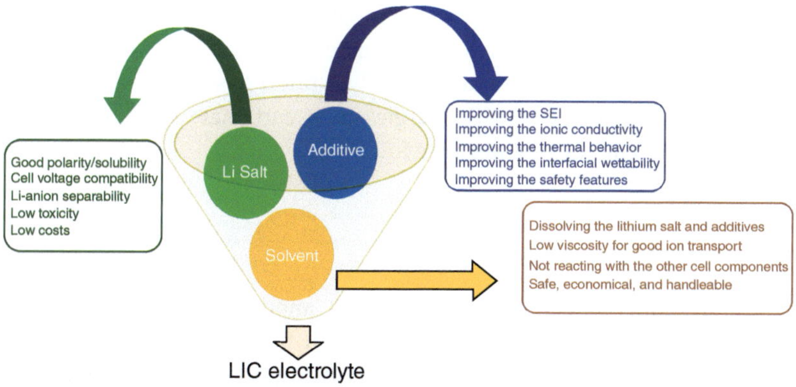

Figure 10.5 Major characteristics of LIC electrolyte components. Source: Reproduced with permission from Soltani et al. [3]. © 2020/Elsevier.

processes, the electrolyte determines the interfacial structure at the surface of the electrode and the internal ion transport inside the electrode. Therefore, the choice of electrolyte is key to improving the safety and performance of energy storage devices. Compared with aqueous electrolytes, organic electrolytes, which consist primarily of organic solvents and lithium salts, have a wider potential window for higher operating voltages. An ideal solvent should usually have good ionic solvent interactions, high safety, and a high potential window. Carbonates are organic liquids that are fairly good solvents for lithium salts. In addition, they have a relatively low viscosity, which results in a low activation energy for lithium-ion diffusion. Commonly used organic solvents for electrolytes are propylene carbonate, ethylene carbonate, diethyl carbonate, dimethyl carbonate, ethyl methyl carbonate, gamma-butyrolactone, and so on.

The most commonly used electrolytes in the market today primarily use $LiPF_6$ dissolved in carbonate solvents. Their widespread use is largely due to their wide thermodynamic electrochemical stability window. However, the major reason for the degradation in battery performance and reduced lifetime is the prevalence of side reactions, especially at transition metal oxides of the cathode operating at high potentials, leading to the loss of active lithium and an increase in cell impedance. Currently, the use of additives in electrolytes to form surface films has been reported to be one of the most effective strategies to stabilize the electrode–electrolyte interface. Among them, lithium bis(oxalate)borate (LiBOB) has been recognized as one of the effective additives and salts for the formation of a stable (solid electrolyte interphase) SEI, and the use of LiBOB as an additive to form the SEI protects the graphite anode material from the damage by the propylene carbonate-rich electrolyte. However, at the end of the charging and discharging process, the slightly higher SEI-induced polarization due to LiBOB may decrease the cycling performance. Therefore, designing new, more suitable high-voltage electrolytes is a direct solution for high-performance devices based on high-voltage electrodes. Molecular orbital calculations have demonstrated that the energy of the highest occupied molecular orbital (HOMO) correlates with the oxidative decomposition potential and that the energy of the lowest unoccupied molecular orbital (LUMO) corresponds to the reductive decomposition potential. Specifically, functional additives with higher HOMO energies than carbonate solvents will be preferentially oxidized to modify the positive (cathode electrolyte interphase) CEI layer, while functional additives with lower LUMO energies than carbonate solvents will be preferentially reduced to modify the negative SEI layer.

10.3.3 Pre-lithium Technologies

During the lithium-ion embedding process, if the anode potential is low, the electrolyte will be simultaneously reduced to form a SEI layer due to parasitic reactions on the anode surface. This SEI layer plays an important role in preventing the electrolyte ions from co-insertion into the electrode, as well as improving the conductivity and chemical activity of the electrode surface. However, during this process, a large amount of Li^+ is consumed and the concentration of lithium ions and electrolytes can change drastically, resulting in irreversible losses. It is estimated that 5–20% lithium depletion in the cathode is responsible for the loss of initial capacity and low initial coulombic efficiency. Therefore, it is necessary to pre-lithiate the LIC. This not only reduces the negative potential of lithium storage and increases the operating voltage and energy density of the device, but also generates a SEI on the negative surface of the capacitor. This surface layer can improve the performance and stability of the capacitor, thus reducing the internal resistance and capacitance loss of the capacitor in the early stage of use and extending the service life by pre-embedding lithium in the capacitor. Common negative electrode pre-embedding methods include internal short-circuit method, lithium compound doping method, electrochemical lithium insertion method, and lithium powder with lithium charging method. The internal short-circuit method is currently the most widely used one. In this method, the lithium metal foil and capacitor anode material are short-circuited in the electrolyte containing Li^+, thus generating a SEI film on the electrode surface. The amount of lithium embedded is controlled by controlling the quality of the lithium metal foil. The lithium compound doping method uses a "rocking chair mechanism" to mix lithium compounds into the anode of the LIC.

During the charging process, lithium ions can be removed from the positive electrode and embedded into the negative electrode to make up for lithium-ion loss and to achieve the effect of lithium doping. Electrochemical pre-embedding method uses the three-electrode system, that is lithium foil as a reference electrode the negative electrode as the working electrode, lithium metal as the counter electrode. Then the three-electrode system is electrochemical charge and discharge after the electrolyte is injected. Lithium ions transfer from the lithium metal counter electrode to the negative electrode in the discharge process, to achieve the purpose of pre-embedding lithium. In the lithium powder with lithium charging method, passivated lithium powder or lithium metal foil is used on the surface of the negative electrode to achieve the effect of lithium doping, but the practical large-scale application of this method is highly challenging due to safety risks;

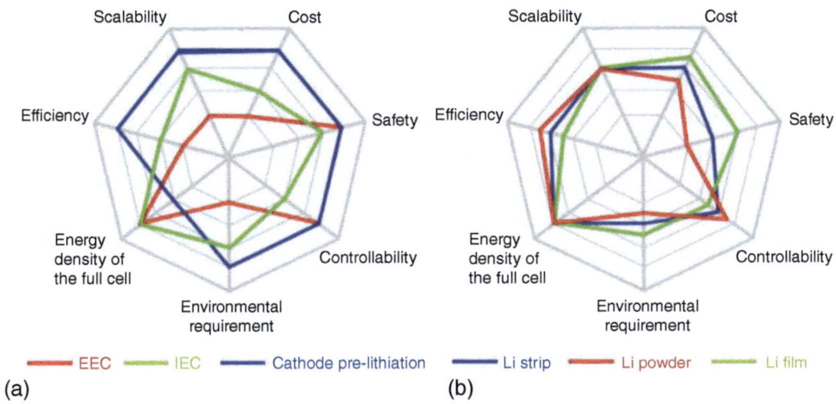

Figure 10.6 Comparison of the scalability, efficiency, cost, safety, controllability, environmental requirements, and the resulting energy density of the full cell for different types of pre-lithiation methods: (a) IEC (in situ electrochemical) and cathode pre-lithiation technologies; and (b) comparison using various lithium sources in the (in situ short circuit) ISC pre-lithiation method. Source: Reproduced with permission from Jin et al. [5]/Royal Society of Chemistry.

thus, the amount of lithium powder and the thickness of lithium foil need to be controlled carefully.

In summary, pre-lithiation exerts considerable influence on LIBs and LICs (Figure 10.6). Through pre-lithiation, the issue of high-capacity irreversibility of LIBs can be effectively addressed, and improved cyclic stability can be achieved. The electrochemical performance, including the energy density, power density, and cycle life can be enhanced. In short, pre-lithiation is a prerequisite for realizing high-performance LIBs and LICs.

10.3.4 Capacity Ratio of Anode and Cathode

However, the kinetics of LICs with battery-type electrodes with an intercalation mechanism is much slower than that of capacitor-type electrodes with a bilayer formation mechanism compared with conventional EDLC, resulting in a lower power of LICs. In addition, LICs retain the high energy decay characteristics of LIBs due to the limitations caused by capacitor electrodes. A key approach to improving LICs is capacity and potential matching between positive and negative electrodes. By combining suitable battery-type electrodes with fast kinetics, LICs can reach power densities comparable with those of SCs. Furthermore, the cycling performance of LICs is also important for their practical applications. Therefore, in addition to the intrinsic behavior of the cell and capacitor electrodes, the mass ratio between the cell and capacitor electrodes has an important influence on the performance of LICs. For asymmetric cells, the charges at the two electrodes should be balanced and the stored charge is related to the specific capacity (C) and mass (m) of the electrodes. Thus, the mass ratio between capacitive and battery electrodes is optimized; however, the capacitance of capacitive electrodes is lower than that of battery electrodes. In summary, since the positive and negative electrodes in

LICs provide different specific capacities and potential windows, capacity and potential matching of electrode materials is critical to fully utilize the performance of capacitors and battery materials in their optimal potential windows.

In general, the energy storage mechanism of electrochemical energy storage devices is based on the migration of ions in electrolytes and chemical reactions in electrode materials. The mass ratio of positive and negative electrodes is usually determined by the specific capacity and specific gravity of electrode materials. The positive and negative electrode mass ratio of electrochemical energy storage devices is generally considered 1 : 1, i.e. the mass of the positive electrode is equal to that of the negative electrode. For example, if an electrode material has a higher specific capacity, then the mass of the positive material can be lower, while for a heavier negative material, the mass of the negative material needs to be increased appropriately to maintain the balance between the positive and negative electrodes; otherwise, the energy density and power density may be limited and even affect the cycle life of the capacitor. However, for Li-ion hybrid capacitors, the mass ratio of the anode and cathode needs to be adjusted according to specific application scenarios and design requirements. For example, for applications requiring high energy density and high power density, the specific gravity of capacitors needs to be increased to improve their power density and energy density. On the other hand, for applications requiring long cycle life and high safety, the proportion of the positive electrode material of LICs should be reduced appropriately to improve the stability and safety of capacitors.

10.3.5 Conductive Electron/Ion Networks

A conductive electron/ion network is a network structure composed of ions in the electrolyte and electrode materials. It has the dual functions of conducting electricity and storing energy. The electrons/ions in the conductive ion network can be involved in ion migration and energy storage processes between the electrolyte and electrode material of the hybrid capacitor. Therefore, the quality and structure of the conductive ion network play a crucial role in the energy density of the capacitor. Generally, the energy storage density and power density of a capacitor can be directly changed by adjusting the structure and composition of the conductive ion network. In addition, the conductive network also affects the cycle life of the capacitor. If the structure and composition of the conductive ion network are not suitable, irreversible chemical reactions will occur during migration and storage of electrons/ions, resulting in a reduction in the cycle life of the capacitor. Electrons/ions in the conductive ion network undergo redox reaction during the charging and discharging process of the capacitor, thus producing by-products such as gas and heat. Furthermore, if the structure and composition of the conductive ion network are not suitable, electrons/ions will generate too much gas and heat during the reaction process, resulting in less safety of the capacitor. Therefore, when designing lithium-ion hybrid capacitors, it is necessary to optimize the structure and composition of the conductive ion network so that the best performance and stability can be achieved. Furthermore, the quality and structure of the conductive ion network should be strictly controlled and tested to ensure that the performance and safety of the capacitor meet the requirements. Therefore, the conductive ion network has an important influence on the performance and stability of LIC devices.

Silicon, as a new type of lithium storage anode, can be used as raw materials in the form of commercial nanosilicon and inexpensive gelatin, by directly mixing and carbonizing to realize continuous conductive network construction and dual-interfacial bonding effect. Such a prepared silicon anode shows high coulombic efficiency, high reversible specific capacity, and excellent long cycle stability. The specific surface area and capacity of the silicon anode are significantly improved without an additional binder and conductive agent. LICs assembled with these silicon anodes have high operating voltage, low self-discharge rate, high energy density, high power density, and long cycle life. The simple and inexpensive preparation method also makes this silicon anode practical in applications in LICs and LIBs. In addition to two-dimensional conductive networks, three-dimensional conductive networks can also be constructed. Furthermore, high-performance SCs have been extensively studied, but little work has focused on the optimization of the conductive bonding network in electrodes to improve their electrochemical performance. A rationally designed three-dimensional conductive bonding network consisting of one-dimensional and two-dimensional structures can be vacuum-filtered, which has the following advantages: three-dimensional structure, electrical conductivity, and electrolyte wettability. Therefore, it has ideal electron transport and ion diffusion properties, which promote the effective use of active materials when introduced into the electrode, thus greatly enhancing the overall performance.

10.3.6 Dense, Continuous, and Robust SEI

For LICs, the inherent high reactivity of lithium and its disordered deposition during electrochemical stripping/sedimentation causes repeated fracturing/reconstruction of the SEI. The term “SEI film” generally refers to a passivation layer formed by the reduction reaction of the electrolyte (including solvents, salts, and additives) on the surface of the negative electrode. It prevents further decomposition of the electrolyte and is an integral part of LICs. An ideal SEI film should have high ionic and negligible electronic conductivity, should be tightly bonded to the surface of the active material (e.g. graphite), and should be thin, strong, and tough.

Due to the inhomogeneous SEI, cation transport also becomes inhomogeneous, leading to uncontrollable dendrite growth and side reactions, especially at high current densities. On the one hand, this leads to continuous parasitic side reactions and SEI formation, which consumes electrolytes and reactive lithium and hinders lithium-ion transport. On the other hand, the repeated and random deposition of lithium results in crystal formation and pulverization, leading to the production of dead lithium and membrane puncture, which in turn leads to serious safety problems in the battery. Artificial SEI can inhibit dendrite growth by enhancing cation diffusion. However, this strategy is ineffective due to insufficient ionic conductivity and inhomogeneous distribution of solid components between the electrolyte and metal surface. Therefore, SEI layers with high ionic conductivity need to be designed to effectively inhibit the dendrite growth and side reactions. An ideal SEI should not only have a specific structure and composition but, more importantly, realize a tight bonding with the metal electrode to facilitate a rapid and dense lithium electroplating/ionization process. Moreover, by constructing, tuning, and optimizing the SEI layer, there is a positive impact on the cycle life of the LIC.

In situ modulation of SEI membranes refers to the formation of specific SEI membranes on the electrode surface through electrochemical cycling by optimizing the electrolyte components or by adding a film-forming agent to the electrolyte or electrode. This method is simple and, to a certain extent, can repair the SEI membranes that may be damaged during the cycling process. It is one of the most common and effective means to regulate the structure and components of SEI membranes. In addition to the in situ modulation of SEI membranes, artificial SEI membranes can be constructed by pretreating the graphite surface with coatings.

10.4 Electrode Materials

10.4.1 Lithium-Ion Capacitor Positive Electrode

Lithium-ion SCs are hybrid electrochemical energy storage devices combining the intercalation mechanism of a LIB with the positive electrode of a double-layer electrocapacitor. Lithium-ion SCs are superior to LIBs and other SCs in terms of capacity, voltage, self-discharge, safety, lifetime, and high-temperature resistance. They have the characteristics of high use voltage, energy density, and power density. However, the development of lithium-ion SCs has some challenges, especially the capacity mismatch between the positive and negative electrodes. The specific capacity of the negative electrode is often higher than that of the positive electrode when the reaction of lithium embedded in the bulk phase occurs, so traditional materials need to be modified to meet the requirements of consumer electronics and electric vehicles that require high energy and peak power. In addition, efficient use of energy from renewable sources also needs to be considered while meeting the needs of economic development. This issue is generally thought of in terms of modifying conventional electrode materials with other conductive materials or using metal compounds for material compounding. The principle behind LIC anodes is discussed below.

10.4.1.1 Carbon-Based Materials

Carbon materials have been widely studied and used in energy storage devices due to their low cost, abundant supply, isotope and conversion abundance, and physical/chemical stability. They have also been inserted into electrodes due to their large specific surface area, high electrical conductivity, and good availability to electrolytes. In addition, carbon compounds can be used as active compounds for Li-ion battery-type electrodes as they have active Li^+ insertion/de-insertion sites.

Currently, the primary material used in the cathode of LIBs is carbon materials. Generally, an ideal LIC capacitive cathode material requires a high surface area, good material and electrode conductivity, good pore structure, and high electrolyte ion accessibility. In general, the higher the specific surface area of porous carbon, the higher the specific capacitance of the material and the higher the energy it can store. However, numerous experimental studies have shown that the relationship between the specific capacity of porous carbon and its specific surface area is not always linear. Besides the specific surface area, the microstructure and surface properties of porous carbon also influence the performance of LICs. AC materials have a very wide pore size distribution, with most of the pores contributing to the specific surface area. The pores contributing the most to the specific surface area are often below 0.5 nm in size. Electrolyte ions can hardly break the potential barrier of pores below 0.5 nm and enter them. As a result, the ion transport rate in the pore channel is significantly reduced, limiting the energy and power density of LICs. In organic electrolyte systems, this is exacerbated by the larger size of electrolyte ions, which makes the situation even more severe. Surface functionalization is another effective way to improve the capacitive properties of carbon materials by increasing Faraday pseudocapacitance. For example, nitrogen-containing groups can generate or increase redox reactions, electron donor capacity, and electrode wettability.

Amatucci et al. developed a LIC system with AC as the cathode and nanostructured $Li_4Ti_5O_{12}$ (LTO) as the anode, the first time AC was used in a LIC. In an organic electrolyte dependent on AC cathodes, the voltage window of the LIC was designed to be in the range of 1.5–4.5 V to ensure high energy density and long cycle stability of the LIC [1]. In addition, metal-organic framework (MOF)-derived carbon with different structures has been extensively studied in LICs. For example, carbon rectangles with a large surface area (2714 $m^2\ g^{-1}$) were synthesized by pyrolysis of Zn-based MOF-5, which showed a special wrinkled sheet-like porous morphology with micro- and mesoporous values. MOF-dependent LICs show a maximum effective energy density of 65 $Wh\ kg^{-1}$ with an excellent power capacity depending on the advanced structure. Similarly, polyhedral hollow carbon derived from MOF has been produced and used in LICs [7]. It is worth noting that the removal of metal ions is a necessary step in the preparation of MOF-derived carbon due to the presence of metal ions in MOF precursors. The unique two-dimensional structure and physical properties of graphene and its derivatives make graphene a unique building block for the fabrication of a wide range of porous three-dimensional structures. Due to the combination of the porous structure and the excellent inherent properties of graphene and its derivatives, these 3D structures have excellent chemical stability and a robust specific surface area as well as rapid electron transfer kinetics. These high-performance 3D architectures are therefore strong candidates for use as cathodes in LICs. Various techniques for building graphene-based LIC architectures have been comprehensively documented in recent years. Among them, integration of microporous carbon with surface connectivity into graphene structures is an important technique to further increase the strength and energy density of LICs. In order to achieve higher performance, LICs require a proper mass balance, and therefore, Li^+ insertion compounds (lithium metal oxides) are used in LIC compositions.

These compounds have high theoretical capacity, initial capacity, and potential, but their low electrical conductivity and low lithium-ion diffusion coefficient ultimately limit their specific power. To exploit the advantages of carbonaceous and intercalated compounds, composite structures may be used. In recent years, biomass carbon has become a research hotspot due to its tunable chemical/physical properties, eco-friendliness, abundant resources, and high economic value. Using natural leek as the biomass carbon precursor, Lv et al. prepared porous AC for the cathode of a hybrid LIC using a two-step method. They designed a hybrid LIC using this cathode material and silicon/flake graphite/carbon nanocomposite as the anode material, which showed high energy density and excellent cycling stability. Xu et al. used sorghum-core-derived carbon flakes as the negative and positive electrodes of a LIC. They achieved high energy density of the LIC with good cycling stability by adjusting the potential of the electrodes so that the negative and positive electrodes were of equal capacity [8]. Gokhale et al. prepared 3D high-surface-area porous carbon by direct pyrolysis of oligomeric salts. They carried out in situ activation at the molecular level and the formation of graphene-like sheets using specific functional groups during the preparation process. The 3D structure was a hexahedral cage formed by individual graphene-like sheets. LICs were assembled using this oligomer-derived carbon as the positive electrode and lithium titanium phosphate as the negative electrode, with a negative-to-positive-electrode mass ratio of 1 : 3.04. The LIC showed a maximum energy density of 63 Wh kg^{-1} and a maximum power density of 10 kW kg^{-1}. In addition, the capacitance retention rate was 81% after 1000 cycles at a current density of 2 A g^{-1}. The performance of this LIC is significantly better than that of commercial LICs (AC/LTO) and symmetrical SCs [9]. Graphene is considered one of the best LIC anode materials due to its excellent electrical conductivity, high surface area, and high mechanical strength. Due to the limited exfoliation process of monolayer graphene, in practice, graphene oxide is generally used, and then reduced graphene oxide is used. Aravindan et al. prepared trimethylene-glycol-reduced graphene oxide that showed good PF_6^- adsorption/desorption properties and excellent performance when used as an anode for LICs, showing an energy density of 45 Wh kg^{-1} and significant cycling stability [10]. The energy density performance of AC in general is relatively poor, and the primary focus of current research has been on improving the specific capacitance and energy density performance of AC. Yu et al. prepared nitrogen-rich mesoporous carbon nanospheres/graphene nanocomposites using triblock copolymer F127 as a soft template and graphene oxide as a matrix. The carbon nanospheres with a diameter of about 70 nm were well distributed in the crumpled graphene network. Even at high current densities of 80 A g^{-1}, these nanocomposites showed excellent multiplicative capabilities and capacitance retention of 48% [11].

10.4.1.2 Metal Oxide Materials

Current global research on LICs is focused on the development of high-capacity alternating electrode materials such as metals, metal oxides, and sulfides with theoretical capacities much higher than that of graphitic carbon. Metals such as Si,

Sn, Al, and Mg undergo reversible alloying/de-alloying reactions with lithium, while metal oxides such as MoS_2, FeO, Li_3VO_4, Co_3O_4, and sulfides undergo reversible conversion/reconversion with lithium. These reactions are quite promising for a wide range of applications. Transition metal oxide materials should meet the following characteristics in order to be used in LICs: (i) they should have excellent electrical conductivity; (ii) metal ions in the oxide should be present in multiple oxidation states and should not cause phase changes; and (iii) lithium ions can move freely in and out of the oxide lattice through reduction reactions.

The first metal oxide electrode was RuO_2; however, since RuO_2 is expensive and toxic, thus limiting its practical application, some inexpensive metal compounds have been developed and applied. Li_3VO_4 (LVO) is an ideal electrode material for LICs due to its high capacity, moderate voltage plateau, and stable structure. Liu Wenjie et al. used carbon-covered Li_3VO_4 (LVO/C) layered structure prepared using the solid phase method. In addition, a LIC device with AC as the cathode and LVO/C as the anode has high energy density and long cycle life [12]. Metal phosphide materials with high capacity, moderate redox voltage, and low polarization are ideal as LIC materials; however, the irreversibility of the discharge product (Li_3P) leads to severe capacity degradation and inhibits their practical application.

10.4.1.3 MXene-Based Materials

Recently, two-dimensional MXenes and MXene-based nanocomposites have become the most important electrode materials due to their unique physical and chemical properties. As electrodes for LICs, MXenes have metallic conductivity and plastic layer structure that provides more chemically active interfaces and shorter ion diffusion lengths, which can effectively mitigate ion kinetic imbalances between the anode and cathode. To further improve the electrochemical performance of MXenes, their composition, morphology, surface chemistry, and structural configuration have been extensively investigated. MXenes, including $Ti_3C_2T_x$, Ti_2CT_x, Nb_2CT_x, and V_2CT_x, have novel ultrathin and open-layer structures that alleviate the slow insertion and removal kinetics of Li ions. As a result, MXenes exhibit nonideal battery behavior, but are similar to capacitors, suggesting their application in LICs. Furthermore, the functional groups (–F, –OH, –O, –Cl) on the MXene layer surface have a significant effect on their energy storage performance, electron transport rate, and stability. According to experimental data from half-lithium batteries, the specific capacity of $Ti_3C_2T_x$ and Ti_2CT_x is lower than that of Nb_2CT_x and V_2CT_x. The delamination process of Nb_2CT_x is usually carried out using a different chemical method. Olha et al. reported the delamination of Nb_2CT_x with amines instead of DMSO. The delaminated Nb_2CT_x suspension was then mixed with CNTs and filtered through a polyester membrane to obtain Nb_2CT_x paper. Results show that the Nb_2CT_x/CNT paper electrode has a high volumetric capacity of $325\,F\,cm^{-3}$. Nb_2CT_x/CNT can be paired with graphite anodes or $LiFePO_4$ cathodes. In addition, symmetrical capacitors can be assembled using Nb_2CT_x/CNT electrodes as the anode and cathode. These three types of LIC devices were able to operate within a 3-V voltage window and provided stable capacities of 43, 24, and $36\,mA\,g^{-1}$, respectively. Continuous cycling showed that all three types of LICs had long cycle life and volumetric energy densities of ~50–70 Wh l^{-1} (based on electrode volume), exceeding the energy density of conventional LTO/AC LICs [13].

10.4.2 Lithium-Ion Supercapacitor Negative Electrodes

10.4.2.1 Silicon-Based Material

Anode materials commonly used in lithium storage include graphite, tin oxide, and silica compounds with layered structures. Silicon-based materials are preferred in LIBs due to their excellent lithium storage performance and low charge and discharge potential. However, the process of lithium storage involves the alloying and dealloying of silicon and lithium, resulting in a volume expansion problem of up to three times. This expansion can lead to issues such as pulverization, rupture, and capacity attenuation of silicon electrodes, thereby adversely affecting their cycle performance. To enhance the cycling performance of silicon, researchers have explored various nanosilicon electrodes for lithium storage, including porous core–shell and microdome-shaped electrodes. The use of nanostructured electrodes increases the specific surface area and creep of the electrode material, thereby reducing the ion diffusion path during the electrochemical process. Additionally, nanostructured electrodes allow for volume expansion of the silicon material during lithium storage, effectively mitigating the stress caused by expansion during lithium removal. Furthermore, incorporating a buffer matrix into the silicon material partially alleviates the volume expansion problem caused by lithium-ion detachment. Cu/Si, Mn/Si, and Fe/Si electrode materials are composed of silicon material and a buffer matrix. These materials have been found to effectively improve the cycling performance of the electrode. Recently, carbon materials have been increasingly used in the buffer matrix of silicon-based electrodes as they show stable electrochemical performance and good electrical conductivity and effectively inhibit the volume expansion of silicon material.

10.4.2.2 Graphite-Based Materials

Voltage can be increased by combining a graphite negative insertion electrode and a positive EDL electrode of AC, using lithium salts in an organic electrolyte. In this method, the graphite negative electrode is initially pre-lithiated by an auxiliary lithium metal electrode to form a stable insertion compound. After this loading step, the battery can be charged/discharged in the voltage range of 1.9–3.6 V. Based on this principle, Aida et al. [14] proposed a system using AC and non-graphitized carbon as positive and negative electrodes, respectively. Without pre-lithiation of the negative electrode, the capacitance of the hybrid device decreased within a few tens of charge/discharge cycles. LICs typically utilize high-power SC electrodes (e.g. AC and porous carbon) as the cathode, high-energy LIB electrodes (graphite, Si/C, and metal oxides) as the anode, and lithium salts containing organic electrolytes. One of the key difficulties in developing high-performance LICs is the imbalance in power capability and charge storage capacity between the two electrodes. For high-capacity lithium-embedded anodes, the greatest challenge is to suppress cycling-induced electrode degradation (volume change, pulverization, and deactivation). A wide range of low-power IC systems have been reported with high performance in terms of optimizing the combination of anode and cathode. Jain et al. [15] used a hybrid structure with AC as the cathode and nanostructured $Li_4Ti_5O_{12}$ as the anode with an energy density of >20 Wh kg^{-1}. The hybrid SC showed the highest energy density to date (257 Wh kg^{-1} at 867 W kg^{-1}). In addition, other LIC systems including AC//B-Si/SiO_2/C, TiC//PHPNC, 3D graphene//Fe_3O_4-graphene, SnO_2-C//porous carbon, AC//hard carbon, AC//soft carbon, AC/LTO, SWNT//V_2O_5, porous graphene//$Li_4Ti_5O_{12}$/C, and AC//graphite have also been successfully explored. Despite these efforts, there is much room to explore and evaluate new LIC configurations to avoid kinetic and capacity differences between the cathode and anode. Of course, the key to high-performance ICs is coupling a high-rate performance anode and a high-capacity cathode. Metal tin (Sn), with high theoretical lithium storage capacity (about 992 mAh g^{-1} for $Li_{4.4}Sn$) and suitable low discharge potential relative to Li/Li^+, is considered one of the most promising alternative cathode materials for next-generation LIBs. Meanwhile, biomass AC is considered the most promising electrode material for SCs because of its large specific surface area, abundant resources, controllable microstructure, easy scale-up, etc. Notably, some anode materials with good nanostructures, such as VN (vanadium nitride) and TiC, have been widely used in LIC systems due to their rapid Li^+ diffusion kinetics that can further overcome the kinetic imbalance between the slow Faraday-inserted anode and the fast capacitive cathode. Of late, transition metal oxides such as Nb_2O_5, Fe_3O_4 based on the conversion reaction mechanism have been explored as high-specific-capacity, high-rate anodes for LIBs. In these studies, well-defined nanostructured anode materials are essential to obtaining stable cycling performance. On the cathode side, the primary limitation is the low specific capacitance of conventional AC. Despite the significant surface area of AC, the unconnected carbon particles result in a low electrochemically active surface area, which limits electron/ion transfer during the charge/discharge cycle.

10.5 Microgrid Energy Storage

Energy storage plays an essential role in modern power systems. The increasing use of renewable energy sources in power systems raises several challenges in coping with power imbalances and ensuring that standards are maintained. Backup supply and resilience are also primary concerns. Energy storage systems also provide ancillary services to the grid, such as frequency regulation, peak shaving, and energy arbitrage. There are several technologies for storing energy at different stages, but there are both benefits and drawbacks in how each one is suited to a particular situation. Energy storage systems play an essential role in modern grids such as addressing the need for the modernization of power systems and energy transition to a decarbonized grid that involves more renewable sources. Renewable energy intermittency requires flexible ancillary services to smoothen the variability in power production, on both large and small scale, e.g. interconnected bulk power systems and microgrids. Energy storage systems may be used to cater to these needs. They also provide peak shaving, backup power, and energy arbitrage services, improving reliability and power quality. Currently, the two primary approaches used in the grid are grid-connected and standalone renewable energy systems. However, new energy sources are unstable, making the final output power highly unreliable and unable to meet the requirements of standard power quality grids, greatly limiting their integration capabilities. In addition, fluctuating power usually causes harmful frequency and/or voltage deviations that may reduce the lifespan of load appliances or damage them. Various technologies including renewable power output smoothing and its optimal dispatch capability, load profile smoothing, overcoming demand-generation mismatches, power quality assurance during grid faults, support for distributed generation during black-start, contribution to system inertia, storage enhancements for electric vehicles, improvements in regenerative braking extraction, and wireless power transmission utilize battery-ultracapacitor hybrid devices. The application of batteries has come a long way since their inception, ranging from commercial electronic equipment, satellite applications, medical instrumentation, and substation installation to electrical circuitry of conventional vehicles. Conventionally, the principle of application of batteries was based on its capability to allow devices to be electrically autonomous. Furthermore, since the green revolution, due to the rapid growth as a result of ever-increasing energy consumption and increase in population and urbanization, battery technologies have witnessed major advances, hence establishing their prominence in the power sector. In addition, most electronic devices, electric vehicles, and telecommunication systems share common load profiles, which are characterized by relatively high peak-to-peak average power demands. These types of loads can be closely represented as constant-current characteristics, i.e. pulsed load profiles, and therefore sulfurized batteries in robust and stand-alone operations can result in increased system losses and the constant need for battery replacement due to the high current profiles. SCs, also known as electric double-layer capacitors, have 20 times the energy storage capacity of conventional electrolytic capacitors and can support higher power pulses than batteries. However, they do not have the same energy storage capacity as batteries. Therefore, combining a battery energy storage system with a SC storage system will lead to a hybrid energy storage system (HESS) with high energy density and high power density (Figure 10.7) with increased operational flexibility. Moreover, the requirements of such a HESS can be decoupled such that each energy storage system involved can perform to its characteristic potential and hence achieve reduced sizing of the energy storage system and cost, which can optimize the efficiency of the overall ESS. Energy-oriented storage devices can usually have huge energy storage capacity and long continuous periods of charging and discharging time, but their response speed is slow and the number of cycle times is small. In contrast, power-oriented storage devices have fast response speed and a large number of cycle times, with small energy storage capacity and short continuous charging and discharging time. It is difficult to design a single energy storage device with (i) both power density and energy density; (ii) superior high- and low-temperature performance; and (iii) efficient charging/discharging

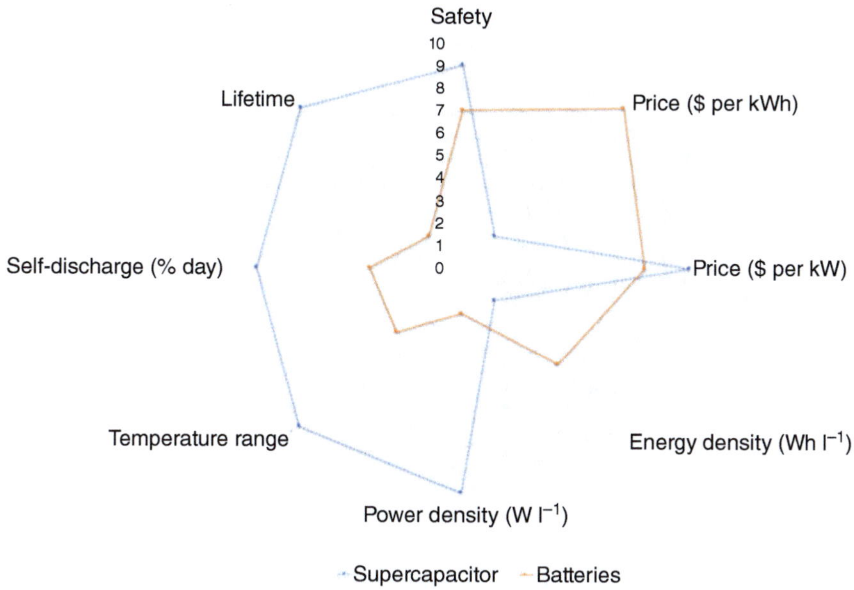

Figure 10.7 Qualitative comparison of batteries and supercapacitors. Source: Reproduced with permission from Khalid [16]/MDPI.

rate and number of cycle times. A HESS is usually composed of an energy-oriented storage device and a power-oriented storage device in a certain capacity ratio, taking into account both the power density and the energy density. Consequently, it has the advantages of each energy storage device, without their respective drawbacks. The performance of energy storage devices, both technically and economically, are expected to be improved. Nowadays, renewable energies, especially photovoltaic and wind power, have become the most attractive solution to address the threat of climate change. However, these two technologies can hardly supply power continuously by themselves due to their intermittent nature. An effective solution that includes energy storage in renewable energy generation systems is proposed to stabilize the power supply of renewable energy. Figure 10.8 depicts the applications of the BS-HESS (battery-supercapacitor hybrid energy storage system) in a new power system, where the proportion of renewable energy is gradually increasing.

10.6 Summary and Perspectives

The advantages of hybrid SCs lie in their high energy density, high power density, long cycle life, and fast charging and discharging, which have greatly contributed to advances in the field of energy storage. This review systematically addresses the current development of hybrid capacitors, encompassing the historical background, charge storage and transfer mechanisms, electrode materials, and full-cell design. However, the existing hybrid capacitors have not yet achieved the excellent

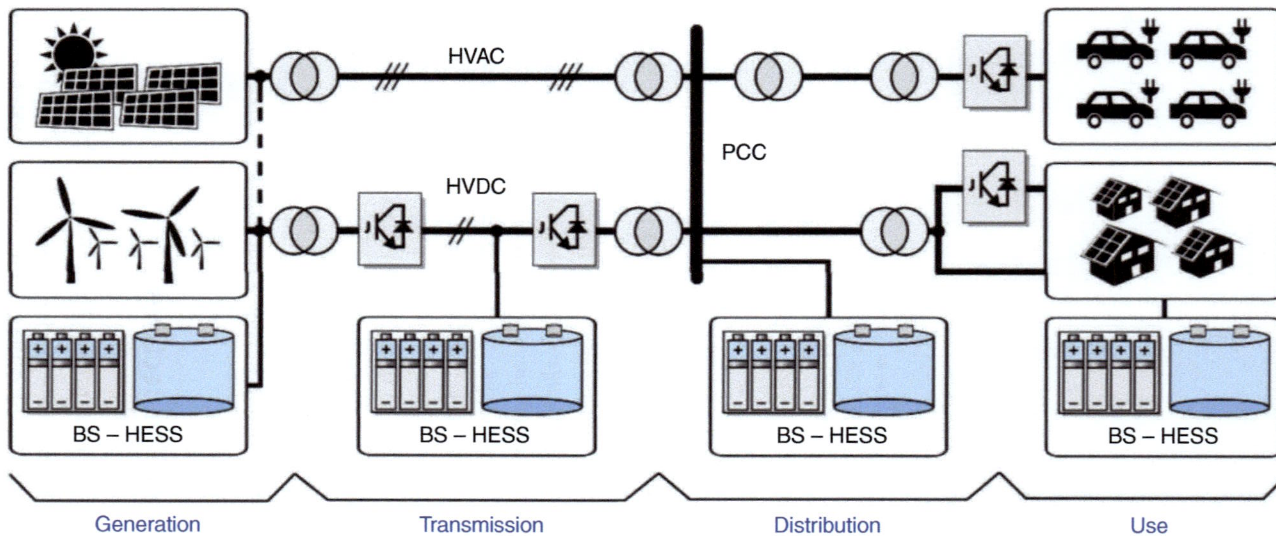

Figure 10.8 Applications of BS-HESS in a new power system. Qualitative comparison of batteries and supercapacitors. Source: Dong et al. [15]/MDPI/CC BY 4.0.

performance of high energy, high power, and cycle stability similar to those of LIBs, and it is expected that LIBs will dominate the market for the foreseeable future. Currently, the energy density of commercial LICs is only about 20 Wh kg^{-1}, and the excessively short discharge times associated with large magnification make them far from the ideal target. Unlike commercially available lithium-ion analogues, NICs are in the early stages of academic validation and are not ready for mass market introduction. In most cases, the superior performance claimed for advanced electrodes fabricated in hybrid capacitors does not generally translate into assembled commercial devices. Coin cells based on the effective mass of two electrodes have been reported to perform very well in parameters such as energy density (100–200 Wh kg^{-1}) and power density (10–20 kW kg^{-1}). In reality, the electrode material represents only a small fraction ($\approx$30%) of the total weight of the entire system. Therefore, further optimization of the overall performance indexes of hybrid capacitors to achieve the desired indexes remains a direction of development in both academia and industry. Potential future research directions are as follows:

1. Screening of Battery Materials with High Electrochemical Properties
   To date, ion/electron hybrid conductors, using carbon materials, doping, and nanoengineering to build three-dimensional electrode structures, remain the most widely used techniques to improve electrical conductivity and compensate for the relatively slow diffusion of ions from battery-type electrodes. These techniques will inevitably result in high manufacturing costs, at the expense of bulk packing density, stability, and sustainability. Therefore, focusing on novel materials with appropriate main lattices and high conductivity will be a key direction for future research, e.g. MXenes, conductive MOFs, and conductive intercalation pseudocapacitive materials.
2. Optimization of Capacitor Materials with High Capacity
   The specific capacity of the most commonly used capacitive materials (such as AC) is usually lower than that of battery-like materials. Therefore, increasing the specific capacity of AC is an important aspect. Doping of heteroatoms and optimization of their pore structure, crystallinity, and connectivity are effective strategies. In addition, new materials with tailored structures such as MXenes, metal carbides/nitrides, conducting MOFs, conducting conjugated microporous polymers, and conducting covalent organic frameworks will be highly promising in enhancing the electrochemical performance of capacitive materials.
3. Exploring the Integrated Materials Engineering
   From a practical point of view, an ideal high-performance hybrid capacitor depends not only on electrode materials, but also on the combination of other components, such as electrolytes ("salt-in-water" electrolytes, "salt-in-solvent" electrolytes, dual redox electrolytes, gel polymer electrolytes, etc.), diaphragms, collectors, and other components that affect the overall performance of the device. Therefore, in-depth research on integrated materials engineering is needed.
4. Advanced Characterization Technology
   Most of the previous characterization techniques focused on secondary batteries, such as LIBs. Hybrid capacitors beyond lithium ions are significant, emerging

energy storage systems. However, the charge storage mechanism of electrode materials, especially the phase between the electrode and electrolyte, is of great research significance and requires more in-depth mechanistic studies combining electrochemistry and various in situ characterization techniques. In addition, the synergistic effect and mechanism of capacitive materials and battery-like materials are yet to be understood thoroughly. This is an important but often neglected research direction. In situ scanning electrochemical probe microscopy, X-rays, electron microscopy, and spectroscopy may be powerful methods to gain insight into electrochemical processes. Techniques such as electrochemical quartz crystal microbalance, scanning electrochemical cell microscopy, spectroscopy with frequency generation, neutron reflectometry, and solid-state NMR, as well as their combinations, are important tools for the elucidation of mechanisms of action.

5. Evaluating the Implementation Performance
   SCs show satisfactory performance at extreme temperatures, but the rapid self-discharge behavior should not be ignored. In contrast, secondary batteries have no significant self-discharge behavior, but their performance at extreme temperatures is not as good as that of SCs. Therefore, research on the self-discharge behavior of hybrid capacitors and their high- and low-temperature characteristics is extremely important for evaluating their performance. In fact, research on hybrid systems has a long way to go. As more and more researchers enter the field of hybrid capacitors, it is believed that hybrid devices with excellent overall performance will be developed in the near future.

References

1 Amatucci, G., Badway, F., Pasquier, A. et al. (2001). An asymmetric hybrid nonaqueous energy storage cell. *Journal of Electrochemical Society* 148: A930–A939.

2 Xing, F., Bi, Z., Su, F. et al. (2022). Unraveling the design principles of battery-supercapacitor hybrid devices: from fundamental mechanisms to microstructure engineering and challenging perspectives. *Advanced Energy Materials* 12 (26): 2200594.

3 Soltani, M. and Beheshti, S.H. (2021). A comprehensive review of lithium ion capacitor: development, modelling, thermal management and applications. *Journal of Energy Storage* 34: 102019.

4 Li, J., Zhang, X., Peng, R. et al. (2016). $LiMn_2O_4$/graphene composites as cathodes with enhanced electrochemical performance for lithium-ion capacitors. *RSC Advances* 6: 54866.

5 Jin, L., Shen, C., Shellikeri, A. et al. (2020). Progress and perspectives on pre-lithiation technologies for lithium ion capacitors. *Energy & Environmental Science* 13 (8): 2341–2362.

6 Jin, L., Yuan, J., Shellikeri, A. et al. (2021). An overview on design parameters of practical lithium-ion capacitors. *Batteries & Supercaps* 4 (5): 749–757.

7 Cai, M., Sun, X., Chen, W. et al. (2017). Performance of lithium-ion capacitors using pre-lithiated multiwalled carbon nanotubes/graphite composite as negative electrode. *Journal of Materials Science* 53: 749–758.

8 Xu, X., Cui, Y., Shi, J. et al. Sorghum core-derived carbon sheets as electrodes for a lithium-ion capacitor. *RSC Advances* 7 (28): 17178–17183.

9 Gokhale, R., Aravindan, V., Yadav, P. et al. (2014). Oligomer-salt derived 3D, heavily nitrogen doped, porous carbon for Li-ion hybrid electrochemical capacitors application. *Carbon* 80: 462–471.

10 Dattakumar Mhamane, Anil Suryawanshi, Abhik Banerjee, Vanchiappan Aravindan, et al. Non-aqueous energy storage devices using graphene nanosheets synthesized by green route[J]. AIP Advances, 2013, 3(4).

11 Yu, X., Zhan, C., Lv, R. et al. (2015). Ultrahigh-rate and high-density lithium-ion capacitors through hybriding nitrogen-enriched hierarchical porous carbon cathode with prelithiated microcrystalline graphite anode. *Nano Energy* 15: 43–53.

12 Liu, W., Zhang, X., Li, C. et al. (2020). Carbon-coated Li_3VO_4 with optimized structure as high capacity anode material for lithium-ion capacitors. *Chinese Chemical Letters* 31 (9): 2225–2229.

13 Mashtalir, O., Lukatskaya, M., Zhao, M. et al. (2015). Amine-assisted delamination of Nb_2C MXene for Li-ion energy storage devices. *Advanced Materials* 27: 3501–3506.

14 Aida, M., Juan, C., Liliana, G. et al. (2020). Adsorption of Triton X-100 in aqueous solution on activated carbon obtained from waste tires for wastewater decontamination. *Adsorption* 26: 303–316.

15 Jain, A., Aravindan, V., Jayaraman, S. et al. (2013). Activated carbons derived from coconut shells as high energy density cathode material for Li-ion capacitors[J]. *Scientific Reports* 3 (1): 3002.

16 Khalid, M. (2019). A review on the selected applications of battery-supercapacitor hybrid energy storage systems for microgrids. *Energies* 12 (23): 4559. https://doi.org/10.3390/en12234559.

11

Fuel Cells

Yuqing Wang[1] *and Minghua Chen*[2]

[1] *Harbin University of Science and Technology, School of Electrical and Electronic Engineering, Harbin, 150080, China*
[2] *Harbin University of Science and Technology, School of Electrical and Electronic Engineering, Key Laboratory of Engineering Dielectric and Applications (Ministry of Education), No. 52 Xuefu Road, Harbin, 150080, China*

In recent years, as energy shortage and environmental degradation have become more and more serious, the development of the fuel cell technology, as an efficient and clean power generation technology, has attracted great attention. The fuel cell technology is regarded as the fourth type of power generation after thermal power generation, hydropower generation, and nuclear power generation. It is considered the preferred clean and efficient power generation technology. Research and development of the fuel cell technology has made great progress in recent years, and fuel cell systems as small as a few watts and as large as megawatts have been successfully investigated and applied in power stations, transportation tools, and portable electronic devices. Fuel cells are also suitable for the construction of substations, which can help us deal with the problem of widespread paralysis caused by centralized grid transmission failures or damage.

11.1 Overview

A fuel cell is an energy conversion device that converts the chemical energy stored in the fuel directly into electrical energy through electrochemical reactions that occur at its cathode and anode. The energy carried by the fuel and the oxidizer is continuously fed into the fuel cell and converted into electrical energy via electrochemical reactions, and the products are continuously discharged. The electrodes of the fuel cell do not change during this process, they simply provide a place for the electrochemical reactions. Therefore, as long as the fuel and the oxidant are continuously supplied, the fuel cell can generate electricity continuously and its electrodes are not consumed. The conversion of chemical energy to electrical energy does not involve combustion or a heat engine. Therefore, the fuel cell has the outstanding advantages of high efficiency, low pollution, and low noise.

Towards Next Generation Energy Storage Technologies: From Fundamentals to Commercial Applications, First Edition. Edited by Minghua Chen.

11.1.1 Development

In 1838, Christian Friedrich Schönbein discovered the principle of the fuel cell. Then, inspired by experiments with water electrolysis, William Robert Grove reported the first fuel cell in 1839 (Figure 11.1) [1]. In 1889, L. Mond and C. Langer assembled a fuel cell using a porous non-conducting material impregnated with an electrolyte as the diaphragm of the cell, Pt black as the electrocatalyst, and Pt plates as the current collector. With an output voltage of 0.73 V at an operating current density of 3.5 mA cm^{-2}, the structure of the cell was close to that of a modern fuel cell. In 1893, Friedrich Wilhelm Ostwald established the basic theory of fuel cells, which laid the foundation for later fuel cell researchers.

In 1932, G.W. Heise prepared a water-repellent electrode with wax as the waterproofing agent, the prototype of the gas diffusion electrode (GDE) currently in use. In the 1950s, General Electric Company and Union Carbide Company developed porous gas diffusion water-repellent electrodes. In 1959, Dr. F.T. Bacon developed the first 5-kW fuel cell stack that could power a welding machine, the first alkaline fuel cell (AFC). In the 1960s, fuel cells became the first choice for NASA's spacecraft power source due to their high energy density and high power density.

Since the 1970s, fuel cell research has entered a rapid development with the introduction of different types of fuel cells and the development of new materials and processes. In 1972, DuPont successfully developed a polymeric proton exchange membrane, Nafion (polytetrafluoroethylene sulfonic acid membrane), which solved the key long-standing problem of solid electrolyte membrane material for fuel cells. In 1993, Ballard Power System launched the first bus powered by a proton exchange membrane fuel cell (PEMFC).

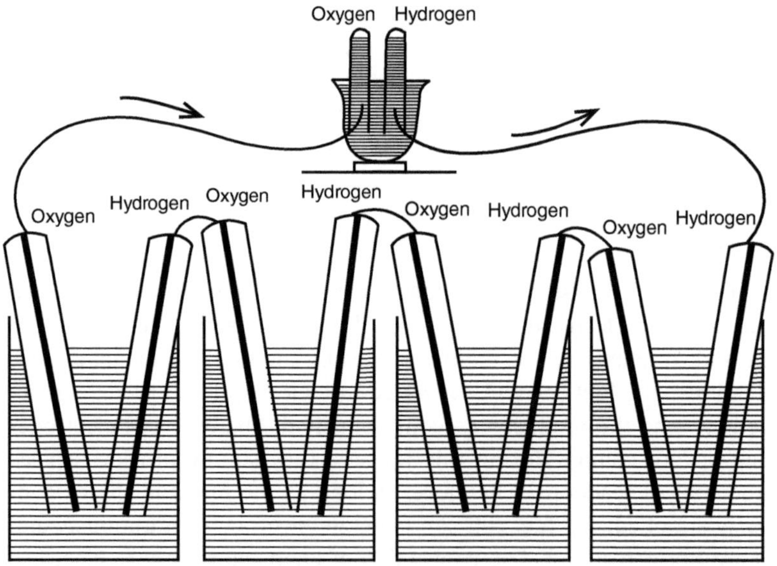

Figure 11.1 Grove gas battery. Source: Reproduced with permission from Grove [1]. © 1842/Elsevier.

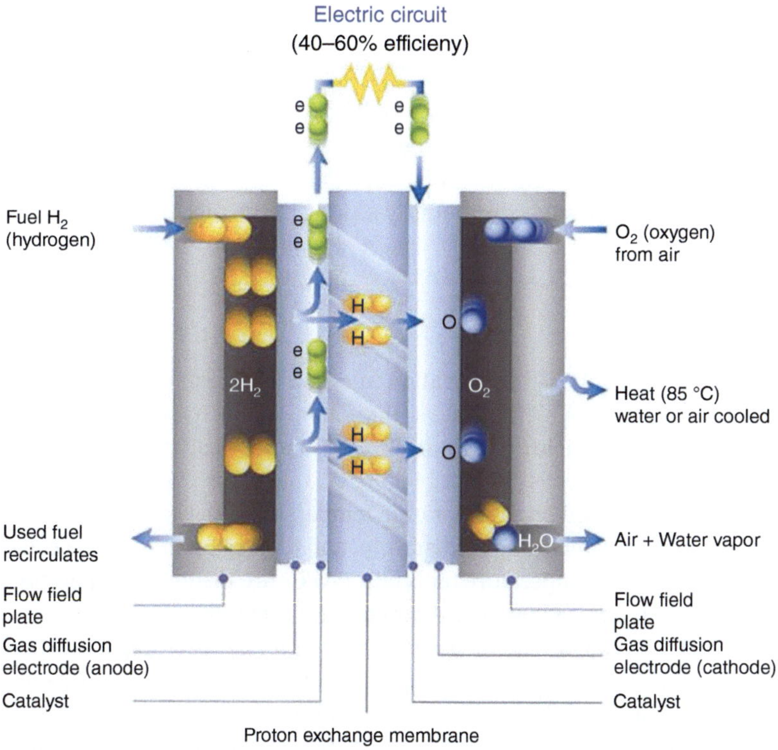

Figure 11.2 Operating principle of a fuel cell. Source: Taa'mneh et al. [2]/Polish Society of Ecological Engineering/CC BY 4.0.

11.1.2 Operating Principle

Fuel cells consist of an anode, a cathode, and an electrolyte. The electrolyte conducts ions and prevents direct contact between the fuel and the oxidizer. The anode is the site of the oxidation reaction of the fuel, whereas the cathode is the site of the reduction reaction of the oxidizer. Consider a hydrogen–oxygen fuel with an acidic electrolyte as an example to illustrate the working principle of fuel cells. As shown in Figure 11.2 [2], hydrogen is continuously delivered as the fuel to the anode of the fuel cell, and the following electrochemical oxidation reaction occurs at the anode under the action of an electrocatalyst:

$$H_2 \rightarrow 2H^+ + 2e^- \qquad (E^\circ = 0\,V) \tag{11.1}$$

Protons are generated and two free electrons are released. The protons pass from the anode to the cathode through the acidic electrolyte, while the free electrons move from the anode to the cathode through the electron conductor and the load. At the cathode, oxygen undergoes the following electrochemical reduction reaction in the presence of a catalyst:

$$1/2O_2 + 2H_2 + 2e^- \rightarrow H_2O \quad (E^\circ = 1.23\,V) \tag{11.2}$$

The total reaction is as follows:

$$H_2 + 1/2O_2 \rightarrow H_2O \qquad (E^\circ_{cell} = 1.23\,V) \tag{11.3}$$

In a fuel cell, hydrogen and oxygen are not in direct contact with each other, but the oxidation and reduction take place at the respective electrodes. Based on the operating principle of the fuel cell, it can be considered as an energy conversion device and can produce electricity continuously as long as the fuel and the oxidizer are supplied.

11.1.3 Classifications

At present, the widely adopted classification method is based on the type of electrolyte used in fuel cells, which proposes the following five types of fuel cells: AFCs, PEMFCs, phosphoric acid fuel cells (PAFCs), molten carbonate fuel cells (MCFCs), and solid oxide fuel cells (SOFCs). Figure 11.3 shows an overview of different types of fuel cells [3]. Based on the operating temperature, fuel cells can be classified into low-temperature, medium-temperature, and high-temperature fuel cells. AFCs and PEMFCs are low-temperature fuel cells, PAFCs are medium-temperature fuel cells, and MCFCs and SOFCs are high-temperature fuel cells.

Furthermore, fuel cells can be classified into three categories according to the fuel type and the usage. The first type is direct fuel cells, where the fuel is oxidized directly on the anode catalyst of the cell, e.g. direct methanol fuel cells, direct carbon fuel cells, and direct borohydride fuel cells. The second type is indirect fuel cells, in which

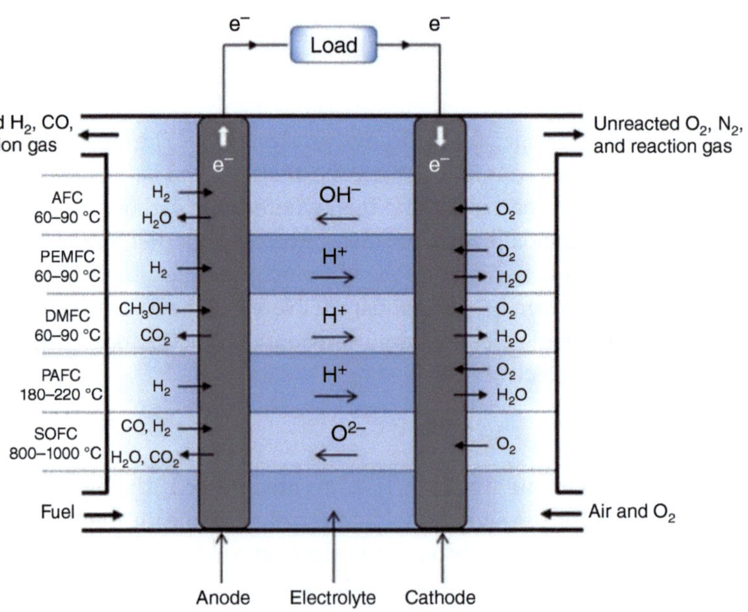

Figure 11.3 Schematic overview of different types of fuel cells. Source: Reproduced with permission from Seselj et al. [3]. © 2015/Elsevier.

the fuel is not used directly but is converted and then used. The third type is regenerative fuel cells, in which the water generated by the fuel cell reaction is decomposed into hydrogen and oxygen and then the hydrogen and oxygen are reintroduced into the fuel cell to generate electricity.

11.1.4 Applications

Fuel cells are likely to become the fourth type of power generation after thermal power, hydropower, and nuclear power generation. The new type of power generation can effectively save energy and reduce air pollution on a large scale. The development of fuel cell vehicles (FCVs) has attracted much attention, and various demonstration zero-emission FCVs have been unveiled to show the appeal and prospect of the new technology. Fuel cells can be scaled from a few watts to megawatts and have a wide range of applications, including stationary power stations, powering vehicles, and powering portable electronic devices.

Fuel cell power generation systems are modular in design, easy to adjust the scale, and highly efficient, generate low noise, and have combined heat and power supply. Thus, they can be used as medium-sized decentralized power stations or as large centralized power plants for grid-connected power generation. Among the five major types of fuel cells, except for AFCs, the other types can be used as fixed power stations with different power generation capacities.

Using a fuel cell system as the power source, an energy conversion efficiency of about 50% can be achieved, which is much higher than that of the conventional internal combustion engine. In addition, the exhaust emissions of fuel cell systems are much lower than those of conventional internal combustion engines, and the noise is low. The development of fuel cells for transportation has been currently focused on PEMFCs, which have the characteristics necessary for transportation, such as fast start at low temperatures and no electrolyte corrosion and spillage problems.

With the increasing complexity of portable electronic products, traditional batteries have become increasingly unable to keep up with the needs of electronic products. Therefore, people have begun to turn their attention to fuel cells. The energy density of fuel cells can be several times higher than that of lithium-ion batteries, and the batteries can be charged by filling with fuel, so they have a long operating time and an extremely short charging time. If technical and cost problems can be overcome, fuel cells will become the next generation of mobile power.

11.2 Thermodynamics and Kinetics

A fuel cell is an energy conversion device that uses an electrochemical reaction to convert the chemical energy stored in the fuel directly into electricity. In this energy conversion process, the power generation efficiency is high because it is not limited by the Carano cycle of traditional heat engine conversion. The theoretical efficiency of the fuel cell is determined by the thermodynamics of the reaction, which is related to the working parameters of the fuel cell, such as temperature and pressure, and also depends on the physical state of the reactants and products. The actual efficiency of the cells is closely related to the electrode process. When the fuel cell is working, current flows through the two electrodes, the electrolyte, and the current output carrier, which leads to the polarization phenomenon, resulting in the fuel cell voltage being less than its equilibrium voltage, thus resulting in the reduction in the fuel cell efficiency.

11.2.1 Thermodynamics of Fuel Cells

Fuel cells usually work at constant temperature and pressure, so the reaction can be regarded as taking place in a constant temperature and pressure system, and the Gibbs free energy change can be expressed as follows:

$$\Delta G = \Delta H - T\Delta S \tag{11.4}$$

where ΔH (enthalpy change) is the total energy released by the fuel cell. When the system is under reversible conditions, i.e. the electrode reactions are reversible, in essence, there is no net current through, and the Gibbs free energy change is the maximum non-volume work that the system can do:

$$\Delta G_{\mathrm{r}} = -W_{\mathrm{r}} \tag{11.5}$$

where W_{r} is the maximum non-volume work of the system. During the electrochemical reaction occurring in the fuel cell, the maximum non-volume work is the maximum electrical work. Therefore, the theoretical efficiency of the fuel cell is as follows:

$$\eta_{\mathrm{r}} = \frac{W_{\mathrm{r}}}{-\Delta H_{\mathrm{r}}} = \frac{\Delta G_{\mathrm{r}}}{\Delta H_{\mathrm{r}}} = 1 - T\frac{\Delta S_{\mathrm{r}}}{\Delta H_{\mathrm{r}}} \tag{11.6}$$

where η_{r} is the theoretical efficiency of the fuel cell, namely the thermodynamic efficiency, the maximum efficiency possible that the fuel cell can achieve.

Under standard conditions (25 °C, 0.1 MPa), the theoretical efficiency of a hydrogen–oxygen fuel cell can be calculated as follows:

$$\mathrm{H_2\,(g)} + \frac{1}{2}\mathrm{O_2\,(g)} = \mathrm{H_2O\,(l)} \tag{11.7}$$

$$\eta_{\mathrm{r}} = \frac{\Delta G_{\mathrm{r}}^{0}}{\Delta H_{\mathrm{r}}^{0}} = \frac{-237.2\ \mathrm{kJ\ mol^{-1}}}{-285.1\ \mathrm{kJ\ mol^{-1}}} = 0.83 \tag{11.8}$$

It can be seen that the maximum efficiency of the fuel cell is nearly 100%. In the actual working process, due to different temperatures and pressures, the maximum efficiency of the fuel cell will change accordingly.

The heat machine conversion process is first combustion, converting the chemical energy of the fuel into heat energy, then converting the heat energy into mechanical energy through the heat machine, and then converting the mechanical energy into electric energy through the generator. During the conversion process, the conversion efficiency is controlled by the Carnot cycle:

$$\eta_{\mathrm{r}} = \frac{W_{\mathrm{r}}}{-\Delta H} = 1 - \frac{T_2}{T_1} \tag{11.9}$$

where T_1 and T_2 are thermodynamic temperatures (K) at the inlet and outlet of the heat machine, respectively. It can be seen from the equation that only when $T_1 \rightarrow \infty$ K or $T_2 \rightarrow 0$ K, $\eta \rightarrow 1$. In practice, such conditions cannot be achieved, so the conversion efficiency of the heat engine will not exceed 50%. The fuel cell directly realizes energy conversion through an electrochemical reaction, as shown in Eq. (11.6), which shows that during the reaction, the heat exchange ($T\ \Delta\ S_r$) between the system and the environment is very small and the conversion efficiency tends to be 100%.

When the fuel cell is discharged under constant temperature and reversible pressure, the following relationship exists between the electromotive force of the fuel cell and Gibbs free energy:

$$\Delta G_r = -W_r = -nFE_r \tag{11.10}$$

where E_r is the electromotive potential of the fuel cell or the equilibrium cell voltage, the reversible cell voltage, and the ideal cell voltage, F is Faraday's constant, transferring the charge carried by 1 mol electron, $F = 96{,}485\ \text{C mol}^{-1}$, and n (mol) is the number of electrons transferred by the fuel cell reaction. Equation (11.10) is a bridge between thermodynamics and electrochemistry, through which the electrochemical parameters (E) of the fuel cell can be calculated using the thermodynamics function of the fuel cell reaction (ΔG).

For example, consider the hydrogen–oxygen fuel cell with an acid electrolyte:

The anode process is

$$H_2 \rightarrow 2H^+ + 2e^- \tag{11.11}$$

The cathode process is

$$\frac{1}{2}O_2 + 2H^+ + 2e^- \rightarrow H_2O(l) \tag{11.12}$$

The number of electron transfers during the reaction is $n = 2$; if the reaction is conducted under the standard conditions of 25 °C, 0.1 MPa, the Gibbs free energy is 237.2 kJ mol^{-1}, so

$$E_r^0 = \frac{-\Delta G_r^0}{nF} = \frac{237.2\ \text{kJ mol}^{-1}}{2 \times 96{,}485\ \text{C mol}^{-1}} = 1.229\ \text{V} \tag{11.13}$$

where E_r^0 is a standard electromotive potential or standard cell voltage.

11.2.2 Electrode Kinetics

The electrode in equilibrium, calculated by the thermodynamic function of fuel cell electric momentum is the maximum potential, the potential can only be on the electrode without the current being achieved. When the current passes through, a series of physical and chemical processes occur inside the fuel cell with high or low resistance. Therefore, the voltage of the fuel cell will be lower than the theoretical equilibrium voltage, and the efficiency will also be lower than the maximum efficiency. From a kinetics point of view, the voltage drop is called the fuel cell polarization phenomenon, i.e. the fuel cell voltage deviates from the equilibrium voltage. Based on the different causes, polarization can be classified into activation, concentration, and ohmic polarization.

11.2.2.1 Polarization and Overpotential

The potential when no current passes on the electrode and when the electrode process is at equilibrium is called equilibrium electrode potential φ_{eq}. When current flows through the electrode and causes the electrode process to deviate from equilibrium, electrode potential deviates from the equilibrium value, which is called actual potential φ. The phenomenon of deviation from the equilibrium potential is called electrode polarization, and the amount that quantitatively represents the degree of polarization is called overpotential, usually expressed by η:

$$\eta = |\varphi - \varphi_{eq}| \tag{11.14}$$

When polarization occurs on the anode, electrode potential moves in the positive direction, $\varphi > \varphi_{eq}$; when it occurs on the cathode, electrode potential moves in the negative direction, $\varphi < \varphi_{eq}$; hence, the actual potential of the anode and cathode is as follows:

Anode (negative electrode)

$$\varphi_a = \varphi_{eq,a} + \eta_a \tag{11.15}$$

Cathode (positive electrode)

$$\varphi_c = \varphi_{eq,c} - \eta_c \tag{11.16}$$

The actual voltage of the fuel cell is the difference between the voltage drop caused by the potential difference between the positive and negative poles and the resistance between the two poles (Eq. 11.17). The electrode polarization degree (η) determines the actual output voltage of the fuel cell. The higher the overpotential, the higher the resistance and the lower the voltage.

$$E = |\varphi_c - \varphi_a| - IR = (\varphi_{eq,c} - \varphi_{eq,a}) - (\eta_c + \eta_a) - IR = E_r - (\eta_c + \eta_a) - IR \tag{11.17}$$

11.2.2.2 Activation Overpotential

Activation polarization is directly related to the rate of the electrochemical reaction, so it is also called electrochemical polarization. For a given reaction, the magnitude of activation overpotential is closely related to the catalytic activity of the electrode. The relationship between activation overpotential (η_{act}) and current density (j) can be expressed by the Tafel equation (Eq. 11.18). The theoretical form of this empirical equation is derived from Butler and Volmer. The cathode reaction, as shown in Eq. (11.19), is called the Butler–Volmer equation. a and b in Eqs. (11.18) and (11.19) can be derived from Eqs. (11.20) and (11.21):

$$\eta_{act} = a + b \ln j \tag{11.18}$$

where a and b are the empirical constants, and b is commonly known as the Tafel slope.

$$\eta = -\frac{RT}{\alpha nF}\ln j_0 + \frac{RT}{\alpha nF}\ln j \tag{11.19}$$

where a is the transfer coefficient and j_0 is the exchange current density.

$$a = -\frac{RT}{\alpha nF}\ln j_0 \tag{11.20}$$

$$b = \frac{RT}{\alpha nF}. \tag{11.21}$$

According to Eq. (11.18), constant a is equivalent to overpotential per unit current density, the smaller the value of a, the lower the polarization. Under low overpotential, a large current output can be obtained. The value of a depends on the electrode material, electrode surface state, solution composition, temperature, etc. According to Eq. (11.20), for an electrode reaction, n is certain and a is in the range of 0–1 and depends on the exchange current density j_0. The higher the exchange current density, the smaller the a. Therefore, the exchange current density of the electrochemical reaction on the cathode and anode of the fuel cell is crucial to the performance, the higher the exchange current density, the more favorable it is.

Given the cathode reaction of oxygen reduction, only the activation overpotential on the electrode is assumed. The potential of the oxygen cathode can be calculated according to Eqs. (11.16) and (11.19):

$$\varphi_c = \varphi_{eq,c} - b_c \ln\left(\frac{j}{j_0}\right) \tag{11.22}$$

where Tafel slope $b_c = RT/(4\alpha F)$.

Exchange current density j_0 can be determined experimentally. First, the Tafel curve is determined to extend the straight part when the current density is high, and the intercept of the extension on the *lgj* axis is the logarithmic value lgj_0 of exchange current density. Reducing the Tafel slope b is an important way to reduce the activation overpotential, which depends on the electrode catalyst.

The relationship between cathode activation overpotential and current density can be calculated from Eqs. (11.16) and (11.22):

$$\eta_{act,c} = b_c \ln\left(\frac{j}{j_{0,c}}\right) \tag{11.23}$$

Similarly, the relationship between anode activation overpotential and current density is as follows:

$$\eta_{act,a} = b_a \ln\left(\frac{j}{j_{0,a}}\right) \tag{11.24}$$

The total activation overpotential (η_{act}) of the fuel cell is equal to the sum of anode overpotential ($\eta_{act,\,a}$) and cathode overpotential ($\eta_{act,\,c}$):

$$\eta_{act} = \eta_{act,a} + \eta_{act,c} = b_a \ln\left(\frac{j}{j_{0,a}}\right) + b_c \ln\left(\frac{j}{j_{0,c}}\right) \tag{11.25}$$

Increasing the exchange current density and reducing the Tafel slope can effectively reduce the activation polarization, thus improving the voltage efficiency of the fuel cell. Especially for complex multi-electron and multistep electrode reactions, such as the electrochemical reduction of oxygen and electrochemical oxidation of methanol. The following methods are used to improve the exchange current density and reduce the impact of electrode activation polarization on the performance: using electrocatalysts with high activity and large specific surface, increasing the operating temperature, and increasing the concentration of the reactants.

Limited by the working conditions of the fuel cell, especially low-temperature fuel cells, such as PEMFCs and direct methanol fuel cells, the improvement due to increasing temperature and concentration is limited. Therefore, it is particularly important to develop electrode catalysts with high activity and large specific surfaces. Furthermore, the research on electrocatalysts accounts for a large proportion of the research on low-temperature fuel cells.

11.2.2.3 Concentration Overpotential

When the fuel cell is discharged, the fuel and the oxidant undergo electrochemical oxidation and reduction reactions in the anode and the cathode, respectively, and are constantly consumed, while the product is continuously generated on the two electrodes. If the rate of the reactant delivery to the electrode is less than the consumption rate or the rate of the product leaving the electrode is less than the generation rate, the concentration of the reactant on the surface of the electrode is lower than the bulk concentration and the product concentration is higher than the bulk concentration; i.e. a concentration difference is formed between surface concentration and bulk concentration. The polarization formed by this difference in concentration is called concentration polarization, and the resulting overpotential is concentration overpotential η_{conc}.

To simplify the discussion, concentration polarization caused by the product is ignored, and only that caused by the slow diffusion of the reactants is considered. Assuming that the reactant surface concentration is c_s and the bulk concentration is c_b, the relationship between concentration overpotential and concentration is as follows:

$$\eta_{\text{conc}} = |\varphi - \varphi_{\text{eq}}| = \frac{RT}{nF} \ln \frac{c_b}{c_s} \tag{11.26}$$

The transfer process of the reactants to the electrode surface follows Fick's first law:

$$\frac{j}{nF} = D\left[\frac{\partial c(x,t)}{\partial x}\right]_{x=0} \approx D\frac{c_b - c_s}{\delta} \tag{11.27}$$

where D is the diffusion coefficient of the reactor and δ is the thickness of the reactor on the electrode surface.

When the electrode potential deviates significantly from the equilibrium value, the electrode reaction is sufficiently impelled and occurs rapidly enough for the reactant to reach the surface of the reaction immediately. In this case, the reactant concentration on the electrode surface is zero. This is known as the limiting current density,

j_L, which is the current density at which the electrode reaction reaches its maximum rate.

$$\frac{j_L}{nF} = D\frac{c_b}{\delta} \tag{11.28}$$

From Eq. (11.28), we know that limiting current density j_L is directly proportional to the diffusion coefficient and bulk concentration and inversely proportional to the thickness of the diffusion layer. Increasing the temperature increases the diffusion coefficient, and increasing the flow rate of the reactants reduces the thickness of the diffusion layer. Therefore, limiting current can be increased by increasing the reaction temperature, increasing the flow rate of the reactant, and increasing the concentration of the reactant.

From Eqs. (11.27) and (11.28),

$$\frac{c_s}{c_b} = 1 - \frac{j}{j_L} \tag{11.29}$$

Substituting Eq. (11.29) into Eq. (11.26),

$$\eta_{conc} = -\frac{RT}{nF}\ln\left(1 - \frac{j}{j_L}\right) = \frac{RT}{nF}\ln\left(\frac{j_L}{j_L - j}\right) \tag{11.30}$$

$B = \frac{RT}{nF}$, and then,

$$\eta_{conc} = B\ln\left(\frac{j_L}{j_L - j}\right) \tag{11.31}$$

Equation (11.31) shows the relationship between concentration overpotential and current density. It shows that as current density increases, the degree of concentration polarization increases. When $j \ll j_L$, concentration polarization is not obvious; when j is close to j_L, concentration overpotential increases sharply.

When the fuel cell discharges, as long as the working current density of the fuel cell reaches the limiting current density of either of the electrodes, the voltage of the whole fuel cell will drop rapidly to zero. Because of the limiting current density of the other electrode, the concentration difference overpotential of the whole battery can also be expressed by Eq. (11.31). Assuming that only the concentration polarization causes voltage loss, the output voltage of the fuel cell can be represented as follows:

$$E = E_r - \eta_{conc} = E_r - B\ln\left(\frac{j_L}{j_L - j}\right) \tag{11.32}$$

In the design process of the fuel cell, limiting current density should be increased as much as possible to avoid the fuel cell entering the concentration polarization control zone prematurely. The ultimate current density can be increased by optimizing the electrode structure and operating parameters, such as increasing the working temperature, increasing the reactant concentration, and increasing the flow rate of the reactants.

11.2.2.4 Ohmic Overpotential

When the fuel cell works, electrons flow through electronic conductors such as electrodes and fluid collectors, and ions move through the electrolyte between the two electrodes. The flow of both electrons and ions is subject to resistance, resulting in a voltage drop, which is called ohmic overpotential η_{ohm}. Since both electronic and ionic conductors obey Ohm's law, the ohmic overpotential η_{ohm} can be expressed as follows:

$$\eta_{ohm} = iR_{\Omega} = jR_{ASR} \tag{11.33}$$

where i is the current passing through the fuel cell and R_{Ω} is the total resistance of the fuel cell, including electronic conductor resistance, ion conductor resistance, contact resistance, etc. Since the current generated by the fuel cell flows through all the conductors, the total resistance is the sum of the resistance of each part. In addition, j is the current density (A cm^{-2}) passing through the fuel cell, and R_{ASR} is the area-specific resistance (Ω cm^2), which is equal to the product of the area of the fuel cell and ohm resistance:

$$R_{ASR} = A_{\text{fuel cell}} \times R_{\Omega} \tag{11.34}$$

Ohmic overpotential should strictly be called the ohm voltage drop of the fuel cell. Obviously, its presence will increase the voltage loss of the fuel cell and reduce the voltage efficiency, especially when the operating current of the fuel cell is large. Therefore, reducing Ohm losses is an important aspect of fuel cell development. Resistance of the electrolyte usually plays a crucial role in the fuel cell, so reducing ohmic polarization usually starts with using ionic conductors and including measures of reducing the distance between the cathode and the anode and improving the conductivity of the electrolyte.

Through thermodynamic analysis, when there is no net current in the fuel cell, cathodes and anodes are in equilibrium and there are no concentration polarization and ohmic polarization. Therefore, the voltage of the fuel cell is determined by the thermodynamic function Gibbs free energy to change the balance of the electromotive force E_r. It is equal to the difference between the equilibrium electrode potential of the cathode and the anode. E_r, the maximum possible voltage of a fuel cell, can be expressed as follows:

$$E_r = -\frac{\Delta G_r}{nF} = \varphi_{eq,c} - \varphi_{eq,a} \tag{11.35}$$

According to kinetics analysis, it is known that when the output current of the fuel cell is worked externally, activation polarization, concentration polarization, and ohmic polarization will be generated, resulting in the actual output voltage E of the cell being lower than its thermodynamic equilibrium voltage E_r. The voltage loss caused by various polarizations is superimposed, so the actual output voltage of the fuel cell can be expressed as follows:

$$E = E_r - (\eta_{act} + \eta_{conc} + \eta_{ohm}) \tag{11.36}$$

Using Eqs. (11.25), (11.32), and (11.33) in Eq. (11.36), the mathematical expression of the current density–voltage ($j - E$) polarization curve of the fuel cell is obtained:

$$E = E_r - \left[b_a \ln\left(\frac{j}{j_{0,a}}\right) + b_c \ln\left(\frac{j}{j_{0,c}}\right)\right] - B\ln\left(\frac{j_L}{j_L - j}\right) - jR_{ASR} \tag{11.37}$$

11.2.3 Fuel Cell Efficiency

From the previous thermodynamics analysis, we know that the theoretical efficiency of the fuel cell is

$$\eta_r = \frac{\Delta G_r}{\Delta H_r} \tag{11.38}$$

It means that when the fuel cell performs a reversible electrochemical reaction, the maximum energy (ΔG_r) that can be used for electrical work is the fraction of the total energy (ΔH_r) released by the fuel cell reaction. However, there are many irreversible factors in fuel cell practice, so the actual efficiency of the fuel cell is always lower than the ideal reversible thermodynamic efficiency. There are two primary reasons why the actual efficiency of fuel cells is lower than the theoretical efficiency: voltage loss due to different polarization and fuel utilization loss due to insufficient fuel utilization.

11.2.3.1 Actual Efficiency of Fuel Cells

Electrochemical Efficiency of Fuel Cells In the actual operation of fuel cells, activation polarization, concentration polarization, and ohmic polarization occur. The overpotential generated by these polarization phenomena reduces the actual output voltage of the fuel cell to be lower than the theoretical equilibrium voltage, i.e. polarization leading to voltage loss. Electrical work $W = ZFE$, and voltage loss is the loss of electrical output energy. The ratio of the actual output voltage (E) of the fuel cell to the thermodynamically reversible voltage (E_r) is defined as the voltage efficiency of the fuel cell (η_{voltage}), also known as the electrochemical efficiency:

$$\eta_{\text{voltage}} = \frac{E}{E_r} \tag{11.39}$$

Equation (11.37) shows that the operating voltage depends on the output current. The higher the current load, the lower the voltage. Therefore, fuel cells have higher efficiency at low loads. This is in stark contrast with an internal combustion engine, which typically has its highest efficiency at the maximum load.

Fuel Utilization Ratio Not all the fuel supplied to the fuel cell participates in the electrochemical reaction. Some fuel may participate in other reactions without generating electrical work. For example, in direct methanol fuel cells, methanol permeates through a proton exchange membrane to the cathode, where oxidation occurs, or is discharged directly with air. Some fuel may flow through the fuel cell without participating in electrochemical oxidation reactions. In this case, the fuel provided to the fuel cell is not fully utilized. The ratio of the fuel that generates electricity to the total fuel provided to the fuel cell is called the fuel utilization ratio η_{fuel}:

$$\eta_{\text{fuel}} = \frac{i/nF}{v_{\text{fuel}}} \tag{11.40}$$

where i is the current generated by the fuel cell and v_{fuel} is the fuel supply rate (mol s^{-1}). For fuel cells, the primary cycle utilization rate of fuel cannot reach 100% because when the pressure of a gaseous fuel or the concentration of a liquid fuel decreases to a certain extent, concentration polarization increases, making stable functioning of the battery difficult. Usually, to improve fuel utilization, fuel cycle methods or special fuel cell designs are used.

Considering dynamic losses and fuel utilization losses, the actual efficiency of a fuel cell can be expressed as follows:

$$\eta = (\eta_{\mathrm{r}}) \times (\eta_{\mathrm{voltage}}) \times (\eta_{\mathrm{fuel}}) = \left(\frac{\Delta G_{\mathrm{r}}}{\Delta H_{\mathrm{r}}}\right) \times \left(\frac{E}{E_{\mathrm{r}}}\right) \times \left(\frac{i/nF}{v_{\mathrm{fuel}}}\right) \tag{11.41}$$

One of the important goals of fuel cell research and development is to improve the actual working efficiency. From the above analysis, it can be seen that the actual efficiency of fuel cells can be improved by focusing on aspects such as thermodynamics, dynamics, and fuel utilization. Optimizing electrode structure, increasing electrode reaction area, and developing highly active electrocatalysts can effectively reduce activation overpotential and thus improve fuel utilization. Ohmic overpotential can be reduced by increasing the ionic conductivity of the electrolyte, reducing the thickness of the electrolyte membrane, and increasing the conductivity and contact of various conductive components within the fuel cell. Reversible voltage and thermodynamic efficiency can be improved by selecting the appropriate operating temperature and pressure, changing the composition of fuel gas, and reducing impurities in fuel gas.

11.2.3.2 Actual Efficiency of Fuel Cell Systems

A fuel cell power generation system comprises fuel cells and the complete set of ancillary equipment. In this system, the fuel cell is the core, which converts the chemical energy of the fuel into electrical energy. Other components perform a series of auxiliary functions, such as continuously supplying fuel and oxidants to the fuel cell (pumps, compressors, blowers, etc.); stabilizing and adjusting the output voltage of the fuel cell (DC/DC converter); converting DC power into AC power to meet the needs of AC users; stabilizing the operating temperature of the fuel cell, and monitoring the fuel cell. Therefore, the actual efficiency of a power generation system is not only related to the efficiency of the fuel cell but also depends on the efficiency of all auxiliary equipment. Only when the actual efficiency of the entire system is high, the advantages of fuel cell efficiency can be achieved. When only the electrical energy output from a fuel cell system is used without utilizing the thermal energy generated, the efficiency of the fuel cell system is usually expressed as electricity efficiency, which is expressed as follows:

$$\eta_{\mathrm{sys}} = \frac{itE_{\mathrm{cell}} - W_{\mathrm{loss}}}{|\Delta H|} \tag{11.42}$$

where itE_{cell} is the electrical energy output of the fuel cell, W_{loss} is the electrical energy consumed by all auxiliary equipment, and ΔH is the chemical energy released by the reacted fuel. The consumption of electrical energy by auxiliary systems reduces the actual electrical efficiency of the entire fuel cell power generation system. The electrical efficiency of the fuel cell is higher at low loads and lower at high loads, whereas for auxiliary devices, the opposite is true: it is lower at low loads and higher at high loads. Therefore, the overall electrical efficiency of the fuel cell system remains roughly unchanged. This is different from most other energy conversion systems, including heat engines, which tend to be less efficient when operated at low loads.

Medium- and high-temperature fuel cells, such as PAFCs, MCFCs, and SOFCs, can generate considerable residual heat during the power generation process. If the residual heat is recycled, the energy conversion efficiency of the fuel cell system can be improved. If the residual heat can be further converted into electrical energy, the power generation efficiency can be improved. As in the case of using the high-temperature residual heat generated by SOFCs for power generation to drive a steam engine. The efficiency of a fuel cell system during waste heat recovery and utilization is commonly referred to as the combined heat and power efficiency η_{Hp}, which can be expressed as follows:

$$\eta_{CHP} = \frac{itE_{cell} - W_{loss} + Q_{recovered}}{|\Delta H|} \tag{11.43}$$

In currently available fuel cells, the voltage efficiency $\eta_{voltage}$ is 50% ~ 80%, the electric efficiency η is 40% ~ 70%, the power generation efficiency η_{sys} is 35% ~ 65%, and the thermoelectric merger efficiency η_{CHP} is 60% ~ 80%. With technological advances in fuel cells and ancillary equipment, the efficiency of fuel cell systems will be further improved.

11.3 Proton Exchange Membrane Fuel Cells

A PEMFC can be used as a power source for FCVs, distributed power supply, heat source, and portable power supply. Therefore, it is being investigated for its applications in electric vehicles, and tens of watts to kilowatts of small- and medium-sized PEMFC mobile power supply is under development.

11.3.1 Operating Principle

A PEMFC single cell is primarily composed of a hydrogen chamber, anode, proton exchange membrane, cathode, and oxygen chamber. The mechanism of a PEMFC using hydrogen as the fuel is shown in Figure 11.4 [4]. During fuel cell operation, hydrogen and oxygen are supplied to the anode and cathode, respectively.

A PEMFC typically uses hydrogen as the fuel. After entering the hydrogen chamber and reaching the anode, hydrogen loses two electrons and oxidizes to H^+:

$$H_2 \rightarrow 2H^+ + 2e^- \tag{11.44}$$

The generated H^+ reaches the cathode by means of hydration through the proton exchange membrane to the cathode, and the electrons reach the cathode through the external circuit after working on the load, to realize electrical conduction and reduce to form water.

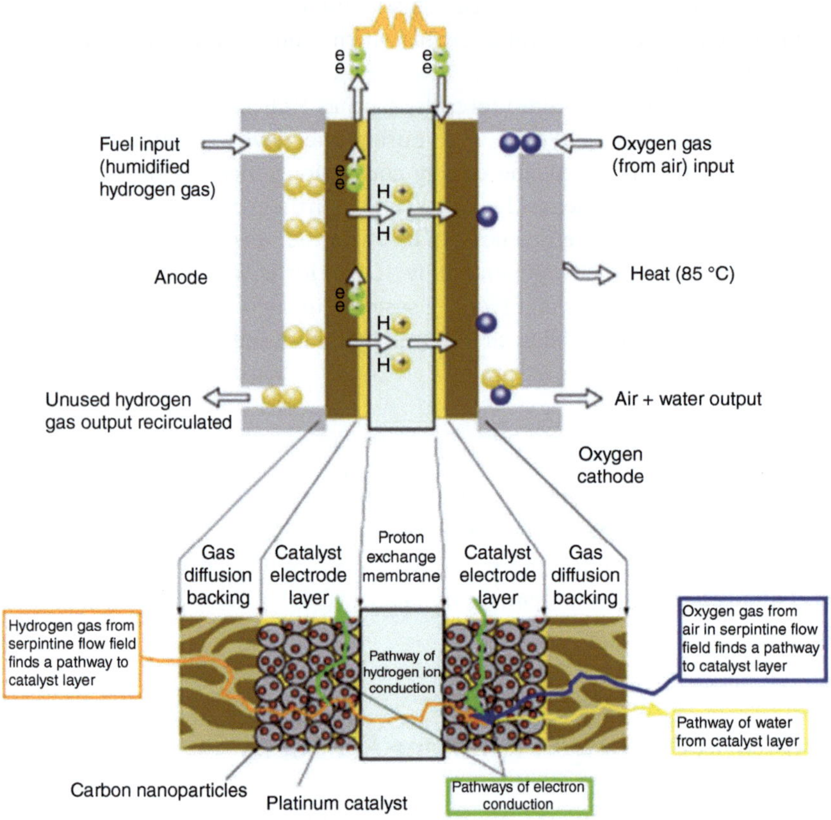

Figure 11.4 Mechanism of a PEMFC using hydrogen as fuel. Source: Reproduced with permission from Ren et al. [4]. © 2017/Elsevier.

Oxygen enters the oxygen chamber to reach the cathode, and under the action of the cathode catalyst, the oxidant combines with H^+; the electrons reaching the cathode generate water, which is exhausted via the tailpipe:

$$\frac{1}{2}O_2 + 2H^+ + 2e^- \rightarrow H_2O \tag{11.45}$$

The total reaction is as follows:

$$\frac{1}{2}O_2 + H_2 \rightarrow H_2O \tag{11.46}$$

The theoretical voltage of a PEMFC in the standard state can be expressed as follows:

$$E^\circ = -\frac{\Delta G^o}{nF} = 1.41\ \text{V} \tag{11.47}$$

11.3.2 Membrane Electrode Assembly

The core component of a PEMFC is the membrane electrode assembly (MEA). The MEA is a single component that combines the anode, proton exchange membrane, and cathode in a sandwich structure. The performance of the MEA varies widely depending on different manufacturing processes. Efforts are being made to optimize the preparation process of MEA while improving the fuel cell performance, reducing the electrode catalyst load, and simplifying the preparation process and equipment to reduce the cost. As high cost is an important factor hindering the commercialization of fuel cells, for commercialization of fuel cells, the production costs such as the platinum load, and MEA should be reduced and the manufacturing process should be simplified while improving the performance.

Generally, electrodes used in PEMFCs can be classified into three types: thick-layer hydrophobic electrodes, thin-layer hydrophilic electrodes, and ultrathin catalyst-layer electrodes. Electrodes can be prepared via two methods (Figure 11.5) [5]: first, preparing the catalyst layer (CL) directly on the electrolyte film, forming the so-called catalyst-coated membrane (CCM), and then combining the CCM and the diffusion layer to form the membrane electrode; and second, forming

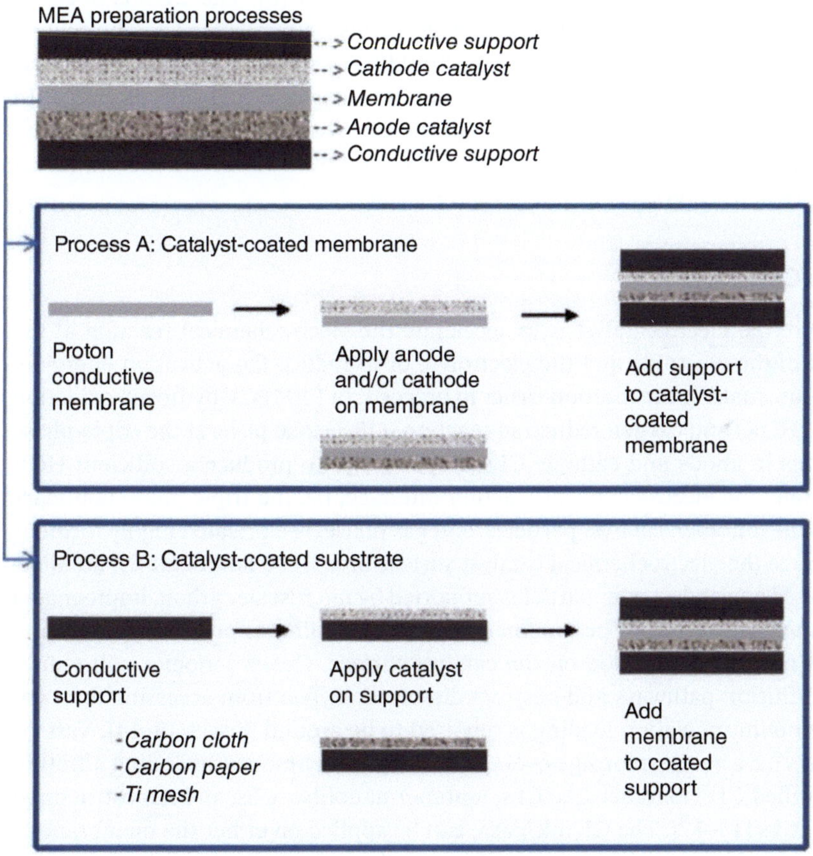

Figure 11.5 Preparation method of MEA. Source: Reproduced with permission from Faid and Sunde [5]/John Wiley & Sons/CC BY 4.0.

the catalyst-coated substrate (CCS), which combines the CL and the conductive support to form a GDE, and then hot-pressing the GDE together with the electrolyte membrane to prepare the membrane electrode.

The overview of progressive improvements in PEMFCs and a schematic of the working principle are as follows [6]. The proton exchange membrane divides the cell into the cathode and the anode, which use the GDE. The GDE is composed of the gas diffusion layer (GDL), the microporous layer (MPL), and the CL. The CL is primarily the oxidation reaction of the fuel and the reduction reaction of the oxidant. The MEA usually consists of five layers: cathode GDL, cathode CL, proton exchange membrane, anode CL, and anode GDL. The MEA is placed between two bipolar plates (BPs), where gas flow channels (GFCs) are placed.

A PEMFC electrolyte is a polymeric membrane (usually Nafion), which plays multiple roles, including proton conduction, electronic insulation, and separation of reactant gases [7]. The role of the CL is to facilitate electrochemical reactions and provide pathways for both reactant transport and electron/proton conduction. A GDL/MPL is placed between the CL and BP to conduct electrons and heat, transport gas reactants, and enable water management [8]. The GDL is not only the place where the reaction material spreads but also supports the CL. The contact resistance between the CL and the GDL should be as small as possible, otherwise, the internal resistance of the whole fuel cell will increase fuel cell discharge. In addition, the GDL is in direct contact with the BP flow field, so the strength of the GDL needs to be observed, especially when the serpentine flow field is used, which is higher than the porous body and mesh flow field.

11.3.3 Catalyst Layers

The role of the electrocatalyst is to accelerate the electrochemical reaction at the interface of the electrode and the electrolyte or to reduce the activation energy of the reaction, making the reaction easier to proceed. In PEMFCs, hydrogen oxidation reaction (HOR) and oxygen reduction reaction (ORR) take place at the triple-phase boundaries in anode and cathode CLs, respectively. To produce a sufficient HOR or ORR rate, the electrochemical reaction interface, i.e. the triple-phase boundary between the ionomer, catalyst particles, and gas phase, is fabricated highly tortuous to maximize the electrochemical catalyst surface area . The dominant CL architecture is based on catalyst nanoparticles supported by microscale carbon, impregnated with ionomer thin films. The ionomer binder also facilitates proton transport and dissolves oxygen for reaction on the catalyst surface. Excess ionomer will reduce the gas diffusion pathways and obstruct dissolved oxygen from accessing the catalyst. The optimum Nafion loading is reported to be around 30 wt% [9–14]. Various developers have been exploring novel CL fabrication techniques, including ultrathin CLs, stratified CLs, ionomer-free CLs, ionomer nanofiber CLs, and nanostructured thin film CLs [15–17]. The CL ink/paste can be applied on either the membrane or GDL/MPL surface using diverse deposition methods, including spraying, painting, screen printing, rolling, descaling, electrodeposition, evaporative deposition, and impregnation reduction [18].

Electrocatalysts should have a high catalytic activity for HOR and ORR and also a good inhibitory effect on the side reactions that can occur during the reaction process. In general, Pt has the highest electrocatalytic activity among metal elements, for both HOR and ORR. The stability of an electrocatalyst depends on its chemical stability and resistance to toxicity. Good chemical stability of an electrocatalyst means that it does not corrode in the electrolyte solution. Toxicity resistance means that the electrocatalyst cannot be easily poisoned by substances. For example, if the hydrogen gas contains CO, it will strongly adsorb on the surface of a Pt catalyst and poison it. To prevent this, a second or third component such as Ru should be added to the Pt catalyst to improve its toxicity resistance.

Electrocatalytic activity is generally related to the specific surface area of the catalyst. Generally, a higher specific surface area leads to increased catalyst activity. The higher the specific surface area of the catalyst, the smaller the particle size and the better the dispersion. This can be achieved with appropriate support. Some of the commonly used supports are activated carbon and carbon black because of the large specific surface area and good electrical conductivity [19–24]. In addition, as electrons from the hydrogen or oxygen have to be transferred through the catalyst after the reaction on the catalyst, the catalyst should have a high electrical conductivity.

11.3.3.1 Electrocatalysts

Anode Electrocatalysts For the oxidation of hydrogen in the anode, Ni and Pd were initially used as catalysts, and then Pt black was used. However, the large granularity and low dispersion of Pt black lead to the low utilization rate of Pt. Pt/C is now the most commonly used catalyst for the anode reaction. Apart from a small amount of hydrogen produced by water electrolysis, the majority of hydrogen is produced from coal, natural gas, petroleum, and organic molecules, which often contain some CO, which poses the following challenge: CO adsorption at the Pt site causes severe loss in performance – the so-called CO poisoning phenomenon. To solve the problem of CO poisoning, first of all, the CO content in hydrogen needs to be reduced, i.e. to separate CO in hydrogen. The volume fraction of CO is controlled as far as possible below the 10^{-5} level. Secondly, effective measures should be taken to improve the tolerance of PEMFCs to CO. If the oxygen bleeding technique is adopted, air/oxygen is injected into the anode fuel gas to oxidize the CO adsorbed on the Pt surface into CO_2, so that CO toxicity can be avoided to a certain extent [25]. In addition, a small amount of hydrogen peroxide can also be added to the fuel gas humidifier, which can also play a similar role to the oxygen infiltration method. Finally, catalysts against CO poisoning should be investigated. Studies show that Pt-based composite catalysts have better effects against CO poisoning. Many Pt group composite catalysts have been studied, such as Pt-Ru, Pt-Sn, Pt-Mo, Pt-Cr, Pt-Mn, Pt-Pd, and Pt-Ir [26–30]. The function of the second element is primarily to reduce the adsorption strength of CO on Pt, make it oxidized to CO_2, and improve the resistance of the catalyst to CO poisoning.

Cathode Electrocatalysts In PEMFCs, the polarization of the cell is primarily derived from the oxygen electrode. Therefore, the performance of the cathode catalyst should be improved. Pt catalysts are generally used in the cathode because Pt has the best electrocatalytic performance for ORR among all elements. The electrocatalytic activity of the composite catalyst with the transition metal and Pt is significantly higher in ORR than that of the Pt catalyst. Many binary and ternary Pt and transition metal alloy catalysts have been investigated for ORR, such as Pt-Co, Pt-Fe, Pt-Ni, Pt-Mn, Pt-Cr-Co, Pt-Fe-Mn, Pt-Fe-Co, Pt-Fe-Ni, and Pt-Cr-Cu. In addition, some composite catalysts of Pt with transition metal oxides, e.g. Pt-WO_3, Pt-TiO_2, and Pt-Cu-MO_x, have high electrocatalytic activity.

11.3.3.2 Preparation

Sol-Gel Method Bonnemann et al. proposed the reaction between $PtCl_2$ and a special organic reducing agent in an organic solvent to produce Pt sol. Then, an activated carbon support was mixed with Pt sol, washed, and dried. Finally, a Pt/C catalyst was obtained. A series of polymetallic sols such as Pt-Ru, Pd-Au, Pt-Ru-Sn, Pt-Ru-Mo, and Pt-Ru-W have been prepared using this method. The average size of metal particles in the resulting catalysts is small, around 1.7 nm. However, the sol-gel method is extremely complex, the conditions are harsh, and the cost of raw materials is high, thus making this approach suitable only for laboratory research. Furthermore, the catalysts obtained using this method often contain some impurities.

Liquid-Phase Reduction Method Liquid phase reduction is one of the most widely used methods for the preparation of Pt/C and Pt-Ru/C catalysts. In this method, the catalyst precursor and H_2PtCl_6 are dissolved and mixed with the activated carbon support, and then the reducing agent, such as $NaBH_4$, formaldehyde, sodium citrate, sodium formate, and hydrazide, is added to reduce Pt and deposit on the activated carbon. The structure and properties of the resulting catalyst vary depending on the reducing agent used. The advantage of this method is simplicity. However, the disadvantages are that the dispersion of the catalyst produced is relatively poor and the average particle size of the metal particles is large. Furthermore, for multi–component composite catalysts, uneven distribution of the components is observed.

Impregnation Reduction Method Impregnation reduction is another common method for preparing Pt/C catalysts. In this method, the support is dispersed in a solvent, and then, an appropriate amount of precious metal precursor is added. Based on the pH of the reducing solution, an excess reducing agent is added at an appropriate temperature. After allowing the reaction to proceed for some time, the desired Pt/C electrocatalyst can be achieved through filtration, washing, drying, and other steps. Generally, the catalyst produced has a small particle size, about a few nanometers. The preparation conditions (such as solvent, pH value, and reaction temperature) are the key factors affecting the catalyst performance, so they should be strictly controlled.

Pre-precipitation Method The pre-precipitation method can be used to prepare the catalyst with small metal particles. For example, an aqueous solution of H_2PtCl_6 and activated carbon were mixed evenly, and then, ammonia salt was added under constant stirring. Since $(NH_4)_2PtCl_6$ is insoluble in water, it was uniformly deposited onto activated carbon. After reduction with the reducing agent, the Pt/C catalyst with smaller Pt particles was obtained.

Colloidal Method In this method, a reducing agent is used to reduce the catalyst precursor into a metal colloid and evenly and stably disperse it in the solvent. The solvent is dispersed into a slurry so that the metal can adsorb or deposit on carbon to yield carbon-supported metal catalysts. Or to synthesize precious metal oxide colloids, which are then reduced and adsorbed on the carbon carrier to prepare the catalyst.

Solid-Phase Reaction Method Due to the low probability of collision between particles in solid-phase systems, the average particle size of the metal particles generated by the reaction is small and the crystallinity is lower, so the electrocatalytic performance of the prepared catalyst is better. For example, under solid-phase conditions, with H_2PtCl_6, the average particle size of Pt particles in the Pt/C catalyst synthesized by polyformaldehyde and activated carbon is about 3 nm, while the average particle size of the Pt/C catalyst in the general liquid-phase reduction is about 8 nm. Therefore, the electrocatalytic activity of the catalyst achieved by the solid-phase reaction is much better than that of the Pt/C catalyst obtained by the liquid-phase reaction.

11.3.3.3 Structure and Characterization

Theoretically, to increase the current density, the specific surface area of the catalyst must be increased by reducing the size of Pt particles. However, if Pt particles are too small, they become unstable. Research has shown that Pt catalysts have the highest mass-specific activity at a particle size of 3 nm.

The crystal properties of metal particles are related to the performance of the catalyst. Two aspects of crystal properties affect the performance of catalysts: the first is crystal structure because the catalytic activity of catalysts is different on different crystal planes. It is presumed that the catalytic activity of Pt atoms is higher on (111) and (100) planes. The different degrees of exposure of (111) and (100) planes lead to different catalytic activities. The second is crystallinity. The crystallinity of metal particles has a significant influence on the performance of the catalyst. In general, a catalyst with low crystallinity shows good electrocatalytic performance.

The surface roughness of metal particles also affects the performance of the catalyst. With an increase in roughness, surface defects increase and the atoms at surface defects often have a stronger ability to dissociate and adsorb hydrogen than general surface atoms, which can greatly improve the catalytic activity of the catalyst.

Using transmission electron microscopy to observe the morphology of metal particles in the catalyst, the dispersion of metal particles can be studied, and the average particle size and particle size distribution can be calculated. From the X-ray diffraction spectrum of the catalyst, the crystal form of metal particles in the catalyst can be obtained, and the average particle size and relative crystallinity of metal particles can be calculated. The electrochemical activity surface area of the catalyst can be determined by measuring the specific surface area of the catalyst using cyclic voltammetry in a sulfuric acid solution.

11.3.4 Gas Diffusion Layer

The GDL typically consists of a carbon black MPL on top of carbonized fibers. The whole layer is teflonated to form a mixed wettable material. GDLs tend to show decreasing hydrophobicity with operating time, which changes the ability to remove water, thereby affecting performance. To improve multiphase, particularly liquid, flow characteristics, the MPL is usually integrated between the GDL and the CL [31–37]. The layer is comprised of carbon black powder with a fine pore structure and can be beneficial for water management and fuel cell performance. Due to the finer pore structure than the GDL, the MPL provides better protection for catalyst nanoparticles and better physical contact with the CL. The thickness of an MPL is typically around 50 mm. A thick MPL raises concerns about increased transport resistance for liquid water removal and oxygen supply.

11.3.4.1 Function of the Gas Diffusion Layer

Supporting the CL and the membrane: The contact resistance between the CL and the GDL needs to be low. In addition, the GDL should provide mechanical support to the CL and protect it from erosion. The thickness of the present commercial GDLs ranges from 100 to 400 mm. The Toyota Mirai's GDL is around 100 mm thick, possibly because it uses a porous media flow field, which provides practical mechanical support and heat/electron pathway via its solid matrix [38].

Diffusion of gas and heat/water removal: To facilitate the diffusion of gas, the GDL should have high porosity. Besides, it must be thick to resist high thermal resistance to the in-plane heat removal under the GFC. A GDL with a gradient pore size, possibly realized via manipulation of carbon fiber arrangement, is expected to establish a more effective bridge between the flow field and the MEA and to improve the mass transfer ability [39–49]. For example, decreasing the porosity on one or both sides of the GDL may reduce the contact resistance and generate pore gradients inside the GDL to facilitate reactant supply and water removal.

Transfer of the electric current between the BP and the CL: The GDL needs to transmit the electrons produced by the oxidation of hydrogen on the CL, and the electrons required for the reduction of oxygen are transmitted to the cathode CL. Therefore, the GDL should be a good conductor.

Protection of the CL from corrosion or erosion: The GDL should be hydrophobic for better waterproofing and good chemical stability.

11.3.4.2 Requirements for Gas Diffusion Layer Materials

GDL materials should meet the following requirements: uniform porous structure with good air permeability; low resistivity and high electron-conducting capacity;

able to reduce contact resistance and improve conductive performance; appropriate mechanical strength, rigidity, and flexibility, which are conducive to the protection of the electrode and stability of the electrode structure under long-term operation conditions; appropriate hydrophilic/hydrophobic balance, which prevents excessive water from blocking pores and reduces gas penetration performance; chemical and thermal stability; low manufacturing cost; and a high performance–price ratio.

11.3.4.3 Gas Diffusion Layer Materials

Carbon paper: Carbon paper has advantages in terms of electrical conductivity, mechanical strength, chemical resistance, and fabrication cost [50]. Carbon fiber paper is widely used in the GDL in PEMFCs. The advantages of this material are uniform porous thin layer structure, excellent electrical conductivity, chemical stability, and thermal stability.

Carbon cloth: Due to the brittleness and lack of flexibility of carbon fiber paper, it can be easily damaged while preparing electrodes. Therefore, carbon fiber cloth is used as a GDL in PEMFC electrodes. The advantages of carbon fiber cloth are its bending ability and resistance to compression.

Nonwoven fabrics: Because carbon fiber paper lacks flexibility and carbon fiber cloth lacks stability, nonwoven fabrics have been developed. They have the following advantages: good mechanical strength, high flexibility, and dimensional stability. Fibers suitable for nonwoven fabrics include carbon fibers, glass fibers, ceramic fibers, and fibers containing organic polymers. These organic polymers can be polypropylene, polyester, polyethylene terephthalate, polyphenyl sulfur polyether ketone, etc. The fiber diameter is $0.2 \sim 50\,\mu m$, and the length is $0.5 \sim 150\,mm$, which is made into nonwoven fabrics by melt-blown or papermaking process.

Carbon black paper: Carbon black paper is a sheet made of carbon powder and polymer binder and is formed on a flat surface. The mass ratio of the polymer and carbon powder is between 20 : 80 and 45 : 55. The following can be used as carbon powder: activated carbon, carbon black, acetylene black, or its mixture. The polymer can be a fluorine resin, such as PTFE and poly 1,1-difluoride ethylene. Fluorine resin can also be used as a water treatment agent for carbon black paper, thus simplifying the water treatment process and reducing the cost.

11.3.4.4 Preparation of the Gas Diffusion Layer

Given the function of the GDL, the hydrophobic reactive gas channel and the hydrophilic liquid water transport channel should be formed. At present, carbon paper or carbon cloth is generally used as the GDL. Considering the drawback of strength, the thickness of a GDL is generally $100 \sim 300\,\mu m$. To increase the hydrophobicity of carbon paper and carbon cloth, it is necessary to immerse them in PTFE emulsion and make them evenly dispersed. Figure 11.6 shows the preparation of a water-repellent substrate by Yongjin Qiao of Beijing Nonferrous Metals Research Institute (BNMRI) and Zhicheng Pang of Beijing Institute of Technology (BIT).

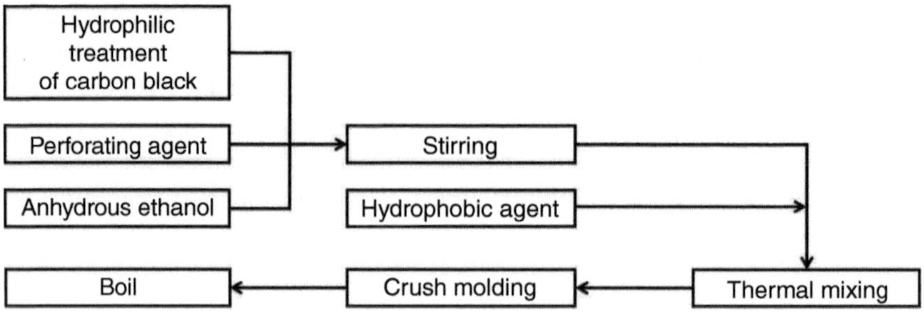

Figure 11.6 Preparation of a water-repellent substrate.

11.3.5 Proton Exchange Membrane

The proton exchange membrane is the core component of PEMFCs. It delivers the H^+ generated by hydrogen during anodic oxidation to the cathode, provides the H^+ needed for the cathode reaction, and enables the fuel cell to form an electric loop. Materials that are desirable as the proton exchange membrane include those that exhibit high ionic conductivity and simultaneously prevent electron transport and crossover of hydrogen and oxygen gases. In addition, they should be chemically stable in an environment with HO and HOO radicals, thermally stable throughout the entire operating temperature range, and mechanically robust.

The perfluoro sulfonic acid (PFSA) membrane comprises a carbon backbone and ether-branched chains with sulfonic acid groups, has high conductivity and chemical stability, and is one of the most suitable electrolytes for PEMFCs. The hydrophobic perfluorinated backbone provides mechanical support and chemical stability, while the hydrophilic sulfonated side chains promote water absorption, forming hydrated clusters [51]. The most commonly used PFSA membrane is Nafion developed by Du Pont. Nafion offers high proton conductivity, durability above 60,000 hours, and chemical stability. However, it is expensive to fabricate and requires hydration and hence humidification to conduct protons [52–57].

The Nafion membrane has good proton conductivity, but the migration of H^+ in the Nafion membrane should be accompanied by water migration. Hence, it is important to keep the membrane moderately moist. However, too much water in the fuel cell is not conducive to the diffusion of gas reactants to the catalyst, which will reduce the performance. In addition, other ions in water such as Na, Ni, Cr, and Fe will also reduce the H^+ mobility of the Nafion membrane. A proton exchange membrane should also have good mechanical strength and flexibility. A dry Nafion film has good mechanical strength, but as the water content increases, the mechanical strength decreases. Therefore, controlling the water content of the PEMFC is an important issue.

The lower the thickness of the proton exchange membrane, the more conducive it is to reducing the internal resistance of the fuel cell and improving the migration rate of H^+. However, if the membrane is too thin, hydrogen and oxygen can easily pass through it, i.e. the transmittance of gas is inversely proportional to the thickness of the membrane. Thus, the thickness of the Nafion membrane should be maintained within an appropriate range. Another advantage of the Nafion membrane is good chemical stability, the reduction of oxygen produces an intermediate, H_2O_2, which is a strong oxidant, but the Nafion membrane has good resistance to hydroperoxide oxidation.

The disadvantages of the Nafion membrane are as follows: first, its price is high, US$ 500 ~ 800 per m^2; second, the diffusion of H^+ in the Nafion membrane should be accompanied by the movement of the water, which makes controlling the water quantity in the membrane an important issue. Moreover, it also requires that the operating temperature of PEMFCs should not exceed 100 °C, generally about 80 °C. If the fuel cell is below 0 °C, water will freeze and the structure of the membrane will be destroyed.

11.3.6 Bipolar Plate and Flow Field

11.3.6.1 Bipolar Plate

The operating voltage of a fuel cell is approximately 0.7 V. To achieve the desired voltage, multiple individual cells should be linked via a conductive partition to create a fuel cell stack. The BP, also known as the collector plate, is a crucial component of the fuel cell. In PEMFCs, each BP has two gas chambers on both sides, with one chamber containing hydrogen and the other containing oxygen. The BP performs several important functions, primarily separating the reaction gas and introducing it into the fuel cell via the flow field. In addition, it collects and conducts current, supports the membrane electrode, and manages the heat dissipation and drainage needs of the entire fuel cell system.

The functions and requirements of the BPs are as follows: providing gas channels (the BP should have a suitable flow field structure) to make the reaction gas evenly distributed in the gas chamber and bring out the water and gas generated in the fuel cell; separating hydrogen and oxygen (the BP is required to have a gas resistance function, not using porous material); easy to process and shape; low price; collector current (the fuel cell is electrically connected by the BP, so the BP should have good electrical conductivity); controlling the temperature (the middle of the BP is designed with a cooling water channel to control the temperature, so the BP should be a good heat conductor to ensure uniform temperature distribution); supporting the diaphragm and the electrode (the BP also plays the role of supporting the diaphragm and electrode and maintaining the stability of the fuel cell stack structure, therefore, it should have appropriate strength); good corrosion resistance (as the electrolyte of PEMFCs is acidic, the environment of the BP also contains an oxidation and reduction medium, which can corrode the BP. In general, PEMFCs operate for tens of thousands of hours, so the BP material should have good corrosion resistance); light in quality; low resistance; and high mechanical strength.

11.3.6.2 Materials for Bipolar Plates

Metal Bipolar Plates Metal BPs have the advantages of good electrical conductivity, thermal conductivity, density, airtight nature, and good processing performance. Using metal materials, BPs can be easily mass-produced. Metal BPs can be applied to a variety of flow fields, and there are also stamping methods to process the flow fields. Thin metal plates can be prepared using stamping technology. However, metal BPs can be easily corroded. The metal ions generated by the corrosion of metal plates enter the battery, poison the catalyst, and reduce the H^+ conductivity of the proton exchange membrane. The corrosion of metal plates may also lead to the formation of a passivation film on the surface of metal plates, resulting in lower electrical conductivity and lower performance. Therefore, good corrosion-resistant materials need to be used, such as stainless steel, titanium, and nickel. Furthermore, the surface of a metal BP can be coated with precious metal or its oxide for better conductivity and anti-corrosion properties. In addition, the contact resistance between the metal BP and the electrode diffusion layer is high. Therefore, good conductive metals such as Au should be placed on the metal BP.

Graphite Bipolar Plates Nonporous graphite BPs and injection-molded graphite BPs are two types of graphite BPs. Nonporous graphite BPs are generally prepared from carbon powder or graphite powder and graphitized resin. Their advantages in a fuel cell environment include good chemical stability, high conductivity, and high gas resistance. However, they have some disadvantages, such as a long graphitization processing cycle and difficult flow field processing, which result in high manufacturing costs, thus, their application is limited. Injection-molded graphite BPs are made of graphite powder or carbon powder and resin (phenolic resin, epoxy resin, etc.) via conductive adhesive molding and then graphitization. The molding method can be injection molding, slurry injection, etc. Injection molding of a BP can be directly performed on the surface of the BP, which has the advantages of low cost and short cycle compared with machining. However, after the BP is prepared, it should be graphitized, which increases the production costs. In addition, injection-molded graphite BPs also have another disadvantage: the resin in the graphitization process will shrink, leading to the deformation of the graphite plate. Thus, while preparing a graphite BP, temperature should be controlled diligently. The requirements for the processing technology are also very strict:

Nonporous graphite bipolar plate: Graphite material is a traditional material developed earlier and applied to the BP. The earlier developed BP generally adopts non-porous graphite, and the graphite powder and resin mixed compaction through machining groove, graphitization (2500 °C heating), after cutting and grinding, made of 2 ~ 5 mm thick graphite plate. The porous graphite plate formed after the machining hole and drawing flow field, the porous graphite plate. For this type of graphite plate, the preparation process is complex, involves high cost, and is not suitable for mass production.

Expanded graphite bipolar plate: Expanded graphite, also known as flexible graphite, is compressed to the desired density and then molded into a plate, which can be machined into various sealed products. Expanded graphite has low air permeability and good electrical and thermal conductivity. Ballard Company uses the stamping method to prepare expanded graphite BPs, in which a small amount of resin can be added to improve the strength and gas resistance ability of BPs.

Molded graphite bipolar plate: Molded graphite BPs are formed by mixing graphite powder and resin at an appropriate temperature. The graphite BP is not graphitized and has a high resistance. However, because of the molding process, the production cost is low, thus making it suitable for large-scale production.

Composite Bipolar Plates Due to the disadvantages of metal BPs and graphite BPs, composite BPs are now widely being used. Generally, the metal is the inner partition of the BP, and graphite is the outer flow field plate. However, while processing, attention should be paid to the electrical contact between the metal plate and the graphite plate. Composite BPs can be primarily classified into the following three categories:

Carbon/carbon composite bipolar plate: The advantages of carbon/carbon composite materials are as follows: high conductivity, high thermal conductivity, light quality, high temperature resistance, high strength, high corrosion resistance, high chemical resistance, and no pollution to the MEA. However, the production process of this BP is more complex and requires high-temperature conditions; hence, their price is high.

Polymer/filler composite bipolar plate: This composite BP is prepared as follows: the conductive filler, resin, and conductive adhesive are mixed and then placed in an appropriate mold. Under suitable polymer melting temperature, pressure, and molding conditions, polymer/filler composite BPs can be prepared via injection. They have corrosion resistance similar to that of graphite, and their production cost is low, making them suitable for large-scale production, with the possibility of commercialization.

Metal-based composite bipolar plate: A metal-based composite BP comprises the following: a thin metal plate or any other conductive material as the partition plate; plastic, polysulfone carbonate, or any other material to make the frame; conductive glue to bond the frame and metal plates; and graphite as the flow field board. Metal-based composite BPs show good corrosion resistance, the quality is relatively light, and the strength is also high. However, they have the disadvantage of having a complex structure.

11.3.6.3 Flow Field

The function of the flow field is to guide the flow direction of the reaction gas and ensure that the reaction gas is evenly distributed throughout the electrode. A reasonable flow field structure can lead to the electrode obtaining sufficient reactants uniformly everywhere and timely discharge of the fuel-cell-generated water, ensuring that the fuel cell has good performance and stability. GFCs are typically embedded in BPs, thus, the dimensions of the GFC, specifically its depth, limit the minimal thickness of the BP. A small GFC may raise multiple concerns, including a large pressure drop, leading to high pumping power consumption, liquid blockage, and GDL intrusion. Many types of flow fields are used in PEMFCs, such as pin-type, network, multichannel, serpentine, alternating finger, spiral, parallel, parallel serpentine, and interdigitated flow field (Figure 11.7) [58].

The serpentine flow field is the most applied and most studied flow field type, including single-channel and multiple-channel structures. In a serpentine flow field,

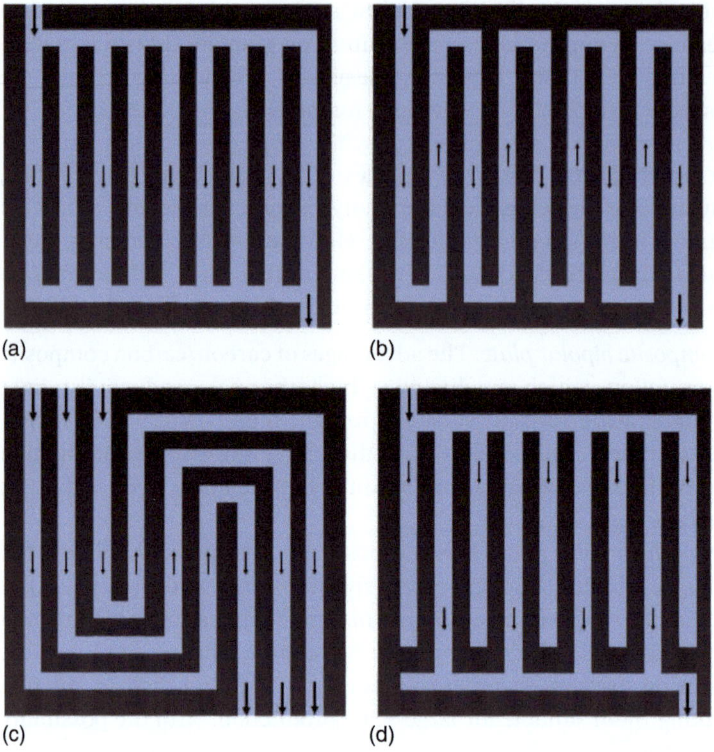

Figure 11.7 Schematic of various flow fields: (a) parallel flow field; (b) serpentine flow field; (c) parallel serpentine flow field; and (d) interdigital flow field. Source: Su et al. [58]/MDPI/CC BY 4.0.

the fuel and the oxidant flow through the electrode surface in the gas chamber and then diffuse into the catalytic layer inside the electrode. The reaction products also diffuse from the inside of the catalytic layer to the gas chamber, and the mass transfer speed is controlled by diffusion, so the rate is relatively low. Its advantage is that it can quickly discharge the water generated by the fuel cell, thus avoiding the blockage of the water convection channel. However, for large plates, this flow field will cause a large pressure drop and uneven current density distribution due to the long flow channel.

In the interdigitated flow field structure, the afferents and efflux of reactants and products on the electrode change from diffusion control to forced convection control. Hence, the mass transfer rate is relatively high, and the performance of the fuel cell is better, especially in large current discharges, the difference is more obvious. Most importantly, the structure of the interdigitated flow field easily takes away the product CO_2 generated by the anode and the water produced by the cathode, so that the anode reaction is not affected by the formation of CO_2 bubbles, but is also conducive to the drainage of the cathode.

The structural design of the parallel flow field makes it achieve a relatively uniform current density distribution. However, in this flow field design, water is prone to aggregation and obstructing the flow channel. In addition, if the design is not reasonable, it will lead to small fluid flow or even no flow, so the performance of the fuel cell is reduced due to concentration polarization.

An alternating finger flow field can improve the power density of the fuel cell. The reactant can fully pass through the flow channel with good drainage capacity, but at the same time, due to the high resistance of the diffusion layer, the pressure drop of the flow field will increase, which can lead to short circuits or groove flow.

11.3.7 Commercial Applications of PEMFCs

The primary applications of PEMFCs are in the fields of transport, distributed/ stationary, and portable power generation. Several FCVs have been developed and demonstrated, e.g. Honda Clarity, Toyota Mirai, GM ZH2, and Hyundai Tucson FCV [59]. Distributed PEMFC power systems primarily focus on small power demands [60]. Early designs have considered fuel cells for residential applications, where the waste heat from the fuel cell can be used for domestic purposes, significantly increasing the overall efficiency. Another promising area is portable electronics and aircraft, where the limited energy capacity of batteries is unlikely to meet the rapidly growing energy demands of modern portable electrical devices such as laptops, mobile phones, and military radiocommunications equipment. PEMFCs provide continuous power as long as hydrogen fuel is available and can be manufactured in small sizes without the loss of efficiency. Regarding aircraft, PEMFCs have advantages in providing direct electrical power, high power density, and large energy capacity. Both Boeing and Airbus are considering PEMFCs as auxiliary power units for their aircraft.

Transportation is a primary area of application for PEMFCs because of the zero-emission, high energy conversion efficiency, and high power density. Major motor companies have been extensively investigating PEMFCs to overcome the primary barriers to commercialization, including cost, durability, and cold-start capability [61, 62]. Although many metrics have been met, significant efforts are still needed to reduce the production cost and improve the durability of PEMFCs. Figure 11.8 shows the power and energy properties of several energy sources for PEMFCs [63].

11.4 Alkaline Fuel Cells

The AFC is the first type of fuel cell to be developed and successfully applied. In the 1950s, Francis Thomas Bacon used a KOH solution instead of the more corrosive acidic electrolyte solution that had been used since Grove's time and developed a double-hole electrode based on the concept of a GDE with a porous structure proposed by Schmid. This diffusion electrode increased the reaction interface of the electrode. He improved the device and resulted in the first AFC, which was also known as the Bacon cell. By the 1960s, NASA began funding fuel cell research programs to promote space technology. Small AFCs with long lifespan were successfully developed and used on the Apollo moon spacecraft, indicating

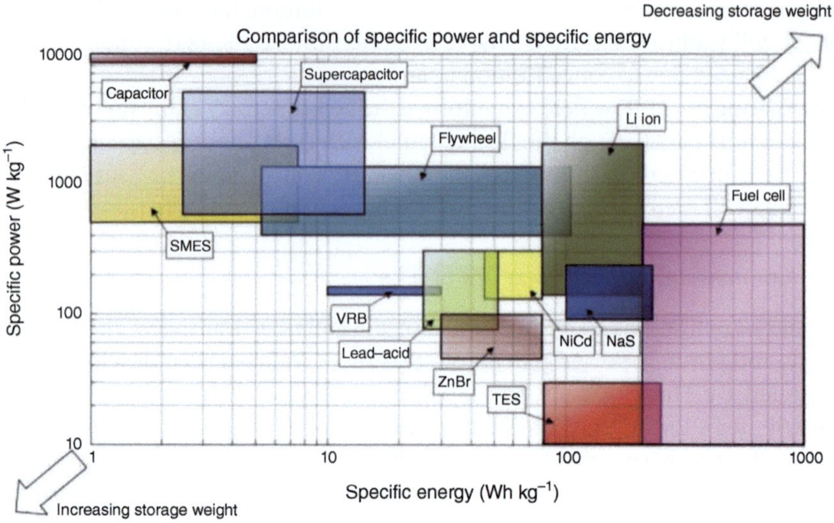

Figure 11.8 Specific energy versus specific power. PTFE: polytetrafluoroethylene; SMES: superconducting magnetic energy storage; VRB: vanadium redox flow battery; TES: thermal energy storage. Source: Reproduced with permission from Luo et al. [63]. © 2014/Elsevier.

that AFCs had reached a practical level, thus setting off the first wave of fuel cell research worldwide.

11.4.1 Operating Principle

AFCs usually use aqueous KOH or NaOH as the electrolyte solution, and the conducting ions transported in the electrolyte are OH^- ions. A more typical electrolyte solution is a KOH solution with a mass fraction of 35–50%, which can be used at lower temperatures (<120 °C). At higher temperatures, such as 200 °C, a high mass fraction (85%) of KOH can be used. Generally, porous GDEs prepared by Pt-Pd/C, Pt/C, Ni, and other electrocatalysts with better catalytic activity for hydrogen are used as hydrogen electrodes. In contrast, porous GDEs prepared by Pt/C, Ag, and other catalysts with better catalytic activity for electrochemical oxygen reduction are used as oxygen electrodes. The fuel reacts at the anode and air or oxidant at the cathode of an electrochemical system, which produces electric power and some excess heat [64]. The working principle of an AFC is shown in Figure 11.9 [2].

A humidified oxygen source, typically purified air/oxygen, is supplied to the cathode together with water. Oxygen gas, solvated in water, is reduced at the cathode to form hydroxide ions, which diffuse through the electrolyte to participate in the HOR that takes place on the anode. The reaction is as follows:

$$H_2 + 2OH^- \rightarrow 2H_2O + O_2 \;\; (\varphi^0 = -0.828\,\text{V}). \tag{11.48}$$

Humidified hydrogen gas is supplied to the anode, which reacts with the hydroxide ions in the electrolyte to produce water and electrons. The ORR is a complicated process involving intermediates and the transfer of four coupled proton and electron

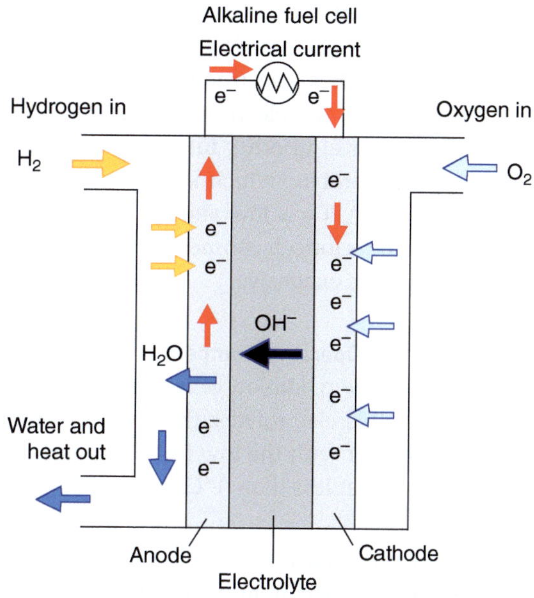

Figure 11.9 Alkaline fuel cell working principle. Source: Adapted from [2].

steps, which leads to several viable reaction pathways. The cathodic reaction is as follows:

$$\frac{1}{2}O_2 + H_2O + 2e^- \rightarrow 2OH^- \quad (\varphi^0 = 0.401\ V) \tag{11.49}$$

The total reaction is as follows:

$$\frac{1}{2}O_2 + H_2 \rightarrow H_2O \tag{11.50}$$

The theoretical electric potential at 25 °C can be expressed as follows:

$$E = 0.401 - (-0.828) = 1.229\ V \tag{11.51}$$

In fact, due to the irreversibility of the reaction, the open circuit voltage is usually below 1.1 V. The operating conditions of AFCs are as follows: the temperature should be 60 ~ 80 °C and the operating pressure of the cell should be 0.4 ~ 0.5 MPa.

11.4.2 Characteristics

11.4.2.1 Good Performance

In general, the ORR is more difficult than the hydroxide reduction, so the improvement in fuel cell performance primarily depends on the improvement in oxygen reduction performance. Since the reduction performance of oxygen in an alkaline solution is much better than that in an acidic solution, the performance of AFCs is generally better than that of PAFCs and PEMFCs using an acidic electrolyte.

11.4.2.2 High Energy Conversion Efficiency

In AFCs, the activation overvoltage of the ORR is low, the operating voltage can be as high as 0.875 V, and the electrical energy conversion efficiency can reach 60–70%.

11.4.2.3 Low Cost

The electrodes of AFCs can be made of non-precious metal materials, and the production costs of these electrodes are much lower than those of other types of fuel cells. In addition, the electrolyte used in AFCs is potassium hydroxide, which is highly inexpensive. In other fuel cells, such as PEMFCs, the Nafion membrane is used as the electrolyte, which is particularly expensive. In addition, the operating temperature of AFCs is low and the electrolyte is relatively less corrosive. Hence, the requirements for cell composition materials are low, the choice is wide, and the materials are less expensive.

11.4.2.4 Wide Operating Temperature Range

Since the electric oxidation of hydrogen and the electric reduction of oxygen are better in alkaline electrolytes, the low-temperature working performance of AFCs is good. Together with the low freezing point of the concentrated KOH electrolyte, AFCs can work at less than 0 °C.

11.4.2.5 Easy to Start

Due to the low working temperature of AFCs and good low-temperature working performance, AFCs can start quickly and can be started at room temperature.

11.4.2.6 Catalyst Poisoning

The KOH electrolyte used in AFCs will react with CO_2 to form carbonates such as potassium carbonate and sodium carbonate with low solubility, which will reduce the conductivity and degrade the performance of the fuel cell. In particular, when the mass fraction of K_2CO_3 reaches more than 30%, the fuel cell performance drops sharply. Moreover, K_2CO_3 is easy to crystallize in the electrode and destroys the electrode structure. Therefore, pure hydrogen and pure oxygen are better suitable for AFCs, which have higher operation costs and can only be used in some special cases, such as in spacecraft.

$$CO_3^{2-} + 2K^+ \rightarrow K_2CO_3 \tag{11.52}$$

11.4.3 Basic Structure

11.4.3.1 Fuel and Oxidizer

AFCs generally use pure hydrogen as fuel, but there are problems such as the high price of pure hydrogen and the low hydrogen storage capacity of the hydrogen source when an AFC is used on the ground. In addition, the use of organic matter pyrolysis to make hydrogen as fuel poses many problems as it contains CO. Under such circumstances, liquid fuels have been considered to replace hydrogen as they are more convenient and safer to store and transport. Liquid fuels that have been studied so far include hydrazine, liquid ammonia, methanol, and hydrocarbons.

Hydrazine is highly susceptible to decomposition at the anode:

$$N_2H_4 \rightarrow N_2 + 2H_2 \tag{11.53}$$

In addition, ammonia–air fuel cells have also been studied. The ideal anode reaction is as follows:

$$4N_3 + 3O_2 \rightarrow 2N_2 + 6H_2O \quad (11.54)$$

The oxidant for AFCs can be either air or oxygen. At a certain voltage, the output current density of AFCs with air as an oxidant is 50% lower than that with pure oxygen. The most severe problem is that air contains impurities such as CO_2. Although CO_2 is removed by pretreatment, there are some other impurities, such as SO_2, that are also detrimental to the fuel cell.

11.4.3.2 Electrode

The electrode generally consists of three parts. First, it has a diffusion layer, which is in contact with the gas. It is therefore required to have a large pore size, generally higher than 30 μm, to facilitate gas diffusion. Second, it should have good water repellency, which is achieved by adding PTFE. Finally, it should have good electrical conductivity. Therefore, the primary material to prepare the diffusion layer is usually a porous metal powder, such as sintered nickel powder and Raney nickel. The second is the catalytic layer, which is obtained by mixing the catalyst and PTFE so that the catalytic layer also has a certain degree of water resistance. The mass ratio of PTFE is generally 30–50% of the catalyst. The third is the collector fluid, which is generally made of nickel mesh, which cannot be easily corroded in KOH and has good electrical conductivity. For good performance of the catalyst, it should have good electrical conductivity, electrochemical stability, appropriate catalytic activity, and selectivity.

A general electrode creation procedure is shown in Figure 11.10. Electrodes are prepared using either dry or wet methods, depending on whether, for instance, the binder PFTE is used as a powder or a suspension liquid [64, 65]. The ink/paste

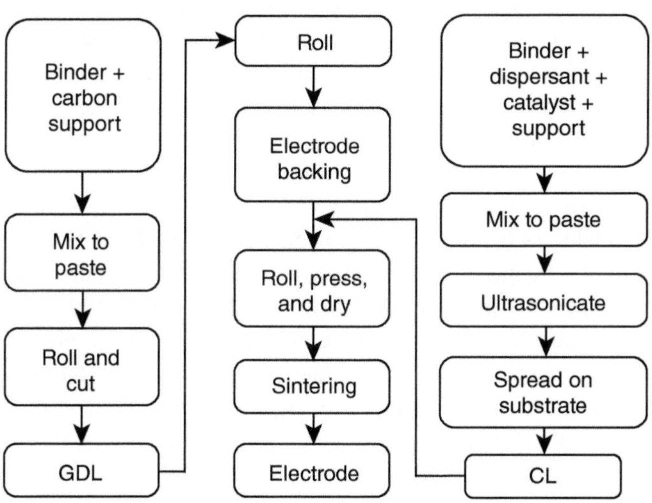

Figure 11.10 Electrode fabrication procedure. Source: Ferriday and Middleton [64]/Elsevier/ CC BY 4.0.

method is a common method for preparing electrodes, which involves mixing a dispersant and a binder and adding a catalyst powder to create a paste/ink, which is subsequently distributed over a substrate by either rolling or screen-printing. Ultrasonication before distribution over the substrate ensures uniform distribution of the catalyst in the paste/ink, decreases diffusion path lengths, and increases the number of gas-dissolving sites in the CL.

The energy required for the decomposition of hydrogen molecules into hydrogen atoms is huge, so only some electrodes with a large affinity for hydrogen atoms and high adsorption of hydrogen atoms can adsorb hydrogen atoms. Electrodes that can meet these conditions are mostly metal electrodes, such as Pt, Pb, Fe, and Ni. Therefore, although many non-precious metal catalysts have been researched, the best performance is achieved by precious metal catalysts. Although non-precious metal catalysts have an advantage in terms of price, precious metal catalysts have an irreplaceable advantage in terms of performance.

In the development of AFCs, pure hydrogen and pure oxygen are generally not used as the fuel and oxidizer, so it is necessary to consider further improving the electrocatalytic activity of the catalyst, improving the anti-toxicant ability of the catalyst, and reducing the amount of precious metal catalyst. Pt-based binary and ternary composite catalysts are generally used to achieve the above requirements.

AFCs have great advantages for commercial fuel cell applications because they operate at room temperature, do not require humidification systems, and have lower costs for electrocatalysts and electrolytes. However, to further expand their applications, the cost of the catalyst should be further reduced. Therefore, another direction of development for AFC catalysts is to use nickel or Raney nickel as an anode catalyst [66, 67]. Nickel-based materials have been comprehensively studied as alternatives for Pt group metals (PGM) due to their high abundance, moderate cost, and immunity toward corrosion at high pH. Raney nickel catalysts were an early recipient of attention as a non-PGM HOR catalyst due to their inherent high surface area. Raney nickel is one of the most active non-PGM catalysts for the HOR and is a preferred non-PGM catalyst in AFCs as they can endure the severity of 6.0 M KOH solutions. To ensure the liquid permeability and gas barrier of the electrode, the nickel electrode should be made into two layers so that it forms a wetted porous structure on the liquid side and has more micropores on the gas side. The pore size on the near-gas side is larger than 30 μm, whereas that on the near-liquid side is less than 16 μm, and the electrode thickness is about 1.6 mm to facilitate the absorption of electrolytes. However, to make the gas–liquid interface at an appropriate position, the pressure difference between the gas and electrolyte needs to be strictly controlled. With proper control, the reaction zone can be effectively stabilized within the coarse pore layer. In the reaction of H_2 oxide, if nickel alone is used, its activity is lower than that of Pt [67–69], which can be improved by adding a co-catalyst.

A co-catalyst is a small amount of material added to the catalyst that has little or no activity, which can improve the activity and selectivity of the main catalyst and the properties of the catalyst such as heat resistance, toxicity resistance, mechanical strength, and lifespan. Co-catalysts can be classified into structural co-catalysts and electronic co-catalysts.

Initially, precious metals are used as cathode catalysts in AFCs, but they are expensive and the resources are limited. Because O_2 has a faster reaction rate in alkaline media and can be used without precious metal catalysts, studies have been conducted on non-precious metal catalysts such as composite oxides, metallic porphyrins, and activated carbon.

11.4.3.3 Catalyst Poisoning

If the AFC uses air as an oxidant, CO_2 in the air will enter the electrolyte and electrode together with oxygen, which will form carbonate, reduce the conductivity of the electrolyte, and also lead to cathode catalyst poisoning.

Cathode Catalyst Poisoning Carbonate production results in blockage. CO_2 in the air will react with OH^-:

$$CO_2 + 2OH^- \rightarrow CO_3^{2-} + H_2O \tag{11.55}$$

The generated CO_3^{2-} is deposited in the micropores of the catalyst, resulting in the blockage of micropores, loss of catalyst activity, and decline in fuel cell performance. In addition, the reaction reduces the carrier OH^- concentration in the electrolyte, which affects the conductivity of the electrolyte.

Anode Catalyst Poisoning

Effect of toxic metal impurities: Some impurities such as Hg and Pd are highly toxic to the catalyst, so attention should be paid to prevent the introduction of these impurities in the preparation of catalysts or purification of fuel as the primary source of impurities is reaction raw materials, chemical drugs, and equipment materials.

CO poisoning: It is well known that CO impurities in the anode fuel can poison the catalyst. CO will adsorb on the surface of the catalyst and occupy the active sites, so that the effective surface area of the catalyst is reduced and the catalytic effect of the catalyst on the H_2 oxidation reaction is weakened, resulting in catalyst poisoning.

Adsorption of anions: Similar to the toxicity of CO to the catalyst, the adsorption of the anions present in the electrolyte on the electrode surface can also poison the catalyst. The anion with the strongest adsorption on the electrode surface is Cl^-, followed by SO_4^{2-}, and the anion with the weakest adsorption is ClO_4^-.

Methods to Prevent Catalyst Poisoning Catalyst poisoning can be prevented using the following approaches: chemical absorption of CO_2; molecular sieve screening; electrochemical removal of CO_2; use of liquid hydrogen; use of circulating electrolyte; and optimizing the electrode preparation method.

11.4.4 Electrolyte

AFCs generally use KOH or NaOH as the electrolyte, and the conducting ion transported inside the electrolyte is OH^-. A more typical electrolyte solution is a KOH solution with a mass fraction of 35–50%, which can be used at lower temperatures (<120 °C). NaOH can also be used as an electrolyte, which has the advantage of being less expensive than KOH. However, if CO_2 is present in the reaction gas, Na_2CO_3 is produced, which has a lower solubility than K_2CO_3 and tends to block the gas channels of the cell. Therefore, KOH is the most commonly used electrolyte.

11.4.4.1 Circulating Electrolytes

The majority of AFCs use circulating electrolytes. These electrolytes facilitate the replacement of electrolytes because during the operation of the cell, in addition to the primary reaction at the two electrodes, there is the following side reaction:

$$2KOH + CO_2 \rightarrow K_2CO_3 + H_2O \tag{11.56}$$

The electrolyte potassium hydroxide and carbon dioxide react to form potassium carbonate, so that as the reaction proceeds, the concentration of potassium hydroxide gradually decreases, which means that OH^- in the solution is gradually replaced by CO_3^{2-}, reducing the efficiency of the cell. To avoid or reduce the effect of CO_2 in the air on cell performance, air must pass through a CO_2 scavenger and an air filter to filter the CO_2 and impurities in it before it enters the fuel cell. Figure 11.11 shows the basic structure of a circulating electrolyte fuel cell.

The advantages of circulating electrolytes are as follows:

(1) A circulating electrolyte system can be used as a cooling system for the fuel cell, reducing the cost and complexity of additional cooling systems. The electrolyte can be continuously stirred and mixed during the circulation process, thus solving the problem of high electrolyte concentration around the cathode and low electrolyte concentration around the anode.

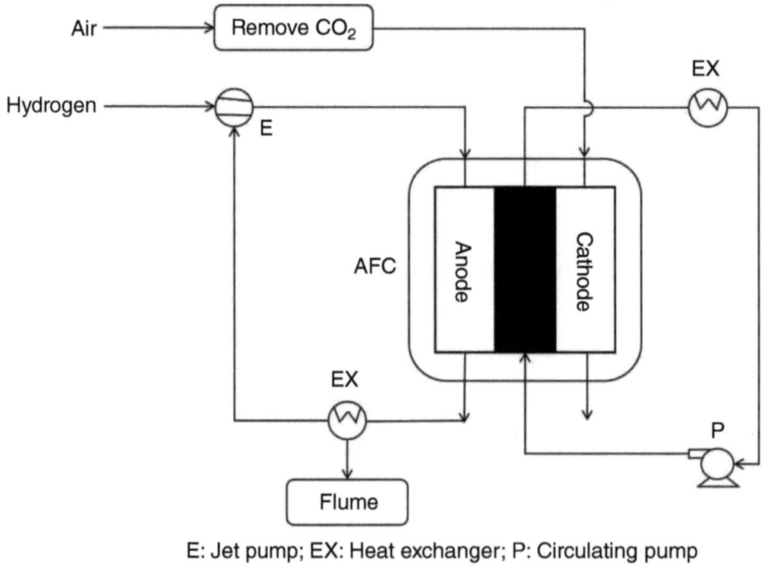

Figure 11.11 Basic structure of a circulating electrolyte fuel cell.

(2) The electrolyte can take away the water generated at the anode while circulating, without requiring an additional evaporation system.
(3) When the reaction time increases, the concentration of electrolyte will change and a fresh solution can be pumped in to replace the old solution.

However, circulating electrolytes have some disadvantages as well:

(1) They need additional devices, such as the pump.
(2) They easily generate parasitic current.
(3) The additional piping increases the possibility of fuel cell system leakage.
(4) Each cell should have its independent electrolyte circulation, otherwise, short circuits can easily occur.

11.4.4.2 Static Electrolyte and Carrier Material

Figure 11.12 shows the basic structure of a static electrolyte fuel cell. In this system, the KOH electrolyte is stabilized in the matrix material. The asbestos diaphragm is the most widely used as the matrix material. It has a porous matrix structure, high strength, and good corrosion resistance. Pure oxygen is used as the cathode oxidizer because if air is used as the oxidizer, once carbonates are produced to poison the electrolyte, replacing the electrolyte solution in this electrolyte-fixed fuel cell system is highly difficult. To discharge the water generated at the anode, hydrogen is circulated. In spacecraft, the water generated by the anode is purified to be used as drinking water for the staff, cooking and wetting the cabin, etc. However, for this purpose, a cooling system is needed. In the cooling system of the Apollo spacecraft, a mixture of ethylene glycol and water was used as the coolant. Each system of this fuel cell stack had a separate electrolyte, thus solving the problem of the so-called short

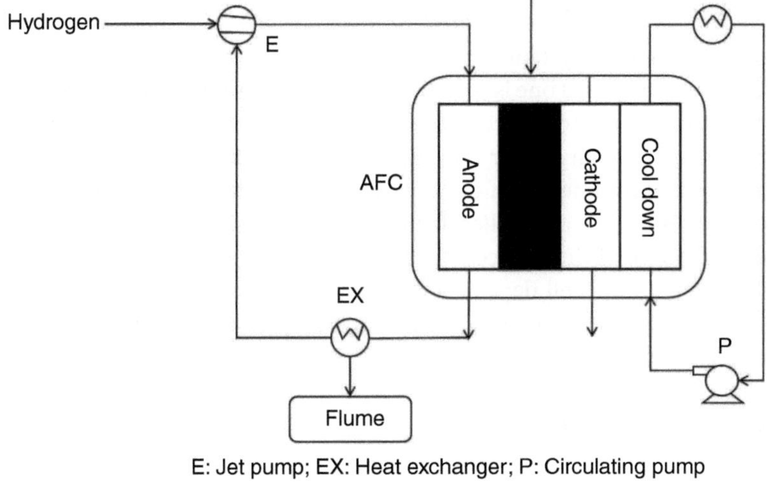

Figure 11.12 Basic structure of a static electrolyte fuel cell.

circuit in the circulating electrolyte system. However, static electrolyte AFCs pose the problem of managing water generation, anode water generation, and cathode water evaporation. In terms of water management, AFCs and PEMFCs are similar. However, the water management problem of AFCs is not as serious as that of PEMFCs because the saturation vapor pressure of KOH solution does not increase as fast as the temperature of pure water, i.e. the water of KOH solution evaporates more slowly.

Alkaline hydrogen–oxygen fuel cells primarily use the asbestos membrane as the electrolyte support material. The role of the asbestos membrane saturated with the alkaline solution is essentially twofold: on the one hand, it acts as a separator, separating the oxidant and reductant to prevent their reaction; on the other hand, it provides a channel for the transfer of OH^-. The major components of the asbestos membrane are magnesium oxide and silicon oxide, with the molecular formula $3MgO{\cdot}2SiO_2{\cdot}2H_2O$. When immersed in an aqueous KOH solution for a long time, the acidic components will react with the alkali to form a micro-soluble silicate (K_2SiO_3) and the asbestos membrane will be lost. To reduce the corrosion of the asbestos membrane in concentrated alkali, asbestos fibers are generally pretreated with concentrated alkali before making the membrane, or a small amount of potassium silicate is added to the concentrated alkali. Another method to increase the lifespan of the asbestos diaphragm is to use other materials in the preparation of the diaphragm, such as potassium titanate (K_2TiO_3). Potassium titanate is synthesized at high temperatures, making it resistant to oxidation and insoluble in KOH solution. The lifespan of the asbestos membrane prepared with potassium titanate membrane is approximately five times that of the asbestos membrane prepared using conventional methods. All of these methods can partially or completely inhibit asbestos film corrosion and reduce the degradation of battery performance due to asbestos film loss during the operation of the fuel cell.

11.4.5 Operating Conditions

The performance of a fuel cell system can be determined by several electrochemical parameters. The commonly used one is specific energy, which is the electrical energy output per unit volume or mass of the cell. It includes two expressions: mass-specific energy, which is the power output per unit mass of the cell ($kW\,kg^{-1}$), also called mass power density; and volume-specific energy, which is the power output per unit volume of the cell ($kW\,dm^{-3}$), also called volume power density. Both parameters take into account the role of the fuel cell stack and auxiliary equipment. The performance and stability of the cell depend on several parameters such as electrode composition, operating temperature of the cell, operating pressure, oxidant, and the concentration of KOH in the electrolyte.

11.4.5.1 Operating Temperature

From the viewpoint of electrode process kinetics, increasing the operating temperature of the cell can improve the electrochemical reaction rate and the mass transfer rate; reduce the concentration polarization; improve the OH^- migration rate, and reduce the ohmic polarization. Thus, the performance of the fuel cell can be improved by increasing the cell temperature.

In AFCs, the electrolyte has good low-temperature conductivity, so the operating temperature of AFCs is about 70 °C. As the temperature decreases, the performance of the cell also decreases, and if it operates at room temperature, the performance is reduced to half of that at 70 °C. At lower temperatures (50–60 °C), an almost linear relationship exists between cell output and the increase in electrolyte temperature. However, as the temperature continues to increase, the linear relationship no longer holds due to the KOH concentration in the electrolyte and other factors.

11.4.5.2 Operating Pressure

Most AFCs operate at pressures higher than atmospheric pressure. This is because when the operating pressure increases, the open circuit voltage of the fuel cell also increases, as can be seen from the following equation:

$$\Delta V = \frac{RT}{4F} \ln \frac{p_2}{p_1} \tag{11.57}$$

In addition, an increase in pressure will also increase the exchange current density. These factors lead to a significant improvement in the performance of the AFC. Therefore, most AFCs operate under high pressure. Working under high pressure requires a high mechanical strength of the materials used. In addition, as the gas is operated at very high pressures, high-pressure equipment is required, which increases the quality and cost of the cell system. There is also the issue of leakage from working under high pressure. Besides wasting gas, leaks can cause gas to enter the electrolyte area, disrupting fuel cell operation and reducing performance. It can also lead to the mixing of two gases, which can lead to an explosion and compromise safety.

11.4.5.3 Effect of Oxidants

The oxidant used in AFCs can be either air or oxygen. It is primarily because when air is used as the oxidizer, it contains a large amount of inert gas, which reduces the performance of the fuel cell. In addition, even though CO_2 in air can be removed via pretreatment, some other impurities are present, which is not good for the cell. Therefore, the best oxidant for AFCs is oxygen.

11.4.5.4 Drainage

Dynamic Drainage Method In AFCs, water is generated at the hydrogen electrode, which is evaporated into hydrogen using the dynamic drainage method, and hydrogen is circulated with a fan so that the water is cooled and separated in a condenser outside the cell. When the hydrogen circulation rate is high, the KOH concentration can be considered constant, but the power consumption of the circulation pump increases, which increases the internal consumption of the cell. Therefore, the hydrogen circulation rate should be controlled. This drainage method is relatively simple but is power-consuming. Moreover, the use of the pump increases the moving parts of the fuel cell system and reduces the operational reliability.

Static Drainage Method In this method, a microporous water-conducting membrane dipped in KOH solution is added to the side of the hydrogen chamber of the fuel cell, and the water vapor chamber outside the membrane is maintained at a negative pressure. Therefore, the water generated at the hydrogen electrode evaporates into hydrogen, which is then absorbed by the KOH membrane, evaporates into the water vapor chamber under vacuum, and then diffuses to the outside of the cell for condensation and separation by pressure difference. This method is easy to implement as the vacuum in the water vapor chamber is the only aspect that needs to be controlled. This method is especially useful in space, where vacuum conditions exist. However, a steam chamber should be added to each cell, making the cell structure more complicated.

11.4.5.5 Heat Extraction

Since the cell reaction is exothermic, the heat generated should be discharged for the cell to operate at a constant temperature. One of the most convenient methods for heat extraction when using a circulating electrolyte is to exhaust the heat. Alternatively, a combination of both air circulation and electrolyte circulation can be used to remove heat.

11.4.5.6 Lifespan

The lifespan of the cell depends primarily on the performance of the catalyst and the stability of the system. The lifespan of hydrogen–oxygen and hydrogen–air AFCs can generally reach about 5000 hours. The lifespan of AFCs can vary based on the applications in which they are used, some applications require AFCs with a lifespan of only 5000 hours or even 1000 hours, whereas some require AFCs with a higher lifespan, up to 25,000 hours. In general, power plants require AFCs with a lifespan of more than 40,000 hours.

11.5 Other Fuel Cells

11.5.1 Phosphoric Acid Fuel Cell

PAFCs use phosphoric acid as the electrolyte. One of the distinctive features of PAFCs is their tolerance to CO, which makes them the first fuel cells to be used on the ground. Since phosphoric acid has good thermal, chemical, and electrochemical stability; little volatility at high temperatures; and unique tolerance to CO_2, it was first successfully applied in fuel cells.

PAFCs use phosphoric acid as the electrolyte and hydrogen as the fuel. Hydrogen enters the gas chamber, reaches the anode, loses two electrons, and oxidizes to H^+ under the action of the anode catalyst.

Anodic reaction:

$$H_2 \rightarrow 2H^+ + 2e^- \qquad (11.58)$$

H^+ reaches the cathode through the phosphoric acid electrolyte, and electrons reach the cathode through the external circuit. Oxygen enters the gas chamber and arrives at the cathode, where it is combined with H^+ and electrons arriving at the cathode by the action of the cathode catalyst and reduced to produce water.

Cathodic reaction:

$$1/2O_2 + 2H^+ + 2e^- \rightarrow H_2O \tag{11.59}$$

Total reaction:

$$H_2 + 1/2O_2 = H_2O \tag{11.60}$$

PAFCs have the following advantages:

(1) The system is simplified and costs are reduced because there is no need to pretreat the gas to remove CO_2.
(2) The operating temperature of the fuel cell is between 180 and 210 °C, which is relatively mild, so the materials required to make up the fuel cell are not high.
(3) The hot water generated by the PAFC during operation can be exploited, i.e. cogeneration is possible.
(4) The start-up time is shorter and the stability is better.

However, PAFCs also have some disadvantages:

(1) The power generation efficiency is low, reaching only 40–45%.
(2) The use of acidic electrolytes necessitates the use of more stable precious metal catalysts, and therefore higher costs.
(3) The 100% phosphoric acid used has a corrosive effect, making it difficult to exceed 40,000 hours of life.
(4) Since Pt, a precious metal, is used as the catalyst, the fuel gas should be purified to prevent catalyst poisoning by CO.

Furthermore, PAFCs require the following operating conditions:

(1) The choice of operating temperature is based on the vapor pressure of the phosphoric acid electrolyte, the corrosion resistance of the material, the ability of the electrocatalyst to resist CO poisoning, and the actual operating requirements. If the temperature increases, the efficiency of the cell will also increase.
(2) The working pressure is generally $0.7 \sim 0.8$ MPa for pressurized working conditions, while for small-capacity fuel cells, the working pressure is often normal. Under pressurization, the reaction rate can be accelerated and the power generation efficiency can be increased.
(3) The fuel utilization rate is generally 70–80%. Fuel utilization refers to the ratio of the amount of hydrogen converted to electricity inside the fuel cell to the amount of hydrogen in the fuel gas.
(4) The utilization rate of oxidants is generally 50–60%. If air is used as the oxidizer, the mass fraction of oxygen in it is 21%, and the utilization rate of 50–60% indicates that 10–12% of the oxygen in air is consumed in fuel cell power generation.
(5) The mass fraction of H_2 in a typical PAFC fuel gas is about 80%, the mass fraction of CO_2 is about 20%, and there are small amounts of CH_4, CO, and sulfides.

11.5.2 Molten Carbonate Fuel Cells

The operating temperature of MCFCs is high (about 873–923 K), their cost and efficiency are more competitive compared with other low-temperature fuel cells, and their advantages are primarily reflected in five aspects:

First, under temperature conditions in which the MCFC operates, reforming of the fuel (e.g. natural gas) can take place inside the cell stack. For example, the reforming of methane can occur in the anode reaction chamber, and the heat required for the reforming reaction is provided by the cell reaction, which reduces the cost and increases the efficiency.

Second, MCFCs operate at temperatures that are sufficient to produce valuable waste heat without excessive free energy loss. The cell can emit waste heat up to 673 K, which can be used to compress the reaction gas to improve cell performance, to absorb heat produced due to reforming of the fuel, or to heat the boiler, resulting in an overall thermal efficiency of 80%.

Third, almost all fuel reforming produces CO. It can poison low-temperature fuel cell electrode catalysts. However, it can be the fuel for MCFCs. Therefore, MCFCs can use fuel with high CO content, such as gas.

Fourth, the high operating temperature and low activation energy of the electrode reaction do not require efficient catalysts for either HOR or ORR. Compared with low-temperature fuel cells that require precious metal catalysts and CO removal in hydrogen-rich fuel reforming, MCFC electrocatalysts are primarily shackled and do not use precious metals.

Fifth, MCFCs can be cooled by air instead of water, which is especially suitable for remote areas where water is scarce. The high reactivity of H_2 is attributable to operating at high temperatures. Although increasing the reaction temperature decreases the theoretical efficiency of the cell, it also decreases the overpotential loss and increases the actual efficiency.

The disadvantages of MCFCs are as follows:

First, the corrosive nature of the electrolyte and high operating temperatures have strict requirements on the long-term corrosion resistance of various materials, and the fuel cell life is somewhat limited.

Secondly, the high-temperature wet sealing technology at the edge of the single cell is difficult, especially in the anode area, where severe corrosion can occur. In addition, there are some inherent problems with molten carbonates, such as rupture due to cooling.

Third, the fuel cell system requires CO cycling to redeliver the CO precipitated from the anode to the cathode, which increases the structural complexity of the system.

Therefore, despite the obvious advantages of MCFCs in terms of reaction kinetics, molten salt corrosion and sealing associated with their high-temperature operation have hindered their rapid development.

11.5.3 Solid Oxide Fuel Cells

The solid oxygen ion conductor was discovered by Nernst at the end of the nineteenth century, and in 1935, Schottky published a paper stating that the substance discovered by Nernst could be used as a solid electrolyte for fuel cells. Among all types of fuel cells, SOFCs have unique advantages and a wide range of applications in large, medium, and small power stations, mobile and portable power supplies, and military and aerospace fields. The advantages of SOFCs are primarily manifested in the following aspects:

SOFCs can use a wide range of fuels, not only H_2 and CO but also natural gas, coal gasification gas, and other hydrocarbons (such as methanol, ethanol, and even high-carbon chain liquid fuels such as gasoline and diesel) directly.

Since SOFCs use ceramics, which are oxygen-ion conductors at high temperatures, as electrolytes, there is no evaporation or precipitation of electrolytes, and problems such as corrosion and electrolyte loss associated with the use of liquid electrolytes are avoided.

SOFCs have high efficiency in comprehensive energy utilization and are one of the more efficient fuel cells using hydrocarbons as fuel, with a primary power generation efficiency of over 65%.

The electrode reaction process of SOFCs is quite rapid because they operate at medium to high temperatures and can withstand high concentrations of sulfide and CO, so electrode requirements are greatly reduced and there is no need to use precious metal electrodes, thus reducing the cost.

The high operating temperature reduces the possibility of catalyst poisoning to some extent, and the low fuel purity requirements make SOFCs attractive for operation with high-carbon-chain fumes such as diesel and even kerosene, while natural-gas-fueled plants do not require desulfurization systems.

In addition, SOFCs use electrocatalytic anodes to produce chemicals while generating electricity. More importantly, because of the advantages of high efficiency, high power density, simple structure, and long lifespan, SOFCs can be used to replace large-scale thermal power. Since a solid oxide fuel has an all-solid structure and small size, it is highly suitable for modular design and scaling up, and the capacity can be flexible, either large or small. Therefore, SOFCs are known as the fourth type of power generation devices after hydropower, thermal power, and nuclear power.

11.6 Fuel Cell Systems

The function of a fuel cell is to convert the chemical energy of the fuel into electrical energy and ultimately to provide the required power to the load. To achieve this goal, a single fuel cell is not enough, but usually, several single fuel cells are combined to form a fuel cell stack, together with a set of ancillary devices to form a complex fuel cell system. These auxiliary devices perform a series of tasks, including fuel storage, processing, continuous and stable delivery of the fuel and oxidizer to the cell, conversion, power regulation of the cell output, heat management of the cell, and monitoring of all parts of the system. Usually, these auxiliary devices occupy a large space in the whole system, are an important part of the system cost, and also consume electrical energy, which leads to the large size, high cost, and reduced efficiency of the fuel cell system. Therefore, the design of the fuel cell system should be considered in the context of fuel cell application goals. When a fuel cell system is used as a mobile or portable power source and high energy density is required, it is important to simplify the ancillary components.

The fuel cell system generally consists of five subsystems: fuel subsystem (fuel storage, handling, and delivery); fuel cell subsystem (fuel cell stack); thermal management subsystem (cooling and waste heat recovery); power conversion subsystem (DC conversion, adjustment, and voltage stabilization); and automatic fuel cell monitoring system. The relationship between each system is shown in Figure 11.13 [70].

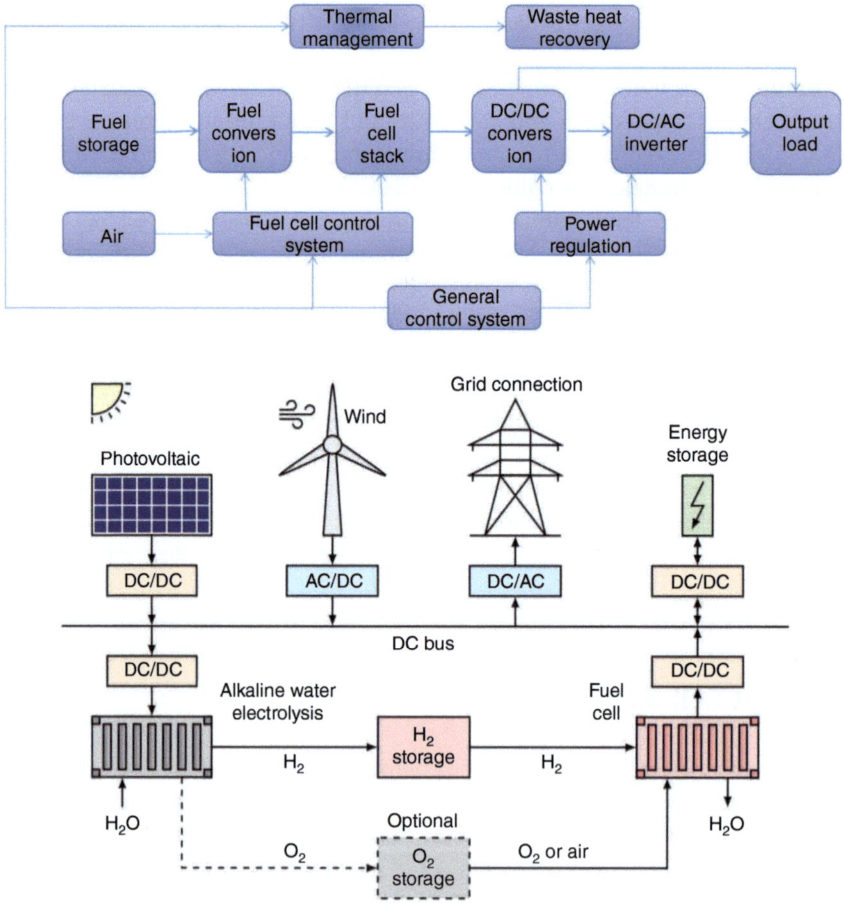

Figure 11.13 Schematic of a fuel cell power generation system. Source: Brauns and Turek [70]/MDPI/CC BY 4.0.

11.6.1 Fuel Cell Stack

In the case of hydrogen–oxygen fuel cells, the operating voltage of a single cell is usually in the range of 0.6–0.7 V. It is an optimal voltage range in which the output power of the fuel cell is close to its maximum value and the electrical efficiency is suitable (about 45%). However, the voltage required in practical applications is often higher than the operating voltage of a single cell, which usually ranges from several volts to several hundred volts. Therefore, to meet the high voltage requirements, several single cells are usually connected in series to form a fuel cell stack. The voltage of the stack is the sum of individual cell voltages. Figure 11.14 shows a schematic of a fuel cell stack [71].

The series connection of fuel cells is realized by "BPs," where one side of the BP is the anode of a single cell and the other side is the cathode of the adjacent single cell. By stacking several BPs, the sandwich structure of the "anode–electrolyte–cathode" of every single cell, with one end plate at each end as the anode and cathode plates, a cell stack is formed. The BP is an important component of the fuel cell stack and is responsible for several tasks: (i) the BP transports the reaction gas, and the two sides are processed with different forms of gas channels (flow fields) to ensure the uniform distribution of the reaction gas throughout the electrode so that it has good processing performance; (ii) it collects current, making it a good electron conductor; (iii) it isolates the fuel from the oxidizer, so it should not be air permeable; and (iv) it plays a role in supporting the electrodes, so it should be of high strength. In addition, BPs should be light in mass and as thin as possible to reduce the mass and volume of the cell stack; they should be good conductors of heat to facilitate the dissipation of reaction heat and ensure the uniform temperature of the cell stack; furthermore, they should be stable and corrosion resistant under the working environment of the cell.

The size of the fuel cell stack depends on the user's needs. The area of the electrode and the number of series connections of a single cell can be determined according to the usage requirements and cell performance. The area of the electrode determines the operating current of the fuel cell stack, and the number of series connections determines the operating voltage of the fuel cell stack.

The design of the fuel cell stack should meet the voltage and power requirements, but also meet other requirements as follows: the distribution of the reaction gas among individual cells in the stack must be uniform, and the resistance to flow through each cell should be equal to ensure that each cell performs at its best efficiency uniformly. In addition, it is important to ensure that the reactive gases on both sides of the anode and cathode do not cascade and leak out, i.e. the sealing should be good. The contact resistance between the BP and the electrode assembly should be low to reduce the loss of electrical energy. These properties are closely related to BPs, sealing materials, and structural design and are often the key factors that determine the performance of fuel cell stacks.

11.6.2 Thermal Management System

The electrical efficiency of fuel cells cannot reach 100%, which is currently only 40–60%. The energy that is not converted into electricity is released in the form of

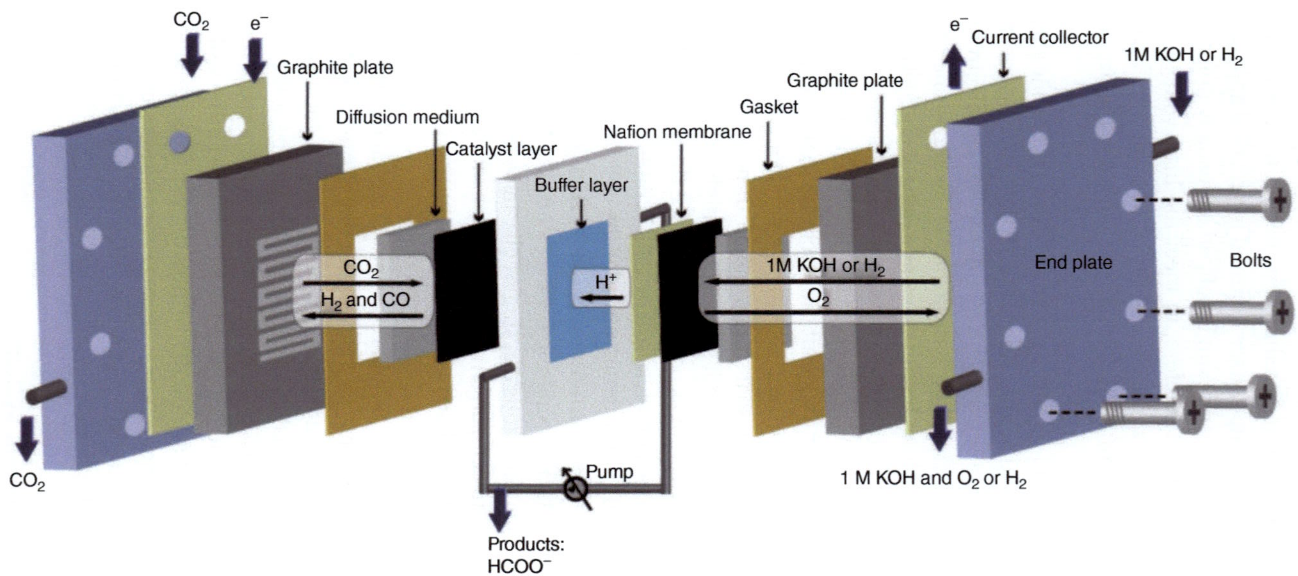

Figure 11.14 Schematic of a fuel cell stack. Source: Reproduced with permission from Wu et al. [71]/The Electrochemical Society.

heat. If the heat is not removed from the stack, it may lead to overheating of the entire fuel cell stack or a large temperature difference between different parts of the stack. This is detrimental to the smooth operation of the fuel cell stack. In addition, to improve the overall energy utilization, the released heat should be recycled. Therefore, the thermal management subsystem plays an important role in the fuel cell system. The heat extraction and cooling method of the fuel cell stack depends on the type and size of the fuel cell. Small low-temperature fuel cells (e.g. PEMFCs) are typically cooled by "passive" cooling, i.e. by natural convection, and heat is often not recovered. Medium- and high-temperature fuel cells (e.g. PAFCs, MCFCs, and SOFCs) and large low-temperature fuel cells (e.g. PEMFCs) use "active" cooling to recycle waste heat.

Small portable low-power PEMFC systems (e.g. <100 W) usually do not require special cooling equipment, but require careful heat transfer design of the fuel cell stack structure using the natural convection of ambient air to achieve the heat balance inside and outside the fuel cell stack. By this approach, the reaction temperature of the fuel cell stack can be maintained without overheating. Larger low-temperature PEMFC systems and medium-temperature PAFC systems usually require forced air convection or heat transfer using a coolant to maintain the cell temperature. This requires auxiliary equipment such as blowers and pumps to move the air or liquid through the cooling ducts. These auxiliary devices consume a portion of the electrical energy generated by the fuel cell. Due to space constraints, automotive fuel cell systems are often cooled by active liquid circulation because the heat capacity of the liquid is much higher than that of the gas, thus reducing the size of the system, but often increasing the complexity of the system. High-temperature fuel cells, such as MCFCs and SOFCs, are not highly sensitive to overheating due to their high operating temperatures, and cooling is not an important issue. However, the heat generated by high-temperature fuel cells is of high value and can be used to preheat the fuel and the oxidizer and to provide heat for fuel pretreatment (hydrocarbon water–gas conversion, hydrogen reforming, etc.). The remaining heat can also be used, for instance, to drive a steam engine to generate electricity, thereby increasing the electrical efficiency of the fuel cell system.

11.6.3 Power Conditioning and Conversion Systems

The function of a fuel cell power conditioning and conversion system is usually twofold: first, to regulate the power, i.e. to ensure an exact and stable voltage output from the fuel cell system; and second, to convert the power, i.e. to convert the direct current output from the fuel cell to alternating current. Almost all fuel cells require power regulation, power conversion is also necessary for fuel cell power plants and automotive fuel cell systems.

11.6.3.1 Power Regulation

Most of the practical applications require a power supply system that can provide a stable voltage over a long period. However, the output voltage of fuel cells is often unstable and depends heavily on operating conditions such as cell temperature, reaction gas pressure, and flow rate. Most notably, the voltage can change drastically when load variations increase. In addition, even when fuel cell stacks are designed in line with the user's voltage requirements, the design voltage often does not meet the exact requirements of the specific application. To solve these problems and to make the fuel cell system deliver a fixed and stable DC voltage, a DC/DC converter (DC/DC converter) is used between the fuel cell stack and the user load, which converts the input voltage (fuel cell stack voltage), and varies within a certain range, into a fixed and stable output voltage. This conversion requires some power consumption. The ratio of the output power to the input power of a DC/DC converter is usually defined as its conversion efficiency. Generally, DC/DC converters can have an efficiency higher than 85%. The DC/DC converter can increase or decrease the voltage and is usually used to increase the voltage. The increase in output voltage comes at the cost of a decrease in output current.

11.6.3.2 Power Conversion

The fuel cell output is direct current. If the fuel cell system has an AC user or if the electricity generated by the fuel cell has to be integrated into the AC grid (e.g. a fuel cell power station), DC power from the fuel cell stack needs to be converted to AC power. This conversion process is usually performed by a DC/AC converter, so a DC/AC converter needs to be installed between the fuel cell stack and the AC user or the grid. Currently, DC/AC converters are usually capable of achieving efficiencies of 85% or higher.

11.6.4 Control System

The load of the fuel cell system changes frequently, especially the fuel cell for vehicles, which have to start, stop, and change speed, resulting in very large load changes. Therefore, each part of the fuel cell system should be monitored in real-time and continuously adjusted according to the load changes, which is the control subsystem. The control subsystem detects, adjusts, and controls key control parameters in other subsystems of the fuel cell system to ensure stable and reliable operation. This system also includes the start-up procedure, stopping procedure, and fault detection procedure of the fuel cell system. The control system consists of various sensing elements (e.g. temperature and pressure sensors), actuators (e.g. solenoid valves and air valves), and computer programs. It can finally realize the fully automatic operation of the fuel cell system.

References

1 Grove, W.R. (2012). On a gaseous voltaic battery. *Philosophical Magazine* 92 (31): 3753–3756.

2 Taa'mneh, T., Ayadi, O., and Al, A.J. (2019). Design of stand-alone proton exchange membrane fuel cell hybrid system under Amman climate. *Journal of Ecological Engineering* 20 (9): 1–10.

3 Seselj, N., Engelbrekt, C., and Zhang, J. (2015). Graphene-supported platinum catalysts for fuel cells. *Science Bulletin* 60 (9): 864–876.

4 Ren, X., Zhang, X., Teng, T. et al. (2020). Research on estimation method of fuel cell health state based on lumped parameter model. *Energies* 13 (23): 6425.

5 Faid, A.Y. and Sunde, S. (2022). Anion exchange membrane water electrolysis from catalyst design to the membrane electrode assembly. *Energy Technology* 10 (9): 2200506.

6 Jiao, K., Xuan, J., Du, Q. et al. (2021). Designing the next generation of proton-exchange membrane fuel cells. *Nature* 595 (7867): 361–369.

7 Peighambardoust, S.J., Rowshanzamir, S., and Amjadi, M. (2010). Review of the proton exchange membranes for fuel cell applications. *International Journal of Hydrogen Energy* 35 (17): 9349–9384.

8 Dicks, A.L. and Rand, D.A.J. (2018). *Fuel Cell Systems Explained*. Wiley.

9 Passalacqua, E., Lufrano, F., Squadrito, G. et al. (2001). Nafion content in the catalyst layer of polymer electrolyte fuel cells: effects on structure and performance. *Electrochimica Acta* 46 (6): 799–805.

10 Xie, J., Xu, F., Wood, D.L. III et al. (2010). Influence of ionomer content on the structure and performance of PEFC membrane electrode assemblies. *Electrochimica Acta* 55 (24): 7404–7412.

11 Wang, Y., Yu, J., Wu, J. et al. (2017). Rapid analysis of platinum and Nafion loadings using laser induced breakdown spectroscopy. *Journal of the Electrochemical Society* 164 (13): 1294–1300.

12 Snyder, S.C., Wickun, W.G., Mode, J.M. et al. (2011). The detection of palladium particles in proton exchange membrane fuel-cell water by laser-induced breakdown spectroscopy (LIBS). *Applied Spectroscopy* 65 (6): 642–647.

13 Wang, Y. and Feng, X. (2009). Analysis of the reaction rates in the cathode electrode of polymer electrolyte fuel cells: II. Dual-layer electrodes. *Journal of the Electrochemical Society* 156 (3): B403.

14 Feng, X. and Wang, Y. (2010). Multi-layer configuration for the cathode electrode of polymer electrolyte fuel cell. *Electrochimica Acta* 55 (15): 4579–4586.

15 Debe, M.K. (2010). Electrocatalyst approaches and challenges for automotive fuel cells. *Nature* 486 (7401): 43–51.

16 Shao, M., Chang, Q., Dodelet, J.P. et al. (2010). Recent advances in electrocatalysts for oxygen reduction reaction. *Chemical Reviews* 116 (6): 3594–3657.

17 Fernández, J.L., Walsh, D.A., and Bard, A.J. (2005). Thermodynamic guidelines for the design of bimetallic catalysts for oxygen electroreduction and rapid screening by scanning electrochemical microscopy. M–Co (M: Pd, Ag, Au). *Journal of the American Chemical Society* 127 (1): 357–365.

18 Kunusch, C., Puleston, P., and Mayosky, M. (2012). *Sliding-Mode Control of PEM Fuel Cells*. Springer Science and Business Media.

19 Xie, X., Yu, G., Liu, N. et al. (2012). Graphene–sponges as high-performance low-cost anodes for microbial fuel cells. *Energy and Environmental Science* 5 (5): 6862–6866.

20 Jha, N., Ramesh, P., Bekyarova, E. et al. (2013). Functionalized single-walled carbon nanotube-based fuel cell benchmarked against US DOE 2017 technical targets. *Scientific Reports* 3 (1): 2257.

21 Proietti, E., Jaouen, F., Lefèvre, M. et al. (2011). Iron-based cathode catalyst with enhanced power density in polymer electrolyte membrane fuel cells. *Nature Communications* 2 (1): 416.

22 Nabae, Y. and Ishihara, A. (2019). Recent progress in non-precious metal fuel cell catalysts. In: *Nanocarbons for Energy Conversion: Supramolecular Approaches. Nanostructure Science and Technology* (ed. N. Nakashima). Cham: Springer https://doi.org/10.1007/978-3-319-92917-0_11.

23 Wang, Y. (2007). Analysis of the key parameters in the cold start of polymer electrolyte fuel cells. *Journal of the Electrochemical Society* 154 (10): 1041.

24 Wang, Y. and Feng, X. (2008). Analysis of reaction rates in the cathode electrode of polymer electrolyte fuel cell I. Single-layer electrodes. *Journal of the Electrochemical Society* 155 (12): 1289.

25 Gottesfeld, S. and Pafford, J. (1988). A new approach to the problem of carbon monoxide poisoning in fuel cells operating at low temperatures. *Journal of the Electrochemical Society* 135 (10): 2651–2652.

26 Ismagilov, Z.R., Kerzhentsev, M.A., Shikina, N.V. et al. (2005). Development of active catalysts for low Pt loading cathodes of PEMFC by surface tailoring of nanocarbon materials. *Catalysis Today* 102: 58–66.

27 Yu, P., Pemberton, M., and Plasse, P. (2005). PtCo/C cathode catalyst for improved durability in PEMFCs. *Journal of Power Sources* 144 (1): 11–20.

28 Xie, J., Wood, D.L., Wayne, D.M. et al. (2005). Durability of PEFCs at high humidity conditions. *Journal of the Electrochemical Society* 152 (1): A104–A113.

29 González-Huerta, R.G., Chávez-Carvayar, J.A., and Solorza-Feria, O. (2006). Electrocatalysis of oxygen reduction on carbon supported Ru-based catalysts in a polymer electrolyte fuel cell. *Journal of Power Sources* 153 (1): 11–17.

30 Eisman, G.A. (1986). The physical and mechanical properties of a new perfluorosulfonic acid ionomer for use as a separator/membrane in proton exchange processes. *In Proceedings of the Electrochemical Society* 86 (13): 156.

31 Benziger, J., Nehlsen, J., Blackwell, D. et al. (2005). Water flow in the gas diffusion layer of PEM fuel cells. *Journal of Membrane Science* 261 (1–2): 98–106.

32 Nam, J.H., Lee, K.J., Hwang, G.S. et al. (2009). Microporous layer for water morphology control in PEMFC. *International Journal of Heat and Mass Transfer* 52 (11–12): 2779–2791.

33 Phillips, R.K., Friess, B.R., Hicks, A.D. et al. (2012). Ex-situ measurement of properties of gas diffusion layers of PEM fuel cells. *Energy Procedia* 29: 486–495.

34 Che, J., Cagin, T., and Goddard, W.A. III (2000). Thermal conductivity of carbon nanotubes. *Nanotechnology* 11 (2): 65.

35 Balandin, A.A., Ghosh, S., Bao, W. et al. (2008). Superior thermal conductivity of single-layer graphene. *Nano Letters* 8 (3): 902–907.

36 Biercuk, M.J., Llaguno, M.C., Radosavljevic, M. et al. (2002). Carbon nanotube composites for thermal management. *Applied Physics Letters* 80 (15): 2767–2769.

37 Stankovich, S., Dikin, D.A., Dommett, G.H. et al. (2006). Graphene-based composite materials. *Nature* 442 (7100): 282–286.

38 Wang, Y. (2009). Porous-media flow fields for polymer electrolyte fuel cells: II. Analysis of channel two-phase flow. *Journal of the Electrochemical Society* 156 (10): B1134–B1141.

39 Tian, X., Zhao, X., Su, Y.Q. et al. (2019). Engineering bunched Pt-Ni alloy nanocages for efficient oxygen reduction in practical fuel cells. *Science* 366 (6467): 850–856.

40 Chong, L., Wen, J., Kubal, J. et al. (2018). Ultralow-loading platinum-cobalt fuel cell catalysts derived from imidazolate frameworks. *Science* 362 (6420): 1276–1281.

41 Chen, C., Kang, Y., Huo, Z. et al. (2014). Highly crystalline multimetallic nanoframes with three-dimensional electrocatalytic surfaces. *Science* 343 (6177): 1339–1343.

42 Li, M., Zhao, Z., Cheng, T. et al. (2016). Ultrafine jagged platinum nanowires enable ultrahigh mass activity for the oxygen reduction reaction. *Science* 354 (6318): 1414–1419.

43 Huang, X., Zhao, Z., Cao, L. et al. (2015). High-performance transition metal-doped Pt3Ni octahedra for oxygen reduction reaction. *Science* 348 (6240): 1230–1234.

44 Ott, S., Orfanidi, A., Schmies, H. et al. (2020). Ionomer distribution control in porous carbon-supported catalyst layers for high-power and low Pt-loaded proton exchange membrane fuel cells. *Nature Materials* 19 (1): 77–85.

45 Yarlagadda, V., Carpenter, M., Moylan, T.E. et al. (2018). Boosting fuel cell performance with accessible carbon mesopores. *ACS Energy Letters* 3 (3): 618–621.

46 Poojary, S., Islam, M.N., Shrivastava, U.N. et al. (2020). Transport and electrochemical interface properties of ionomers in low-Pt loading catalyst layers: effect of ionomer equivalent weight and relative humidity. *Molecules* 25 (15): 3387.

47 Yoon, K.R., Lee, K., Jo, S. et al. (2019). Mussel-inspired polydopamine-treated reinforced composite membranes with self-supported CeO_x radical scavengers for highly stable pem fuel cells. *Advanced Functional Materials* 29 (3): 1806929.

48 Park, C.H., Lee, Y., Huang, S. et al. (2016). Nanocrack-regulated self-humidifying membranes. *Nature* 532 (7600): 480–483.

49 Liu, X., Li, Y., Xue, J. et al. (2019). Magnetic field alignment of stable proton-conducting channels in an electrolyte membrane. *Nature Communications* 10 (1): 842.

50 Ozden, A., Shahgaldi, S. et al. (2019). A review of gas diffusion layers for proton exchange membrane fuel cells-with a focus on characteristics, characterization techniques, materials and designs. *Progress in Energy and Combustion Science* 74: 50–102.

51 Haile, S.M. (2003). Fuel cell materials and components. *Acta Materialia* 51 (19): 5981–6000.

52 Bauer, F., Denneler, S. et al. (2005). Influence of temperature and humidity on the mechanical properties of Nafion® 117 polymer electrolyte membrane. *Journal of Polymer Science Part B: Polymer Physics* 43 (7): 786–795.

53 Devanathan, R. (2008). Recent developments in proton exchange membranes for fuel cells. *Energy & Environmental Science* 1 (1): 101–119.

54 Jones, D. (2015). Perfluorosulfonic acid membranes for fuel cell and electrolyser applications. *Material Matters* 10 (3): 42–43.

55 Kienitz, B., Kolde, J., Priester, S. et al. (2011). Ultra-thin reinforced ionomer membranes to meet next generation fuel cell targets. *ECS Transactions* 41 (1): 1521–1530.

56 Jalani, N.H., Dunn, K., and Datta, R. (2005). Synthesis and characterization of Nafion®-MO_2 (M = Zr, Si, Ti) nanocomposite membranes for higher temperature PEM fuel cells. *Electrochimica Acta* 51 (3): 553–560.

57 Chen, S.Y., Han, C.C., Tsai, C.H. et al. (2007). Effect of morphological properties of ionic liquid-templated mesoporous anatase TiO_2 on performance of PEMFC with Nafion/TiO_2 composite membrane at elevated temperature and low relative humidity. *Journal of Power Sources* 171 (2): 363–372.

58 Su, D., Zheng, J., Ma, J. et al. (2023). Application of machine learning in fuel cell research. *Energies* 16 (11): 4390.

59 Wang, Y., Diaz, D.F.R., Chen, K.S. et al. (2020). Materials, technological status, and fundamentals of PEM fuel cells–a review[J]. *Materials today* 32: 178–203.

60 Wang, Y., Chen, K.S., Mishler, J. et al. (2011). A review of polymer electrolyte membrane fuel cells: technology, applications, and needs on fundamental research. *Applied Energy* 88 (4): 981–1007.

61 Stumper, J. and Stone, C. (2008). Recent advances in fuel cell technology at Ballard. *Journal of Power Sources* 176 (2): 468–476.

62 Wang, Y., Basu, S., and Wang, C.Y. (2008). Modeling two-phase flow in PEM fuel cell channels. *Journal of Power Sources* 179 (2): 603–617.

63 Luo, X., Wang, J., Dooner, M. et al. (2015). Overview of current development in electrical energy storage technologies and the application potential in power system operation. *Applied Energy* 137: 511–536.

64 Ferriday, T.B. and Middleton, P.H. (2021). Alkaline fuel cell technology – a review. *International Journal of Hydrogen Energy* 46 (35): 18489–18510.

65 Badea, N.I. (2021). Hydrogen as energy sources – basic concepts. *Energies* 14 (18): 5783.

66 Schultze, K. and Bartelt, H. (1992). Formation and reduction of nickel (II) hydroxide on Raney-nickel electrodes and its influence on the hydrogen overvoltage. *International Journal of Hydrogen Energy* 17 (9): 711–718.

67 Davydova, E.S., Mukerjee, S., Jaouen, F. et al. (2018). Electrocatalysts for hydrogen oxidation reaction in alkaline electrolytes. *ACS Catalysis* 8 (7): 6665–6690.

68 Sheng, W., Gasteiger, H.A., and Shao-Horn, Y. (2010). Hydrogen oxidation and evolution reaction kinetics on platinum: acid vs alkaline electrolytes. *Journal of the Electrochemical Society* 157 (11): B1529.

69 Schmidt, T.J., Ross, P.N. Jr., and Markovic, N.M. (2002). Temperature dependent surface electrochemistry on Pt single crystals in alkaline electrolytes: part 2. The hydrogen evolution/oxidation reaction. *Journal of Electroanalytical Chemistry* 524: 252–260.

70 Brauns, J. and Turek, T. (2020). Alkaline water electrolysis powered by renewable energy: a review. *Processes* 8 (2): 248.

71 Wu, J., Risalvato, F.G., Sharma, P.P. et al. (2013). Electrochemical reduction of carbon dioxide: II. Design, assembly, and performance of low temperature full electrochemical cells. *Journal of the Electrochemical Society* 160 (9): F953.

Index

Towards Next Generation Energy Storage Technologies: From Fundamentals to Commercial Applications, First Edition. Edited by Minghua Chen.

M